FIAS Interdisciplinary Science Series

More information about this series at http://www.springer.com/series/10781

Johannes Kirsch · Stefan Schramm (deceased) ·
Jan Steinheimer-Froschauer ·
Horst Stöcker
Editors

Discoveries at the Frontiers of Science

From Nuclear Astrophysics to Relativistic Heavy Ion Collisions

Springer

Editors
Johannes Kirsch
Frankfurt Institute for Advanced Studies
Frankfurt am Main, Hessen, Germany

Jan Steinheimer-Froschauer
Frankfurt Institute for Advanced Studies
Frankfurt am Main, Hessen, Germany

Stefan Schramm (deceased)
Stefan Schramm wrote this book while at
Frankfurt Institute for Advanced Studies
Frankfurt am Main, Hessen, Germany

Horst Stöcker
Frankfurt Institute for Advanced Studies
Frankfurt am Main, Hessen, Germany

ISSN 2522-8900 ISSN 2522-8919 (electronic)
FIAS Interdisciplinary Science Series
ISBN 978-3-030-34233-3 ISBN 978-3-030-34234-0 (eBook)
https://doi.org/10.1007/978-3-030-34234-0

This Springer imprint is published by the registered company Springer Nature Switzerland AG
The registered company address is: Gewerbestrasse 11, 6330 Cham, Switzerland

Preface

This volume addresses problems and current research at the forefront of theoretical physics. It also mirrors the deep and broad interests of Professor Walter Greiner (1935–2016), the founder and senior editor of Springer Nature's 'FIAS Interdisciplinary Science Series'.

In June 2017, the Frankfurt Institute for Advances Studies FIAS hosted the 'International Symposium on Discoveries at the Frontiers of Science' to honor and commemorate the scientific legacy of Walter Greiner. The Symposium succeeded in bringing together many world-class scientists for a lively and inspiring exchange of ideas.

Subsequently, the participants have worked diligently to prepare this collection of overview articles. Their contributions cover and connect together, topics ranging from atomic and molecular physics to quantum field theory and nuclear physics; and from relativistic heavy ion collisions and the Equation of State EoS of hot dense nuclear QCD-matter to general relativistic binary black hole—and neutron star mergers, which are being tested experimentally by gravitational wave observatories like LIGO/VIRGO. Further, forefront topics include fundamental quantum mechanics and possible modifications of Einstein's GR.

The Equation of State (EoS) and the quest for phase transitions of nuclear and neutron star matter to quark matter, or from a pure Yang-Mills gluon plasma to a quark-gluon plasma (QGP) are among the great challenges that can be addressed both in studies of the core of neutron stars or neutron star mergers and in high energy heavy ion collisions, during the ultradense, hot matter formation, expansion, and hadronization.

Related topics include the interaction of heavy quarks with QGP partons, heavy ion collision experiments at GSI, FAIR, NICA, SPS, LHC, and RHIC beam energies to investigate the transition from baryon-dominated to meson-dominated matter, the analysis of baryon, pion, and kaon flow as well as the formation of D- and B-mesonic states in heavy ion collisions.

The volume includes reviews addressing photon scattering experiments, superheavy elements, extended versions of general relativity, new experimental developments in heavy ion collisions, and renewable energy networks, and—finally—tributes to Walter Greiner's scientific career.

Many of these contributions have roots in Walter's work and bear witness to his remarkably productive life as a brilliant scientist.

Walter will continue to be present in our hearts and in our minds.

Frankfurt am Main, Germany
September 2019

Johannes Kirsch
Stefan Schramm (Deceased)
Jan Steinheimer-Froschauer
Horst Stöcker

Contents

Contributors

Joerg Aichelin SUBATECH, UMR 6457, Université de Nantes, Ecole des Mines de Nantes, IN2P3/CNRS, Nantes Cedex 3, France

Miron Ya. Amusia Racah Institute of Physics, The Hebrew University, Jerusalem, Israel;
A. F. Ioffe Physical-Technical Institute, St. Petersburg, Russia

Hartmuth Arenhövel Institut für Kernphysik Johannes Gutenberg-Universität, Mainz, Germany

Debades Bandyopadhyay Astroparticle Physics and Cosmology Division, Centre for Astroparticle Physics, Saha Institute of Nuclear Physics, HBNI, Kolkata, India

Sarmistha Banik Department of Physics, Birla Institute of Technology and Science, Pilani, Medchal District, Hyderabad, India

Jean Cleymans Physics Department, UCT-CERN Research Centre, University of Cape Town, Rondebosch, South Africa

Gustavo A. Contrera Facultad de Ciencias Exactas, IFLP, UNLP, CONICET, La Plata, Argentina;
Grupo de Gravitación, Astrofísica y Cosmología, Facultad de Ciencias Astronómicas y Geofísicas, Universidad Nacional de La Plata, La Plata, Argentina;
CONICET, Buenos Aires, Argentina

Hans-Thomas Elze Dipartimento di Fisica "Enrico Fermi", Universitá di Pisa, Pisa, Italy

Amand Faessler Institute of Theoretical Physics, University of Tuebingen, Tübingen, Germany

Leonid Frankfurt Sackler School of Exact Sciences, Tel Aviv University, Tel Aviv, Israel

Avraham Gal Racah Institute of Physics, The Hebrew University, Jerusalem, Israel

Hans Geissel Gesellschaft für Schwerionenforschung GSI, Darmstadt, Germany

Mark I. Gorenstein Bogolyubov Institute for Theoretical Physics, Kyiv, Ukraine; Giersch Science Center, Frankfurt Institute for Advanced Studies, Frankfurt, Germany

Pol B. Gossiaux SUBATECH, UMR 6457, Université de Nantes, Ecole des Mines de Nantes, IN2P3/CNRS, Nantes Cedex 3, France

Robert Greifenhagen Helmholtz-Zentrum Dresden-Rossendorf, Dresden, Germany

Martin Greiner Department of Engineering, Aarhus University, Aarhus C, Denmark

Matthias Hanauske Institut für Theoretische Physik, Frankfurt, Germany;
Frankfurt Institute for Advanced Studies, Frankfurt, Germany

John W. Harris D. Allan Bromley Professor of Physics, Department of Physics, Yale University, New Haven, CT, USA

Christoph Hartnack SUBATECH, UMR 6457 IMT Atlantique, IN2P3/CNRS, Université de Nantes, Nantes, France

Peter O. Hess Instituto de Ciencias Nucleares, UNAM, Circuito Exterior, C.U., Mexico City, Mexico;
Frankfurt Institute for Advanced Studies, Johann Wolfgang Goethe Universität, Frankfurt am Main, Germany

Burkhard Kämpfer Helmholtz-Zentrum Dresden-Rossendorf, Dresden, Germany;
TU Dresden, Institute of Theoretical Physics, Dresden, Germany

Leonid P. Kaptari Helmholtz-Zentrum Dresden-Rossendorf, Dresden, Germany;
JINR, Bogoliubov Institute for Theoretical Physics, Dubna, Russia

Johannes Kirsch Frankfurt Institute for Advanced Studies (FIAS), Frankfurt am Main, Germany

Alexei Larionov Institut für Kernphysik, Forschungszentrum Jülich, Jülich, Germany;
National Research Center "Kurchatov Institute", Moscow, Russia;
Frankfurt Institute for Advanced Studies, Giersch Science Center, Frankfurt am Main, Germany

Andreas Lehrach Institut für Kernphysik, Forschungszentrum Jülich, Jülich, Germany;
JARA-FAME (Forces and Matter Experiments), Forschungszentrum Jülich and RWTH Aachen University, Aachen, Germany

Rudolf Maier Institut für Kernphysik, Forschungszentrum Jülich, Jülich, Germany;
JARA-FAME (Forces and Matter Experiments), Forschungszentrum Jülich and RWTH Aachen University, Aachen, Germany

Gottfried Münzenberg GSI Helmholtzzentrum für Schwerionenforschung mbH, Darmstadt, Germany;
Johannes Gutenberg-Universität Mainz, Mainz, Germany;
Manipal Centre for Natural Sciences, Manipal University, Manipal, Karnataka, India

Marlene Nahrgang SUBATECH, UMR 6457, Université de Nantes, Ecole des Mines de Nantes, IN2P3/CNRS, Nantes Cedex 3, France

Milva G. Orsaria Grupo de Gravitación, Astrofísica y Cosmología, Facultad de Ciencias Astronómicas y Geofísicas, Universidad Nacional de La Plata, La Plata, Argentina;
CONICET, Buenos Aires, Argentina

Johann Rafelski Department of Physics, The University of Arizona, Tucson, AZ, USA

Mirko Schäfer INATECH, University of Freiburg, Freiburg, Germany;
Department of Engineering, Aarhus University, Aarhus C, Denmark

Christoph Scheidenberger GSI Helmholtzzentrum für Schwerionenforschung mbH, Darmstadt, Germany;
Justus-Liebig-Universität Gießen, Gießen, Germany

Christian Spieles Institut für Theoretische Physik, Goethe Universität Frankfurt, Frankfurt am Main, Germany

William M. Spinella School of Physical Sciences and Technologies, Irvine Valley College, Irvine, CA, USA

Horst Stöcker Institut für Theoretische Physik, Frankfurt, Germany;
Frankfurt Institute for Advanced Studies, Frankfurt, Germany;
GSI Helmholtzzentrum für Schwerionenforschung GmbH, Darmstadt, Germany

Mark Strikman Pennsylvania State University, University Park, PA, USA

Jürgen Struckmeier Frankfurt Institute for Advanced Studies (FIAS), Frankfurt am Main, Germany;
Goethe-Universität, Frankfurt am Main, Germany;
GSI Helmholtzzentrum für Schwerionenforschung GmbH, Darmstadt, Germany

Laura Tolos Institut für Theoretische Physik, University of Frankfurt, Frankfurt am Main, Germany;
Frankfurt Institute for Advanced Studies, University of Frankfurt, Frankfurt am Main, Germany;
Institute of Space Sciences (ICE, CSIC), Campus UAB, Barcelona, Spain;
Institut d'Estudis Espacials de Catalunya (IEEC), Barcelona, Spain

Bo Tranberg Ento Labs ApS, Aarhus C, Denmark;
Department of Engineering, Aarhus University, Aarhus C, Denmark

Wolfgang Trautmann GSI Helmholtzzentrum, Darmstadt, Germany

Hendrik van Hees Institut für Theoretische Physik, Goethe Universität Frankfurt, Frankfurt am Main, Germany

David Vasak Frankfurt Institute for Advanced Studies (FIAS), Frankfurt am Main, Germany

Volodymyr Vovchenko Frankfurt Institute for Advanced Studies, Giersch Science Center, Frankfurt am Main, Germany;
Institut für Theoretische Physik, Goethe Universität Frankfurt, Frankfurt am Main, Germany

Fridolin Weber Department of Physics, San Diego State University, San Diego, CA, USA;
Center for Astrophysics and Space Sciences, University of California, San Diego, La Jolla, CA, USA

Klaus Werner SUBATECH, UMR 6457, Université de Nantes, Ecole des Mines de Nantes, IN2P3/CNRS, Nantes Cedex 3, France

What the Azimuthal Distribution of Heavy Mesons Tells Us About the Quark Gluon Plasma?

Joerg Aichelin, Pol B. Gossiaux, Marlene Nahrgang and Klaus Werner

Abstract Heavy mesons (charm and bottom) are one of the few probes which are sensitive to the time evolution of a Quark Gluon Plasma (QGP), light mesons come to a statistical equilibrium latest at the end of the QGP expansion and do therefore not carry information on the QGP properties during the expansion. We discuss here the interaction of the heavy quarks with the QGP partons and how this interaction influences the azimuthal distribution of the heavy mesons. We will argue that there are indications that small p_T heavy quarks equilibrate in the QGP whereas those with a high p_T create a finite azimuthal flow due to the different path lengths in the QGP. These are results of the pQCD based Monte-Carlo (MC@sHQ) approach which is coupled to EPOS 2 modeling the expansion of the QGP.

1 Introduction

I have neither studied in Frankfurt nor have I been employed by Frankfurt. Nevertheless, I owe a lot to Walter. When I appeared the first time in Frankfurt, innocently invited to present my work, which I did together with Horst Stöcker, in an one hour seminar called "Palaver" Walter bombarded me with questions for more than 2 1/2 h. At the end I was completely exhausted and Walter had knowledge about all the details of our work. Obviously me answers satisfied him because from then on we had many discussions, supported also by the red wine which he asked me regularly to bring to him from my French hometown. Among many things I learned from him is that a physicist should work at the frontiers to unknown land, what he always demonstrate by his own work. So I started a couple of years ago to work on heavy quarks and especially on what one can learn from heavy quarks about the quark gluons plasma which is created in heavy ion collisions.

To understand the formation, the expansion and the hadronization of a quark gluon plasma (QGP) created in ultra-relativistic heavy ion collisions is the ultimate

J. Aichelin (✉) · P. B. Gossiaux · M. Nahrgang · K. Werner
SUBATECH, UMR 6457, Université de Nantes, Ecole des Mines de Nantes, IN2P3/CNRS, 4 rue Alfred Kastler, 44307 Nantes Cedex 3, France
e-mail: Jorg.Aichelin@univ-nantes.fr

J. Kirsch et al. (eds.), *Discoveries at the Frontiers of Science*,
FIAS Interdisciplinary Science Series, https://doi.org/10.1007/978-3-030-34234-0_1

objective of the heavy ion experiments at the Large Hadron Collider (LHC) at CERN and the Relativistic Heavy Ion Collider (RHIC) at Brookhaven. To achieve this objective turned out to be more complicated than expected because the multiplicities of the observed hadrons indicate that in the end of the expansion the QGP is in statistical equilibrium and therefore their multiplicity follows statistical laws. In addition, the temperature at which the transition to hadrons takes place, extracted from this statistical analysis, is about the same as that obtained by lattice gauge calculation in which the Quantumchromodynamics, the underlying theory for strong interactions, is solved on a computer [1, 2]. Once statistical equilibrium is obtained the information on how the system approaches this equilibrium is lost. Hadrons which are created from the QGP can therefore tell little, in order not to say nothing, about the time evolution of the QGP before hadronization.

If one wants to study the time evolution of the QGP one has to rely on probes which pass the QGP but which do not come to an equilibrium with the QGP. Such probes exist. One may study electromagnetic probes like photons or dileptons or hard probes which are created in the first interaction between projectile and target nucleons in hard processes, means in those in which the momentum transfer is large. Both of these probes have their advantages and inconveniences. Electromagnetic probes are rare and many processes contribute, before and after the hadronization, to the measured spectra. Many of them are hard to assess because the production cross sections are unknown or only vaguely known and the composition of the hadron gas after the hadronization is debated, depending on how the hadron properties change in a dense environment at high temperature. Despite of these difficulties many interesting features have been discovered by analyzing the electromagnetic probes [3]. As an example, the large anisotropy in the azimuthal plane has for long been a surprise because early produced photons and dileptons should not show such a feature. In the meantime it has been discovered that bremsstrahlung and hadronic interactions are the origin of this anisotropy.

Hard probes, heavy quarks and energetic light quarks or gluons are created in initial hard collisions between projectile and target and have to traverse the plasma. We concentrate here on heavy quarks. The interpretation of jet observables is more complicated because the leading jet parton may change its identity has just started and interesting results are expected for the near future. Being colored objects the interaction of heavy quarks with plasma particles is strong. Due to the propagation through the colored partonic medium high-p_T heavy quarks suffer from a substantial energy loss, while low-p_T heavy quarks are expected to thermalize at least partially within the medium. The nuclear modification factor, R_{AA}, which is the ratio of the spectra measured in heavy-ion collisions to the scaled proton-proton reference, and the elliptic flow, v_2, are traditional observables of heavy-flavor hadrons and decay leptons. A suppression of high-p_T D mesons, heavy-flavor decay electrons and muons has been measured by the STAR [4, 5] and Phenix [6] collaborations at RHIC and the ALICE [7–9] and CMS [10] collaborations at LHC. The v_2 of D mesons, heavy-flavor decay electrons and muons was found to be nonvanishing both at RHIC [11] and at LHC [12].

2 The MC@sHQ Approach

In the MC@sHQ approach [13, 14] the heavy quark ($Q\bar{Q}$) pairs are initialized according to the p_T distribution from FONLL [15–17]. We assume the LO production processes, i.e., an azimuthally back-to-back initialization of the $Q\bar{Q}$ pairs with $\mathbf{p}_{T,\bar{Q}} = -\mathbf{p}_{T,Q}$. The heavy quarks can interact with the plasma constituents purely elastically or in a combination of elastic and inelastic collisions. The elastic cross sections in Born approximation are obtained within a hard thermal loop (HTL) calculation, including a running coupling constant α_s [13, 18]. The contribution from the t-channel is regularized by a reduced Debye screening mass κm_D^2, which is calculated self-consistently [13, 19], yielding a gluon propagator with

$$1/Q^2 \rightarrow 1/(Q^2 - \kappa \tilde{m}_D^2(T)) \tag{1}$$

for a momentum transfer Q^2. In this HTL+semihard approach [13], κ is determined such that the average energy loss is maximally insensitive to the intermediate scale between soft (with a HTL gluon propagator) and hard (with a free gluon propagator) processes. The inelastic cross sections include both, the incoherent gluon radiation [20] and the effect of coherence, i.e. the Landaul-Pomeranchuk-Migdal (LPM) effect [21]. In this approach the incoming light partons are considered as massless [22–25].

The fluid dynamical evolution is used as a background providing us with the temperature and velocity fields necessary to sample thermal scattering partners for the heavy quarks. The MC@sHQ approach couples the Monte-Carlo treatment of the Boltzmann equation of heavy quarks (MC@sHQ) [13] to the $3+1$ dimensional fluid dynamical evolution of the locally thermalized QGP following the initial conditions from EPOS2 [26, 27]. EPOS2 is a multiple scattering approach which combines pQCD calculations for the hard scatterings with Gribov-Regge theory for the phenomenological, soft initial interactions. Jet components are identified and subtracted while the soft contributions are mapped to initial fluid dynamical fields. By enhancing the initial flux tube radii viscosity effects are mimicked, while the subsequent $3+1$ dimensional fluid dynamical expansion itself is ideal. Including final hadronic interactions the EPOS2 event generator has successfully described a variety of bulk and jet observables, both at RHIC and at LHC [26, 27]. For details we refer to the references.

Including elastic and inelastic collisions this approach reproduces quite well the experimental D-meson and non photonic electron data at RHIC and LHC. As an example we display in Fig. 1 the D meson R_{AA} as dashed line, for elastic (coll) as well as for elastic+inelastic collisions (coll+rad) in comparison with ALICE data. Elastic cross sections alone give not sufficient stopping in this approach.

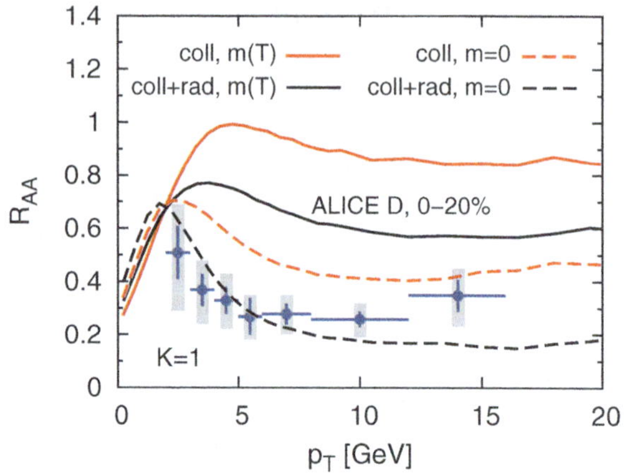

Fig. 1 Comparison of the D meson $R_{\rm AA}$ for a QGP consisting of massive quasiparticles (solid lines) and massless partons (dashed lines). Purely collisional (orange, light) and collisional + radiative (LPM) (blackline) energy loss scenarios are shown

3 Azimuthal Distribution of a Back-to-Back Emitted $Q\bar{Q}$ Pair

The heavy (anti)quarks are propagated through the QGP by means of the coupled MC@sHQ+EPOS approach, which was described above. Here, we track the evolution of the heavy (anti)quark until it leaves the QGP [22]. At this transition point we extract the difference of the azimuthal angles, $\Delta\phi$, of those $Q\bar{Q}$ pairs which were initially produced together. The distributions of $\Delta\phi$ are shown in Fig. 2 for $c\bar{c}$ pairs in the left column and for $b\bar{b}$ pairs in the right column. These pairs are taken into account if both the quark and the antiquark are finally at a rapidity $|y_Q| < 1$ and $|y_{\bar{Q}}| < 1$. The results for the 0–20% most central collisions are plotted in the upper row, while in the middle row we see results for 20–40% centrality and in the lowest row for peripheral collisions (40–60% most central). In each individual plot we show the distribution of azimuthal correlations for three different classes of p_T. The lowest p_T class collects all $Q\bar{Q}$ pairs, where both, the quark and the antiquark, have a final p_T between 1 and 4 GeV. In the intermediate-p_T class quark and antiquark have a final p_T between 4 and 10 GeV and in the high p_T-class the final p_T of the quark and antiquark is in between 10 and 20 GeV. For these calculations we have multiplied our cross section with a K-factor to obtain the best agreement with the experimental R_{AA}. This is due to the fact that the purely collisional interaction mechanism produces a larger average $p_\perp^2$ per unit time [25].

We see, first of all, that the initial correlations are broadened and they are broadened more strongly for the purely collisional interaction mechanism than for the mechanism including radiative corrections.

The systems that are created in the most central collisions are the largest and reach the highest temperatures and densities. Therefore the broadening of the initial delta-function-like correlations is most efficient. We find a substantial broadening of these correlations for all p_T classes and both interaction mechanisms for $c\bar{c}$ pairs in Fig. 2a

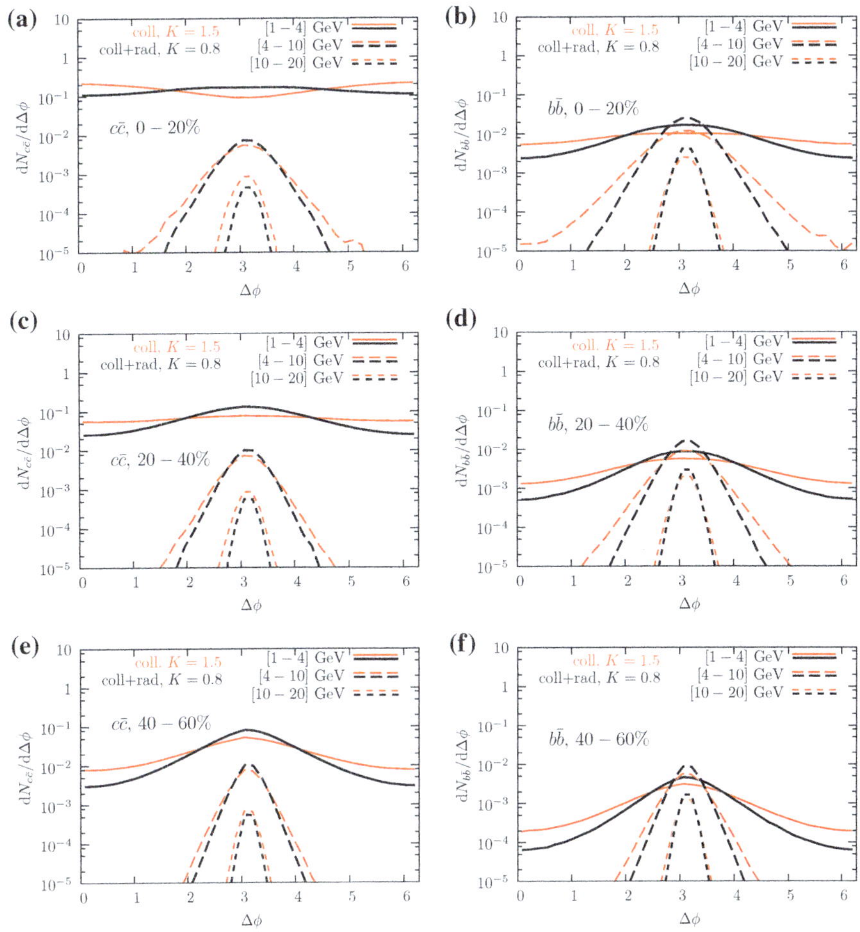

Fig. 2 Azimuthal correlations of initially correlated $Q\bar{Q}$ pairs at the transition temperature. In the left column the azimuthal distributions of $c\bar{c}$ pairs are shown, in the right column those of $b\bar{b}$ pairs at midrapidity. The centralities are 0–20% (upper row), 20–40% (middle row) and 40–60% (lower row). In each plot we compare the purely collisional (orange/light) to the collisional plus radiative (black/dark) interaction mechanism for different classes of final p_T. See text for more details

and for $b\bar{b}$ pairs in Fig. 2b. For the quark with lowest p_T the initial correlations are almost completely washed out. This almost flat $dN_{Q\bar{Q}}/d\Delta\phi$ distribution is a strong hint that the heavy quarks are in equilibrium with their environment.

4 v_2 as a Measure of Approaching Equilibrium

This question, whether the heavy quarks approach equilibrium with the QPG partons can be studied by measuring the azimuthal anisotropy of the heavy quarks. Initially, produced in a hard collisions the heavy quarks have no preferred direction in the transverse plane whereas the QGP partons have a fluid dynamical flow. This fluid dynamical flow is the response to the eccentricity in the initial geometry

$$\varepsilon_n = \frac{\sqrt{\langle r^n \cos(n\phi)\rangle^2 + \langle r^n \sin(n\phi)\rangle^2}}{\langle r^n \rangle} \tag{2}$$

where ϕ is the spatial azimuthal angle and $r = \sqrt{x^2 + y^2}$ the distance from the center. The average $\langle\cdot\rangle$ is weighted by the local energy density. Similarly the n-th order angles of the participant plane[1] can be obtained from the initial state via

$$\psi_n^{\mathrm{PP}} = \frac{1}{n} \arctan \frac{\langle r^n \cos(n\phi)\rangle}{\langle r^n \sin(n\phi)\rangle} . \tag{3}$$

It has been shown that the flow coefficients

$$v_n^{\mathrm{EP}} = \frac{\int \mathrm{d}\phi \cos\left[n(\phi - \psi_n^{\mathrm{EP}})\right] \frac{\mathrm{d}N}{\mathrm{d}y\mathrm{d}\phi}}{\int \mathrm{d}\phi \frac{\mathrm{d}N}{\mathrm{d}y\mathrm{d}\phi}} , \tag{4}$$

taken as the Fourier coefficients of the single-particle azimuthal distribution with respect to the event-plane angle $\psi_n^{\mathrm{EP}} = (1/n)\arctan(\langle p_T \sin(n\phi)\rangle / \langle p_T \cos(n\phi)\rangle)$, where ϕ is the azimuthal angle of the transverse momentum of the measured particles, corresponds very well to the flow coefficients v_n^{PP} obtained from correlating the single particles with the initial participant plane [28].

We now investigate the centrality dependence of the heavy-flavor flow further by plotting the integrated v_n/ε_n, which is dominated by low p_t quarks as a function of the centrality in Fig. 3. We concentrate on the collisional + radiative(LPM) energy loss model at $\sqrt{s} = 2.76$ TeV. Under the assumption that $v_n \propto \varepsilon_n$ (which holds for small and intermediate centralities) the plotted quantity can be identified with the efficiency of the medium to transform an initial geometry into an anisotropy in momentum space. By comparing the D mesons flow to the flow of the light charged hadrons from the bulk and the heavy B mesons we can make the following observations. For all particles we see that the efficiency of the system to respond to the initial geometry decreases toward more peripheral collisions and a mass hierarchy can be observed in the slopes of this decrease. For v_2/ε_2, D mesons and light charged hadrons show a very similar behavior in both the magnitude and the slope, which as such would imply that the overall efficiency of transferring an initial ellipticity to bulk flow

[1] The term "participant plane" is commonly used for the following definition. We would like to point out though, that the initial conditions used here, do not rely on a participant picture.

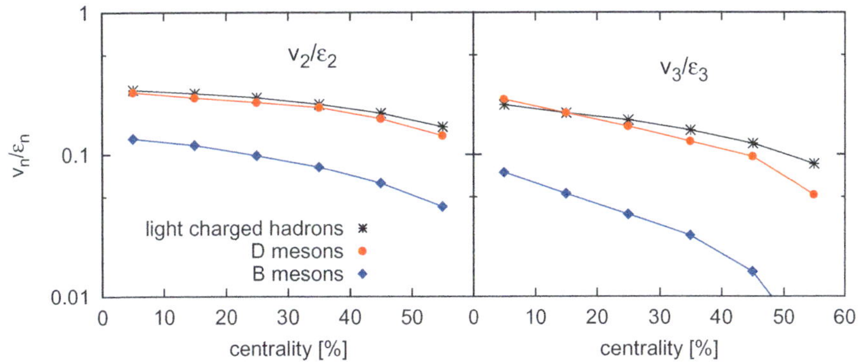

Fig. 3 The centrality dependence of the ratios v_n/ε_n for $n = 2, 3$ in the collisional + radiative (LPM) energy loss model at $\sqrt{s} = 2.76$ TeV collision energies for the light charged hadron, the D and B meson flow. The bulk flow is obtained as $v_n(2, |\Delta\eta| > 1)$ from the full EPOS2 model [27]

and to flow of the charm quarks is of the same order suggesting a perfect coupling of the charm quarks to the bulk. We can see, however, that this does not hold for the third-order Fourier coefficient of the flow where, although being of the same magnitude (within expected errors) in the central collisions, the ratio v_3/ε_3 falls off more quickly for D mesons than for the bulk flow toward more peripheral collisions. For B mesons, the flow is smaller in magnitude and by the steeper decrease one can see a more rapid decoupling from the bulk medium that can be understood. This is a consequence of the large mass of the b quark which leads to larger times needed for equilibration. This time is evidently not available in the rapidly expanding system. We can conclude, that the p_T integrated v_2, which is dominated by low p_T heavy mesons, is another hint that c quarks observed as D-mesons came to an equilibrium with the expanding QGP partons.

5 Eccentricity Is only One Reason for a Finite v_2

We continue our study on the information which is contained in the v_2 observable. In theoretical studies we can artificially switch off the bulk flow by assuming that the local rest frame of the fluid is the same as the laboratory frame. This procedure is of course only a first approximation to a scenario without bulk flow as the temperature field is still taken from an evolution that includes bulk flow, yet it gives an idea of how much of the heavy-flavor flow stems from the path length difference due to the initial eccentricity. For this study we calculate peripheral collisions at $\sqrt{s} = 2.76$ TeV [22]. The results are displayed in Fig. 4. We find that around $p_T \sim 2$ GeV both the v_2 and the v_3 of charm quarks are almost entirely due to the bulk flow of the medium. At $p_T \approx 4$ GeV the charm quark v_2 originating from path length differences is $\sim 50\%$ of the charm quark v_2 produced in a medium with bulk flow. This picture is

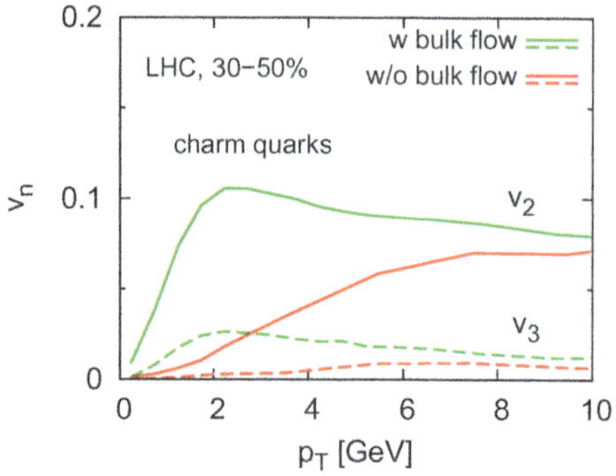

Fig. 4 The contribution of the bulk flow to the charm quark elliptic (solid) and triangular (dashed) flow (right plot) for 30–50% most central Pb+Pb collisions at $\sqrt{s} = 2.76$ TeV

slightly different for the triangular flow v_3. Path length differences seem to be smaller in triangularly shaped event geometries and the corresponding angular sectors are smaller, which diminishes the importance of this contribution to the flow. Up to $p_T \sim 4$ GeV we find that the charm quark v_3 is built up almost exclusively from the bulk flow of the medium, which makes it an excellent probe of the dynamics and interactions of charm quarks in the quark-gluon plasma.

6 Summary

Heavy quarks have been identified as a tool to study the time evolution of a QGP created in ultra-relativistic heavy ion collisions. We demonstrated that calculations using the pQCD based Monte-Carlo (MC@sHQ) approach which is coupled to EPOS 2 for the modeling the expansion of the QGP show that heavy quarks with a small p_T come to an equilibrium with the QPG partons. Therefore their azimuthal distribution is close to that of the light QGP partons. Heavy quarks with a large p_T do not equilibrate. Their azimuthal distribution measures the path length difference of the trajectories of the heavy quarks in the QGP. It depends on the elementary interactions of the heavy quarks with the QGP constituents. The transition between both distributions is smooth, so only theoretical investigations can separate them.

Acknowledgements This work was supported by the project "Together" of the region Pays de la Loire, France.

References

1. S. Borsanyi, Z. Fodor, C. Hoelbling, S.D. Katz, S. Krieg, K.K. Szabo, Phys. Lett. B **730**, 99 (2014)
2. A. Bazavov et al. [HotQCD Collaboration], Phys. Rev. D **90**(9), 094503 (2014)
3. O. Linnyk, E.L. Bratkovskaya, W. Cassing, Prog. Part. Nucl. Phys. **87**, 50 (2016). https://doi.org/10.1016/j.ppnp.2015.12.003, arXiv:1512.08126 [nucl-th]
4. STAR Collaboration, Phys. Rev. Lett. **98**, 192301 (2007); Erratum-ibid. **106**, 159902 (2011)
5. X. Dong [STAR Collaboration], Nucl. Phys. A **904–905**, 19c (2013)
6. PHENIX Collaboration, Phys. Rev. C **84**, 044905 (2011)
7. ALICE Collaboration, JHEP **09**, 112 (2012)
8. A. Dainese [ALICE Collaboration], arXiv:1212.0995 [nucl-ex]
9. B. Abelev et al. [ALICE Collaboration], arXiv:1305.2707 [nucl-ex]
10. CMS Collaboration [CMS Collaboration], CMS-PAS-HIN-15-005
11. L. Adamczyk et al. [STAR Collaboration], arXiv:1405.6348 [hep-ex]
12. J. Adam et al. [ALICE Collaboration], Phys. Lett. B **753**, 41 (2016)
13. P.B. Gossiaux, J. Aichelin, Phys. Rev. C **78**, 014904 (2008), arXiv:0802.2525 [hep-ph]
14. J. Aichelin, P.B. Gossiaux, T. Gousset, Phys. Rev. D **89**(7), 074018 (2014), arXiv:1307.5270 [hep-ph]
15. M. Cacciari, M. Greco, P. Nason, JHEP **9805**, 007 (1998)
16. M. Cacciari, S. Frixione, P. Nason, JHEP **0103**, 006 (2001)
17. M. Cacciari, S. Frixione, N. Houdeau, M.L. Mangano, P. Nason, G. Ridolfi, JHEP **1210**, 137 (2012)
18. A. Peshier, arXiv:hep-ph/0601119
19. S. Peigne, A. Peshier, Phys. Rev. D **77**, 114017 (2008)
20. J.F. Gunion, G. Bertsch, Phys. Rev. D **25**, 746 (1982)
21. P.B. Gossiaux, Nucl. Phys. A **910–911**, 301 (2013), arXiv:1209.0844 [hep-ph]
22. M. Nahrgang, J. Aichelin, S. Bass, P.B. Gossiaux, K. Werner, Phys. Rev. C **91**(1), 014904 (2015). https://doi.org/10.1103/PhysRevC.91.014904, arXiv:1410.5396 [hep-ph]
23. M. Nahrgang, J. Aichelin, S. Bass, P.B. Gossiaux, K. Werner, Nucl. Phys. A **931**, 575 (2014). https://doi.org/10.1016/j.nuclphysa.2014.08.094, arXiv:1409.1464 [hep-ph]
24. M. Nahrgang, J. Aichelin, P.B. Gossiaux , K. Werner, Phys. Rev. C **89**(1), 014905 (2014). https://doi.org/10.1103/PhysRevC.89.014905, arXiv:1305.6544 [hep-ph]
25. M. Nahrgang, J. Aichelin, P.B. Gossiaux, K. Werner, Phys. Rev. C **90**(2), 024907 (2014). https://doi.org/10.1103/PhysRevC.90.024907, arXiv:1305.3823 [hep-ph]
26. K. Werner, I. Karpenko, T. Pierog, M. Bleicher, K. Mikhailov, Phys. Rev. C **82**, 044904 (2010)
27. K. Werner, I. Karpenko, M. Bleicher, T. Pierog, S. Porteboeuf-Houssais, Phys. Rev. C **85**, 064907 (2012)
28. H. Holopainen, H. Niemi, K.J. Eskola, Phys. Rev. C **83**, 034901 (2011). https://doi.org/10.1103/PhysRevC.83.034901, arXiv:1007.0368 [hep-ph]

Limiting Temperature, Phase Transition(s), Crossover, ...

Mark I. Gorenstein

Abstract I present a short review of two physical models devoted to the equation of state at high energy densities: The Hagedorn concept of limiting temperature $T = T_H$ and the statistical bag model of phase transitions at $T = T_c$. The statistical bag model admits the different orders of phase transitions between hadrons (small bags) and quark gluon plasma (infinitely large bags). The crossover transition between hadron resonance gas and cluster quark gluon plasma at $T = T_{\rm cr}$ is also possible within the statistical bag model. For all these different phenomena rather similar values of the temperatures, $T_H \sim T_c \sim T_{\rm cr} = 150$–$160$ MeV, have been assumed.

Keywords Limiting temperature · Quark-gluon bags · Phase transitions · Crossover

1 Introduction

Equation of state and other equilibrium properties of matter are the subjects of statistical mechanics. This physical approach is used to describe a multiparticle system and calculate its partition function, i.e., the sum over all permitted microstates. To make this, one should define particle species, interparticle interactions, conserved charges in the considered system, and the external conditions. All this information requires careful experimental and theoretical investigations, and at present it is rather well known for typical atomic and molecular systems, at least in their gas and liquid phases. These statistical systems are usually defined in terms of several number of atomic and molecular species and two-particle potentials between their different pairs. If chemical reactions are possible, the numbers of conserved atomic species, or corresponding chemical potentials, should be additionally defined. One also chooses an appropriate statistical ensemble which reflects the physical boundary conditions.

M. I. Gorenstein (✉)
Bogolyubov Institute for Theoretical Physics, Kyiv 03680, Ukraine

Giersch Science Center, Frankfurt Institute for Advanced Studies, 60438 Frankfurt, Germany
e-mail: goren@fias.uni-frankfurt.de

J. Kirsch et al. (eds.), *Discoveries at the Frontiers of Science*,
FIAS Interdisciplinary Science Series, https://doi.org/10.1007/978-3-030-34234-0_2

Thus, a statistical description of real gases and liquids becomes a purely mathematical problem. Nevertheless, this mathematical problem remains to be rather difficult. For example, one should still follow some phenomenological models in a description of the liquid-gas phase transition and its critical point as rigorous analytical results in this field are rather poor.

Strongly interacting matter at high energy density is formed at the early stages of nucleus-nucleus collisions and/or in the central regions of neutron stars. What types of particles should be considered as the fundamental ones and what are the composite objects? What are the fundamental forces between the matter constituents? What are the conserved charges? What statistical ensemble should be used for their description? At high energy densities an equilibrium system consists from elementary particles, and the answers to the above questions have been changed in time with increasing of our knowledge about basic physical features of elementary particles.

The fundamental microscopic constituents and basic interactions at high energy density were not well known up to the recent time. First experiments on high energy collisions carried out in the middle of the last century demonstrated large amount of new particle species—hadrons and resonances. These particles are influenced by the strong interactions. And the main feature of the strong interactions is a creation of new and new types of hadrons and resonances with increasing of collision energies. These features of the strong interactions lead Hagedorn to formulation of his statistical model with exponentially increasing spectrum of resonances at large masses. This model considered in Sect. 2 introduced the new hypothetical physical constant—the limiting temperature $T_H \cong 160$ MeV.

Today we know that fundamental constituents of the strongly interacting matter are quarks and gluons, and their interactions are described by the quantum chromodynamics (QCD). This fact is important for our understanding of nature. It becomes clear that hadrons and resonances can not be the point-like objects, as it was assumed in the Hagedorn model. Hadrons and resonances should have an internal quark structure and, thus, non-zero proper volume. This was taken into account by the excluded volume procedure in statistical models with arbitrary number of hadron species discussed in Sect. 3.

An understanding of a fundamental role of quarks and gluons does not make however easier a theoretical description of relativistic nucleus-nucleus collisions. The main feature of these reactions at high collision energy is a production of huge amount of hadrons and resonances. Measured physical quantities are mostly presented in terms of hadrons. One still has to work with composite objects both at the beginning and the end of nucleus-nucleus collisions. Quarks and gluons can appear at the short early stage of nucleus-nucleus reaction and the only straightforward signals of this stage can be its specific electromagnetic radiation. The statistical model discussed in Sect. 4 treats matter at high energy densities in terms of hadron-like degrees of freedom—quark-gluon bags. The notions of hadrons have been changed with increasing of our knowledge on the strong interactions: hadrons, resonances, fireballs, and quark-gluon bags. Quark-gluon bags play a role of the bridge between hadron gas from one side and quarks-gluon plasma (QGP) from the other side. Section 5 summarizes my presentation.

2 Hagedorn Limiting Temperature

The first model of matter at high energy density was formulated in 1950 by Fermi [1]. It was assumed that a system created in high energy proton-proton collisions consists of pions and behaves as the black-body radiation, i.e., pions were considered as non-interacting particles, and the pion mass was neglected as compared to high temperature of the system.

In what follows we consider the system with zero values of the net baryon number, electric charge, and strangeness. These conditions correspond approximately to nucleus-nucleus collisions in Large Hadron Collider at CERN. The temperature T remains then the only independent thermodynamical variable in the thermodynamic limit when the system volume goes to infinity. The system pressure is defined in terms of the grand canonical partition function $Z(T, V)$ as (the system of units with $h/(2\pi) = c = k_B = 1$ will be used),

$$p(T) = T \lim_{V\to\infty} \frac{\ln Z(T, V)}{V} . \tag{1}$$

The pressure function plays a role of the thermodynamical potential in the grand canonical ensemble. The entropy density s and energy density ε can be calculated from $p(T)$ using the thermodynamical identities:

$$s(T) \equiv \frac{dp}{dT}, \qquad \varepsilon(T) \equiv T\frac{dp}{dT} - p . \tag{2}$$

Let us consider the ideal gas of particle with mass m and degeneracy factor g. The partition function can be calculated in the Boltzmann approximation as

$$Z(T, V) = \sum_{N=0}^{\infty} \frac{V^N}{N!} \left[g \int_0^{\infty} \frac{k^2 dk}{2\pi^2} \exp\left(-\frac{\sqrt{k^2 + m^2}}{T} \right) \right]^N \equiv \exp\left[V\phi_m(T) \right] , \tag{3}$$

where

$$\phi_m(T) = \frac{g}{2\pi^2} \int_0^{\infty} k^2 dk \, \exp\left(-\frac{\sqrt{k^2 + m^2}}{T} \right) = g \frac{m^2 T}{2\pi^2} K_2(m/T) , \tag{4}$$

with K_2 being the modified Bessel function. The function $\phi_m(T)$ has physical meaning as particle number density. Using Eqs. (1) and (2) one finds

$$p = T\phi_m(T) , \qquad \varepsilon = T^2 \frac{d\phi_m}{dT} . \tag{5}$$

At $T/m \gg 1$ one obtains

$$p \cong \frac{g}{\pi^2} T^4 , \quad \varepsilon \cong \frac{3g}{\pi^2} T^4 , \tag{6}$$

i.e., in the high temperature limit a behavior of the energy density has the familiar Stephan-Boltzmann (S-B) form, $\varepsilon = \sigma T^4$ with S-B constant $\sigma = 3g/\pi^2 \cong 0.30\,g$. The Bose and Fermi statistics lead to the same behavior $\varepsilon \sim T^4$ with the corresponding S-B constants $\sigma_{\text{Bose}} = \pi^2 g/30 \cong 0.33\,g$ and $\sigma_{\text{Fermi}} = 7\pi^2 g/240 \cong 0.29\,g$ which are only slightly different from their Boltzmann approximation.

In the opposite limit, $m/T \gg 1$, one obtains from Eq. (4)

$$\phi_m(T) \cong g \left(\frac{mT}{2\pi}\right)^{3/2} \exp\left(-\frac{m}{T}\right) . \tag{7}$$

In this large mass limit the quantum statistics effects play no role at all. In what follows we will neglect quantum statistics effects. Note that this Boltzmann approximation is rather reasonable for the analysis of nucleus-nucleus collisions at very high collision energies where the baryon chemical potential is approximately equal to zero. This classical approximation is however violated for systems with large baryon densities and moderate temperatures, e.g., Fermi statistics effects may play a crucial role in a description of nuclear matter created in nucleus-nucleus reactions at small collisions energies or matter formed inside neutron stars.

The number of particles N is a random variable and has the Poisson probability distribution with the average value

$$\langle N \rangle = V\,\phi_m(T) , \tag{8}$$

which leads to a most familiar form of the ideal gas equation of state

$$pV = \langle N \rangle T . \tag{9}$$

The only difference from the school textbook formula is that $\langle N \rangle$ in Eq. (9) is not a constant number but depends on the system volume and temperature according to Eq. (8).

One can generalize the above equations to the system of several particle species with masses $m_1, \ldots, m_n$:

$$p(T; m_1, \ldots, m_n) = \sum_{i=1}^{n} p(T; m_i) , \quad \varepsilon(T; m_1, \ldots, m_n) = \sum_{i=1}^{n} \varepsilon(T; m_i) , \tag{10}$$

where $p(T; m_i)$ and $\varepsilon(T; m_i)$ are given by Eq. (5) with $m = m_i$ and $g = g_i$. These expressions look again similar to the ideal gas formulae for a mixture of different atoms. However, in contrast to non-relativistic physics, the numbers of each particle species marked by the mass m_i are not constant values but are changed with the system temperature according to Eq. (8). This is a feature of relativistic physics. Only conserved charges, not just the number particles, are really conserved. Adding the energy, i.e., by increasing the temperature, one observes more and more new particles in the system. Note again that our discussion does not introduce any types of conserved charges.

Let us fix $m_1 < \cdots < m_N$ in Eq. (10) and extend the sums in Eq. (10) to infinity. For these series to exist it is necessary that nth term goes to zero at $n \to \infty$, this can be only achieved if $m_n \to \infty$ at $n \to \infty$, otherwise the infinite sums in Eq. (10) would be divergent. The lightest particle in the hadron spectrum is the pion with $m_\pi \cong 140$ MeV. Then one needs to add all known particles (and antiparticles). The particle degeneracy factor g_m, i.e., the number of the internal degrees of freedom for the particle with mass m, assumes that g_m equal terms are present in Eq. (10) for each m value.

It is convenient to introduce the mass spectrum density $\rho(m)$, i.e., $\rho(m)dm$ gives the number of different particle mass states including their internal degeneracies g_m in the interval $[m, m+dm]$. The partition function of the system takes then the following form

$$Z(T,V) = \sum_{N=0}^{\infty} \frac{V^N}{N!} \int_0^\infty \rho(m_1)dm_1 \int_0^\infty \frac{k_1^2 dk_1}{2\pi^2} \exp\left(-\frac{\sqrt{k_1^2+m_1^2}}{T}\right) \times \cdots \tag{11}$$

$$\times \int_0^\infty \rho(m_N)dm_N \int_0^\infty \frac{k_N^2 dk_N}{2\pi^2} \exp\left(-\frac{\sqrt{k_N^2+m_N^2}}{T}\right) = \exp\left[V \int_0^\infty \rho(m)\,\phi_m(T)\right].$$

The system pressure p, particle number density $n \equiv \langle N \rangle / V$, and energy density ε read

$$p = Tn = T\int_0^\infty dm\,\rho(m)\,\phi_m(T)\,, \qquad \varepsilon = T^2 \int_0^\infty dm\,\rho(m)\,\frac{d\phi_m}{dT}\,. \tag{12}$$

It was suggested by Hagedorn [2] that hadron mass spectrum increases exponentially at $m \to \infty$

$$\rho(m) \sim Cm^{-a}\exp(bm)\,, \tag{13}$$

with model parameters C, a, and b. The spectrum (13) was originally motivated by the data for the spectrum of known particles and resonances, and the exponential behavior of transverse momentum spectra of secondary particles in high energy collisions. Both arguments lead to the same value of $T_H \equiv 1/b \cong 160$ MeV. Later, Eq. (13) was also supported by the theoretical arguments formulated within the statistical bootstrap model [3]. From Eqs. (7) and (13) it follows that the integrands in Eq. (12) behave as $m^{3/2-a+l}\exp[-m(1/T - 1/T_H)]$ at $m \to \infty$ with $l = 0$ for p and $l = 1$ for ε. The values of $T_H = 1/b$ becomes, therefore, a limiting temperature (Hagedorn temperature), i.e., at $T > T_H$ the integrals in Eq. (12) become divergent. At $T \to T_H - 0$ a behaviour of the thermodynamical functions (12) with mass spectrum (13) depends crucially on parameter a in Eq. (13). For $T = T_H$ the exponential part of the integrands in Eq. (12) vanishes and the convergence or divergence of the integrals at their upper limit are defined by the parameter a:

$$p,\ n,\ \varepsilon\ \to \infty\ , \quad \text{for } a \le 5/2\ , \tag{14}$$

$$p,\ n,\ \to \text{const}\ , \quad \varepsilon\ \to \infty\ , \quad \text{for } 5/2 < a \le 7/2\ , \tag{15}$$

$$p,\ n,\ \varepsilon\ \to \text{const}\ , \quad \text{for } a > 7/2\ . \tag{16}$$

The system properties at T near T_H are dependent on contributions of heavy particles with $m \to \infty$. This contribution, in turn, are defined by the value of a. The limiting temperature singularity $T = T_H$ appears because of the exponential increasing factor, $\exp(m/T_H)$, of the mass spectrum (13), but the specific behaviour near this singular point is defined by the power factor, m^{-a}, in Eq. (13). The mass spectrum (13) was suggested by Hagedorn more than 50 years ago. A special name—fireball—was introduced for the heavy hadrons. They were considered as an extensions of hadron resonances to the high mass region. A clear experimental identification of the individual fireball states in the region of very large masses is rather problematic. The mass distinction of these excited states at high m becomes smaller then their expected decay widths. Even today, a presence of the fireball-like states in nature remains as the open question.

One feature of these states does not look however as the physical one. All particles including fireballs with $m \to \infty$ were treated as point-like objects. We overcome this unrealistic feature of the statistical bootstrap model. By taking into account particle proper volumes we hope to transform a limiting temperature T_H into the temperature of a phase transition.

3 Excluded Volume Effects

We introduce now the particle proper volume effects. For fixed particle number, N, Eq. (8), $pV = NT$, will be modified according to the van der Waals excluded volume procedure with v_0 being the proper volume particle parameter

$$p(V - v_0 N)\ =\ NT\ . \tag{17}$$

The grand canonical partition function (3) is then transformed into

$$Z(T, V)\ =\ \sum_{N=0}^{\infty} \frac{[(V - v_0 N)\, \phi_m]^N}{N!}\, \theta(V - v_0 N)\ . \tag{18}$$

To proceed further one can use the Laplace transformation of Eq. (18):

$$\hat{Z}(T,s) \equiv \int_0^\infty ds\, \exp(-sV)\, Z(T,V) = \left[s - \exp(-v_0 s)\phi_m(T)\right]^{-1}. \quad (19)$$

An exponentially increasing part of the partition function behaves as $Z(T,V) \sim \exp(pV/T)$ and generates the singularity of the function $\hat{Z}$ in variable s. The farthest-right singularity s^* gives us the system pressure,

$$p(T) = T\,s^*. \quad (20)$$

This is because at $s^* > p(T)/T$ the V-integral in Eq. (19) diverges at its upper limit. The connection of the farthest-right s-singularity of $\hat{Z}(T,s)$ to the asymptotic $V \to \infty$ behaviour of $Z(T,V)$ given by Eq. (19) is a general mathematical property of the Laplace transform. The farthest-right s-singularity of the function (19) is a simple pole. It leads to the following transcendental equation for the system pressure [4]

$$p(T) = \exp\left(-\frac{v_0 p(T)}{T}\right) T\phi_m(T). \quad (21)$$

At $v_0 = 0$ Eq. (21) is reduced to the ideal gas result of Eq. (5).

Equation (19) can be generalize for an arbitrary number of types of particles $(m_1, v_1), \ldots, (m_n, v_n)$:

$$\hat{Z}(T,s) = \left[s - \sum_{j=1}^{n} \exp(-v_j s)\phi_{m_j}(T)\right]^{-1}. \quad (22)$$

As long as the number n is finite, the farthest-right singularity s^* of the function (22) is always the pole. We denote this pole point as $s_H(T)$. The equation of state of the system reads then as

$$p(T) = \sum_{j=1}^{n} \exp\left(-\frac{v_j\, p(T)}{T}\right) T\phi_{m_j}(T). \quad (23)$$

Our final step in the model formulation is to extend the summation over different (m, v)-types of particles up to infinity. As before, it is convenient to work with an integral over particle mass-volume spectrum. The spectrum function $\rho(m, v)$ is introduced, so that $\rho(m, v)dmdv$ gives the number of different mass-volume states. The sum $\sum_{j=1}^{n} \ldots$ over different particle species in (22) is then replaced by $\int dmdv\rho(m, v)\ldots$ integral.

4 Statistical Models of Bags

The statistical model with $\rho(m, v)$ mass-volume spectrum is defined by the following formulae [5, 6]:

$$\hat{Z}(T, s) \equiv \int_0^\infty dV \exp(-sV)\, Z(T, V) = \left[s - \int dm dv\, \rho(m, v) \exp(-vs) \phi_m(T) \right]^{-1}$$
$$= [s - f(T, s)]^{-1} , \quad (24)$$
$$p(T) \equiv T \lim_{V \to \infty} \frac{\ln Z(T, V)}{V} = T s^*(T) , \quad (25)$$

where $s^*(T)$ is the farthest-right s-singularity of the function $\hat{Z}(T, s)$. Note that all models discussed in the previous sections can be obtained using the particular choices of the $\rho(m, v)$ spectrum when this spectrum is reduced to corresponding special forms of either $\rho(m)$ or to $\rho(v)$ functions.

One possible singular point of the function $\hat{Z}$ in variable s is evidently the pole singularity, $s^* = s_H$ defined by the transcendental equation

$$s_H(T) = f(T, s_H) . \quad (26)$$

If the mass-volume spectrum $\rho(m, v)$ is restricted by a finite number of states, (m_j, v_j), the farthest-right singularity s^* is always the pole s_H given by Eq. (26). For an infinite number of states, another singular point $s^* = s_Q$ can emerge. This is a possible singularity of the function $f(T, s)$ itself, which can appear due to a divergence of the $dmdv$-integrals at their upper limits. The equation of state takes then the form:

$$p(T) = \max\{s_H(T), s_Q(T)\} . \quad (27)$$

The mathematical mechanism for possible phase transitions in our model is the 'coalescence' (coincidence) of the two singularities $s_H(T)$ and $s_Q(T)$. Note that the possible phase transitions reveal themselves as the singularities of the $p(T)$ function, and they can only appear in the thermodynamical limit $V \to \infty$.

The mass-volume spectrum $\rho(m, v)$ is supposed to reproduce the known low-lying hadron states. The region where both m and v are large will be described within the bag model [7]. The density of states of quarks and gluons with the total energy $m - Bv$ ($B = \text{const} > 0$) can be then presented at large m and v as

$$\rho(m, v) \cong C\, v^\gamma\, (m - Bv)^\delta \exp\left[\frac{4}{3} \sigma_Q^{1/4} m^{1/4} (m - Bv)^{3/4} \right] , \quad (28)$$

where C, γ, and δ are the model parameters, and σ_Q is the Stefan-Boltzmann constant counting gluons (spin, color) and (anti-)quarks (spin, color, flavor) states inside the bag.

The position of the singularity s_Q as a function of T can be calculated analytically,

$$s_Q(T) = \frac{1}{3}\sigma_Q T^3 - \frac{B}{T} . \tag{29}$$

The value of s_H given by Eq. (26) is always positive, whereas s_Q given by Eq. (29) becomes negative at small T. Thus, at small T the farthest-right singularity s^* is always the pole s_H, and the equation of state is mainly defined by the contributions of low-laying hadron and resonance states, i.e., by the small bags. If s_Q becomes larger than s_H at large T the system pressure then reads:

$$p(T) = T\, s_Q(T) = \frac{1}{3}\sigma_Q T^4 - B , \tag{30}$$

and corresponds to the behavior of the ideal gas of quarks and gluons corrected by the vacuum pressure B. The energy density in this case equal to

$$\varepsilon = T\frac{dp}{dT} - p = \sigma_Q T^4 + B , \tag{31}$$

and the system behaves as one infinitely large bag.

The s_Q singularity is generated by the exponential factor in mass-volume bag spectrum $\rho(m, v)$ (28) and is dependent on its two parameters, σ_Q and B. However, a possibility and types of phase transitions in the system of bags are defined by the pre-exponential factors in Eq. (28) and depend crucially on the powers γ and δ. This resembles the case of the statistical bootstrap model discussed in Sect. 2. The limiting temperature $T_H = 1/b$ appears because of the exponentially increasing factor $\exp(bm)$ in the mass spectrum (13), but the thermodynamical behavior at $T \to T_H$ depends crucially on the value of the parameter a in the pre-exponential factor, m^{-a}, in the fireball mass spectrum $\rho(m)$. The parameters γ and δ depend significantly on the set of selected constraints. For example, original values $\gamma = -7/4$ and $\delta = -7/4$ are transformed to $\gamma = -9/4$ and $\delta = -19/4$ under an additional selection of only colorless quark-gluon bags. These constraints in statistical mechanics are usually ignored as the related to them pre-exponential factors in the partition function do not contribute to the equation of state in the thermodynamic limit. Because of this possible sets of physical constraints are not well defined and their consequences were not carefully studied. In what follows the values of γ and δ will be considered as the free model parameters.

The function $f(T, s)$ has the singular point $s = s_Q$, and $f(T, s_Q)$ can be either finite or infinite depending on the value of $\gamma + \delta$. The phase transition never exists for

$$\gamma + \delta \geq -3 \tag{32}$$

because of $f(T, s_Q) = \infty$, and this case corresponds to $s_H > s_Q$ at all T. Inequality

$$\gamma + \delta < -3 \tag{33}$$

leads to finite values of $f(T, s_Q)$. An existence of a phase transition becomes then possible and depends on the value of the parameter δ. A phase transition takes place if, additionally to (33), one more inequality takes place,

$$\delta < -7/4 . \tag{34}$$

The phase transition temperature $T = T_c$ corresponds to an intersection of two singularities,

$$s_H(T_c) = s_Q(T_c) , \tag{35}$$

with $s_H > s_Q$ at $T < T_c$ and $s_Q > s_H$ at $T > T_c$. For γ and δ parameters in the region $\gamma + \delta < -4$, the system of bags exhibits the first-order phase transition at $T = T_c$. In this case there are discontinuities of the energy density, $\varepsilon(T_c + 0) - \varepsilon(T_c - 0) > 0$, and entropy density. The physical system at $T = T_c$ is the mixed phase of the finite size bag-hadrons and infinite size bags which correspond to the QGP equation of state (30).

The values of γ and δ in the region $-4 < \gamma + \delta < -3$ lead to the second- and higher-orders phase transition in the system of bags at $T = T_c$. These possibilities were discussed in Ref. [8]. The behavior of thermodynamical functions at the second- or higher-order phase transitions is the following. In a comparison to the first-order phase transition, at the higher-order transitions there is a stronger contribution from large-size bags at $T \to T_c - 0$. This makes the first derivative of the pressure $p(T)$ as a continues function at $T = T_c$. Note also that at $T \to T_c - 0$, a contribution of large size bags to the energy density is much stronger than that to the pressure. This because the large bags contribute strongly to the energy density due to their large masses. The average volume of the bag $\overline{v}$ increases in the vicinity of the phase transition temperature, and and $\overline{v} \to \infty$ at $T \to T_c - 0$.

The basic equations of the bag-gas model include the information of low-mass (hadrons) bag spectrum. This information is important for the low-temperature behavior. However, an absence or presence of the phase transitions (and the order of possible phase transitions) are completely defined by a behavior of the mass-volume spectrum of bags (28) at large $m \to \infty$ and $v \to \infty$. A possibility that the parameters δ and γ may change their values at the non-zero baryon chemical potential was discussed in Ref. [9].

If the inequalities (33) and (34) are not satisfied simultaneously the phase transition does not take place in the statistical bag model. This possibility was considered in Ref. [10]. In this case, inequality $s_H > s_Q$ takes place at all T, thus, no intersection of the singularities s_H and s_Q happens. The average bag volume $\overline{v}$ can either remain finite or increase to infinity at $T \to \infty$. However, in both cases the QGP equation of state

$$p \cong \frac{1}{3}\sigma_Q T^4 , \quad \varepsilon \cong \sigma_Q T^4 , \tag{36}$$

becomes approximately valid at $T \to \infty$. The system of bags transforms smoothly to the QGP equation of state (36) in some region of temperature. Just such a picture is suggested by the present lattice QCD results at zero baryon chemical potential. The crossover region with $T \sim T_{cr}$ can be rather narrow for some types of physical observables, but rather wide for the others.

5 Summary

In this presentation we have studied the gas of the quark-gluon bags at different system temperature T. The mass-volume spectrum function $\rho(m, v)$ for the quark-gluon bags is taken in the form (28). It describes the hypothetical bag states for large m and v. Note that an approximation of point-like bags, $\rho(m, v) = \rho(m)\delta(v)$, with $\rho(m)$ given by Eq. (13) leads to the Hagedorn model and the limiting temperature T_H. In addition to $\rho(m, v)$ (28), the experimental spectrum of low-lying hadron and resonance states should be added. These states describe a behavior of the statistical system at small T. However, a presence and the types of phase transitions in the system of quark-gluon bags are defined by the spectrum (28) and depends crucially on the values of the γ and δ parameters. Different regions of the γ and δ values lead to different behavior of the quark-gluon bags. Particularly, to the different orders of possible phase transitions. On the other hand, the pressure p and energy density ε have always the same asymptotic behavior (36) at high temperature for all different values of γ and δ. This happens even in the absence of the phase transition and corresponds to the equation of state of non-interacting quarks and gluons inside the bags, i.e., to the ideal quark-gluon plasma. The average volume of the bag $\overline{v}$ and its average mass $\overline{m}$ can have however rather different behavior in different regions of the γ and δ parameters. If the system of quark-gluon bags has no phase transition, the values of $\overline{v}$ and $\overline{m}$ can remain finite at high temperature. Such a cluster QGP can be rather different from the ideal QGP despite of the similar to that equation of state.

Several aspects of the statistical model of quark-gluon bags were developed in my collaboration with Walter Greiner [4, 8–10]. Walter liked the idea of the cluster QGP and believed that these colorless clusters can be observed in the future experiments on high energy physics.

Acknowledgements This presentation is dedicated to my friend Walter Greiner. The scientific discussions with Walter and his support of my activity were very important in all aspects of my life during the last three decades.

References

1. E. Fermi, Prog. Theor. Phys. **5**, 570 (1950)
2. R. Hagedorn, Nuovo Cim. Suppl. **3**, 147 (1965)
3. S. Frasustchi, Phys. Rev. D **3**, 2821 (1971)
4. D.H. Rischke, M.I. Gorenstein, H. Stöcker, W. Greiner, Z. Phys. C **51**, 484 (1991)
5. M.I. Gorenstein, V.K. Petrov, G.M. Zinovjev, Phys. Lett. B **106**, 327 (1981)
6. M.I. Gorenstein, V.K. Petrov, V.P. Shelest, G.M. Zinovev, Theor. Math. Phys. **52**, 843 (1982)
7. A. Chodos et al., Phys. Rev. D **9**, 3471 (1974)
8. M.I. Gorenstein, W. Greiner, S.N. Yang, J. Phys. G **24**, 725 (1998)
9. M.I. Gorenstein, M. Gazdzicki, W. Greiner, Phys. Rev. C **72**, 024909 (2005)
10. V.V. Begun, M.I. Gorenstein, W. Greiner, J. Phys. G **36**, 095005 (2009)

The Charming Beauty of the Strong Interaction

Laura Tolos

Abstract Charmed and beauty hadrons in matter are discussed within a unitarized coupled-channel model consistent with heavy-quark spin symmetry. We analyse the formation of D-mesic states as well as the propagation of charmed and beauty hadrons in heavy-ion collisions from LHC to FAIR energies.

1 Introduction

One of the main research activities in nuclear and particle physics is the exploration of the Quantum Chromodynamics (QCD) phase diagram for high density and/or temperature. Up to now the studies have been concentrated on light quarks due to energy constraints of the experimental setups, but with the upcoming research facilities, the goal is to move to the heavy-quark domain, where heavy degrees of freedom, such as charm and beauty, play a crucial role.

In order to understand the QCD phase diagram, one needs first to understand the interaction between heavy hadrons. In particular, the nature of newly discovered heavy excited states is of major concern, whether they can be described within the standard quark model and/or better understood as dynamically generated states via hadron-hadron interactions.

Given the success of unitarized coupled-channel approaches in the description of some of the existing experimental data in the light-quark sector, charmed and beauty degrees of freedom have been recently incorporated in these models and

L. Tolos (✉)
Institut für Theoretische Physik, University of Frankfurt, Max-von-Laue-Str. 1, 60438 Frankfurt am Main, Germany
e-mail: tolos@th.physik.uni-frankfurt.de

Frankfurt Institute for Advanced Studies, University of Frankfurt, Ruth-Moufang-Str. 1, 60438 Frankfurt am Main, Germany

Institute of Space Sciences (ICE, CSIC), Campus UAB, Carrer de Can Magrans, 08193 Barcelona, Spain

Institut d'Estudis Espacials de Catalunya (IEEC), 08034 Barcelona, Spain

J. Kirsch et al. (eds.), *Discoveries at the Frontiers of Science*, FIAS Interdisciplinary Science Series, https://doi.org/10.1007/978-3-030-34234-0_3

several experimental states have been described as excited baryon molecules. Some examples can be found in Refs. [1–26]. However, some of these models are not fully consistent with heavy-quark spin symmetry (HQSS) [27], which is a QCD symmetry that appears when the quark masses become larger than the typical confinement scale. Thus, a model that incorporates HQSS constraints has been developed in the past years [28–35].

Once the interaction between heavy hadrons has been determined, the study of the properties of heavy hadrons in nuclear matter requires the inclusion of nuclear medium modifications. In this way, it is possible to study the formation of heavy mesic states in nuclei [36–38], as well as the propagation of charmed and beauty in heavy-ion collisions [39–43], all of these topics matter of the present paper.

2 Charm Under Extreme Conditions

2.1 Excited Charmed Baryons

Recently a predictive model has been developed for four flavors including all ground-state hadrons (pseudoscalar and vector mesons, and $1/2^+$ and $3/2^+$ baryons). This scheme reduces to the Weinberg-Tomozawa (WT) interaction when Goldstone bosons are involved and includes HQSS in the sector where heavy quarks appear. In fact, this model is justified due to the results of the SU(6) extension in the three-flavor sector [44] and is based on a formal plausibleness on how the interactions between heavy pseudoscalar mesons and baryons emerge in the vector-meson exchange picture. The WT potential can be then used to solve the on-shell Bethe-Salpeter equation in coupled channels so as to calculate the scattering amplitudes. The poles of the scattering amplitudes are the dynamically-generated charmed baryonic resonances (see Ref. [33] for a review).

Dynamically generated states with different charm and strangeness are predicted in Refs. [28–30]. The studies are constrained to the states coming from the most attractive representations of the SU(6) $\times$ HQSS scheme. Some of them can be identified with known states from the PDG [45], by comparing the PDG data on these states with the mass, width and the dominant couplings to the meson-baryon channels.

In this work, as an example, the results in the $C = 1, S = 0, I = 0$ are presented. In Ref. [30] three Λ_c and one Λ_c^* are obtained. A pole around 2618.8 MeV is identified with the experimental $\Lambda_c(2595)$ resonance, while a second broad Λ_c state at 2617 MeV shows a similar two-pole pattern as in the $\Lambda(1405)$ case [46], coupling strongly to $\Sigma_c\pi$. The third spin-1/2 Λ_c around 2828 MeV cannot be assigned to any experimental state. With regards to the spin-3/2 Λ_c state, this is assigned to $\Lambda_c(2625)$.

2.2 Charmed Hadrons in Matter

The inclusion of dense matter effects modifies the mass and width of the dynamically-generated baryonic states. On l.h.s of Fig. 1 the squared amplitude of D^*N-D^*N transition is shown for different partial waves as function of the center-of-mass energy for zero total momentum. In this case, the position of the $\Lambda_c(2595)$ is fitted and, as a result, the masses of the other states are slightly modified as compared to the results reported in the previous section.

The SU(6) × HQSS model in the $I = 0, 1$, $J = 1/2, 3/2$ sectors predicts several states that can have experimental confirmation: ($I = 0$, $J = 1/2$) $\Lambda_c(2595)$, ($I = 1$,$J = 1/2$) $\Sigma_c(2823)$ and $\Sigma_c(2868)$, ($I = 0$, $J = 3/2$) $\Lambda_c(2660)$,($I = 0$, $J = 3/2$) $\Lambda_c(2941)$, ($I = 1$, $J = 3/2$) $\Sigma_c(2554)$ and ($I = 1$, $J = 3/2$) $\Sigma_c(2902)$ resonances. Several medium modifications are considered: no in-medium corrections, the inclusion of Pauli blocking on the nucleon intermediate states at saturation density $\rho_0 = 0.17\ \text{fm}^{-3}$, and the in-medium solution that takes into account Pauli blocking as well as the self-consistent inclusion of the D and D^* self-energies. These states are modified in mass and width depending on the strength of the coupling to meson-baryon channels with D, D^* and N as well as the closeness to the DN or D^*N thresholds.

The knowledge of the in-medium modified dynamically generated excited baryon states is of extreme importance for the determination of the properties of open-charm mesons, such as D and $\bar{D}$ mesons, in the nuclear medium. And the modification of the properties of open-charm mesons is of crucial value because of the implications for charmonium suppression [48] and the possible formation of D-meson bound states in nuclei [49].

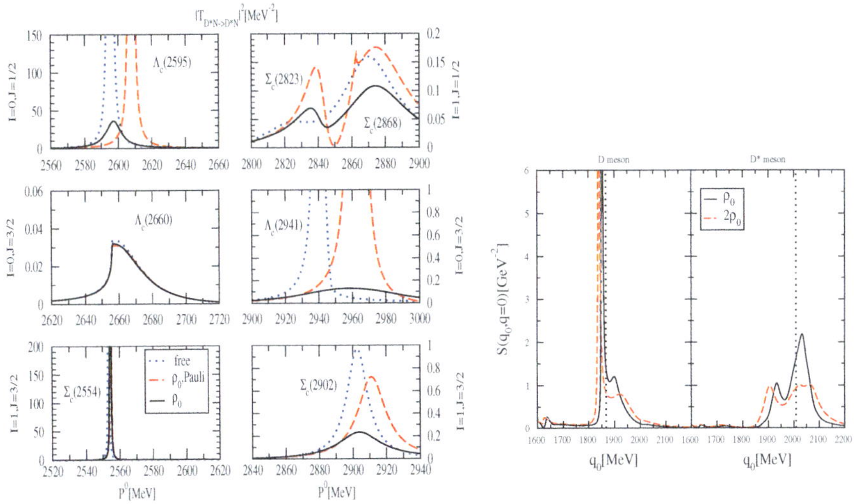

Fig. 1 Left: Charmed baryonic resonances in dense matter (taken from Ref. [47]). Right: The D and D^* spectral functions in dense nuclear matter at zero momentum (taken from Ref. [47])

The properties of open charm in dense matter have been analysed in different schemes: QMC schemes [49], QCD sum-rule approaches [50–52], NJL models [53], chiral effective models in matter [54] or pion-exchange approaches that incorporate heavy-quark symmetry constraints [55]. Nevertheless, the full spectral features of the open-charm mesons in matter have been obtained in self-consistent unitarized coupled-channel models, where the intermediate meson-baryon channels are modified including medium corrections [1, 2, 7–9, 37, 47, 56]. On the r.h.s. of Fig. 1 the D and D^* spectral functions are displayed for two densities at zero momentum. The D-meson quasiparticle peak mixes strongly with $\Sigma_c(2823)N^{-1}$ and $\Sigma_c(2868)N^{-1}$ particle-hole excitations, whereas the $\Lambda_c(2595)N^{-1}$ is visible in the low-energy tail. The D^* spectral function incorporates the $J = 3/2$ resonances, and the quasiparticle peak mixes with the $\Sigma_c(2902)N^{-1}$ and $\Lambda_c(2941)N^{-1}$. For both D and D^* mesons, the particle-hole modes smear out with density while the spectral functions broaden.

2.3 D-Meson Bound States in Nuclei

Since the work of Ref. [49], there have been speculations about the formation of D-meson bound states in nuclei, which are based on the assumption of an attractive interaction between D-mesons and nucleons. Within the model of Ref. [36], D^0-mesons bind in nucleus but very weakly (see Fig. 2), in contradistinction to [49]. Moreover, D^0-mesic states show significant widths. No D^+-nuclear states are found since the Coulomb interaction prevents their formation. As for D^- or $\bar{D}^0$, both mesons bind in nuclei as seen in Fig. 3, though only nuclear states are manifest for $\bar{D}^0$ in nuclei. The atomic states for D^- are less bound as compared to the pure Coulomb levels, whereas the nuclear ones are more bound and might show a significant width, appearing only for low angular momenta [37]. These results are close to [55], but in contrast to [49] for ^{208}Pb.

Fig. 2 D^0-nucleus bound states (taken from [36])

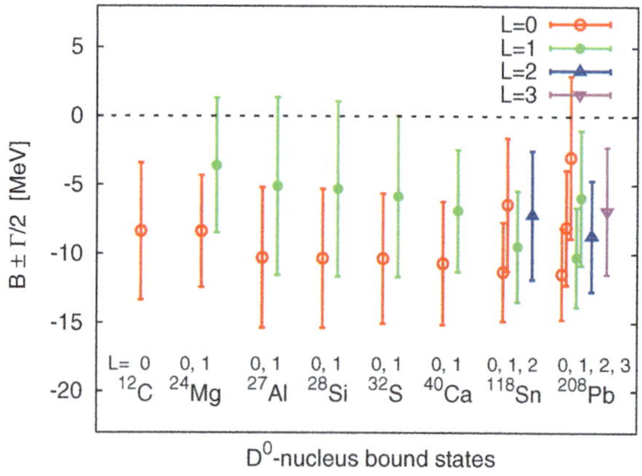

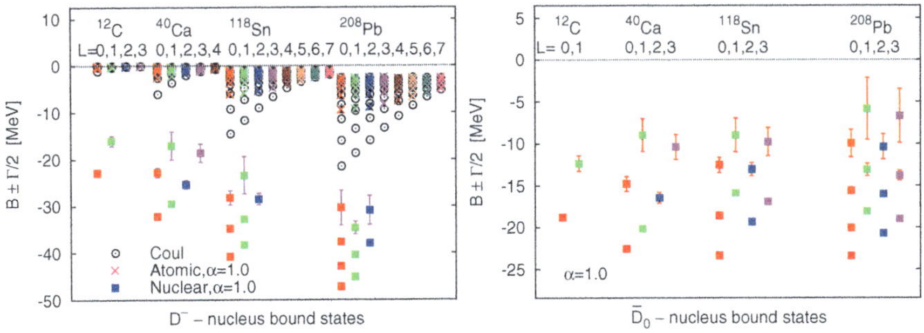

Fig. 3 D^- and $\bar{D}^0$-nucleus bound states (taken from [37])

As shown in Ref. [38], the experimental detection of D and $\bar{D}$-meson bound states is complicated. Reactions of the type ($\bar{p}$, D+N) or ($\bar{p}$, D+2N) may be indeed possible at PANDA with antiproton beams, as long as formation cross sections are not suppressed as well as small or even zero momentum transfer reactions are feasible. More successful mechanisms could involve the emission of pions by intermediate D^* or $\bar{D}^*$, while the resulting open-charm mesons are trapped by the nucleus [38].

2.4 D-Meson Propagation in Hot Matter

The transport coefficients of D mesons in the hot dense medium created in heavy-ion collisions offer the possibility to analyse the interaction of D mesons with light mesons and baryons. Using the Fokker-Planck description, the drag (F_i) and diffusion coefficients (Γ_{ij}) of D mesons in hot dense matter can be obtained using an effective field theory that incorporates both the chiral and HQSS in the meson [57] and baryon sectors [39].

The spatial diffusion coefficient D_x, that appears in Fick's diffusion law, is a relevant quantity that involves both the drag and diffusion coefficients. Within an isotropic bath, the spatial diffusion coefficient reads

$$D_x = \lim_{p \to 0} \frac{\Gamma(p)}{m_D^2 F^2(p)} , \tag{1}$$

as a function of the scalar $F(p)$ and $\Gamma(p)$ coefficients. The interest on the spatial diffusion coefficient relies on the fact that it might show an extremum around the transition temperature between the hadronic and QGP phases, as seen previously for the shear and bulk viscosities [39].

In Fig. 4 the $2\pi T D_x$ is shown around the transition temperature following isentropic trajectories ($s/n_B = $ ct) from RHIC to FAIR energies. The matching between curves in both phases for a given value of s/n_B seems to indicate the possible existence of a minimum in the $2\pi T D_x$ at the phase transition [41, 58].

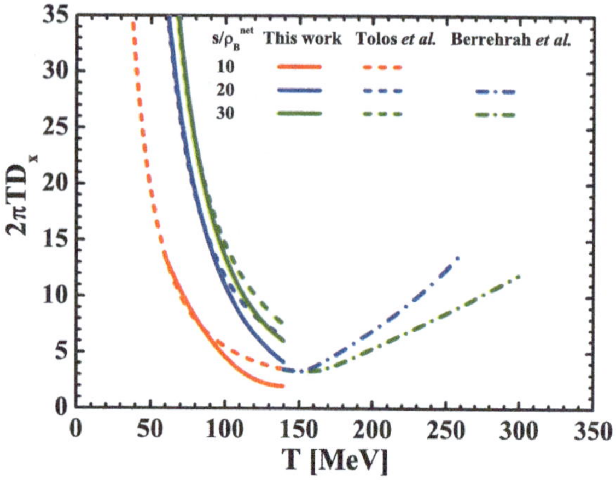

Fig. 4 The coefficient $2\pi T D_x$ around the transition temperature (taken from Ref. [41])

3 Beauty Under Extreme Conditions

3.1 *Excited Beauty Baryons*

The LHCb Collaboration has observed two narrow baryon resonances with beauty, being their masses and decay modes consistent with the quark model orbitally excited states $\Lambda_b(5912)$ and $\Lambda_b^*(5920)$, with $J^P = 1/2^-$ and $3/2^-$, respectively [59].

The existence of these states is predicted within the unitarized meson-baryon coupled-channel dynamical model, which implements HQSS and has been presented in Sect. 2. A summary of the predictions is graphically shown in Fig. 5. Within that scheme, the experimental $\Lambda_b^{(*)}$ states are identified as HQSS partners, explaining their approximate mass degeneracy. An analogy is found between the bottom, charm and strange sectors, given that the $\Lambda_b^0(5920)$ is the bottomed counterpart of the $\Lambda^*(1520)$ and $\Lambda_c^*(2625)$ states. Moreover, the $\Lambda_b^0(5912)$ belongs to the two-pole structure similar as the one seen in the case of the $\Lambda(1405)$ and $\Lambda_c(2595)$.

Mass and decay modes are also predicted for some $\Xi_b(1/2^-)$ and $\Xi_b(3/2^-)$, that belong to the same SU(3) multiplets as the $\Lambda_b(1/2^-)$ and $\Lambda_b(3/2^-)$. Three $\Xi_b(1/2^-)$ and one $\Xi_b(3/2^-)$ states are obtained coming from the most attractive SU(6) $\times$ HQSS representations. Two of these states, $\Xi_b(6035.4)$ and $\Xi_b^*(6043.3)$, form a HQSS doublet similar to that of the experimental $\Lambda_b(5912)$ and $\Lambda_b^*(5920)$. Nevertheless, none of these states have been detected yet.

3.2 $\bar{B}$ *Meson and* Λ_b *Propagation in Hot Matter*

The physical observables in heavy-ion collisions, such as particle ratios, R_{AA} or v_2, are strongly correlated to the behavior of the transport properties of heavy hadrons.

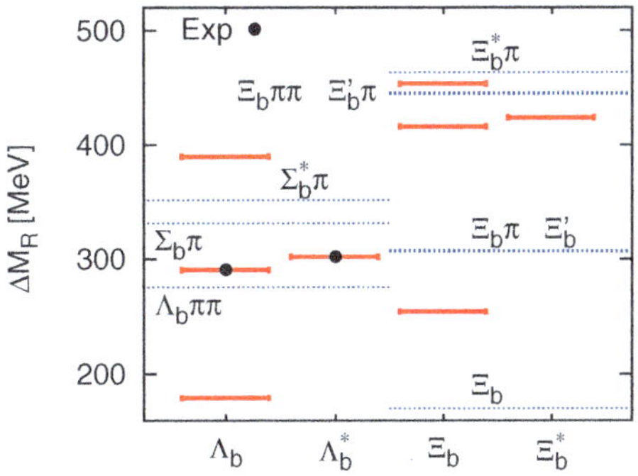

Fig. 5 Summary of the new predicted $\Lambda_b^{(*)}$ and $\Xi_b^{(*)}$ states (red lines). We also show the experimentally observed $\Lambda_b^0(5912)$ and $\Lambda_b^0(5920)$ states (black dots) and some relevant hadronic thresholds (blue dotted lines). These results have been taken from [31]

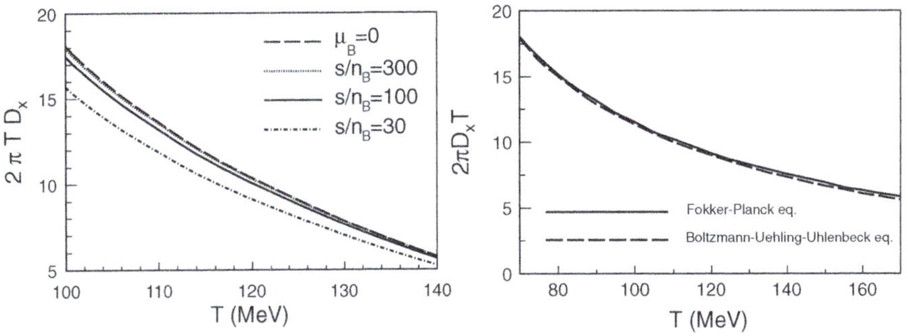

Fig. 6 Spatial diffusion coefficient, D_x, multiplied by the thermal wavenumber ($2\pi T$), for $\bar{B}$ meson (left, taken from [40]) and Λ_b (right, taken [64]). While the $\bar{B}$ spatial diffusion coefficient is shown for different isentropic trajectories, only the $\mu_B = 0$ case is presented for Λ_b

The transport properties depend on the interactions of the heavy particles with the surrounding medium, and these are described by means of effective theories that incorporate the heavy degrees of freedom.

Following the initial works of Refs. [57, 60–63], the effective interaction of heavy mesons, such as D [39] and $\bar{B}$ [40] mesons, with light mesons and baryons has been obtained by exploiting chiral and heavy-quark symmetries (see Sect. 2.1 for the case of baryons with charm). With these interactions, the heavy-meson transport coefficients are obtained as a function of temperature and baryochemical potential of the hadronic bath using the Fokker-Planck equation, as described in Sect. 2.4.

In Fig. 6 the spatial diffusion coefficients D_x, multiplied by the thermal wavenumber ($2\pi T$), for $\bar{B}$ and also Λ_b are presented, as derived in Eq. 1. On the l.h.s the spatial diffusion coefficient for $\bar{B}$ is shown for different isentropic trajectories. The results are quite independent of the entropy per baryon as long as it is high enough, that is, the collision energy is sufficiently high. Thus, these results can be taken as prediction for the hadronic medium created at high energy collisions (like those at the RHIC or the LHC), independently of the precise value of the entropy per baryon of the trajectory. Although the relaxation time is smaller with larger baryonic density, the $\bar{B}$ meson can hardly relax to the equilibrium. Moreover, on the r.h.s of Fig. 6 the analogous coefficient for Λ_b is shown, but only for $\mu_B = 0$. It is obtained that the outcome for the spatial diffusion coefficient from the Fokker-Planck formalism is in good agreement to the one coming from the solution of the Boltzmann-Uehling-Uhlenbeck transport equation. Moreover, a similar behavior for both the $\bar{B}$ and Λ_b spatial diffusion coefficients is observed at $\mu_B = 0$, due to the comparable mass and cross sections [64]. The phenomenological implications of these findings in heavy-ion collisions have been analysed in [65].

Acknowledgements The author warmly thanks Santosh K. Das, Carmen Garcia-Recio, Juan Nieves, Lorenzo L. Salcedo, Olena Romanets and Juan M. Torres-Rincon for their collaboration and discussions that has made this work possible. She also acknowledges support from the Heisenberg Programme of the Deutsche Forschungsgemeinschaft under the Project Nr. 383452331, the Ramón y Cajal research programme, FPA2013-43425-P and FPA2016-81114-P Grants from MINECO, and THOR COST Action CA15213.

References

1. L. Tolos, J. Schaffner-Bielich, A. Mishra, Phys. Rev. C **70**, 025203 (2004). https://doi.org/10.1103/PhysRevC.70.025203
2. L. Tolos, J. Schaffner-Bielich, H. Stoecker, Phys. Lett. B **635**, 85 (2006). https://doi.org/10.1016/j.physletb.2006.02.045
3. M.F.M. Lutz, E.E. Kolomeitsev, Nucl. Phys. A **730**, 110 (2004). https://doi.org/10.1016/j.nuclphysa.2003.10.012
4. M.F.M. Lutz, E.E. Kolomeitsev, Nucl. Phys. A **755**, 29 (2005). https://doi.org/10.1016/j.nuclphysa.2005.03.110
5. J. Hofmann, M.F.M. Lutz, Nucl. Phys. A **763**, 90 (2005). https://doi.org/10.1016/j.nuclphysa.2005.08.022
6. J. Hofmann, M.F.M. Lutz, Nucl. Phys. A **776**, 17 (2006). https://doi.org/10.1016/j.nuclphysa.2006.07.004
7. M.F.M. Lutz, C.L. Korpa, Phys. Lett. B **633**, 43 (2006). https://doi.org/10.1016/j.physletb.2005.11.046
8. T. Mizutani, A. Ramos, Phys. Rev. C **74**, 065201 (2006). https://doi.org/10.1103/PhysRevC.74.065201
9. L. Tolos, A. Ramos, T. Mizutani, Phys. Rev. C **77**, 015207 (2008). https://doi.org/10.1103/PhysRevC.77.015207
10. C.E. Jimenez-Tejero, A. Ramos, I. Vidana, Phys. Rev. C **80**, 055206 (2009). https://doi.org/10.1103/PhysRevC.80.055206
11. J. Haidenbauer, G. Krein, U.G. Meissner, A. Sibirtsev, Eur. Phys. J. A **33**, 107 (2007). https://doi.org/10.1140/epja/i2007-10417-3
12. J. Haidenbauer, G. Krein, U.G. Meissner, A. Sibirtsev, Eur. Phys. J. A **37**, 55 (2008). https://doi.org/10.1140/epja/i2008-10602-x
13. J. Haidenbauer, G. Krein, U.G. Meissner, L. Tolos, Eur. Phys. J. A **47**, 18 (2011). https://doi.org/10.1140/epja/i2011-11018-3
14. J.J. Wu, R. Molina, E. Oset, B.S. Zou, Phys. Rev. Lett. **105**, 232001 (2010). https://doi.org/10.1103/PhysRevLett.105.232001
15. J.J. Wu, R. Molina, E. Oset, B.S. Zou, Phys. Rev. C **84**, 015202 (2011). https://doi.org/10.1103/PhysRevC.84.015202
16. J.J. Wu, T.S.H. Lee, B.S. Zou, Phys. Rev. C **85**, 044002 (2012). https://doi.org/10.1103/PhysRevC.85.044002

17. E. Oset, A. Ramos, E.J. Garzon, R. Molina, L. Tolos, C.W. Xiao, J.J. Wu, B.S. Zou, Int. J. Mod. Phys. E **21**, 1230011 (2012). https://doi.org/10.1142/S0218301312300111
18. W.H. Liang, T. Uchino, C.W. Xiao, E. Oset, Eur. Phys. J. A **51**(2), 16 (2015). https://doi.org/10.1140/epja/i2015-15016-1
19. Y. Yamaguchi, S. Ohkoda, S. Yasui, A. Hosaka, Phys. Rev. D **87**(7), 074019 (2013). https://doi.org/10.1103/PhysRevD.87.074019
20. J.X. Lu, Y. Zhou, H.X. Chen, J.J. Xie, L.S. Geng, Phys. Rev. D **92**(1), 014036 (2015). https://doi.org/10.1103/PhysRevD.92.014036
21. W.H. Liang, C.W. Xiao, E. Oset, Phys. Rev. D **89**(5), 054023 (2014). https://doi.org/10.1103/PhysRevD.89.054023
22. L. Roca, J. Nieves, E. Oset, Phys. Rev. D **92**(9), 094003 (2015). https://doi.org/10.1103/PhysRevD.92.094003
23. J.X. Lu, H.X. Chen, Z.H. Guo, J. Nieves, J.J. Xie, L.S. Geng, Phys. Rev. D **93**(11), 114028 (2016). https://doi.org/10.1103/PhysRevD.93.114028
24. A. Hosaka, T. Hyodo, K. Sudoh, Y. Yamaguchi, S. Yasui, Prog. Part. Nucl. Phys. **96**, 88 (2017). https://doi.org/10.1016/j.ppnp.2017.04.003
25. G. Montana, A. Feijoo, A. Ramos, Eur. Phys. J. A **54**(4), 64 (2018). https://link.springer.com/article/10.1140%2Fepja%2Fi2018-12498-1
26. V.R. Debastiani, J.M. Dias, W.H. Liang, E. Oset, Phys. Rev. D **97**(9), 094035 (2018). https://journals.aps.org/prd/abstract/10.1103/PhysRevD.97.094035
27. N. Isgur, M.B. Wise, Phys. Lett. B **232**, 113 (1989). https://doi.org/10.1016/0370-2693(89)90566-2
28. C. Garcia-Recio, V.K. Magas, T. Mizutani, J. Nieves, A. Ramos, L.L. Salcedo, L. Tolos, Phys. Rev. D **79**, 054004 (2009). https://doi.org/10.1103/PhysRevD.79.054004
29. D. Gamermann, C. Garcia-Recio, J. Nieves, L.L. Salcedo, L. Tolos, Phys. Rev. D **81**, 094016 (2010). https://doi.org/10.1103/PhysRevD.81.094016
30. O. Romanets, L. Tolos, C. Garcia-Recio, J. Nieves, L.L. Salcedo, R.G.E. Timmermans, Phys. Rev. D **85**, 114032 (2012). https://doi.org/10.1103/PhysRevD.85.114032
31. C. Garcia-Recio, J. Nieves, O. Romanets, L.L. Salcedo, L. Tolos, Phys. Rev. D **87**(3), 034032 (2013). https://doi.org/10.1103/PhysRevD.87.034032
32. C. Garcia-Recio, J. Nieves, O. Romanets, L.L. Salcedo, L. Tolos, Phys. Rev. D **87**, 074034 (2013). https://doi.org/10.1103/PhysRevD.87.074034
33. L. Tolos, Int. J. Mod. Phys. E **22**, 1330027 (2013). https://doi.org/10.1142/S0218301313300270
34. C. Garcia-Recio, C. Hidalgo-Duque, J. Nieves, L.L. Salcedo, L. Tolos, Phys. Rev. D **92**(3), 034011 (2015). https://doi.org/10.1103/PhysRevD.92.034011
35. J. Nieves, R. Pavao, L. Tolos, Eur. Phys. J. C **78**(2), 114 (2018). https://link.springer.com/article/10.1140%2Fepjc%2Fs10052-018-5597-3
36. C. Garcia-Recio, J. Nieves, L. Tolos, Phys. Lett. B **690**, 369 (2010). https://doi.org/10.1016/j.physletb.2010.05.056
37. C. Garcia-Recio, J. Nieves, L.L. Salcedo, L. Tolos, Phys. Rev. C **85**, 025203 (2012). https://doi.org/10.1103/PhysRevC.85.025203
38. J. Yamagata-Sekihara, C. Garcia-Recio, J. Nieves, L.L. Salcedo, L. Tolos, Phys. Lett. B **754**, 26 (2016). https://doi.org/10.1016/j.physletb.2016.01.003
39. L. Tolos, J.M. Torres-Rincon, Phys. Rev. D **88**, 074019 (2013). https://doi.org/10.1103/PhysRevD.88.074019
40. J.M. Torres-Rincon, L. Tolos, O. Romanets, Phys. Rev. D **89**(7), 074042 (2014). https://doi.org/10.1103/PhysRevD.89.074042
41. V. Ozvenchuk, J.M. Torres-Rincon, P.B. Gossiaux, L. Tolos, J. Aichelin, Phys. Rev. C **90**, 054909 (2014). https://doi.org/10.1103/PhysRevC.90.054909.
42. T. Song, H. Berrehrah, D. Cabrera, J.M. Torres-Rincon, L. Tolos, W. Cassing, E. Bratkovskaya, Phys. Rev. C **92**(1), 014910 (2015). https://doi.org/10.1103/PhysRevC.92.014910
43. T. Song, H. Berrehrah, J.M. Torres-Rincon, L. Tolos, D. Cabrera, W. Cassing, E. Bratkovskaya, Phys. Rev. C **96**(1), 014905 (2017). https://doi.org/10.1103/PhysRevC.96.014905

44. D. Gamermann, C. Garcia-Recio, J. Nieves, L.L. Salcedo, Phys. Rev. D **84**, 056017 (2011). https://doi.org/10.1103/PhysRevD.84.056017
45. C. Patrignani et al., Chin. Phys. C **40**(10), 100001 (2016). https://doi.org/10.1088/1674-1137/40/10/100001
46. D. Jido, J.A. Oller, E. Oset, A. Ramos, U.G. Meissner, Nucl. Phys. A **725**, 181 (2003). https://doi.org/10.1016/S0375-9474(03)01598-7
47. L. Tolos, C. Garcia-Recio, J. Nieves, Phys. Rev. C **80**, 065202 (2009). https://doi.org/10.1103/PhysRevC.80.065202
48. M. Gonin et al., Nucl. Phys. A **610**, 404C (1996). https://doi.org/10.1016/S0375-9474(96)00373-9
49. K. Tsushima, D.H. Lu, A.W. Thomas, K. Saito, R.H. Landau, Phys. Rev. C **59**, 2824 (1999). https://doi.org/10.1103/PhysRevC.59.2824
50. A. Hayashigaki, Phys. Lett. B **487**, 96 (2000). https://doi.org/10.1016/S0370-2693(00)00760-7
51. T. Hilger, B. Kampfer, S. Leupold, Phys. Rev. C **84**, 045202 (2011). https://doi.org/10.1103/PhysRevC.84.045202
52. K. Suzuki, P. Gubler, M. Oka, Phys. Rev. C **93**(4), 045209 (2016). https://doi.org/10.1103/PhysRevC.93.045209
53. D. Blaschke, P. Costa, Yu.L. Kalinovsky, Phys. Rev. D **85**, 034005 (2012). https://doi.org/10.1103/PhysRevD.85.034005
54. A. Mishra, E.L. Bratkovskaya, J. Schaffner-Bielich, S. Schramm, H. Stoecker, Phys. Rev. C **69**, 015202 (2004). https://doi.org/10.1103/PhysRevC.69.015202
55. S. Yasui, K. Sudoh, Phys. Rev. C **87**(1), 015202 (2013). https://doi.org/10.1103/PhysRevC.87.015202
56. C.E. Jimenez-Tejero, A. Ramos, L. Tolos, I. Vidana, Phys. Rev. C **84**, 015208 (2011). https://doi.org/10.1103/PhysRevC.84.015208
57. L.M. Abreu, D. Cabrera, F.J. Llanes-Estrada, J.M. Torres-Rincon, Ann. Phys. **326**, 2737 (2011). https://doi.org/10.1016/j.aop.2011.06.006
58. H. Berrehrah, P.B. Gossiaux, J. Aichelin, W. Cassing, J.M. Torres-Rincon, E. Bratkovskaya, Phys. Rev. C **90**, 051901 (2014). https://doi.org/10.1103/PhysRevC.90.051901
59. R. Aaij et al., Phys. Rev. Lett. **109**, 172003 (2012). https://doi.org/10.1103/PhysRevLett.109.172003
60. M. Laine, JHEP **04**, 124 (2011). https://doi.org/10.1007/JHEP04(2011)124
61. M. He, R.J. Fries, R. Rapp, Phys. Lett. B **701**, 445 (2011). https://doi.org/10.1016/j.physletb.2011.06.019
62. S. Ghosh, S.K. Das, S. Sarkar, J.E. Alam, Phys. Rev. D **84**, 011503 (2011). https://doi.org/10.1103/PhysRevD.84.011503
63. S.K. Das, S. Ghosh, S. Sarkar, J.E. Alam, Phys. Rev. D **85**, 074017 (2012). https://doi.org/10.1103/PhysRevD.85.074017
64. L. Tolos, J.M. Torres-Rincon, S.K. Das, Phys. Rev. D **94**(3), 034018 (2016). https://doi.org/10.1103/PhysRevD.94.034018
65. S.K. Das, J.M. Torres-Rincon, L. Tolos, V. Minissale, F. Scardina, V. Greco, Phys. Rev. D **94**(11), 114039 (2016). https://doi.org/10.1103/PhysRevD.94.114039

Novel Physics Opportunities at the HESR-Collider with PANDA at FAIR

Leonid Frankfurt, Mark Strikman, Alexei Larionov, Andreas Lehrach, Rudolf Maier, Hendrik van Hees, Christian Spieles, Volodymyr Vovchenko and Horst Stöcker

Abstract Exciting new scientific opportunities are presented for the PANDA detector at the High Energy Storage Ring in the redefined $\bar{p}p(A)$ collider mode, HESR-C, at the Facility for Antiproton and Ion Research (FAIR) in Europe. The high luminosity, $L \sim 10^{31}$ cm^{-2} s^{-1}, and a wide range of intermediate and high energies, $\sqrt{s_{NN}}$ up to 30 GeV for $\bar{p}p(A)$ collisions will allow to explore a wide range of exciting topics in QCD, including the study of the production of excited open charm and bottom states, nuclear bound states containing heavy (anti)quarks, the interplay of hard and soft physics in the dilepton production, and the exploration of the regime where gluons—but not quarks—experience strong interaction.

L. Frankfurt (✉)
Sackler School of Exact Sciences, Tel Aviv University, Tel Aviv, Israel
e-mail: frankfur@tauphy.tau.ac.il

M. Strikman
Pennsylvania State University, University Park, PA, USA

A. Larionov · A. Lehrach · R. Maier
Institut für Kernphysik, Forschungszentrum Jülich, 52425 Jülich, Germany

A. Larionov
National Research Center "Kurchatov Institute", 123182 Moscow, Russia

A. Larionov · V. Vovchenko · H. Stöcker
Frankfurt Institute for Advanced Studies, Giersch Science Center,
60438 Frankfurt am Main, Germany

A. Lehrach · R. Maier
JARA-FAME (Forces and Matter Experiments), Forschungszentrum Jülich
and RWTH Aachen University, Aachen, Germany

H. van Hees · C. Spieles · V. Vovchenko · H. Stöcker
Institut für Theoretische Physik, Goethe Universität Frankfurt, 60438
Frankfurt am Main, Germany

H. Stöcker
GSI Helmholtzzentrum für Schwerionenforschung GmbH,
64291 Darmstadt, Germany

J. Kirsch et al. (eds.), *Discoveries at the Frontiers of Science*,
FIAS Interdisciplinary Science Series, https://doi.org/10.1007/978-3-030-34234-0_4

1 Introduction

The experimental discovery of charmonium [1, 2] and bottomonium [3] in e^+e^- and pA collisions suggests that hadrons containing heavy quarks can be investigated in hadronic processes, where a dense, strongly interacting medium could be formed. It can be particularly useful to study the annihilation of antiprotons on free protons and baryons bound in nuclei in $\bar{p}p(A)$ collisions, in both collider and fixed-target experiments at collision energies of $\sqrt{s} = 2 - 200\,\text{GeV}$.

A unique opportunity to do this in the near future is provided by the Facility for Antiproton and Ion Research (FAIR), with the PANDA detector at the high-energy storage ring (HESR). This concerns both, the presently developed HESR fixed target mode at $\sqrt{s} < 6\,\text{GeV}$, and a future collider mode at $\sqrt{s} < 32\,\text{GeV}$, with PANDA as midrapidity detector. The collider mode would need a dedicated injection-beam transfer line from the SIS 18 directly into HESR, as discussed in Ref. [4]. Additionally, asymmetric HESR collider schemes with somewhat lower center-of-mass energies have been discussed in detail [5–7]. It may also be feasible to study $\bar{p}A$ collisions for $\sqrt{s_{NN}}$ of up to 19 GeV with interesting physics opportunities [8, 9].

Luminosities of up to $5 \cdot 10^{31}\,\text{cm}^{-2}\text{s}^{-1}$ can be reached at $\sqrt{s} \simeq 30\,\text{GeV}$ in the symmetric $\bar{p}p$ collider mode at the HESR [10, 11]. The collision scheme of twelve proton bunches colliding with the same amount of antiproton bunches has to be adapted to the HESR. This modification of the HESR requires a second proton injection, the Recuperated Experimental Storage Ring (RESR), the 8 GeV electron cooler and a modification of the PANDA interaction region.

Besides the deceleration of rare-isotope beams, the RESR storage ring also accumulates high-intensity antiprotons, via the longitudinal momentum stacking with a stochastic cooling system [12]. This is achieved by injecting and pre-cooling the produced antiprotons at 3 GeV in the Collector Ring (CR) storage ring.

The anticipated beam intensities in the HESR proton-antiproton collider version require a full-energy electron cooler (8 MeV) to avoid beam emittance growth, which results in a decreased luminosity during the cycle. The Budker Institute of Nuclear Physics (BINP) presented a feasibility study for magnetized high-energy electron cooling. An electron beam up to 1 A, accelerated in dedicated accelerator columns to energies in the range of 4.5–8 MeV has been proposed. For the FAIR full version, it is planned to install the high-energy electron cooler in one of the HESR straight sections [13–15].

In the fixed-target mode, at $E_{\text{kin}} = 4 - 10\,\text{GeV}$, it will be possible to perform complementary measurements of the cross section of charmonium interaction with nuclear matter with the PANDA detector.

A conservative estimate of the $p\bar{p}$ luminosities which can be reached at the startup phase without RESR is $4 \cdot 10^{30}\,\text{cm}^{-2}\text{s}^{-1}$. We will use it below in our estimates assuming a one year run (10^7s). Energies $\sqrt{s}$ up to 30 GeV could be reached and it may also be feasible to study $\bar{p}$-heavy-nucleus collisions for $\sqrt{s_{NN}}$ of up to 19 GeV. In the present work we outline how the collider at the HESR machine will extend the scope of the PANDA project, with a focus on a few highlights. In Sects. 2 and 3 we discuss the potential of the $\bar{\text{p}}$p collider to provide new information in the field of heavy quark physics, with some attention devoted to the possible discovery of new states. Some other related opportunities are outlined in Sect. 4. Section 5 presents a number of additional physics topics that could be explored in both $\bar{\text{p}}$p and $\bar{\text{p}}A$ modes. These include the production of nuclear fragments containing $\bar{\text{c}}$ and/or c quarks, the c$\bar{\text{c}}$ pair production, color fluctuation effects, probing the pure glue matter, and the production of low-mass dileptons. Concluding remarks in Sect. 6 close the article.

2 Study of Bound States Containing Heavy Quarks

A number of new states containing heavy quarks have been discovered recently. These can be interpreted as pentaquark and tetraquark states containing c$\bar{\text{c}}$ pairs. Some of the states are observed in decays of mesons and baryons containing b-quarks, others in the final states of e^+e^- annihilation; for a review see [16].

It is widely expected that these discoveries represent just the start of the exploration of rich new families of states containing heavy quarks. Understanding the dynamics responsible for the existence of these states would help to clarify many unresolved issues in the spectroscopy of light hadrons.

A unique feature of an intermediate-energy $\bar{\text{p}}$p collider is that it makes possible to study the production of $Q\bar{Q}$ (Q = c, b) pairs and the formation of various hadrons containing heavy quarks rather close to the threshold. The b$\bar{\text{b}}$ pairs are produced mostly in the process of annihilation of valence quarks and antiquarks, i.e. $q\bar{q} \to Q\bar{Q}$. The production of c$\bar{\text{c}}$ pairs in the antiproton fragmentation region also corresponds to this mechanism.

The invariant masses of the produced $Q\bar{Q}$ pairs are much closer to the threshold in the discussed energy range than at the LHC energies. It is natural to expect that the large probability to produce final states with small $Q\bar{Q}$ invariant masses should lead to a higher relative probability to produce pentaquark and tetraquark states compared to the one at the LHC energies. Additionally, the small transverse momenta of the $Q\bar{Q}$ pairs facilitate the pick up of light quarks as compared to Q or $\bar{Q}$ fragmentation. In the antiproton-fragmentation region another $Q\bar{Q}$ production enhancement mechanism, specific for antiproton interactions, is possible: the production of $Q\bar{Q}$ pairs with large $x \sim 0.2 - 0.4$ in the annihilation of $q\bar{q}$, which could merge with a spectator antiquark of the antiproton carrying $x \sim 0.2$.

Another effect which can help to observe new states in medium-energy $\bar{\text{p}}$p collisions is the relatively low spatial density of the system produced at moderate energies. This should suppress final-state interactions, which could possibly hinder the formation of weakly bound clusters of large size. An additional advantage is a relatively small bulk hadron production which reduces the combinatorial background significantly as compared to the LHC.

2.1 The Heavy Quark Production Rates

The need for $t\bar{t}$-production cross sections has stimulated the development of new computational techniques for heavy-quark production in hadron-hadron collisions (see, e.g., Refs. [17, 18]), in particular those which include the effects of threshold resummation.

These calculations, which are currently being validated by comparison to data at high collision energies, predict the following cross section for the $b\bar{b}$ pair production in $\bar{p}p$ collisions [19]:

$$\sigma_{b\bar{b}}(\sqrt{s} = 30\,\text{GeV}) = 1.8 \cdot 10^{-2}\,\mu\text{b}. \tag{1}$$

This calculation has a relative uncertainty of about 30%, and the predicted cross section value is about seven times higher than the corresponding cross section for pp scattering because of the contribution of the valence-quark valence-antiquark annihilation present in $\bar{p}p$. The $b\bar{b}$ cross section per nucleon for the 15 GeV $\times$ 6 GeV kinematics is a factor of 100 smaller. Charm production is dominated by gluon annihilation, $gg \to Q\bar{Q}$. This fact implies that the corresponding cross sections are close in pp and $p\bar{p}$ collisions, with the exception of the fragmentation regions. The experimental data in this case are rather consistent between pp and $\bar{p}p$ and correspond to

$$\sigma_{c\bar{c}} = 30\,\mu\text{b}. \tag{2}$$

For the 15 GeV $\times$ 6 GeV scenario the cross section per nucleon drops by a factor of 3.

2.2 Rate Estimates

The cross sections in Eqs. (1) and (2) correspond to significant event rates for one year (10^7 s) of running at a luminosity of $4 \cdot 10^{30}\,\text{cm}^{-2}\text{s}^{-1}$. We find

$$N_{b\bar{b}} = 10^6, \quad N_{c\bar{c}} = 10^9. \tag{3}$$

These numbers can be easily rescaled for a run at a different luminosity, if required.

At these energies a rearrangement of the light-quark fractions occurs in the final states without significant suppression since heavy quarks are dominantly produced at mid rapidity, and the interaction between a b quark and a light valence quark produces slow light quarks. Thus the overlapping integral between such a $b\bar{q}$ pair and the corresponding hadron wave function should be large. Most likely mesons are produced in excited states. The probability for the formation of $b\bar{q}$ mesons with a given flavor is about 30% of the total cross section because of the competition between different flavors. In the discussed processes gluon radiation is a small correction because of the restricted phase space and the large b-quark mass.

At the collision energies considered here, the combinatorial background is significantly smaller compared to LHC energies, which makes the observation of the excited states containing heavy quarks easier. In particular, the spectator quarks and antiquarks will have quite small velocities relative to the b quark. Indeed, for $\sqrt{s} = 30\,\mathrm{GeV}$ a typical x value for the b-quark is $m_{\mathrm{b}}/\sqrt{s} \gtrsim 0.2$, which is close to the x values of the valence quarks. This fact suggests the possibility of formation of excited states in both the meson and the baryon channels, as well as an enhanced probability of the production of the excited $\mathrm{b}\bar{\mathrm{b}}q\bar{q}$ tetraquark or $(\mathrm{b}\bar{q}) - (\bar{\mathrm{b}}q)$ mesonic molecular states.

2.3 Hidden Beauty Resonance Production

In contrast to charm production, the cross section for hidden beauty resonance production $\mathrm{p}\bar{\mathrm{p}} \to \chi_{\mathrm{b}}$ may be too low for the process to be observed at the HESR. Within the standard quarkonium models, where a $Q\bar{Q}$ pair annihilates into two gluons which subsequently fragment into light quarks, one can estimate that this cross section drops with M_Q as $R_{\mathrm{b/c}} = \Gamma(\chi_{\mathrm{b}} \to \mathrm{p}\bar{\mathrm{p}})/\Gamma_{\mathrm{tot}}(\chi_{\mathrm{b}}) \propto \alpha_s^8/M_Q^8$. This is because this cross section is proportional to the partial width of the decay $\mathrm{p}\bar{\mathrm{p}} \to \chi_{\mathrm{b}}$, which drops with an increase of M_Q. The ratio of cross sections of beauty production through a χ_{b} intermediate state to that for charm is $\approx [\alpha_s(M_Q)/\alpha_s(M_c)]^8/[M_c/M_Q]^{10}$ with an additional factor of M_Q^{-2} stemming from the expression for the resonance cross section. The suppression is due to the necessity of a light-quark rearrangement in the wave function of the proton to obtain decent overlapping with χ_{b} states. In the non-relativistic approximation the wave functions of χ_{b} states vanish at zero inter-quark distance. Thus the overall suppression for the total cross section of χ_{b} production as compared to that of χ_{c} production is approximately 10^{-7}.

3 Potential for Discovery of New States

Investigating $\mathrm{p}\bar{\mathrm{p}}$ collisions at moderate energies carries specific advantages for searches of new states as the $\mathrm{b}\bar{\mathrm{b}}$-production rate is relatively high, while the overall multiplicity, which determines the background level, is rather modest. Also, an equal number of states containing quarks and antiquarks is produced enabling cross checks of observations using conjugated channels. In the following we will discuss resonances containing heavy quarks with the understanding that everything said equally applies to the resonances containing heavy antiquarks. Although the rates in many cases are rather modest, we nevertheless include the discussion of these channels in view of the possibility to have a higher energy collider, as discussed in the final remarks (Sect. 6).

3.1 b*qq*-Baryons and b$\bar{q}$ Mesons

The current knowledge of the spectrum of the excited states containing b- or $\bar{\text{b}}$-quarks is very limited. According to PDG [20], in the $q\bar{\text{b}}$ sector there are two states $\text{B}_\text{J}(5970)^+$ and $\text{B}_\text{J}(5970)^0$ with unknown quantum numbers which could be excited states of the B^+ and B^0, respectively. In the bqq-sector there are two baryons $\Lambda_b(5912)^0$ and $\Lambda_b(5920)^0$ which can be regarded as orbitally excited states of Λ_b^0 and one excited Σ_b^* state. This is much less in comparison with the cqq sector where five excited Λ_c^+ states and two excited Σ_c states (all rated with ***) are observed.

So there are plenty of opportunities here. One noteworthy issue is the comparison of the accuracy with which the heavy-quark limit works for hadrons containing b quarks versus those containing c quarks.

3.2 Excited States Containing $b\bar{b}$

As argued above it is very difficult to produce bound states containing b$\bar{\text{b}}$ in the resonance process of p$\bar{\text{p}}$ annihilation. Nevertheless, many of these states, as well as other states like analogs of X, Y, Z charmonium states, could be produced in inelastic p$\bar{\text{p}}$ interactions. This is because the invariant mass of the produced b$\bar{\text{b}}$ system is rather close to the threshold, and because the b$\bar{\text{b}}$ pair is produced in association with several valence quarks and valence antiquarks which have rather low momenta relative to the b$\bar{\text{b}}$ pair.

3.3 Baryons and Mesons Containing Two Heavy Quarks

Since there are three valence antiquarks colliding with three valence quarks in $\bar{\text{p}}$p, one can produce two pairs of heavy quarks in a double quark-antiquark collision. This entails a possibility for producing the following baryons and mesons containing two heavy quarks:

(i) ccq baryons.

At the collision energies discussed, the contribution of the leading-twist mechanism of 2g $\to Q\bar{Q}Q\bar{Q}$ for the double heavy-quark production should be quite small as it requires very large x of the colliding partons (the situation might be less pronounced for the case of the double c$\bar{\text{c}}$ production than for b$\bar{\text{b}}$). Therefore, the only effective mechanism left is the production of two pairs of heavy quarks in two hard parton-parton collisions (see Fig. 1).

For the case of double c$\bar{\text{c}}$ pair production, one can make estimates by considering the suppression factor for the production of the second c$\bar{\text{c}}$ pair relative to a single c$\bar{\text{c}}$ pair. This factor can be roughly estimated using the high-energy experimental

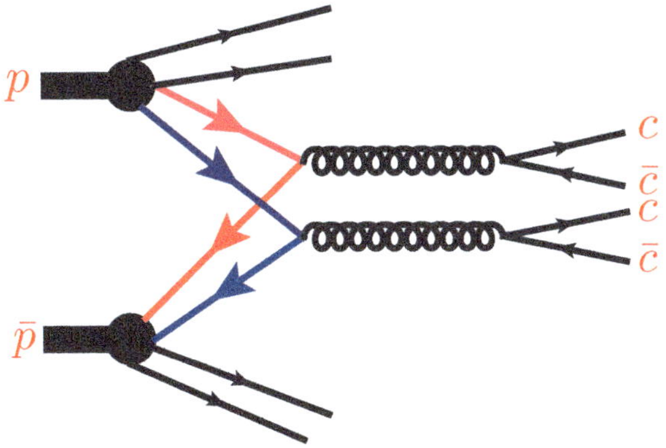

Fig. 1 Double parton interaction mechanism for the production of two pairs of heavy quarks

studies of double-parton collisions at the Tevatron collider. One finds a probability of about 10^{-3} for the ratio of the cross section for producing two $c\bar{c}$ pairs relative to a single pair. (To be conservative we took a factor of two smaller value of the parameter which determines the probability of double collisions ($1/\sigma_{\rm eff}$) than the one measured at Tevatron as in the Tevatron kinematics special small-x effects may enhance this factor.) These considerations result in

$$N_{c\bar{c},c\bar{c}} = 10^6 \tag{4}$$

as an estimate for the yearly number of events with two pairs of $c\bar{c}$. Since the available phase space is rather modest, there is a significant probability that the relative velocity of two c—quarks would be small, and therefore a ccq state would be formed.

Other interesting channels are production of an open charm-anticharm pair plus charmonium, and double charmonium.

(ii) b*cq baryons and* b$\bar{c}$ *mesons.*

The $b\bar{b}c\bar{c}$ pairs are produced pretty close to threshold and have small relative velocities. Hence there is a good chance that they would form a bcq baryon. About 10^3 events per year of running with $b\bar{b}c\bar{c}$ could be expected based on the double parton interaction mechanism.

(iii) bb*q baryons.*

To observe the production of $\sim 10^2$ $b\bar{b}b\bar{b}$ pairs one would need a one-year run at a much higher luminosity of 10^{33} cm^{-2}s^{-1}. The velocities of two b quarks are expected to be close in about 1/2 of the events. So there should be a significant chance for them to form bbq baryons.

(iv) bcc *baryons.*

Whether it is feasible to observe bcc states requires more detailed estimates and may depend on the structure of the three-quark configurations in the nucleon. A naive estimate is that 10^2 events with $bcc\bar{b}\bar{c}\bar{c}$ would be produced in a one-year run at a higher luminosity of $10^{33}\,cm^{-2}s^{-1}$.

Note also that even though a $b\bar{c}$ meson has been observed (although its quantum numbers are not known), there are potentially many other states built of these quarks suggesting a rich spectroscopy (mirroring the spectra of $c\bar{c}$- and $b\bar{b}$-onium states).

3.4 Summary

To summarize, it would be possible with PANDA at the HESR-C $\bar{p}p$ collider to discover and to study properties of meson and baryon states containing one b quark and light (anti) quarks, complementing the charmonium states which are planned to be explored in the PANDA fixed target experiment. There are also good chances to discover the double-heavy-quark baryonic and mesonic states. These new potential observations will allow to achieve a much deeper understanding of the bound-state dynamics in QCD.

4 Other Opportunities

In the last decades much effort has been devoted to studying high-energy properties of QCD in the vacuum channel—the so called perturbative Pomeron. Interactions in non-vacuum channels, on the other hand, have practically not been studied. The PANDA experiment at a collider has a perfect kinematic coverage to study the behavior of Regge trajectories in both the non-perturbative regime (small Mandelstam t) as well as the possible onset of the perturbative regime. One advantage of the antiproton beam is the possibility to study a wide range of baryon and meson Regge trajectories, possibly including Regge trajectories with charmed quarks. The latter will allow to check the non-universality of the slopes of the Regge trajectories which were observed already at positive t values.

Other possibilities include Drell–Yan-pair measurements, which at the lower end of the discussed energy range, may be extended to the limit of exclusive processes like $\bar{p}p \to \mu^+\mu^- +$ meson which are sensitive to generalized parton-distribution functions, etc.

Another direction of studies is the investigation of correlations between valence (anti)quarks in (anti)nucleons using multi-parton interactions (MPI) analogous to those shown in Fig. 1. In the discussed energy range MPI get a significant contribution from collisions of large-x partons (double Drell–Yan, Drell–Yan + charm production, etc.), and the rate of the MPI is inversely proportional to the square of the average distance between the valence quarks. In particular the rates would be strongly enhanced in the case of a large probability of (anti)quark-(anti)diquark configurations in the (anti)nucleon.

The analysis of the production of heavy-quark pairs discussed above would also be of great interest for the study of the multiparton structure of nucleons.

4.1 $\bar{p}A$ Elastic Scattering and Absorption

So far measurements of the antiproton-nucleus elastic scattering were done only at LEAR for $p_{\rm lab} < 1\,{\rm GeV/c}$. It turned out that owing to the forward-peaked $\bar{\rm p}$p elastic scattering amplitude the Glauber model describes LEAR data on the angular differential cross sections of $\bar{\rm p}A$ elastic scattering surprisingly well. This is in contrast to the pA elastic scattering where the Glauber model description starts to work only above $p_{\rm lab} \sim 1.5\,{\rm GeV/c}$ [21].

Glauber theory analysis [22] has shown that the $\bar{\rm p}A$ and pA angular differential elastic scattering cross sections at $p_{\rm lab} = 10\,{\rm GeV/c}$ (fixed target PANDA) strongly differ in the diffraction minima due to the different ratios of the real-to-imaginary parts of $\bar{\rm p}N$ and pN elastic scattering amplitudes. Experimental confirmation of such a behavior would be a good validity test of the Glauber theory, important in view of its broad applications for other reaction channels, and of the input elementary amplitudes which are typically given by Regge-type parameterizations. In particular, the $\bar{\rm p}$n elastic amplitude is accessible only by scattering on complex nuclei. The determination of diffractive structures at $\bar{\rm p}A$-collider energies would require good transverse momentum transfer resolution $\sim$10MeV/c and the capability to trigger on the events where nucleus remains intact (see the discussion in Sect. 5.2). Light nuclear targets are preferred as their diffractive structures are broader in $p_{\rm t}$, and there is a smaller number of possible excited states. Spin 0 targets like ^{4}He are especially good for these purposes.

A related problem is the determination of the antiproton absorption cross section on nuclei (defined as the difference between total and elastic cross sections). Experimental data on the antiproton absorption cross section above LEAR energies are quite scarce, although such data are needed for cosmic ray antiproton flux calculations [23].

4.2 Coherent Hypernuclei Production

While ordinary Λ-hypernuclei were discovered long ago, the Λ_c^+- and Λ_b^0-hypernuclei were predicted in mid-70s [24, 25] but have not been observed so far. However, their existence is expected based on a number of models, e.g. the quark-meson coupling model [26].

The processes $\bar{\rm p}{\rm p} \to \bar{Y}Y$, where $Y = \Lambda, \Lambda_c^+$ or Λ_b^0, have the lowest thresholds among all possible other channels of the respective $\bar{\rm s}$s, $\bar{\rm c}$c or $\bar{\rm b}$b production channels in $\bar{\rm p}$p collisions. Thus, they are preferred for Y-hypernuclei production as the momentum transfer to the hyperon is relatively small.

The coherent reactions $^AZ(\bar{\rm p}, \bar{\Lambda})^A_\Lambda(Z-1)$ for the different states of the hypernucleus have never been studied experimentally. It is expected that these reactions have cross sections of the order of a few 10 nb at $p_{\rm lab} \sim 20\,{\rm GeV/c}$ [27]. Thus, they can serve as a powerful source of Λ-hypernuclei production at the lower end of the

$\bar{\text{p}}A$-collider energies. Here, the amplitude $\bar{\text{p}}\text{p} \to \bar{\Lambda}\Lambda$ should be dominated by the K* (or K* Regge trajectory) exchange.

More challenging is the coherent process $^{A}\text{Z}(\bar{\text{p}}, \bar{\Lambda}_c^-)^{A}_{\Lambda_c^+}\text{Z}$ [28] where the underlying $\bar{\text{p}}\text{p} \to \bar{\Lambda}_c^- \Lambda_c^+$ amplitude is due to D^0 and D^{*0} exchanges. One can also think of the $^{A}\text{Z}(\bar{\text{p}}, \bar{\Lambda}_b^0)^{A}_{\Lambda_b^0}(\text{Z}-1)$ coherent reaction.

5 Unique Opportunities for Probing QCD Properties at the $\bar{\text{p}}A$ Collider

5.1 Space-Time Picture of the Formation of Hadrons Containing Heavy Quarks

The kinematics of heavy-quark-state production in collisions of p($\bar{\text{p}}$) with proton or nuclei (neglecting Fermi motion effects) dictates that heavy states can only be produced with momenta

$$p_Q > M_Q^2/2m_N x_q - m_N x_q/2 \tag{5}$$

in the rest frame of the nucleus. Here x_q is the x of the quark of the nucleus involved in the production of the $Q\bar{Q}$ pair. For $x_q \leq 0.5$ this corresponds to a charm momentum above 4 GeV/c which is much larger than typical momenta of the heavy system embedded in the nucleus.

However, there is a significant probability that D, $\Lambda_c, \ldots$ hadrons slow down due to final-state interactions. Indeed, it is expected in QCD that the interaction strength of a fast hadron with nucleons is determined by the area in which the color is localized. For example, ψ'-N interactions should be comparable to the kaon-nucleon cross section and be much larger than the J/ψ-N cross section, see for example [29]. Also the cross sections of open charm (bottom) interactions should be on the scale $\gtrsim$10 mb.

The formation distance (coherence length) can be estimated as

$$l_{\text{coh}} \simeq \gamma l_0, \tag{6}$$

where $l_0 \simeq 0.5 - 1.0$ fm. For the discussed energies and the case of scattering off heavy nuclei the condition

$$l_{\text{coh}} \leq R_A, \tag{7}$$

is satisfied for hadrons produced in a broad range of momenta including the central and nucleus fragmentation region. So it would be possible to explore the dependence of the formation time and interaction strength on, for example, the orbital angular momentum of D*.

Observing these phenomena and hence exploring QCD dynamics in a new domain could be achieved by studying the A-dependence of charm production at momenta $\lesssim 10\,\mathrm{GeV/c}$ (in the rest frame of the nucleus).

5.2 New Heavy-Quark States

The formation of heavy mesons or baryons, H, inside nuclei implies final-state interactions which slow down these heavy hadrons, leading to the production of hadrons at low momenta forbidden for scattering off a free proton:

$$p_{\mathrm{H}} \leq (m_{\mathrm{H}}^2 - m_N^2)/2m_N. \tag{8}$$

In this kinematics the slow-down may be sufficient to allow for the production of (anti-)charm quarks embedded in nuclear fragments. The collider kinematics would make it easier to detect decays of such nuclei than in fixed-target set ups as these nuclei would be produced with high momenta (velocities comparable to those of ordinary nuclear fragments). Thus the discussed HESR-C collider in the $\bar{\mathrm{p}}A$ mode would have a high discovery potential for observing various nuclear states containing c and/or $\bar{\mathrm{c}}$.

Higher luminosities and higher collider energies will allow search for analogous b and/or $\bar{\mathrm{b}}$ states.

5.3 Color Fluctuations in Nucleons

At high energies hadrons are thought to be interacting with each other in frozen configurations which have different interaction strengths—so-called color fluctuations. One can explore these phenomena in proton-nucleus collisions in a number of ways. Here we give as one example the study of the interaction strength of a hadron in the case of a configuration that contains a large-x ($x \geq 0.4$) parton. One expects that in such configurations the average interaction strength is significantly smaller than on average: in these configurations color screening leads to a suppression of the gluon fields and of the quark-antiquark sea [30]. This picture has allowed to explain [31, 32] strong deviations of the centrality dependence of the leading-jet production from the geometrical picture (Glauber model of inelastic collisions) observed at the LHC in p-Pb collisions and at RHIC in d-Au collisions.

Due to a fast increase of the interaction strength for small-size configurations with increasing energy, the strength of color fluctuations drops at higher energies. Correspondingly color-fluctuation effects are expected to be much enhanced at the HESR-C $\bar{\mathrm{p}}A$-collider energies. For example, for $x \simeq 0.6$, the cross-section ratio, $\sigma_{\mathrm{eff}}(x)/\sigma_{\mathrm{tot}}(NN)$, is expected to be $\sim$0.25, while at the LHC it is $\sim$0.6.

To observe this effect one would need to study Drell–Yan production at large x. A strong drop of hadron production in the nucleus fragmentation region would be a strong signal for the discussed effect. For its detailed study measurements with different nuclei would be desirable.

5.4 Probing Pure Glue Matter

One of the central questions in high-energy hadronic and nuclear collisions is how the initially non-equilibrium system evolves towards a state of apparent (partial) thermodynamic equilibrium at later stages of nuclear collisions. Presently, the community favors a paradigm of an extremely rapid (t_{eq} less than $0.3\,\mathrm{fm/c}$) thermalization and chemical saturation of soft gluons and light quarks.

The large gluon-gluon cross sections lead to the idea [33] that the gluonic components of colliding nucleons interact more strongly than the quark-antiquark ones. The two-step equilibration scenario of the quark-gluon plasma (QGP) was proposed in [34–36]. It was assumed that the gluon thermalization takes place at the proper time $\tau_g < 1$ fm/c and the (anti)quarks equilibration occurs at $\tau_{th} > \tau_g$. The estimates of Refs. [37–39] show that τ_{th} can be of the order of $5\,\mathrm{fm/c}$.

Recently the *pure glue* scenario was proposed for the initial state at midrapidity in Pb+Pb collisions at Relativistic Heavy Ion Collider (RHIC) and Large Hadron Collider (LHC) energies [40, 41]. According to lattice-QCD calculations [42], quarkless purely gluonic matter should undergo a first-order phase transition at a critical temperature $T_c = 270\,\mathrm{MeV}$. At this temperature the deconfined pure glue matter transforms into the confined state of pure Yang–Mills theory, namely into a glueball fluid. This is in stark contrast to full QCD equilibrium with $(2 + 1)$ flavors, where a smooth crossover transition takes place (see Fig. 2 for a comparison of the corresponding equations of state).

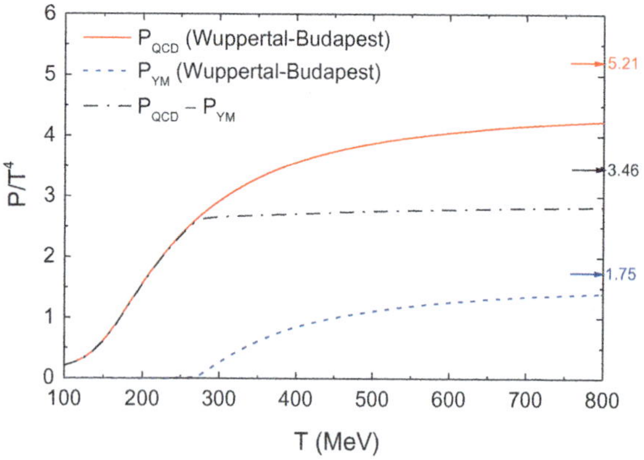

Fig. 2 Temperature dependence of the scaled pressure p/T^4 obtained in lattice QCD calculations of the Wuppertal–Budapest collaboration for $(2 + 1)$-flavor QCD [48] (red line) and for Yang–Mills matter [42] (blue line). The black dash-dotted line depicts the difference between the pressure in full QCD and in Yang–Mills theory

At $\sqrt{s_{NN}} \simeq 30\,\text{GeV}$ $\bar{\text{p}}\text{p}(A)$ collisions can create only small systems. Baryon free matter can be expected if the $\bar{\text{p}}\text{p}$ annihilation occurs briefly in the initial stage of the collision. An enhanced annihilation probability can be expected in $\bar{\text{p}}A$ collisions over the $\bar{\text{p}}\text{p}$ collisions. Therefore, the properties of this baryon-free matter can potentially be studied by looking into the difference in observables between $\bar{\text{p}}\text{p}(A)$ and $\text{pp}(A)$ at the same energy.

If indeed a hot thermalized gluon fluid, initially containing no (anti)quarks, is created in the early stage of a $\bar{\text{p}}\text{p}$ or $\bar{\text{p}}A$ collision at mid rapidity, it will quickly cool and expand until it reaches a mixed-phase region at $T = T_c^{\text{YM}}$. After the initial pure gluon plasma has completely transformed into the glueball fluid, the system will cool down further. These heavy glueballs produced during the Yang–Mills hadronization process where the pure glue plasma forms a glueball fluid. The heavy glueballs will later evolve into lighter states, possibly via a chain of two-body decays [43], and finally decay into hadronic resonances and light hadrons, which may or may not show features of chemical equilibration.

Of course, a more realistic scenario must take into account that some quarks will be produced already before and during the Yang–Mills driven first-order phase transition. This scenario can be modeled by introducing the time-dependent effective number of (anti)quark degrees of freedom, given by the time-dependent absolute quark fugacity λ_q [44]:

$$\lambda_q(\tau) = 1 - \exp\left(\frac{\tau_0 - \tau}{\tau_*}\right). \tag{9}$$

Here τ_* characterizes the quark chemical equilibration time.

To illustrate the above considerations, we apply the (2 + 1)-dimensional relativistic hydrodynamics framework with a time-dependent equation of state, developed in Refs. [45, 46] and implemented in the vHLLE package [47], to $\text{p}\bar{\text{p}}$ collisions at HESR. The equation of state interpolates linearly between the lattice equations of state for the purely gluonic Yang–Mills (YM) theory [42] $P_{\text{YM}}(T)$ at $\lambda_q = 0$ and the full QCD with (2 + 1) quark flavors [48] $P_{\text{QCD}}(T)$ at $\lambda_q = 1$:

$$\begin{aligned} P(T, \lambda_q) &= \lambda_q P_{\text{QCD}}(T) + (1 - \lambda_q) P_{\text{YM}}(T) \\ &= P_{\text{YM}}(T) + \lambda_q \left[P_{\text{QCD}}(T) - P_{\text{YM}}(T)\right]. \end{aligned} \tag{10}$$

$P_{\text{YM}}(T)$ and $P_{\text{QCD}}(T)$ are shown in Fig. 2.

The hydrodynamic simulations of $\text{p}\bar{\text{p}}$ collisions at $\sqrt{s} = 32\,\text{GeV}$ discussed below assume a hard-sphere initial energy density profile with radius $R = 0.6\,\text{fm}$. The normalization of the energy is fixed in order to yield an initial temperature of 273 MeV in the central cell, which is slightly above the critical temperature of 270 MeV. This choice is motivated by Bjorken model based estimates at $\sqrt{s} = 30\,\text{GeV}$ for small systems [49].

Figure 3a shows the τ-dependence of the temperature in the central cell for different quark equilibration times: $\tau_* = 0$ (instant equilibration), 1 fm/c (fast equilibra-

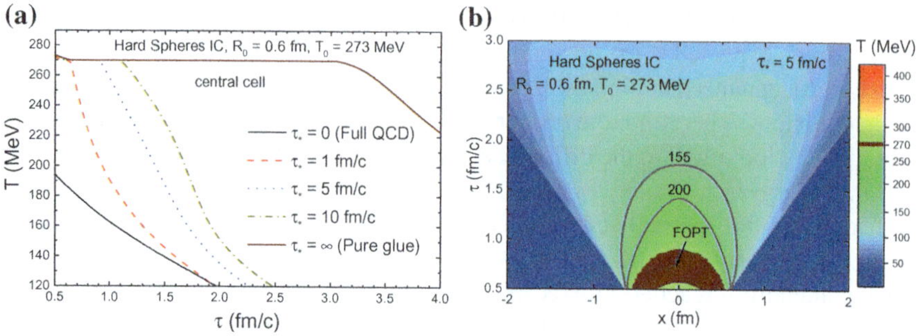

Fig. 3 The temperature profile of the central cell in the longitudinally boost invariant $(2+1)$-dimensional hydro evolution for $p\bar{p}$ collisions in the pure glue initial state, the Yang–Mills scenario. As hard spheres overlap, a transverse density profile with radius $R = 0.6\,\text{fm}$ is used as the initial condition. The normalization is fixed in order to yield the initial temperature of 273 MeV in the central cell. **a** The τ-dependence of the temperature is given for the central cell for different quark equilibration times: $\tau_* = 0$ (instant equilibration), $1\,\text{fm/c}$ (fast equilibration), $5\,\text{fm/c}$ (moderate equilibration), $10\,\text{fm/c}$ (slow equilibration), and for $\tau_* \to \infty$ (pure gluodynamic evolution). **b** The temperature profile in the $x-\tau$ plane for $\tau_* = 5\,\text{fm/c}$

tion), $5\,\text{fm/c}$ (moderate equilibration), $10\,\text{fm/c}$ (slow equilibration), and $\tau_* \to \infty$ (pure gluodynamic evolution). In the pure gluodynamic scenario, $\tau_* \to \infty$, the system spends a very long time in the mixed-phase region. A fast quark equilibration shortens the time period spent in the mixed phase significantly. Nevertheless, a significant fraction of the system evolution takes place in the mixed phase of the gluon-glueball deconfinement phase transition even at presence of a moderately fast quark equilibration ($\tau_* = 5\,\text{fm/c}$), as illustrated by Fig. 3b. Thus, significant effects of the initial pure glue state on electromagnetic and hadronic observables are expected for this collision setup.

These results illuminate the future HESR-collider option with the central PANDA experiment detector as an exciting upgrade for FAIR, a promising option to search for even heavier glueballs and hadrons than envisioned for the fixed target mode, and for other new exotic states of matter.

5.5 *Low- and Large Mass Dilepton Production*

Low-mass lepton pair production has raised the interest in the field for decades. A quite robust theoretical understanding [50–55] of dilepton production in heavy-ion collisions at various energies has been gained. There dileptons play a special role as messengers from the early stages, as penetrating probes.

At large invariant dilepton masses the perturbative Drell–Yan (DY) mechanism of QCD sets in. The minimal $M_{\rm DY}^2$ value, where the DY mechanism dominates, is not well known as it is difficult to separate it experimentally from the contribution

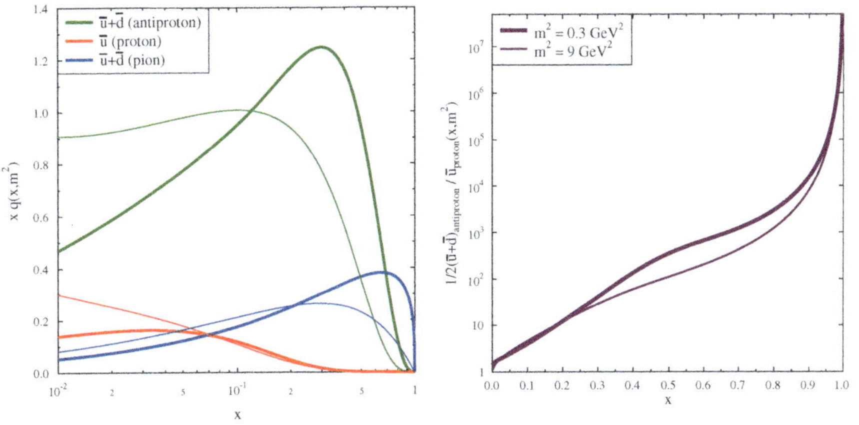

Fig. 4 Left: The light-anti-quark parton-distribution functions (PDFs) at $m^2 = 0.3\ \text{GeV}^2$ (thin lines) and $m^2 = 9\ \text{GeV}^2$ (bold lines) for anti-protons, protons [56], and pions [57]. Right: The ratio of the PDFs for light anti-quarks in anti-protons and in protons

of charm and J/ψ production. It seems that the DY pair production mechanism dominates at $M_{\text{DY}} \geq 2\,\text{GeV}$. For $\bar{\text{p}}$p(A) at intermediate energies, M_{DY} should be lower than for proton projectiles due to presence of abundant valence antiquarks.

Higher-twist mechanisms may delay the inset of the leading-twist contribution in the case of interactions with nuclei. Drell–Yan production of dileptons at intermediate invariant masses by reactions with secondary mesons and (anti-)baryons in pA and AA collisions has been investigated in [58, 59]. At the lower beam energies of the beam-energy scan at RHIC, at FAIR, and at NICA, strong contributions from secondary DY processes are predicted for low- and intermediate-mass dilepton pairs due to the formation of mesons at time scales $\lesssim 1\ \text{fm}/c$, i.e. during the interpenetration stage of projectile and target. This implies that inflying primordial projectile and target nucleons from the interpenetrating nuclei collide with just newly formed mesons (e.g., ρ and ω with constituent-quark and constituent-antiquark masses of $\sim$300 MeV each).

For $\bar{\text{p}}$p and $\bar{\text{p}}A$ reactions, the Drell–Yan production is enhanced already for the primordial collisions due to the presence of *valence* antiquarks of the antiproton, as displayed in Fig. 4 which shows the ratio of light-antiquark PDFs in antiprotons and protons. With the HESR-C $\bar{\text{p}}$p and $\bar{\text{p}}A$ collider discussed here a direct assessment of the valence-quark valence-antiquark parton distributions in $\bar{\text{p}}$p and $\bar{\text{p}}A$ collisions is accessible. The comparison of dilepton production by proton and antiproton induced reactions provides unique information on the relative role of initial- and final-state mechanisms. In particular, by comparing dilepton production in proton and antiproton fragmentation regions one may expect the maximal difference between the two cases.

Calculations of the inclusive (integrated over transverse momenta) DY production can be performed now in the NNLO DGLAP approximation. The techniques

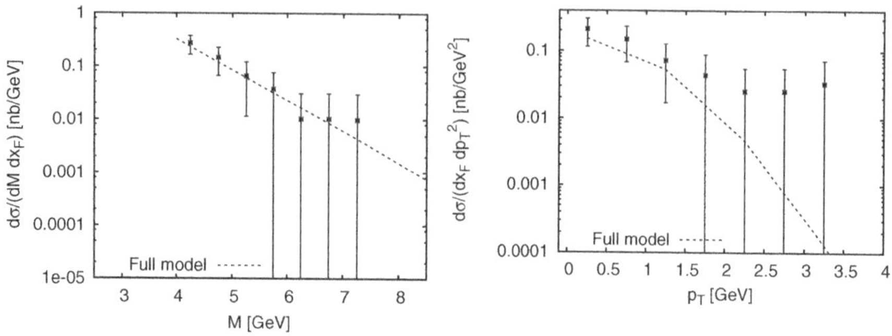

Fig. 5 Invariant-mass (left) and transverse-momentum spectra (right) of DY dimuon pairs in $\bar{p}$W at $\sqrt{s} = 15\,\text{GeV}^2$ collisions. Data are from the E537 collaboration [61]. The used integrated PDF's are the MSTW2008LO68cl set [63]

were developed for calculations of transverse momentum distributions which include effects of multiple gluon emissions (the Sudakov form factor effects) and nonperturbative transverse momentum distributions (TMD), see e.g. [60] and references therein. TMDs contribute to the differential cross section predominantly at small transverse momenta of the DY pairs and for moderate masses of the pairs.

For large masses the theory works now very well providing a parameter free description of the $Z-$boson production at the LHC including the transverse momenta distribution.

Corresponding high precision data for fixed target energies are very limited especially for the antiproton projectiles. Figure 5 shows the result of the model for $\bar{p}$W collisions of the E537 collaboration at $s = 236\,\text{GeV}^2$ [61].

The data are compared with the model calculation [62] which includes pQCD parton evolution, and the TMD effects.

The model parameters have been fixed by using data on dimuon transverse-momentum spectra in pp collisions at $s = 1500\,\text{GeV}^2$ from the E866 collaboration in [64, 65]. This model describes the dimuon production data without any adjustment of model parameters quite well, in particular, the absolute cross section in pd collisions from the E772 collaboration [66], in pCu collisions from the E605 collaboration [67] at the same collision energy, as well as in pA collisions from the E288 collaboration [68] and in pW collisions from the E439 collaboration [69] at $s = 750\,\text{GeV}^2$.

The presence of the abundant antiquarks and the forward kinematics experimental arm of PANDA at the HESR-C $\bar{p}$p collider allows to determine the minimal M^2 for which the DY mechanism works. (Note that the charm contribution is strongly suppressed at $x_p \geq 0.4$ as a lepton in a charm decay carries, on average, only 1/3 of the D-meson momentum (and even less for charm baryons).

The onset of factorization in the DY process with nuclei has still not been explored, but will be very interesting. Indeed, as mentioned already, the formation length in these processes is pretty small. Hence, one could think of the process in a semiclassical way, for a rather broad range of energies. In this picture, in the case of a nuclear target, the antiproton may experience one or more (in)elastic rescatterings on the target nucleons—before it annihilates with a proton into a dilepton pair. This effect can be taken into account within transport models, e.g. GiBUU [70] calculations, which are a good setup for the antiproton-nucleus dynamics. The antiproton stopping shall lead to a softening of the invariant-mass and transverse-momentum spectra of the dilepton pairs. It is also important to include these effects in future calculations, as it influences the conclusions on the in-medium modifications of the nucleon PDFs.

At high enough energies the formation time becomes large, but an antiquark which is involved in the DY process can experience energy losses growing quadratically with the path length [71]. Here one expects p_T broadening $\propto A^{1/3}$. These studies, if performed as a function of the atomic number and of the collision energy at fixed $x_{\bar{q}}$, may allow to explore these important effects in great detail.

6 Final Remarks

The exciting science discussed here could be extended to much higher energies and heavier states if—in a later phase of FAIR—one can manage to reinject antiprotons from the HESR into one or two of the higher energy main FAIR synchrotrons, SIS 100 and SIS 300. Then $\bar{p}A$ collisions can be synchronized to run effectively and with high luminosity in a collider mode too. Also, crossing beams of SIS100 with SIS300, e.g. 45 *A*GeV heavy ions in the SIS 300 colliding with 30 GeV antiprotons (or protons), or with ions of 15 *A*GeV in the SIS100 can be envisioned. Also other asymmetric collisions maybe feasible, e.g., 90 GeV antiprotons/protons (SIS300) colliding with 15 *A*GeV heavy ions (or with 30 GeV protons or antiprotons), in the SIS 100.

Acknowledgements We thank M. Cacciari and R. Vogt for discussions on charm and beauty production in $\bar{p}p$ scattering. Research of L.F. and M.S. was supported by the US Department of Energy Office of Science, Office of Nuclear Physics under Award No. DE-FG02-93ER40771. A.L. acknowledges partial financial support by Helmholtz International Center (HIC) for FAIR. H.v.H. acknowledges the support from Frankfurt Institute for Advanced Studies (FIAS). H.St. acknowledges the support through the Judah M. Eisenberg Laureatus Chair by Goethe University and the Walter Greiner Gesellschaft, Frankfurt.

References

1. J.E. Augustin et al. (SLAC-SP-017 Collaboration), Discovery of a narrow resonance in e^+e^- annihilation. Phys. Rev. Lett. **33**, 1406 (1974). https://doi.org/10.1103/PhysRevLett.33.1406
2. J.J. Aubert et al. (E598 Collaboration), Experimental observation of a heavy particle. J, Phys. Rev. Lett. **33**, 1404 (1974). https://doi.org/10.1103/PhysRevLett.33.1404
3. S.W. Herb et al. Observation of a Dimuon resonance at 9.5 GeV in 400 GeV proton-nucleus collisions. Phys. Rev. Lett. **39**, 252 (1977). https://doi.org/10.1103/PhysRevLett.39.252

4. H. Stöcker, T. Stöhlker, C. Sturm, FAIR - cosmic matter in the laboratory, J. Phys. Conf. Ser. **623**, 012026 (2015). https://doi.org/10.1088/1742-6596/623/1/012026
5. V. Barone et al. (PAX Collaboration), Antiproton-proton scattering experiments with polarization (2005), arXiv: hep-ex/0505054
6. PAX Collaboration, Technical proposal for antiproton-proton scattering experiments with polarization. Technical report, Forschungszentrum Jülich (2006), http://collaborations.fz-juelich.de/ikp/pax/public_files/proposals/techproposal20060125.pdf
7. A. Lehrach, Accelerator configuration for polarized proton-antiproton physics at FAIR. AIP Conf. Proc. **915**, 147 (2007)
8. I.N. Mishustin, L.M. Satarov, J. Schaffner, H. Stöcker, W. Greiner, Baryon anti-baryon pair production in strong meson fields. J. Phys. G **19**, 1303 (1993). https://doi.org/10.1088/0954-3899/19/9/009
9. A.B. Larionov, I.N. Mishustin, L.M. Satarov, W. Greiner, Dynamical simulation of bound antiproton-nuclear systems and observable signals of cold nuclear compression. Phys. Rev. C **78**, 014604 (2008). https://doi.org/10.1103/PhysRevC.78.014604
10. F. Bradamante, I. Koop, A. Otboev, V. Parkhomchuk, V. Reva, P. Shatunov, Y. Shatunov, Conceptual design for a polarized proton-antiproton collider facility at GSI (2005), arXiv: physics/0511252
11. A. Lehrach, O. Boine-Frankenheim, F. Hinterberger, R. Maier, D. Prasuhn, Beam performance and luminosity limitations in the high-energy storage ring (HESR). Nucl. Instrum. Methods A **561**, 289 (2006). https://doi.org/10.1016/j.nima.2006.01.017
12. P. Beller, K. Beckert, C. Dimopoulou, A. Dolinsky, F. Nolden, M. Steck, J. Yang, Layout of an accumulator and decelerator ring for FAIR. Conf. Proc. C **060626**, 199 (2006), http://accelconf.web.cern.ch/AccelConf/e06/PAPERS/MOPCH074.PDF
13. V.V. Parkhomchuk, V.B. Reva, A.N. Skrinsky, V.A. Vostrikov, K. Beckert, P. Beller, A. Dolinskii, B. Franzke, F. Nolden, M. Steck, An electron cooling system for the proposed HESR antiproton storage ring, in *9th European Particle Accelerator Conference (EPAC 2004) Lucerne, Switzerland, 5–9 July 2004* (2004), http://accelconf.web.cern.ch/AccelConf/e04/PAPERS/WEPLT056.PDF
14. D. Reistad et al., Status of the HESR electron cooler design work. Conf. Proc. C **060626**, 1648 (2006)
15. V. Kamerdzhiev et al., 2 MeV electron cooler for COSY and HESR – first results, in *Proceedings, 5th International Particle Accelerator Conference (IPAC 2014): Dresden, Germany*, MOPRI070, 15-20 June 2014 (2014), http://jacow.org/IPAC2014/papers/mopri070.pdf
16. M. Karliner, Heavy exotic quarkonia and doubly heavy baryons. EPJ Web Conf. **96**, 01019 (2015)
17. M. Cacciari, P. Nason, C. Oleari, A study of heavy flavored meson fragmentation functions in e^+e^- annihilation. JHEP **04**, 006 (2006). https://doi.org/10.1088/1126-6708/2006/04/006
18. M. Beneke, A.P. Chapovsky, M. Diehl, T. Feldmann, Soft collinear effective theory and heavy to light currents beyond leading power. Nucl. Phys. B **643**, 431 (2002), arXiv: hep-ph/0206152, https://doi.org/10.1016/S0550-3213(02)00687-9
19. M. Cacciari, R. Vogt, Priv. Commun.
20. M. Tanabashi et al. (Particle Data Group), The review of particle physics. Phys. Rev. D **98**, 030001 (2018), http://pdg.lbl.gov/
21. G.D. Alkhazov, S.L. Belostotsky, A.A. Vorobev, Scattering of 1-GeV protons on nuclei. Phys. Rep. **42**, 89 (1978). https://doi.org/10.1016/0370-1573(78)90083-2
22. A.B. Larionov, H. Lenske, Elastic scattering, polarization and absorption of relativistic antiprotons on nuclei. Nucl. Phys. A **957**, 450 (2017). https://doi.org/10.1016/j.nuclphysa.2016.10.006
23. I.V. Moskalenko, A.W. Strong, J.F. Ormes, M.S. Potgieter, Secondary anti-protons and propagation of cosmic rays in the galaxy and heliosphere. Astrophys. J. **565**, 280 (2002). https://doi.org/10.1086/324402
24. A.A. Tyapkin, A possible way of establishing the existence of charmed particles. Sov. J. Nucl. Phys. **22**, 89 (1976)

25. C.B. Dover, S.H. Kahana, Possibility of charmed hypernuclei. Phys. Rev. Lett. **39**, 1506 (1977). https://doi.org/10.1103/PhysRevLett.39.1506
26. K. Tsushima, F.C. Khanna, Lambda(c)+ and Lambda(b) hypernuclei. Phys. Rev. C **67**, 015211 (2003). https://doi.org/10.1103/PhysRevC.67.015211
27. A.B. Larionov, H. Lenske, Distillation of scalar exchange by coherent hypernucleus production in antiproton-nucleus collisions. Phys. Lett. B **773**, 470 (2017). https://doi.org/10.1016/j.physletb.2017.09.007
28. R. Shyam, K. Tsushima, Production of Λ_c^+ hypernuclei in antiproton - nucleus collisions. Phys. Lett. B **770**, 236 (2017). https://doi.org/10.1016/j.physletb.2017.04.057
29. L. Gerland, L. Frankfurt, M. Strikman, H. Stöcker, W. Greiner, J/ψ production, χ polarization and color fluctuations. Phys. Rev. Lett. **81**, 762 (1998), arXiv:nucl-th/9803034, https://doi.org/10.1103/PhysRevLett.81.762
30. L.L. Frankfurt, M.I. Strikman, Point-like configurations in hadrons and nuclei and deep inelastic reactions with leptons: EMC and EMC-like effects. Nucl. Phys. B **250**, 143 (1985). https://doi.org/10.1016/0550-3213(85)90477-8
31. M. Alvioli, B.A. Cole, L. Frankfurt, D.V. Perepelitsa, M. Strikman, Evidence for x-dependent proton color fluctuations in pA collisions at the CERN large hadron collider. Phys. Rev. C **93**, 011902 (2016). https://doi.org/10.1103/PhysRevC.93.011902
32. M. Alvioli, L. Frankfurt, D. Perepelitsa, M. Strikman, Global analysis of color fluctuation effects in proton- and deuteron-nucleus collisions at RHIC and the LHC (2017), arXiv:1709.04993
33. L. Van Hove, S. Pokorski, High-energy hadron-hadron collisions and internal hadron structure. Nucl. Phys. B **86**, 243 (1975). https://doi.org/10.1016/0550-3213(75)90443-5
34. S. Raha, Dilepton, diphoton and photon production in preequilibrium. Phys. Scr. **T32**, 180 (1990). https://doi.org/10.1088/0031-8949/1990/T32/030
35. E.V. Shuryak, Two stage equilibration in high-energy heavy ion collisions. Phys. Rev. Lett. **68**, 3270 (1992). https://doi.org/10.1103/PhysRevLett.68.3270
36. J. Alam, B. Sinha, S. Raha, Successive equilibration in quark - gluon plasma. Phys. Rev. Lett. **73**, 1895 (1994)
37. T.S. Biro, E. van Doorn, B. Muller, M.H. Thoma, X.N. Wang, Parton equilibration in relativistic heavy ion collisions. Phys. Rev. C **48**, 1275 (1993). https://doi.org/10.1103/PhysRevC.48.1275
38. D.M. Elliott, D.H. Rischke, Chemical equilibration of quarks and gluons at RHIC and LHC energies. Nucl. Phys. A **671**, 583 (2000). https://doi.org/10.1016/S0375-9474(99)00840-4
39. Z. Xu, C. Greiner, Thermalization of gluons in ultrarelativistic heavy ion collisions by including three-body interactions in a parton cascade. Phys. Rev. C **71**, 064901 (2005). https://doi.org/10.1103/PhysRevC.71.064901
40. H. Stöcker et al., Glueballs amass at RHIC and LHC colliders! - the early quarkless 1st order phase transition at $T = 270$ MeV - from pure Yang-Mills glue plasma to GlueBall-Hagedorn states. J. Phys. G **43**, 015105 (2016). https://doi.org/10.1088/0954-3899/43/1/015105
41. H. Stöcker et al., Under-saturation of quarks at early stages of relativistic nuclear collisions: the hot glue initial scenario and its observable signatures. Astron. Nachr. **336** (2015). https://doi.org/10.1002/asna.201512252
42. S. Borsanyi, G. Endrodi, Z. Fodor, S.D. Katz, K.K. Szabo, Precision SU(3) lattice thermodynamics for a large temperature range. JHEP **07**, 056 (2012). 1204.6184, https://doi.org/10.1007/JHEP07(2012)056
43. M. Beitel, C. Greiner, H. Stöcker, Fast dynamical evolution of a hadron resonance gas via Hagedorn states. Phys. Rev. C **94**, 021902 (2016). https://doi.org/10.1103/PhysRevC.94.021902
44. V. Vovchenko, M.I. Gorenstein, L.M. Satarov, I.N. Mishustin, L.P. Csernai, I. Kisel, H. Stöcker, Entropy production in chemically nonequilibrium quark-gluon plasma created in central Pb+Pb collisions at energies available at the CERN large Hadron collider. Phys. Rev. C **93**, 014906 (2016). https://doi.org/10.1103/PhysRevC.93.014906
45. V. Vovchenko, I.A. Karpenko, M.I. Gorenstein, L.M. Satarov, I.N. Mishustin, B. Kämpfer, H. Stöcker, Electromagnetic probes of a pure-glue initial state in nucleus-nucleus collisions at energies available at the CERN large Hadron collider. Phys. Rev. C **94**, 024906 (2016). https://doi.org/10.1103/PhysRevC.94.024906

46. V. Vovchenko, L.-G. Pang, H. Niemi, I.A. Karpenko, M.I. Gorenstein, L.M. Satarov, I.N. Mishustin, B. Kämpfer, H. Stöcker, Hydrodynamic modeling of a pure-glue initial scenario in high-energy hadron and heavy-ion collisions. PoS **BORMIO2016**, 039 (2016)
47. I. Karpenko, P. Huovinen, M. Bleicher, A 3+1 dimensional viscous hydrodynamic code for relativistic heavy ion collisions. Comput. Phys. Commun. **185**, 3016 (2014). https://doi.org/10.1016/j.cpc.2014.07.010
48. S. Borsanyi, Z. Fodor, C. Hoelbling, S.D. Katz, S. Krieg, K.K. Szabo, Full result for the QCD equation of state with 2+1 flavors. Phys. Lett. B **730**, 99 (2014). https://doi.org/10.1016/j.physletb.2014.01.007
49. V. Vovchenko, Quantum statistical van der Waals equation and its QCD applications. Ph.D. thesis, Goethe University Frankfurt (2018)
50. R. Rapp, J. Wambach, Low mass dileptons at the CERN SPS: evidence for chiral restoration? Eur. Phys. J. A **6**, 415 (1999). https://doi.org/10.1007/s100500050364
51. R. Rapp, J. Wambach, H. van Hees, The chiral restoration transition of QCD and low mass dileptons. Landolt-Börnstein **23**, 134 (2010). https://doi.org/10.1007/978-3-642-01539-7_6
52. S. Endres, H. van Hees, M. Bleicher, Photon and dilepton production at the facility for proton and anti-proton research and beam-energy scan at the relativistic heavy-ion collider using coarse-grained microscopic transport simulations. Phys. Rev. C **93**, 054901 (2016). https://doi.org/10.1103/PhysRevC.93.054901
53. T. Galatyuk, P.M. Hohler, R. Rapp, F. Seck, J. Stroth, Thermal dileptons from coarse-grained transport as fireball probes at SIS energies. Eur. Phys. J. A **52**, 131 (2016). https://doi.org/10.1140/epja/i2016-16131-1
54. J. Staudenmaier, J. Weil, V. Steinberg, S. Endres, H. Petersen, Dilepton production and resonance properties within a new hadronic transport approach in the context of the GSI-HADES experimental data (2017), arXiv: 1711.10297 [nucl-th]
55. O. Linnyk, E.L. Bratkovskaya, W. Cassing, Effective QCD and transport description of dilepton and photon production in heavy-ion collisions and elementary processes. Prog. Part. Nucl. Phys. **87**, 50 (2016). https://doi.org/10.1016/j.ppnp.2015.12.003
56. M. Glück, E. Reya, A. Vogt, Dynamical parton distributions of the proton and small x physics. Z. Phys. C **67**, 433 (1995). https://doi.org/10.1007/BF01624586
57. M. Gluck, E. Reya, A. Vogt, Pionic parton distributions. Z. Phys. C **53**, 651 (1992). https://doi.org/10.1007/BF01559743
58. C. Spieles, L. Gerland, N. Hammon, M. Bleicher, S.A. Bass, H. Stöcker, W. Greiner, C. Lourenco, R. Vogt, A microscopic calculation of secondary Drell-Yan production in heavy ion collisions. Eur. Phys. J. C **5**, 349 (1998). https://doi.org/10.1007/s100520050279
59. C. Spieles, L. Gerland, N. Hammon, M. Bleicher, S.A. Bass, H. Stöcker, W. Greiner, C. Lourenco, R. Vogt, Intermediate mass dileptons from secondary Drell-Yan processes. Nucl. Phys. A **638**, 507 (1998). https://doi.org/10.1016/S0375-9474(98)00345-5
60. R. Angeles-Martinez et al., Transverse momentum dependent (TMD) parton distribution functions: status and prospects. Acta Phys. Polon. B **46**, 2501 (2015). https://doi.org/10.5506/APhysPolB.46.2501
61. E. Anassontzis et al., High mass dimuon production in $\bar{p}n$ and $\pi^- n$ interactions at 125-GeV/c. Phys. Rev. D **38**, 1377 (1988). https://doi.org/10.1103/PhysRevD.38.1377
62. F. Eichstaedt, S. Leupold, K. Gallmeister, H. van Hees, U. Mosel, Description of fully differential Drell-Yan pair production. PoS **BORMIO2011**, 042 (2011), arXiv: 1108.5287 [hep-ph]
63. A.D. Martin, W.J. Stirling, R.S. Thorne, G. Watt, Parton distributions for the LHC. Eur. Phys. J. C **63**, 189 (2009). https://doi.org/10.1140/epjc/s10052-009-1072-5
64. J.C. Webb et al. (NuSea Collaboration), Absolute Drell-Yan dimuon cross-sections in 800 GeV/c pp and pd collisions (2003), arXiv: hep-ex/0302019
65. J.C. Webb, Measurement of continuum dimuon production in 800 GeV/c proton nucleon collisions. Ph.D. thesis, New Mexico State U. (2003), arXiv: hep-ex/0301031, https://doi.org/10.2172/1155678

66. P.L. McGaughey et al. (E772 Collaboration), Cross-sections for the production of high mass muon pairs from 800-GeV proton bombardment of H-2. Phys. Rev. D **50**, 3038 (1994), [Erratum: Phys. Rev. D **60**, 119903 (1999)]. https://doi.org/10.1103/PhysRevD.50.3038, https://doi.org/10.1103/PhysRevD.60.119903
67. G. Moreno et al., Dimuon production in proton - copper collisions at $\sqrt{s}$ = 38.8 GeV. Phys. Rev. D **43**, 2815 (1991). https://doi.org/10.1103/PhysRevD.43.2815
68. A.S. Ito et al., Measurement of the continuum of dimuons produced in high-energy proton - nucleus collisions. Phys. Rev. D **23**, 604 (1981). https://doi.org/10.1103/PhysRevD.23.604
69. S.R. Smith et al., Experimental test of the Drell-Yan model in $pW \rightarrow \mu^+\mu^-X$. Phys. Rev. Lett. **46**, 1607 (1981). https://doi.org/10.1103/PhysRevLett.46.1607
70. O. Buss, T. Gaitanos, K. Gallmeister, H. van Hees, M. Kaskulov, et al., Transport-theoretical description of nuclear reactions. Phys. Rep. **512**, 1 (2012). https://doi.org/10.1016/j.physrep.2011.12.001
71. R. Baier, Y.L. Dokshitzer, A.H. Mueller, S. Peigne, D. Schiff, Radiative energy loss and p(T) broadening of high-energy partons in nuclei. Nucl. Phys. B **484**, 265 (1997). https://doi.org/10.1016/S0550-3213(96)00581-0

Regge Trajectories of Radial Meson Excitations: Exploring the Dyson–Schwinger and Bethe–Salpeter Approach

Robert Greifenhagen, Burkhard Kämpfer and Leonid P. Kaptari

Abstract The combined Dyson–Schwinger and Bethe–Salpeter equations in rainbow-ladder approximation are used to search for Regge trajectories of mesons in the pseudo-scalar and vector channels. We focus on the often employed Alkofer–Watson–Weigel kernel which is known to deliver good results for the ground state meson spectra; it provides linear Regge trajectories in the $J^P = 0^-$ channel.

1 Introduction

Despite of the apparent simplicity of the Lagrangian where Quantum Chromodynamics (QCD) is based upon, it encodes an enormous richness of phenomena, most of them related to non-perturbative regime. While lattice QCD allows for an access to many facets of the hadron spectra, the so-called XYZ states pose still a challenge [1]. Apart the quantitatively adequate description of low-lying hadron states in various flavor channels, the higher excitations call also for a description and confrontation with experimentally well-established facts. It is known for a long time that mesons of a given flavor composition can be grouped on Regge trajectories according to $M_n^2 = M_0^2 + n\mu^2$, where M_n stands for the mass (energy) labeled by the radial quantum number $n = 0, 1, 2, \ldots$, M_0 denotes the ground state mass of a respective

R. Greifenhagen (✉) · B. Kämpfer · L. P. Kaptari
Helmholtz-Zentrum Dresden-Rossendorf, Dresden, Germany
e-mail: r.greifenhagen@hzdr.de

B. Kämpfer
e-mail: kaempfer@hzdr.de

L. P. Kaptari
e-mail: kaptari@theor.jinr.ru

B. Kämpfer
TU Dresden, Institute of Theoretical Physics, Dresden, Germany

L. P. Kaptari
JINR, Bogoliubov Institute for Theoretical Physics, Dubna, Russia

J. Kirsch et al. (eds.), *Discoveries at the Frontiers of Science*,
FIAS Interdisciplinary Science Series, https://doi.org/10.1007/978-3-030-34234-0_5

trajectory and $\mu^2 = 1.25\,\mathrm{GeV}^2$ [2] or $1.35\,\mathrm{GeV}^2$ [3] is a universal slope parameter (cf. [3, 4] for a recent account and [5–8] for the discussion of the experimental database). More generally, Ref. [9] advocate an ordering according to $M^2_{n,J} = \hat{a}(n + J) + \hat{c}$, where J stands for the angular momentum and $\hat{a}$ and $\hat{c}$ are appropriate constants, see also [10].

While being a phenomenological ordering scheme, the arrangement of hadron states on Regge trajectories should emerge from QCD, ideally directly without approximations or based on certain symmetries or as result of suitable models. In fact, the relativistic quark model [11, 12] delivers such linear trajectories. Also, holographic models based on the AdS/CFT correspondence (cf. [6, 13]) cope with Regge trajectories [14], or even use them as input for constraining the dilaton dynamics for further investigations [15–17]. Moreover, functional formulations of QCD such as combined Dyson–Schwinger (DS) and Bethe–Salpeter (BS) equations address the issue of recovering Regge trajectories [18, 19] with appropriate interactions kernels and truncation schemes [20]. The latter approach is interesting since it provides the avenue toward addressing the important quest for medium modifications of hadrons in a hot and dense hadron medium [21]. Considering the medium created transiently in the course of relativistic heavy-ion collisions, the interplay of confinement and chiral symmetry restoration poses further challenges [22].

Here, we focus on the question whether the DS-BS approach in rainbow-ladder approximation is capable to deliver Regge trajectories when using simple interaction kernels. To be specific, we employ the Alkofer–Watson–Weigel (AWW) kernel [23] in the pseudo-scalar and vector channels and search for the first excited states. Such a study is a prerequisite for the extension to nonzero temperatures [24]. The AWW kernel is known to provide a good description of meson ground states supposed the analytic properties of the quark propagators are properly dealt with [25–27]. However, in the literature, one finds some side remarks that AWW is less appropriate for a description of excitations [23]. Since details are lacking we feel that a dedicated investigation is timely, in particular in respect with the above stressed importance of Regge trajectories as a central feature of the meson spectrum. For the search of meson excitations, we employ a novel method of finding solution of the BS equation.

Our paper is organized as follows. In Sect. 2, we recall the DS and BS equations as well as the AWW kernel. Numerical results are described in Sect. 3. We summarize in Sect. 4. The appendix contains some technicalities.

2 Recalling the DS and BS Equations in Rainbow-Ladder Approximation

The DS equation (also dubbed gap equation) aims at solving

$$S^{-1}(p) = S_0^{-1}(p) - \int \frac{d^4k}{(2\pi)^4} \left[-ig^2 \gamma^\nu \frac{\tau^a}{2} \right] \mathscr{D}_{\mu\nu}(p,k) \Gamma^{\mu,a}(p,k) S(k), \quad (1)$$

for the dressed quark propagator S, where S_0 is the undressed quark propagator, γ^ν are the Dirac matrices with $\{\gamma^\mu, \gamma^\nu\} = 2g^{\mu\nu}$, τ^a are color matrices, p and k are four-momenta, g is the QCD coupling constant, and $\mathscr{D}_{\mu\nu}$ stands for the gluon propagator. In Euclidean space, the rainbow-ladder approximation means $\Gamma^{\mu,a}(p,k) \Rightarrow -i\gamma^\mu \frac{\tau^a}{2}$. The final equation to be solved reads

$$S^{-1}(p) = S_0^{-1}(p) + \frac{4}{3}\int \frac{d^4k}{(2\pi)^4}[g^2\mathscr{D}_{\mu\nu}(p-k)]\gamma^\mu S(k)\gamma^\nu \tag{2}$$

with $S_0^{-1}(p) = i\not{p} + m_q$, where the parameter m_q is flavor-dependent.

The dressed quark propagator $S(p)$ enters the BS equation for the vertex function

$$\Gamma(P,p) = \int \frac{d^4k}{(2\pi)^4} K(P,p,k)S(k_1)\Gamma(P,k)S(k_2), \tag{3}$$

with the quark (antiquark) momentum k_1 (k_2), total momentum $P = k_1 + k_2$ ($\sim (M_{12}, \mathbf{0})$ for a meson at rest), the relative momenta p and $k = (k_1 - k_2)/2$ and the rainbow-ladder approximation for the kernel function

$$K(P,p,k) = -g^2\mathscr{D}_{\mu\nu}(p-k)\left(\gamma^\mu \frac{\tau^a}{2}\right)\left(\gamma^\nu \frac{\tau^a}{2}\right). \tag{4}$$

In the Euclidean space, the used BS becomes then

$$\Gamma(P,p) = -\frac{4}{3}\int \frac{d^4k}{(2\pi)^4}\gamma^\mu S(k_1)\Gamma(P,k)S(k_2)\gamma^\nu[g^2\mathscr{D}_{\mu\nu}(p-k)]. \tag{5}$$

We employ here the AWW kernel, i.e. $D(k^2) \Rightarrow D^{\mathrm{AWW}}(k^2)$ in the decomposition of the gluon propagator in Landau gauge, $g^2\mathscr{D}_{\mu\nu}(k) = (g_{\mu\nu} - k_\mu k_\nu k^{-2})\, D(k^2)$ with

$$D^{\mathrm{AWW}}(k^2) = \frac{4\pi^2 Dk^2}{\omega^2} e^{-\frac{k^2}{\omega^2}}, \tag{6}$$

with the interaction strength parameter D and the interaction range parameter ω. (In what follows, we employ the standard model parameters $\omega = 0.5$ GeV and $D = 16\,\mathrm{GeV}^{-2}$, unless explicitly noted.) It is the IR part of the Maris–Tandy kernel [28].

3 Numerical Methods and Results

The numerical details for solving the above quoted DS and BS equations with given truncations and approximations are described in [29]. In vacuum, the quark propagator can be decomposed as $S^{-1}(p) = i\gamma \cdot pA(p) + B(p)$ to split (2) into two coupled integral equations for A and B which are needed for complex arguments p. Figure 1

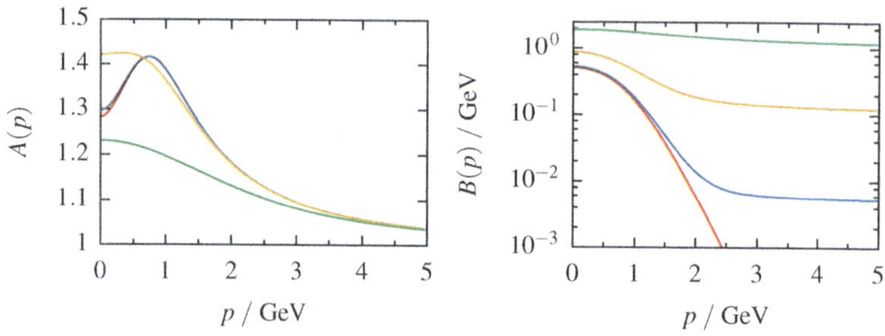

Fig. 1 Propagator functions $A(p)$ (left) and $B(p)$ (right) on the posive real axis. Red curve: $m_q = 0\,\text{MeV}$ (chiral limit), blue curve: $m_q = 5\,\text{MeV}$, yellow curve: $m_q = 115\,\text{MeV}$, and green curve: $m_q = 1130\,\text{MeV}$

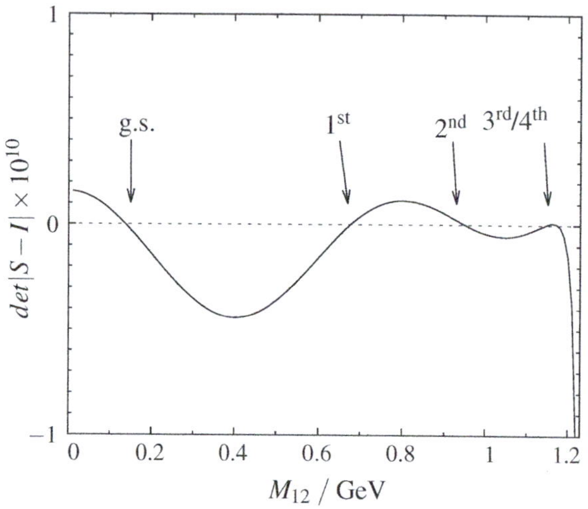

Fig. 2 Smooth determinant function $det(S - \mathbb{1})$ as a function of M_{12} for the pion channel ($m_{1,2} = m_q = m_u = 5\,\text{MeV}$). For $\omega = 0.3\,\text{GeV}$ and $D = 205.761\,\text{GeV}^{-2}$. The arrows denote the masses of ground state (g.s.), first excited state (1st), second excited state (2nd), and third and fourth excited state (3rd, 4th)

exhibits examples for these functions for positive, real values of the momentum p. Note the nonlinear dependence on the quark mass parameter m_q.

Making an expansion of the BS vertex function (5) into spin-angular functions, spherical harmonics, and Gegenbauer polynomials, one arrives at a matrix equation $X_\alpha = S_{\alpha\beta} X_\beta$ with $\alpha, \beta = 1, \ldots N$, where $N = \alpha_{\max} \cdot N_{\text{Gegenbauer}} \cdot N_{\text{Gauss}}$. Here, $\alpha_{\max}$ denotes the number of spin-angular harmonics, $N_{\text{Gegenbauer}}$ is the number of included Gegenbauer polynomials, and N_{Gauss} stands for the mesh number of a Gaussian integration over internal momenta. The chain of manipulations that leads to the quantity $S_{\alpha\beta}$ is recalled in the Appendix, where also the elements of X are defined.

The energy of mesons as $\bar{q}q$ bound states is determined by $\det|S - \mathbb{1}| = 0$ with S being a function of M_{12}. An example is exhibited in Fig. 2.

The AWW kernel depends on two parameters, D and ω; in addition, the quark masses $m_{1,2}$ ($m_1 = m_2 = m_q$ for equal quark mass mesons) must be adjusted. Figure 3 exhibits examples for the pseudo-scalar (left) and vector (right) channels.

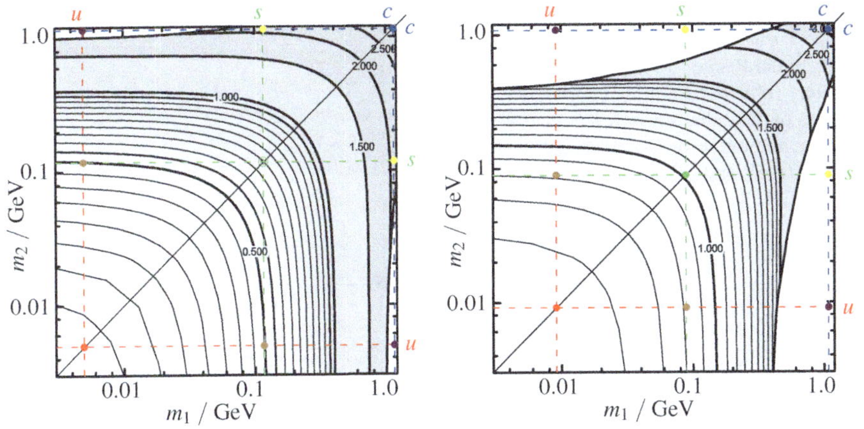

Fig. 3 Contour plot of pseudo-scalar (left) and vector (right) meson ground state masses in units of GeV for varying quark masses m_1 and m_2. The colored bullets denote the experimental values of meson ground states (red: π/ρ, green: K/ϕ, violet: η_c/J_ψ) which could be used for extracting the bare quark mass parameters $m_{1,2}$ (vertical and horizontal dashed lines, labeled by the corresponding quark flavor); the corresponding value of $m_c = 1.110\,\text{GeV}$ in the vector channel can be compared with one suggested in the pseudo-scalar channel, with optimum value $1.130\,\text{GeV}$. In the white region, no solutions of the BS equation could be found w/o accounting explicitly for the pole structure of S in the complex momentum plane

Freezing in $m_u = m_d = 5\,\text{MeV}$, $m_s = 115\,\text{MeV}$, $m_c = 1130\,\text{MeV}$ and keeping our standard parameters $\{\omega, D\}$ we find in the $J^P = 0^-$ channel (in units of GeV, experimental values in parentheses) $M_\pi = 0.137\,(0.140), 0.986\,(1.300), 1.369\,(1.812)$, $M_K = 0.492\,(0.494), 1.162\,(1.460)$, $M_{\bar{s}s} = 0.693\,(-), 1.278\,(-), 1.572\,(-)$, $M_D = -\,(1.870)$, $M_{D_s} = 2.075\,(1.968), 2.313\,(-)$, $M_{\eta_c} = 2.984\,(2.984), 3.278\,(3.639), 3.557\,\,(-)$, where a "–" means either no state found or missing information. The results for the $J^P = 1^-$ states are $M_\rho = 0.758\,(0.775), 1.041\,(1.465), 1.287\,(1.720)$, $M_{K^*} = 0.945\,(894), 1.264\,(1.414)$, $M_\phi = 1.077\,(1.019), 1.402\,(1.680), 1.598\,(2.175)$, $M_{D^*} = -\,(2.010)$, $M_{D_s^*} = -\,(2.112)$, $M_{J/\psi} = 3.136\,(3.097), 3.346\,(3.686), 3.593\,\,(3.773)$. In addition, we note $f_\pi = 0.133\,\text{GeV}$ and $\langle\bar{q}q\rangle = (-0.255\,\text{GeV})^3$ [29]. The overwhelming impression is that, despite of the truncation and the simple AWW kernel, quite reasonable numbers are delivered, however, with some drastic deviations, e.g., the pure pseudo-scalar $\bar{s}s$ states do not appear in nature.

Another issue is the discrepancy of calculated and experimental values of excitations, e.g., in the pion channel. Instead of disputing the impact of mixing effects, we take our calculated values and check the arising Regge trajectories for linearity, see Fig. 4 for an example with apparently linear trajectory. Some survey is exhibited in Fig. 5, where a few 0^- states are depicted (left column) and the Regge slope parameters as well as a linearity measure are displayed too (right column). In that study, we freeze in $a = \omega^5 D$ and vary the parameter ω. As known, the ground state masses are kept constant under such a variation, but evidently, the excited states depend on

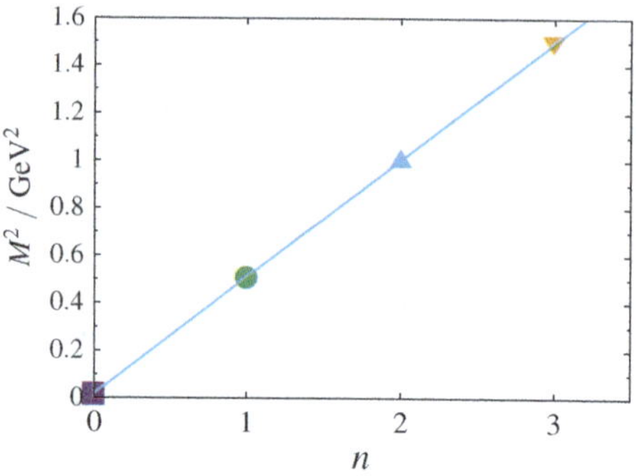

Fig. 4 Pion bound states (symbols, $m_{1,2} = m_q = m_u$) and the fitted Regge trajectory (blue line) as a function of the radial quantum number n for $\omega = 0.32\,\text{GeV}$ and $a = 0.5\,\text{GeV}^3$

ω, even up to a disappearance of certain states, e.g., π and $\bar{s}s$. The slope changes with ω, while the linearity is strikingly good. This is in contrast to the 1^- channel (not displayed; for details cf. [29]), where, at fixed values of $\omega^5 D$, also the ground states vary with changing ω; linear Regge trajectories are hardly accessible within the preferred parameter range adjusted to 0^- states.

4 Summary

In summary, we test the capability to catch the first excited states of mesons (pseudo-scalar and vector channels) by using the Dyson–Schwinger and Bethe–Salpeter equations in rainbow-ladder approximation equipped with the Alkofer–Watson–Weigel kernel. This is the first step of an attempt to describe hadron properties, and thus implicitly confinement and relevant scales, together with the extension to finite temperatures and baryon densities in follow-up investigations. According to our contemporary understanding, at some temperature and at small baryon density, hadrons as quasi-particle degrees of QCD should disappear in favor of quasi-quark and quasi-gluon degrees of freedom. The transition happens gradually and may depend on the flavor channel under consideration. For larger baryon densities, the transition could be abrupt, supposed a critical point occurs in the phase diagram of strongly interacting matter. A few large-scale heavy-ion experiments, e.g., the beam-energy scan at RHIC, NA61/SHINE at SPS, CBM at SIS100, as well at NICA and J-PARC address in their physics programs the critical point search. For that, both the properties of hadrons as individual entities and the behavior of hadron matter are key quantities in reconstructing the final state of strong-interaction matter in collision.

Acknowledgements The authors gratefully acknowledge the collaboration with S. M. Dorkin, T. Hilger, and M. Viebach on the topic.

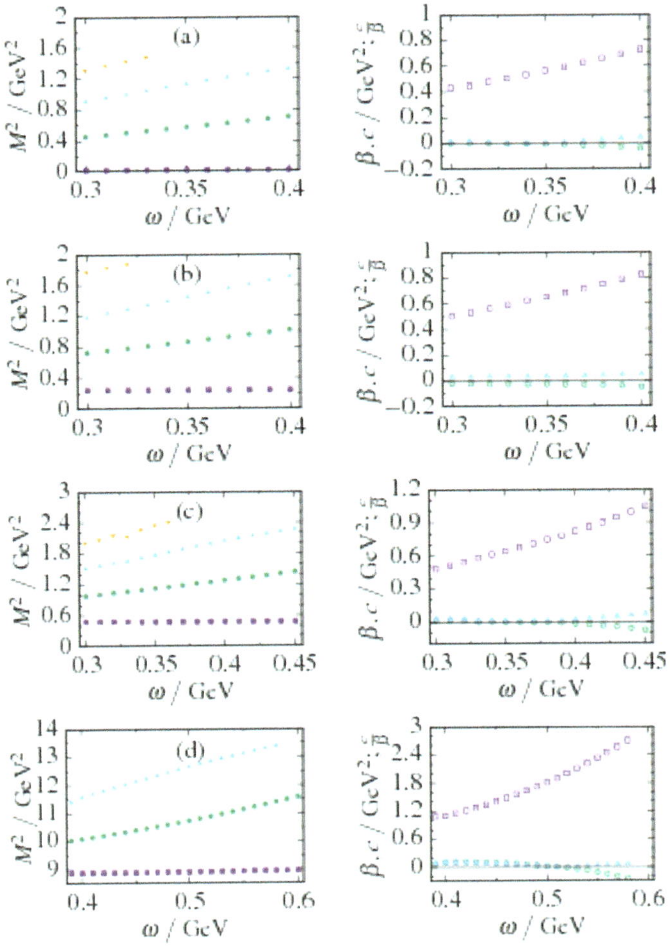

Fig. 5 Dependence of the spectrum of pions (a: $m_q = m_u$), kaons (b: $m_1 = m_u, m_2 = m_s$), fictitious pseudo-scalar $\bar{s}s$ states (c: $m_q = m_s$), and η_c (d: $m_q = m_c$) as a function of ω for $a = 0.5\,\text{GeV}^3$ (left column). The right column depicts the corresponding Regge slope coefficients β (empty violet squares), the quadratic term c (empty green squares) in fits of the spectra by $M_n^2 = M_0 + \beta n + cn^2$, $\beta \equiv \mu^2$, and the deviation measured from linear behavior $|c/\beta|$ (filled blue squares)

Appendix: Spin-Angular Harmonics

The BS vertex function Γ can be expanded into spin-angular harmonics:

$$\Gamma(p) = \sum_{\alpha=1}^{\alpha_{max}} \Gamma_\alpha(p) = \sum_{\alpha=1}^{\alpha_{max}} g_\alpha(p)\, \mathscr{T}_\alpha(\mathbf{p}) \tag{7}$$

with functions g_α fulfilling the orthogonality relation $g_\alpha(p) = \int d\Omega_{\mathbf{p}}\, \mathrm{Tr}[\,\Gamma(p)\, \mathscr{T}_\alpha^\dagger(\mathbf{p})\,]$. For pseudo-scalar mesons ($J^{PC} = 0^-$), the number of independent spin-angular harmonics $\alpha_{max} = 4$, and the set is chosen as

$$
\begin{aligned}
\mathscr{T}_1(\mathbf{p}) &= \frac{1}{\sqrt{16\pi}}\gamma^5 = \mathscr{T}_1^\dagger(\mathbf{p}), & \mathscr{T}_2(\mathbf{p}) &= \frac{1}{\sqrt{16\pi}}\gamma^0\gamma^5 = -\mathscr{T}_2^\dagger(\mathbf{p}),\\
\mathscr{T}_3(\mathbf{p}) &= -\frac{1}{\sqrt{16\pi}}\not{n}_{\mathbf{p}}\gamma^0\gamma^5 = \mathscr{T}_3^\dagger(\mathbf{p}), & \mathscr{T}_4(\mathbf{p}) &= -\frac{1}{\sqrt{16\pi}}\not{n}_{\mathbf{p}}\gamma^5 = \mathscr{T}_4^\dagger(\mathbf{p}),
\end{aligned}
\tag{8}
$$

and for vector mesons ($J^{PC} = 1^-$), $\alpha_{max} = 8$ with

$$
\begin{aligned}
\mathscr{T}_1(\mathbf{p}) &= \sqrt{\frac{1}{16\pi}}\not{\xi}_{\mathscr{M}} = \mathscr{T}_1^\dagger(\mathbf{p}), \quad \mathscr{T}_2(\mathbf{p}) = -\sqrt{\frac{1}{16\pi}}\gamma^0\not{\xi}_{\mathscr{M}} = \mathscr{T}_2^\dagger(\mathbf{p}),\\
\mathscr{T}_3(\mathbf{p}) &= -\sqrt{\frac{3}{16\pi}}(n_{\mathbf{p}}\,\xi_{\mathscr{M}}) = \mathscr{T}_3^\dagger(\mathbf{p}),\\
\mathscr{T}_4(\mathbf{p}) &= \sqrt{\frac{3}{32\pi}}\gamma^0[-(n_{\mathbf{p}}\xi_{\mathscr{M}}) + \not{n}_{\mathbf{p}}\not{\xi}_{\mathscr{M}}] = -\mathscr{T}_4^\dagger(\mathbf{p}),\\
\mathscr{T}_5(\mathbf{p}) &= \sqrt{\frac{1}{32\pi}}[\not{\xi}_{\mathscr{M}} + 3(n_{\mathbf{p}}\,\xi_{\mathscr{M}})\not{n}_{\mathbf{p}}] = -\mathscr{T}_5^\dagger(\mathbf{p}),\\
\mathscr{T}_6(\mathbf{p}) &= \sqrt{\frac{1}{32\pi}}\gamma^0[\not{\xi}_{\mathscr{M}} + 3(n_{\mathbf{p}}\,\xi_{\mathscr{M}})\not{n}_{\mathbf{p}}] = \mathscr{T}_6^\dagger(\mathbf{p}),\\
\mathscr{T}_7(\mathbf{p}) &= -\sqrt{\frac{3}{16\pi}}\gamma^0(n_{\mathbf{p}}\,\xi_{\mathscr{M}}) = \mathscr{T}_7^\dagger(\mathbf{p}),\\
\mathscr{T}_8(\mathbf{p}) &= \sqrt{\frac{3}{32\pi}}[-(n_{\mathbf{p}}\,\xi_{\mathscr{M}}) + \not{n}_{\mathbf{p}}\not{\xi}_{\mathscr{M}}] = -\mathscr{T}_8^\dagger(\mathbf{p}).
\end{aligned}
\tag{9}
$$

Scalar products are displayed here in Minkowski space; $n_{\mathbf{p}}$ is the unit vector $n_{\mathbf{p}} = (0, \mathbf{p}/|\mathbf{p}|)$, $\xi_{\mathscr{M}} = (0, \boldsymbol{\xi}_{\mathscr{M}})$ is the polarization vector fixed by $\boldsymbol{\xi}_{+1} = -(1, i, 0)/\sqrt{2}$, $\boldsymbol{\xi}_{-1} = (1, -i, 0)/\sqrt{2}$, $\boldsymbol{\xi}_0 = (0, 0, 1)$ and slashed quantities such as $\not{x}$ represent $\gamma^\mu x_\mu$.

The partial amplitudes $\Gamma_\alpha(p)$ and the interaction kernel (6) are decomposed over the basis of spherical harmonics $Y_{lm}(\theta, \phi)$ and normalized Gegenbauer polynomials $X_{nl}(\chi)$,

$$
Z_{nlm} = X_{nl}(\chi)Y_{lm}(\theta, \phi) \tag{10}
$$

$$
= \sqrt{\frac{2^{2l+1}}{\pi}\frac{(n+1)(n-l)!l!^2}{(n+l+1)!}}\sin^l\chi\, G_{n-l}^{l+1}(\cos\chi)Y_{lm}(\theta, \phi), \tag{11}
$$

with familiar Gegenbauer polynomials $G_{n-l}^{l+1}(\cos\chi)$. The hyper angle χ is defined by $\cos\chi = p_4/\tilde{p}$ and $\sin\chi = |\mathbf{p}|/\tilde{p}$, where $\tilde{p} = (p_4^2 + \mathbf{p}^2)^{1/2}$ is the modulus for an Euclidean four-vector $p = (p_4, \mathbf{p})$. The partial decompositions of $\Gamma_\alpha(p)$ and $D^{\mathrm{AWW}}(p-k)$ read

$$\Gamma_\alpha(p) = \sum_n \varphi^n_{\alpha,l_\alpha}(\tilde{p}) X_{nl_\alpha}(\chi_p) \mathscr{T}_\alpha(\mathbf{p}), \tag{12}$$

$$D^{\rm AWW}(p-k) = 2\pi^2 \sum_{\kappa\lambda\mu} \frac{1}{\kappa+1} V_\kappa(\tilde{p},\tilde{k}) X_{\kappa\lambda}(\chi_p) X_{\kappa\lambda}(\chi_k) Y_{\lambda\mu}(\Omega_p) Y^*_{\lambda\mu}(\Omega_k), \tag{13}$$

where $V_\kappa(\tilde{p},\tilde{k})$ are the partial kernels and $\varphi^n_{\alpha,l_\alpha}(\tilde{p})$ are the expansion coefficients of the partial amplitudes. Actually, l_α is restricted by the corresponding orbital momentum encoded in $\mathscr{T}_\alpha(\mathbf{p})$. For $\mathscr{T}_{1,2}(\mathbf{p})$ from Eq. (8) $l_\alpha = 0$, while for $\mathscr{T}_{3,4}(\mathbf{p})$ $l_\alpha = 1$. In analogy for vector mesons (see (9)), $l_\alpha = 0$ for $\mathscr{T}_{1,2}(\mathbf{p})$, $l_\alpha = 1$ for $\mathscr{T}_{3,4,7,8}(\mathbf{p})$ and $l_\alpha = 2$ for $\mathscr{T}_{5,6}(\mathbf{p})$.

Changing the integration variables to the hyperspace, $d^4k = \tilde{k}^3 \sin^2\chi_k \sin\theta_k d\tilde{k} \times d\chi_k d\theta_k d\phi_k$, inserting Eqs. (12) and (13) into (5) and performing the necessary angular integration, a system of integral equations for the expansion coefficients $\varphi^n_{\alpha,l_\alpha}(\tilde{p})$ as the N elements of X remains:

$$\varphi^n_{\alpha,l_\alpha}(\tilde{p}) = \sum_\beta \sum_{m=1}^{\infty} \int d\tilde{k}\tilde{k}^3 S_{\alpha\beta}(\tilde{p},\tilde{k},m,n) \varphi^m_{\beta,l_\beta}(\tilde{k}). \tag{14}$$

The explicit expression for $S_{\alpha\beta}(\tilde{p},\tilde{k},m,n)$ reads

$$\begin{aligned} S_{\alpha\beta}(\tilde{p},\tilde{k},m,n) = \sum_\kappa \int & \sin^2\chi_k d\chi_k X_{ml_\beta}(\chi_k) X_{\kappa\lambda}(\chi_k) \sigma_{s,v}(\tilde{k}_1^2) \sigma_{s,v}(\tilde{k}_2^2) \\ & \times A_{\alpha\beta}(\tilde{p},\tilde{k},\kappa,\chi_k,n), \end{aligned} \tag{15}$$

where $\tilde{k}^2_{1,2}$ is given by $\tilde{k}^2_{1,2} = \tilde{k}^2 - \eta^2_{1,2} M^2_{12} \pm \eta_{1,2} M_{12} \cos\chi$ with momentum partitioning parameters $\eta_1 + \eta_2 = 1$ and $A_{\alpha\beta}(\tilde{p},\tilde{k},\kappa,\chi_k,n)$ results from calculations of traces and angular integrations as

$$\begin{aligned} A_{\alpha\beta}(\tilde{p},\tilde{k},\kappa,\chi_k,n) = \int & \sin^2\chi_p d\chi_p d\Omega_p d\Omega_k V_\kappa(\tilde{p},\tilde{k}) X_{nl_\alpha}(\chi_p) X_{\kappa\lambda}(\chi_p) Y_{\lambda\mu}(\Omega_p) Y^*_{\lambda\mu}(\Omega_k) \\ & \times {\rm Tr}[d_{\mu\nu}((p-k)^2)\gamma^\mu \ldots \mathscr{T}_\alpha(\mathbf{p}) \ldots \mathscr{T}_\alpha(\mathbf{p})\gamma^\nu]. \end{aligned} \tag{16}$$

References

1. Z. Yang, Q. Wang, U.G. Meissner, Phys. Lett. B **767**, 470 (2017). https://doi.org/10.1016/j.physletb.2017.01.023, arXiv:1609.08807 [hep-ph]
2. A.V. Anisovich, V.V. Anisovich, A.V. Sarantsev, Phys. Rev. D **62**, 051502 (2000). https://doi.org/10.1103/PhysRevD.62.051502, arXiv:hep-ph/0003113
3. P. Masjuan, E.R. Arriola, W. Broniowski, Phys. Rev. D **85**, 094006 (2012). https://doi.org/10.1103/PhysRevD.85.094006, arXiv:1203.4782 [hep-ph]

4. P. Masjuan, E.R. Arriola, W. Broniowski, Phys. Rev. D **87**(11), 118502 (2013). https://doi.org/10.1103/PhysRevD.87.118502, arXiv:1305.3493 [hep-ph]
5. E. Klempt, A. Zaitsev, Phys. Rept. **454**, 1 (2007). https://doi.org/10.1016/j.physrep.2007.07.006, arXiv:0708.4016 [hep-ph]
6. S.J. Brodsky, G.F. de Teramond, H.G. Dosch, J. Erlich, Phys. Rept. **584**, 1 (2015). https://doi.org/10.1016/j.physrep.2015.05.001, arXiv:1407.8131 [hep-ph]
7. D.V. Bugg, Phys. Rev. D **87**(11), 118501 (2013). https://doi.org/10.1103/PhysRevD.87.118501, arXiv:1209.3481 [hep-ph]
8. D.V. Bugg, Phys. Rep. **397**, 257 (2004). https://doi.org/10.1016/j.physrep.2004.03.008, arXiv:hep-ex/0412045
9. S.S. Afonin, Eur. Phys. J. A **29**, 327 (2006). https://doi.org/10.1140/epja/i2006-10080-2, arXiv:hep-ph/0606310
10. S.S. Afonin, I.V. Pusenkov, EPJ Web Conf. **125**, 04006 (2016). https://doi.org/10.1051/epjconf/201612504006, arXiv:1606.05218 [hep-ph]
11. D. Ebert, R.N. Faustov, V.O. Galkin, Eur. Phys. J. C **66**, 197 (2010). https://doi.org/10.1140/epjc/s10052-010-1233-6, arXiv:0910.5612 [hep-ph]
12. D. Ebert, R.N. Faustov, V.O. Galkin, Phys. Rev. D **79**, 114029 (2009). https://doi.org/10.1103/PhysRevD.79.114029, arXiv:0903.5183 [hep-ph]
13. S.S. Afonin, Phys. Lett. B **675**, 54 (2009). https://doi.org/10.1016/j.physletb.2009.03.073, arXiv:0903.0322 [hep-ph]
14. S.S. Afonin, Phys. Lett. B **678**, 477 (2009). https://doi.org/10.1016/j.physletb.2009.06.071, arXiv:0902.3959 [hep-ph]
15. D. Li and M. Huang, JHEP **1311**, 088 (2013). https://doi.org/10.1007/JHEP11(2013)088,arXiv:1303.6929 [hep-ph]
16. D. Li, M. Huang, Q.S. Yan, Eur. Phys. J. C **73**, 2615 (2013). https://doi.org/10.1140/epjc/s10052-013-2615-3, arXiv:1206.2824 [hep-th]
17. R. Zollner, B. Kampfer, arXiv:1708.05833 [hep-th]
18. C.S. Fischer, S. Kubrak, R. Williams, Eur. Phys. J. A **50**, 126 (2014). https://doi.org/10.1140/epja/i2014-14126-6, arXiv:1406.4370 [hep-ph]
19. T. Hilger, M. Gomez-Rocha, A. Krassnigg and W. Lucha, Eur. Phys. J. A **53**(10), 213 (2017). https://doi.org/10.1140/epja/i2017-12384-4, arXiv:1702.06262 [hep-ph]
20. D. Binosi, L. Chang, J. Papavassiliou, S.X. Qin, C.D. Roberts, Phys. Rev. D **93**(9), 096010 (2016). https://doi.org/10.1103/PhysRevD.93.096010, arXiv:1601.05441 [nucl-th]
21. K.l. Wang, Y.X. Liu, L. Chang, C.D. Roberts, S.M. Schmidt, Phys. Rev. D **87**(7), 074038 (2013). https://doi.org/10.1103/PhysRevD.87.074038, arXiv:1301.6762 [nucl-th]
22. H. Suganuma, T.M. Doi, K. Redlich, C. Sasaki, J. Phys. G **44**, 124001 (2017). https://doi.org/10.1088/1361-6471/aa8e2f, arXiv:1709.05981 [hep-lat]
23. R. Alkofer, P. Watson, H. Weigel, Phys. Rev. D **65**, 094026 (2002). https://doi.org/10.1103/PhysRevD.65.094026, arXiv:hep-ph/0202053
24. S.M. Dorkin, M. Viebach, L.P. Kaptari, B. Kampfer, J. Mod. Phys. **7**, 2071 (2016). https://doi.org/10.4236/jmp.2016.715182, arXiv:1512.06596 [nucl-th]
25. S.M. Dorkin, L.P. Kaptari, B. Kampfer, Phys. Rev. C **91**(5), 055201 (2015). https://doi.org/10.1103/PhysRevC.91.055201, arXiv:1412.3345 [hep-ph]
26. S.M. Dorkin, L.P. Kaptari, T. Hilger, B. Kampfer, Phys. Rev. C **89**, 034005 (2014). https://doi.org/10.1103/PhysRevC.89.034005, arXiv:1312.2721 [hep-ph]
27. S.M. Dorkin, T. Hilger, L.P. Kaptari, B. Kampfer, Few Body Syst. **49**, 247 (2011). https://doi.org/10.1007/s00601-010-0108-6, arXiv:1008.2135 [nucl-th]
28. P. Maris, P.C. Tandy, Phys. Rev. C **60**, 055214 (1999). https://doi.org/10.1103/PhysRevC.60.055214, arXiv:nucl-th/9905056
29. R. Greifenhagen, Investigation of the AWW kernel for describing the excited meson spectrum in a combined Dyson-Schwinger– Bethe-Salpeter approach. Master thesis, TU Dresden (2016)

The Physics Case for the FAIR/NICA Energy Region

Jean Cleymans

Abstract The beam energy region $\sqrt{s_{NN}} \approx 10$ GeV for heavy-ion collisions is a very interesting one. The final state has the highest net baryon density at this beam energy. A transition from a baryon dominated to a meson dominated final state takes place around this beam energy. Ratios of strange particles to mesons show clear and pronounced maxima around this beam energy. The theoretical interpretation can be clarified by covering fully this energy region. In particular the strangeness content needs to be determined, data covering the full phase space (4π) would be helpful to establish the properties of this energy region.

1 Introduction

Heavy-ion collisions [1] at high energies produce a large numbers of secondaries. At the LHC the number of charged particles produced in Pb–Pb collisions at 5.02 TeV [2] is shown in Fig. 1, thus, including neutral particles, a total of approximately 30 000 particles is being produced on average in such a collision. It is natural to try a statistical-thermal model to analyze these. As it turns out such an analysis is useful for a very wide range of beam energies, stretching from 1 GeV all the way up to the highest energies available at the LHC. For such an analysis one has to keep in mind that a relativistic heavy-ion collision passes through several stages. At one of the later, hadronic, stages, the system is assumed to be dominated by hadronic resonances, on which the thermal model focuses. The identifying feature of the thermal model is that all the resonances as listed in [3] are assumed to be in thermal and chemical equilibrium. This assumption drastically reduces the number of free parameters and thus this stage is determined by just a few thermodynamic variables namely, the chemical freeze-out temperature T, the various chemical potentials μ determined

To the memory of Professor Dr. Walter Greiner.

J. Cleymans (✉)
Physics Department, UCT-CERN Research Centre, University of Cape Town, Rondebosch 7701, South Africa
e-mail: jean.cleymans@uct.ac.za

J. Kirsch et al. (eds.), *Discoveries at the Frontiers of Science*,
FIAS Interdisciplinary Science Series, https://doi.org/10.1007/978-3-030-34234-0_6

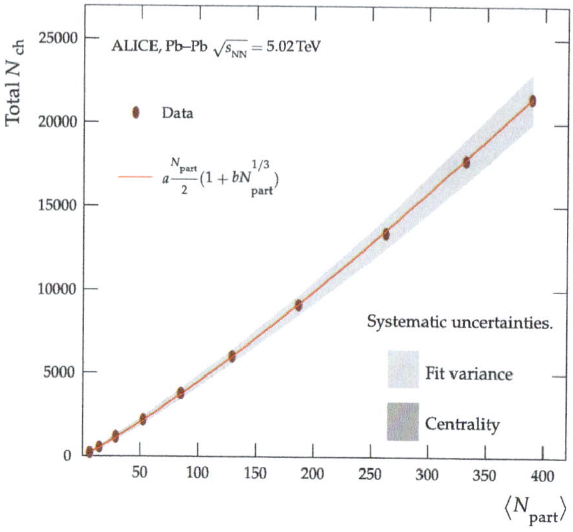

Fig. 1 Number of charged particles produced in a Pb–Pb collision as a function of beam energy as measured by the ALICE collaboration [2]

by the conserved quantum numbers and by the volume V of the system. It has been shown that this description is also the correct one [4–6] for a scaling expansion as first discussed by Bjorken [7].

In relativistic heavy ion collisions a new dimension was given to the model by the highly successful analysis of particle yields, leading to the notion of chemical equilibrium which is now a well-established one in the analysis of relativistic heavy ion collisions, see e.g. [8–10]. In view of the success of chemical freeze-out in relativistic heavy ion collisions, much effort has gone into finding models that describe this chemical freeze-out, a comparison [11] of three parameterizations is shown in Fig. 2.

There are of course uncertainties in the thermal model, one of these is about the decays of resonances, another one is whether some resonances exist or not [3]. Particle yields are determined from:

$$N_i = \sum_j N_j Br(j \to i).$$

Hence a lack of knowledge of branching ratios affects the quality of results obtained from the thermal model.
As an example, the final yield of π^+'s is given by

$$N_{\pi^+} = N_{\pi^+}(\text{thermal}) + N_{\pi^+}(\text{resonance decays})$$

and, depending on the temperature, over 80% of observed pions could be due to resonance decays. Hence the crucial importance of these decays. Various theoretical uncertainties have been recently discussed in [13].

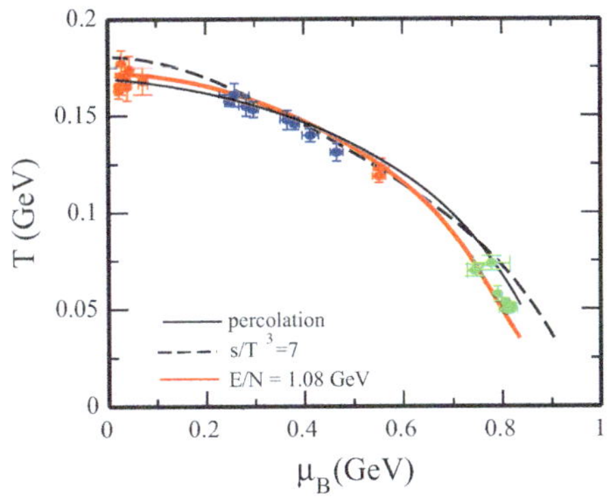

Fig. 2 Chemical freeze-out temperature T versus the baryon chemical potential at different beam energies together with curves corresponding to a fixed ratio of energy per hadron divided by total number of hadrons in the resonance gas before decay of resonances [11]. Also shown are calculations based on the percolation model [12] and for a fixed value of the entropy density divided by T^3

2 What Makes the Beam Energy $\sqrt{s_{NN}} \approx 10$ GeV Special?

2.1 *Maximum Net Baryon Density*

The resulting freeze-out curve in the $T - \mu_B$ plane, shown in Fig. 2, can also be drawn in the temperature T versus net baryon density plane as was done in [14]. The resulting curve is shown in Fig. 3. At very high beam energies the net baryon density is zero because equal numbers of particles and antiparticles are being produced while at low temperatures the net baryon density is very high. Figure 3 shows that a clear maximum exists just below the $\sqrt{s_{NN}} = 10$ GeV beam energy region.

2.2 *Transition from a Baryon Dominated to a Meson Dominated Final State*

A fairly good criterium for chemical freeze-out is the constant value of the entropy density divided $s/T^3 = 7$ ratio as can be seen from Fig. 2. The components that make up the entropy density are shown in Fig. 4, the change from a baryon-dominated to a meson-dominated final state also happens around a beam energy of $\sqrt{s_{NN}} \approx 10$ GeV.

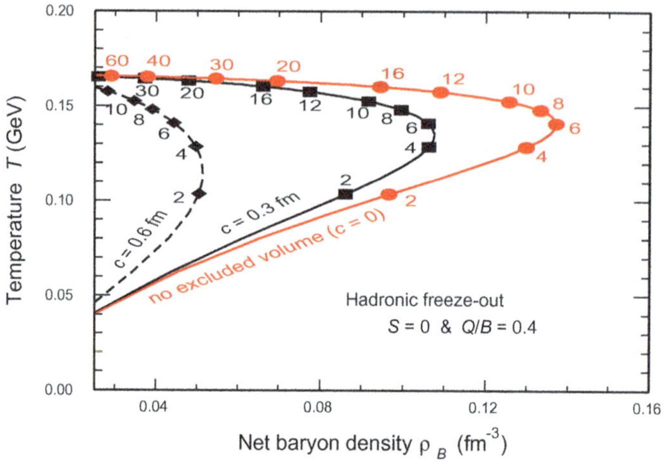

Fig. 3 The hadronic freeze-out line in the $\rho_B - T$ phase plane as obtained from the values of μ_B and T that have been extracted from the experimental data in [11]. The calculation employs values of μ_Q and μ_S that ensure $\langle S \rangle = 0$ and $\langle Q \rangle = 0.4 \langle B \rangle$ for each value of μ_B [14]. Also indicated are the beam energies for which the particular freeze-out conditions are expected. The dependence on a hard-core radius is indicated

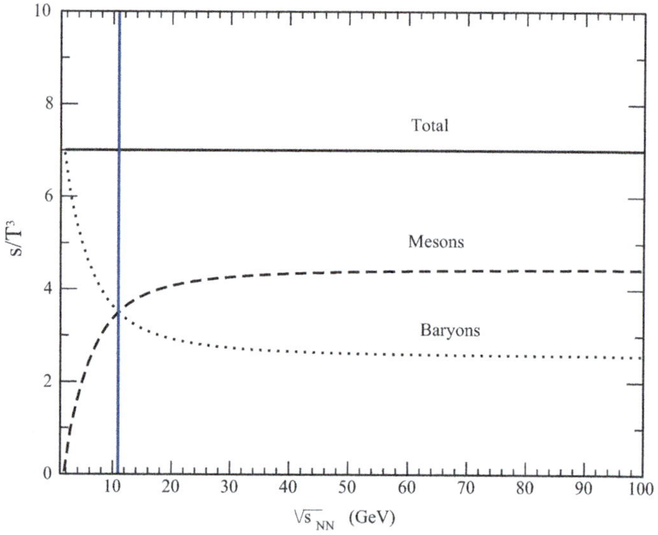

Fig. 4 The s/T^3 ratio calculated in the thermal model along the constant value consistent with chemical freeze-out. Also shown are the contributions from the mesons and the baryons

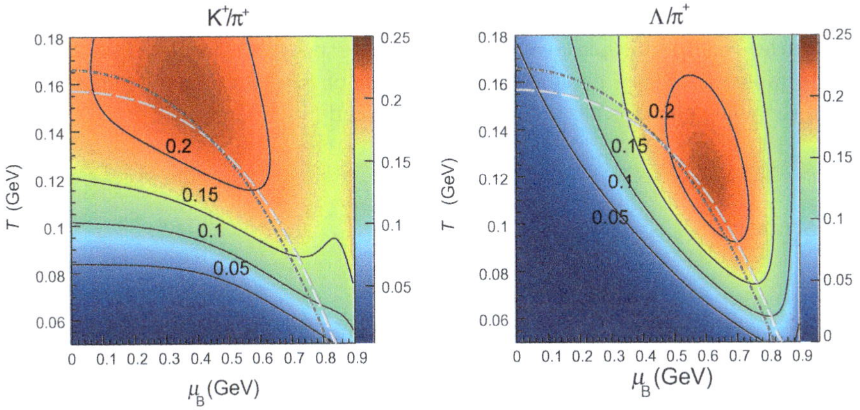

Fig. 5 Lines of constant values of the K^+/π^+ (left panel) and the Λ/π^+ (right panel) ratios in the $T - \mu_B$ plane showing a clear maximum in each ratio close to the boundary given by the chemical freeze-out line but in a different position [17]

2.3 *Ratios of Strange Hadrons to Pions*

Despite the smoothness in the thermal freeze-out parameters as a function of beam energy, strong changes are observed in several particle ratios, e.g. the horn in the K^+/π^+ ratio and a similar strong variation in the Λ/π ratio [15]. These are not observed in $p - p$ collisions, in Pb–Pb collisions they happen at a beam energy of around $\sqrt{s_{NN}} \approx 10$ GeV. Within the framework of thermal models this variation has been connected to a change from a baryon dominated to a meson dominated hadron gas [16]. The values of the K^+/π^+ and Λ/π^+ ratios [17] are shown in Fig. 5. From the lines of constant values for these ratios it can be seen that the maxima in the thermal model hug the chemical freeze-out line. It is also important to note that the maxima occur for different values of T and μ_B.

3 Conclusions

In the thermal model a change is expected as the hadronic gas undergoes a transition from a baryon-dominated to a meson-dominated gas. The strong variations seen in the particle ratios coincide with this transition. This transition occurs at a

- temperature $T = 151$ MeV,
- baryon chemical potential $\mu_B = 327$ MeV,
- energy $\sqrt{s_{NN}} = 11$ GeV.

There are thus several indications that the energy region around 10 GeV, covered by proposed new facilities, is an extremely interesting one. The theoretical interpretation can be clarified by covering this energy region. In particular the strangeness content needs to be determined, data covering the full phase space (4π) would be very helpful to determine the thermal parameters of a possible phase transition and the existence of a quarkyonic phase as has been discussed recently in [18].

References

1. H. Stoecker, W. Greiner, Phys. Rep. **137**, 277 (1986)
2. ALICE Collaboration, J. Adam et al., Phys. Lett. B **772**, 567 (2017)
3. Particle Data Group Collaboration, C. Patrignani et al., Chin. Phys. C **40**, 100001 (2016)
4. J. Cleymans, K. Redlich, Phys. Rev. C **60**, 054908 (1999)
5. W. Broniowski, W. Florkowski, Phys. Rev. Lett. **87**, 272302 (2001)
6. S. Akkelin, P. Braun-Munzinger, Y.M. Sinyukov, Nucl. Phys. A **710**, 439 (2002)
7. J. Bjorken, Phys. Rev. D **27**, 140 (1983)
8. S.T.A.R. Collaboration, L. Adamczyk et al., Phys. Rev. C **96**, 044904 (2017)
9. ALICE Collaboration, M. Floris et al., Quark matter 2014, in *The 24th International Conference on Ultrarelativistic Nucleus-Nucleus Collisions* (Darmstadt, Germany, 2014), pp. c103–c112
10. HADES Collaboration, M.M. Lorenz et al., SQM2015, in *15th International Conference on Strangeness in Quark Matter* (Dubna, Russia, 2015), p. 012022
11. J. Cleymans, H. Oeschler, K. Redlich, S. Wheaton, Phys. Rev. C **73**, 034905 (2006)
12. V. Magas, H. Satz, Eur. Phys. J. C **32**, 115 (2003)
13. L.M. Satarov, V. Vovchenko, P. Alba, M.I. Gorenstein, H. Stoecker, Phys. Rev. C **95**, 024902 (2017)
14. J. Randrup, J. Cleymans, Eur. Phys. J. C **52**, 218 (2016)
15. NA49 Collaboration, C. Alt et al., Phys. Rev. C **77**, 024903 (2008)
16. J. Cleymans, H. Oeschler, K. Redlich, S. Wheaton, Phys. Lett. B **615**, 50 (2005)
17. H. Oeschler, J. Cleymans, B. Hippolyte, K. Redlich, N. Sharma, Eur. Phys. J. C **77**, 584 (2017)
18. A. Andronic et al., Nucl. Phys. A **837**, 65 (2010)

Structure and Width of the d*(2380) Dibaryon

Avraham Gal

Abstract In this contribution, dedicated to the memory of Walter Greiner, we discuss the structure and width of the recently established d*(2380) dibaryon, confronting the consequences of our Pion Assisted Dibaryons hadronic model with those of quark motivated calculations. In particular, the relatively small width $\Gamma_{d^*} \approx 70\,\text{MeV}$ favors hadronic structure for the d*(2380) dibaryon rather than a six-quark structure.

1 Walter Greiner: Recollections

This contribution is dedicated to the memory of Walter Greiner whose wide-ranging interests included exotic phases of matter. My first physics encounter with Walter was in Fall 1983 in a joint physics symposium hosted by him, see Fig. 1.

Greiner's wide-ranging interests included also superheavy elements, so it was quite natural for him to ask my good colleague Eli Friedman, a leading figure in exotic atoms, whether extrapolating pionic atoms to superheavy elements would shed light on a then-speculated pion condensation phase. Subsequently in 1984 Eli spent one month in Frankfurt at Greiner's invitation, concluding together with Gerhard Soff [1] that the strong-interaction π^-_{1s} repulsive energy shift known from light pionic atoms persists also in superheavy elements, as shown in Fig. 2-left, thereby ruling out pion condensation for large Z. However, quite surprisingly, they also found that 1s and 2p π^- atomic states in normal heavy elements up to $Z \approx 100$ have abnormally small widths of less than 1 MeV owing to the repulsive π-nucleus strong interaction within the nuclear volume. Hence 'deeply bound' states (DBS) in pionic atoms are experimentally resolvable, although they cannot be populated radiatively as in light pionic atoms because the absorption width in the higher 3d state exceeds its radiative width by almost two orders of magnitude, as shown in Fig. 2-right.

Friedman and Soff's 1985 prediction of DBS was repeated three years later by Toki and Yamazaki [2], who apparently were not aware of it, and verified experimentally in 1996 at GSI in a (d, ^{3}He) reaction on ^{208}Pb [3]. Subsequent experiments on

A. Gal (✉)
Racah Institute of Physics, The Hebrew University, 91904 Jerusalem, Israel
e-mail: avragal@savion.huji.ac.il

J. Kirsch et al. (eds.), *Discoveries at the Frontiers of Science*,
FIAS Interdisciplinary Science Series, https://doi.org/10.1007/978-3-030-34234-0_7

Pb and Sn isotopes have yielded accurate data on several other pionic-atom DBS [4], showing clear evidence in support of Weise's 1990 conjecture of partial chiral symmetry restoration in the nuclear medium due to a renormalized isovector s-wave πN interaction through the decrease of the pion-decay constant f_π [5]. However, the few DBS established so far are still short of providing on their own the precision reached by comprehensive fits to *all* (of order 100) pionic atom data, dominantly in higher atomic orbits, in substantiating this conjecture; for a recent review on the state of the art in pionic atoms see Ref. [6].

My own encounters with Walter Greiner and several of his colleagues in Frankfurt during several Humboldt-Prize periods in the 1990s focused on developing the concept of Strange Hadronic Matter [8–10] and also on studying Kaon Condensation [11]. I recall fondly that period. Here I highlight another exotic phase of matter: non-strange Pion Assisted Dibaryons, reviewed by me recently in Ref. [12].

Fig. 1 Participants of the Frankfurt–Jerusalem Symposium in Frankfurt, 1983. Walter Greiner is 3rd left on the 1st row. Eli Friedman is 5th right on the 2nd row. I'm missing in this photo

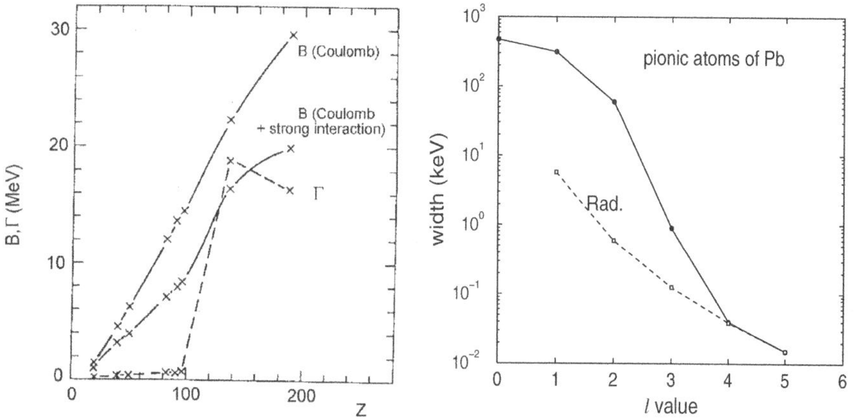

Fig. 2 First prediction of deeply bound pionic atom states [1]. The left panel shows binding energies and widths in 1s deeply bound pionic states [1], and the right panel shows the width saturation in circular states of pionic atoms of Pb [7]

2 Pion Assisted $N\Delta$ and $\Delta\Delta$ Dibaryons

2.1 The Dyson–Xuong 1964 Prediction

Non-strange s-wave dibaryon resonances $\mathscr{D}_{IS}$ with isospin I and spin S were predicted by Dyson and Xuong in 1964 [13] as early as SU(6) symmetry proved successful, placing the nucleon $N(939)$ and its P_{33} πN resonance $\Delta(1232)$ in the same **56** multiplet. These authors chose the **490** lowest-dimension SU(6) multiplet in the **56** $\times$ **56** direct product containing the flavor-SU(3) $\overline{\mathbf{10}}$ and **27** multiplets in which the deuteron, $\mathscr{D}_{01}$, and NN virtual state, $\mathscr{D}_{10}$, are classified. Four more non-strange dibaryons emerged in this scheme, with masses listed in Table 1 in terms of constants A and B. Identifying A with the NN threshold mass 1878 MeV, the value $B \approx 47$ MeV was derived by assigning $\mathscr{D}_{12}$ to the $pp \leftrightarrow \pi^+ d$ coupled-channel resonance behavior noted then at 2160 MeV, near the $N\Delta$ threshold (2.171 MeV). This led in particular to a predicted mass $M = 2350$ MeV for the $\Delta\Delta$ dibaryon candidate $\mathscr{D}_{03}$ assigned at present to the recently established d*(2380) resonance [14].

In retrospect, the choice of the **490** lowest-dimension SU(6) multiplet, with Young tableau denoted [0, 3, 3], is not accidental. This [0, 3, 3] is the one adjoint to [2, 2, 2] for color-SU(3) singlet six-quark (6q) state, thereby ensuring a totally antisymmetric color-flavor-spin-space 6q wavefunction, assuming a totally symmetric $L = 0$ orbital component. For non-strange dibaryons, flavor-SU(3) reduces to isospin-SU(2), whence flavor-spin SU(6) reduces to isospin-spin SU(4) in which the [0, 3, 3] representation corresponds to a **50** dimensional representation consisting of precisely the I, S values of the dibaryon candidates listed in Table 1, as also noted recently in Ref. [15]. Since the **27** and $\overline{\mathbf{10}}$ flavor-SU(3) multiplets accommodate NN s-wave states that are close to binding ($^1S_0$) or weakly bound ($^3S_1$), we focus here on the $\mathscr{D}_{12}$ and $\mathscr{D}_{03}$ dibaryon candidates assigned to these flavor-SU(3) multiplets.

2.2 Pion Assisted Dibaryons

The pion plays a major role as a virtual particle in binding or almost binding NN s-wave states. The pion as a real particle interacts strongly with nucleons, giving rise

Table 1 Predicted masses of non-strange $L = 0$ dibaryons $\mathscr{D}_{IS}$ with isospin I and spin S, using the Dyson–Xuong SU(6)→SU(4) mass formula $M = A + B[I(I+1) + S(S+1) - 2]$ [13]

$\mathscr{D}_{IS}$	$\mathscr{D}_{01}$	$\mathscr{D}_{10}$	$\mathscr{D}_{12}$	$\mathscr{D}_{21}$	$\mathscr{D}_{03}$	$\mathscr{D}_{30}$
BB'	NN	NN	$N\Delta$	$N\Delta$	$\Delta\Delta$	$\Delta\Delta$
SU(3)$_f$	$\overline{\mathbf{10}}$	**27**	**27**	**35**	$\overline{\mathbf{10}}$	**28**
$M(\mathscr{D}_{IS})$	A	A	$A + 6B$	$A + 6B$	$A + 10B$	$A + 10B$

to the πN P_{33} p-wave $\Delta(1232)$ resonance. Can it also assist in binding two nucleons into s-wave $N\Delta$ states? And once we have such $N\Delta$ states, can the pion assist in binding them into s-wave $\Delta\Delta$ states? This is the idea behind the concept developed by Garcilazo and me of pion assisted dibaryons [16, 17], or more generally meson assisted dibaryons to go beyond the non-strange sector, see Ref. [12] for review.

As discussed in the next subsection, describing $N\Delta$ systems in terms of a stable nucleon (N) and a two-body πN resonance (Δ) leads to a well defined πNN three-body model in which $IJ = 12$ and 21 resonances identified with the $\mathcal{D}_{12}$ and $\mathcal{D}_{21}$ dibaryons of Table 1 are generated. This relationship between $N\Delta$ and πNN may be generalized into relationship between a two-body $B\Delta$ system and a three-body πNB system, where the baryon B stands for N, Δ, Y (hyperon) etc. In order to stay within a three-body formulation one needs to assume that the baryon B is stable. For $B = N$, this formulation relates the $N\Delta$ system to the three-body πNN system. For $B = \Delta$, once properly formulated, it relates the $\Delta\Delta$ system to the three-body $\pi N\Delta$ system, suggesting to seek $\Delta\Delta$ dibaryon resonances by solving $\pi N\Delta$ Faddeev equations, with a stable Δ. The decay width of the Δ resonance is considered then at the penultimate stage of the calculation. In terms of two-body isobars we have then a coupled-channel problem $B\Delta \leftrightarrow \pi D$, where D stands generically for appropriate dibaryon isobars: (i) $\mathcal{D}_{01}$ and $\mathcal{D}_{10}$, which are the NN isobars identified with the deuteron and virtual state respectively, for $B = N$, and (ii) $\mathcal{D}_{12}$ and $\mathcal{D}_{21}$ for $B = \Delta$.

Within this model, and using separable pairwise interactions, the coupled-channel $B\Delta - \pi D$ eigenvalue problem reduces to a single integral equation for the $B\Delta$ T matrix shown diagrammatically in Fig. 3, where starting with a $B\Delta$ configuration the Δ-resonance isobar decays into πN, followed by $NB \to NB$ scattering through the D-isobar with a spectator pion, and ultimately by means of the inverse decay $\pi N \to \Delta$ back into the $B\Delta$ configuration. We note that the interaction between the π meson and B is neglected for $B = \Delta$, for lack of known $\pi\Delta$ isobar resonances in the relevant energy range.

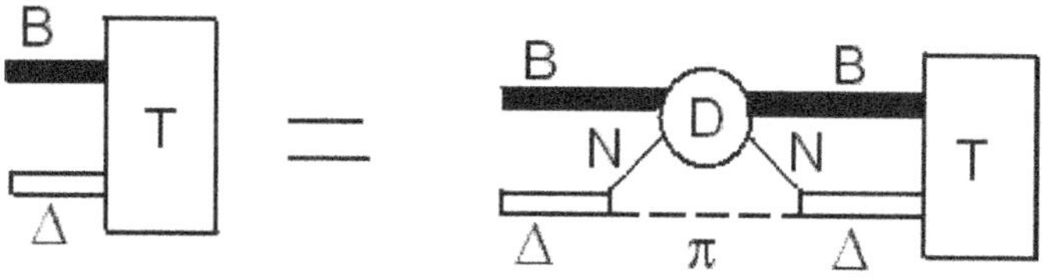

Fig. 3 Diagrammatic representation of the integral equation for the $B\Delta$ T matrix, derived by using separable pairwise interactions in πNB Faddeev equations [17] and solved numerically to calculate $B\Delta$ dibaryon resonance poles for $B = N,\ \Delta$

2.3 $N\Delta$ Dibaryons

The $\mathcal{D}_{12}$ dibaryon of Table 1 shows up clearly in the Argand diagram of the NN $^1D_2$ partial wave which is coupled above the $NN\pi$ threshold to the $I = 1$ s-wave $N\Delta$ channel. Values of $\mathcal{D}_{12}$ and $\mathcal{D}_{21}$ pole positions $W = M - i\Gamma/2$ from our hadronic-model three-body πNN Faddeev calculations [16, 17] described in the previous subsection are listed in Table 2 together with results of phenomenological studies that include (i) early NN phase shift analyses [18] and (ii) $pp \leftrightarrow np\pi^+$ coupled-channels analyses [19]. The $\mathcal{D}_{12}$ mass and width values calculated in the Faddeev hadronic model version using $r_\Delta \approx 1.3$ fm are remarkably close to the phenomenologically derived ones, whereas the mass evaluated in the version using $r_\Delta \approx 0.9$ fm agrees with that assumed in the Dyson–Xuong pioneering discussion [13].

Recent $pp \to pp\pi^+\pi^-$ production data [21] locate the $\mathcal{D}_{21}$ dibaryon resonance almost degenerate with the $\mathcal{D}_{12}$. Our πNN Faddeev calculations produce it about 10–20 MeV higher than the $\mathcal{D}_{12}$, see Table 2. The widths of these near-threshold $N\Delta$ dibaryons are, naturally, close to that of the Δ resonance. We note that only $^3S_1$ NN enters the calculation of the $\mathcal{D}_{12}$ resonance, while for the $\mathcal{D}_{21}$ resonance calculation only $^1S_0$ NN enters, both with maximal strength. Obviously, with the $^1S_0$ interaction the weaker of the two, one expects indeed that the $\mathcal{D}_{21}$ resonance lies above the $\mathcal{D}_{12}$ resonance. Moreover, these two dibaryon resonances differ also in their flavor-SU(3) classification, see Table 1, which is likely to push up the $\mathcal{D}_{21}$ further away from the $\mathcal{D}_{12}$. Finally, the $N\Delta$ s-wave states with $IJ = 11$ and 22 are found not to resonate in the πNN Faddeev calculations [17].

2.4 $\Delta\Delta$ Dibaryons

The $\mathcal{D}_{03}$ dibaryon of Table 1 shows up in the $^3D_3$ nucleon-nucleon partial wave above the $NN\pi\pi$ threshold owing to the coupling between the $I = 0$ $^3D_3$ NN channel and the $I = 0$ $^7S_3$ $\Delta\Delta$ channel, i.e. the coupling between the two-body NN channel and the four-body $NN\pi\pi$ channel. Indeed its best demonstration is by the relatively narrow peak about 80 MeV above the $\pi^0\pi^0$ production threshold and

Table 2 $\mathcal{D}_{12}$ and $\mathcal{D}_{21}$ $N\Delta$ dibaryon S-matrix pole positions $W = M - i\frac{\Gamma}{2}$ (in MeV), obtained by solving the $N\Delta$ T-matrix integral equation of Fig. 3 [17], are listed for two choices of the πN P_{33} form factor specified by a radius parameter r_Δ (in fm) together with two phenomenological values. The last column lists the results of a nonrelativistic meson-exchange Faddeev calculation

$N\Delta$	Phenomenological		Faddeev (present)		Faddeev (non rel.)
$\mathcal{D}_{IS}$	Ref. [18]	Ref. [19]	$r_\Delta \approx 1.3$	$r_\Delta \approx 0.9$	Ref. [20]
$\mathcal{D}_{12}$	2148 − i63	2144 − i55	2147 − i60	2159 − i70	2116 − i61
$\mathcal{D}_{21}$	–	–	2165 − i64	2169 − i69	–

80 MeV below the $\Delta\Delta$ threshold, with $\Gamma_{d^*} \approx 70$ MeV, observed in $pn \to d\pi^0\pi^0$ by the WASA-at-COSY Collaboration [22] and shown in Fig. 4-left. The $I = 0$ isospin assignment follows from the isospin balance in $pn \to d\pi^0\pi^0$, and the $J^P = 3^+$ spin-parity assignment follows from the measured deuteron angular distribution. The $d^*(2380)$ was also observed in $pn \to d\pi^+\pi^-$ [23], with cross section consistent with that measured in $pn \to d\pi^0\pi^0$, and studied in several $pn \to NN\pi\pi$ reactions [24–26]. Recent measurements of pn scattering and analyzing power [27] have led to the $pn\ {}^3D_3$ partial-wave Argand diagram shown in Fig. 4-right, supporting the $\mathcal{D}_{03}$ dibaryon resonance interpretation.

The history and state of the art of the $\mathcal{D}_{03}$ dibaryon, now denoted d*(2380), were reviewed recently by Clement [14]. In particular, its mass was predicted in several quark-based calculations, as listed in Table 3 in the columns following the symmetry-based value predicted first by Dyson and Xuong [13]. Also listed are $\mathcal{D}_{12}$ mass values, wherever available from such calculations. Remarkably, none of these quark-based predictions managed to reproduce the empirical mass values listed in the last column for *both* $\mathcal{D}_{12}$ and $\mathcal{D}_{03}$. More recent quark-based calculations, following the 2008 first announcement of observing the $\mathcal{D}_{03}$ [35], are discussed below.

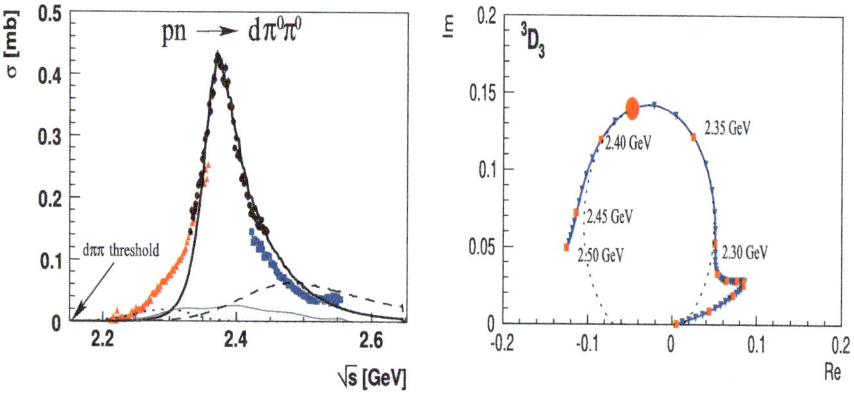

Fig. 4 $d^*(2380)$ dibaryon resonance signatures in recent WASA-at-COSY Collaboration experiments. Left: from the peak observed in the $pn \to d\pi^0\pi^0$ reaction [22]. Right: from the Argand diagram of the ${}^3D_3$ partial wave in pn scattering [27]

Table 3 $\mathcal{D}_{03}$ mass (in GeV) predicted in several quark-based calculations prior to 2008. Wherever calculated, the mass of $\mathcal{D}_{12}$ is also listed

$\mathcal{D}_{IS}$ (BB')	Ref. [13]	Ref. [28]	Ref. [29]	Ref. [30]	Ref. [31]	Ref. [32]	Ref. [33]	Ref. [34]	Exp./phen.
$\mathcal{D}_{03}$ $(\Delta\Delta)$	2.35	2.36	2.46	2.42	2.38	$\leq$2.26	2.40	2.46	2.38
$\mathcal{D}_{12}$ $(N\Delta)$	2.16	2.36	–	–	2.36	–	–	2.17	$\approx$2.15

Table 4 $\mathcal{D}_{03}$ and $\mathcal{D}_{30}$ $\Delta\Delta$ dibaryon S-matrix pole position $W = M - i\frac{\Gamma}{2}$ (in MeV), obtained in our hadronic model by solving the $\Delta\Delta$ T-matrix integral equation of Fig. 3, are listed for two choices of the πN P_{33} form factor specified by a radius parameter r_Δ (in fm). The last two columns list results of post 2008 quark-based RGM calculations with hidden-color $\Delta_8\Delta_8$ components

$\Delta\Delta$	Faddeev (present)		Recent quark-based	
$\mathcal{D}_{IS}$	$r_\Delta \approx 1.3$	$r_\Delta \approx 0.9$	Ref. [36]	Ref. [37]
$\mathcal{D}_{03}$	2383 – i41	2343 – i24	2393 – i75	2380 – i36
$\mathcal{D}_{30}$	2411 – i41	2370 – i22	2440 – i100	–

Values of $\mathcal{D}_{03}$ and $\mathcal{D}_{30}$ pole positions $W = M - i\Gamma/2$ from our hadronic-model three-body $\pi N\Delta$ Faddeev calculations [16, 17] are listed in Table 4. The $\mathcal{D}_{03}$ mass and width values calculated in the Faddeev hadronic model version using $r_\Delta \approx 1.3$ fm are remarkably close to the experimentally reported ones, whereas the mass evaluated in the model version using $r_\Delta \approx 0.9$ fm agrees, perhaps fortuitously so, with that derived in the Dyson–Xuong pioneering discussion [13]. For smaller values of r_Δ one needs to introduce explicit vector-meson and/or quark-gluon degrees of freedom which are outside the scope of the present model. In contrast, the calculated widths Γ are determined primarily by the phase space available for decay, displaying little sensitivity to the radius r_Δ of the πN P_{33} form factor.

The $\mathcal{D}_{30}$ dibaryon resonance is found in our $\pi N\Delta$ Faddeev calculations to lie about 30 MeV above the $\mathcal{D}_{03}$. These two states are degenerate in the limit of equal $D = \mathcal{D}_{12}$ and $D = \mathcal{D}_{21}$ isobar propagators in Fig. 3. Since $\mathcal{D}_{12}$ was found to lie lower than $\mathcal{D}_{21}$, we expect also $\mathcal{D}_{03}$ to lie lower than $\mathcal{D}_{30}$ as satisfied in our Faddeev calculations. Moreover, here too the difference in their flavor-SU(3) classification will push the $\mathcal{D}_{30}$ further apart from the $\mathcal{D}_{03}$. The $\mathcal{D}_{30}$ has not been observed and only upper limits for its production in $pp \to pp\pi^+\pi^+\pi^-\pi^-$ are available [38].

Finally, we briefly discuss the $\mathcal{D}_{03}$ mass and width values, listed in the last two columns of Table 4, from two recent quark-based resonating-group-method (RGM) calculations [36, 37] that add $\Delta_8\Delta_8$ hidden-color (CC) components to a $\Delta_1\Delta_1$ cluster. Interestingly, the authors of Ref. [36] have just questioned the applicability of admixing CC components in dibaryon calculations [39]. The two listed calculations generate mass values that are close to the mass of the d*(2380). The calculated widths, however, differ a lot from each other: one calculation generates a width of 150 MeV [36], exceeding substantially the reported value $\Gamma_{d^*(2380)} = 80 \pm 10$ MeV [27], the other one generates a width of 72 MeV [37], thereby reproducing the d*(2380) width. While the introduction of CC components has moderate effect on the resulting mass and width in the chiral version of the first calculation, lowering the mass by 20 MeV and the width by 25 MeV, it leads to substantial reduction of the width in the second (also chiral) calculation from 133 to 72 MeV. The reason is that the dominant CC $\Delta_8\Delta_8$ components, with 68% weight [37], cannot decay through single-fermion transitions $\Delta_8 \to N_1\pi_1$ to asymptotically free color-singlet hadrons. However, as argued in the next section, these quark-based width calculations miss important kinematical ingredients that make the width of a single compact $\Delta_1\Delta_1$ cluster considerably smaller than $\Gamma_{d^*(2380)}$. The introduction of substantial $\Delta_8\Delta_8$ components only aggravates the disagreement.

3 The Width of d*(2380), Small or Large?

The width derived for the $d^*(2380)$ dibaryon resonance by the WASA-at-COSY Collaboration and the SAID Data-Analysis-Center is $\Gamma_{d^*(2380)} = 80 \pm 10\,\text{MeV}$ [27]. It is much smaller than 230 MeV, twice the width $\Gamma_\Delta \approx 115\,\text{MeV}$ [40, 41] of a single free-space Δ, expected naively for a $\Delta\Delta$ quasibound configuration. However, considering the reduced phase space, $M_\Delta = 1232 \Rightarrow E_\Delta = 1232 - B_{\Delta\Delta}/2\,\text{MeV}$ in a bound-Δ decay, where $B_{\Delta\Delta} = 2 \times 1232 - 2380 = 84\,\text{MeV}$ is the $\Delta\Delta$ binding energy, the free-space Δ width gets reduced to 81 MeV using the in-medium single-Δ width $\Gamma_{\Delta\to N\pi}$ expression obtained from the empirical Δ-decay momentum dependence

$$\Gamma_{\Delta\to N\pi}(q_{\Delta\to N\pi}) = \gamma\,\frac{q^3_{\Delta\to N\pi}}{q_0^2 + q^2_{\Delta\to N\pi}}, \tag{1}$$

with $\gamma = 0.74$ and $q_0 = 159\,\text{MeV}$ [42]. Yet, this simple estimate is incomplete since neither of the two Δs is at rest in a deeply bound $\Delta\Delta$ state. To take account of the $\Delta\Delta$ momentum distribution, we evaluate the bound-Δ decay width $\overline{\Gamma}_{\Delta\to N\pi}$ by averaging $\Gamma_{\Delta\to N\pi}(\sqrt{s_\Delta})$ over the $\Delta\Delta$ bound-state momentum-space distribution,

$$\overline{\Gamma}_{\Delta\to N\pi} \equiv \langle \Psi^*(p_{\Delta\Delta})|\Gamma_{\Delta\to N\pi}(\sqrt{s_\Delta})|\Psi(p_{\Delta\Delta})\rangle \approx \Gamma_{\Delta\to N\pi}(\sqrt{\overline{s}_\Delta}), \tag{2}$$

where $\Psi(p_{\Delta\Delta})$ is the $\Delta\Delta$ momentum-space wavefunction and the dependence of $\Gamma_{\Delta\to N\pi}$ on $q_{\Delta\to N\pi}$ for on-mass-shell nucleons and pions was replaced by dependence on $\sqrt{s_\Delta}$. The averaged bound-Δ invariant energy squared $\overline{s}_\Delta$ is defined by

$$\overline{s}_\Delta = (1232 - B_{\Delta\Delta}/2)^2 - P^2_{\Delta\Delta}, \tag{3}$$

in terms of a $\Delta\Delta$ bound-state r.m.s. momentum $P_{\Delta\Delta} \equiv \langle p^2_{\Delta\Delta}\rangle^{1/2}$.

In Table 5, taken from my recent work [43], we list values of $\sqrt{\overline{s}_\Delta}$ and the associated in-medium decay-pion momentum $\overline{q}_{\Delta\to N\pi}$ for several representative values of the r.m.s. radius $R_{\Delta\Delta} \equiv \langle r^2_{\Delta\Delta}\rangle^{1/2}$ of the bound $\Delta\Delta$ wavefunction, using the equality sign in the uncertainty relationship $P_{\Delta\Delta}R_{\Delta\Delta} \geq 3/2$. Listed also are values of the in-medium single-Δ width $\overline{\Gamma}_{\Delta\to N\pi}$ obtained from Eq. (2). It is implicitly assumed here that the empirical momentum dependence (1) provides a good approximation also for off-mass-shell Δs. Finally, The last column of the table lists values of $\overline{\Gamma}_{\Delta\Delta\to NN\pi\pi}$ obtained by multiplying $\overline{\Gamma}_{\Delta\to N\pi}$ by two, for the two Δs, while applying to one of them the isospin projection factor 2/3 introduced in the Gal-Garcilazo hadronic model [16, 17] to obey the quantum statistics requirements in the leading final $NN\pi\pi$ decay channels. The large spread of $\overline{\Gamma}_{\Delta\Delta\to NN\pi\pi}$ width values exhibited in the table, all of which are much smaller than the 162 MeV obtained by ignoring in Eq. (3) the bound-state momentum distribution, demonstrates the importance of momentum-dependent contributions. It is seen that a compact d*(2380) with r.m.s. radius $R_{\Delta\Delta}$ less than 0.8 fm is incompatible with the experimental value

Table 5 Values of $\sqrt{s_\Delta}$, of the corresponding decay-pion momentum $\overline{q}_{\Delta\to N\pi}$ and of $\overline{\Gamma}_{\Delta\to N\pi}$ (2), listed as a function of $R_{\Delta\Delta}$ using $P_{\Delta\Delta}R_{\Delta\Delta}=\frac{3}{2}$ in Eq. (3). The last column lists deduced values of $\overline{\Gamma}_{\Delta\Delta\to NN\pi\pi}$, approximating it by $\frac{5}{3}\overline{\Gamma}_{\Delta\to N\pi}$ (see text)

$R_{\Delta\Delta}$ (fm)	$\sqrt{s_\Delta}$ (MeV)	$\overline{q}_{\Delta\to N\pi}$ (MeV)	$\overline{\Gamma}_{\Delta\to N\pi}$ (MeV)	$\overline{\Gamma}_{\Delta\Delta\to NN\pi\pi}$ (MeV)
0.6	1083	38.3	1.6	2.6
0.7	1112	96.6	19.3	32.1
0.8	1131	122.0	33.5	55.8
1.0	1153	147.7	50.6	84.4
1.5	1174	170.4	67.4	112.3
2.0	1181	177.9	73.2	122.0

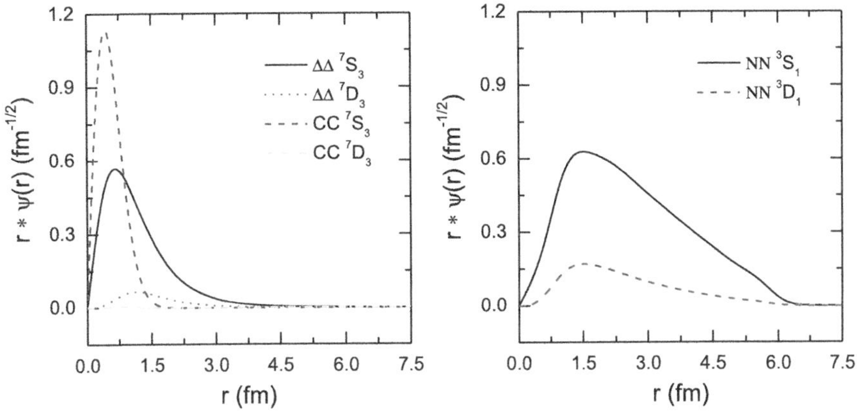

Fig. 5 d*(2380) $\Delta\Delta$ wavefunction with r.m.s. radius $R_{\Delta\Delta}=0.76$ fm (Left) and deuteron wavefunction with r.m.s. radius $R_d \approx 2$ fm (Right) from recent quark-based RGM calculations [37, 44]. Figure adapted from Ref. [44]

$\Gamma_{d^*(2380)} = 80 \pm 10$ MeV from WASA-at-COSY and SAID even upon adding a non-pionic partial width $\Gamma_{\Delta\Delta\to NN} \sim 10$ MeV [27].

Figure 5 shows d*(2380) and d(1876) wavefunctions from quark-based RGM calculations [37]. The d*(2380) appears quite squeezed compared to the diffuse deuteron. Its size, $R_{\Delta\Delta} = 0.76$ fm, leads to unacceptably small upper limit of about 47 MeV for the d*(2380) pionic width. This drastic effect of momentum dependence is missing in quark-based width calculations dealing with pionic decay modes of $\Delta_1\Delta_1$ components, e.g. Ref. [37]. Practitioners of quark-based models ought therefore to ask "what makes $\Gamma_{d^*(2380)}$ so much larger than the width calculated for a compact $\Delta\Delta$ dibaryon?" rather than "what makes $\Gamma_{d^*(2380)}$ so much smaller than twice a free-space Δ width?"

The preceding discussion of $\Gamma_{d^*(2380)}$ suggests that the quark-based model's finding of a tightly bound $\Delta\Delta$ s-wave configuration is in conflict with the observed

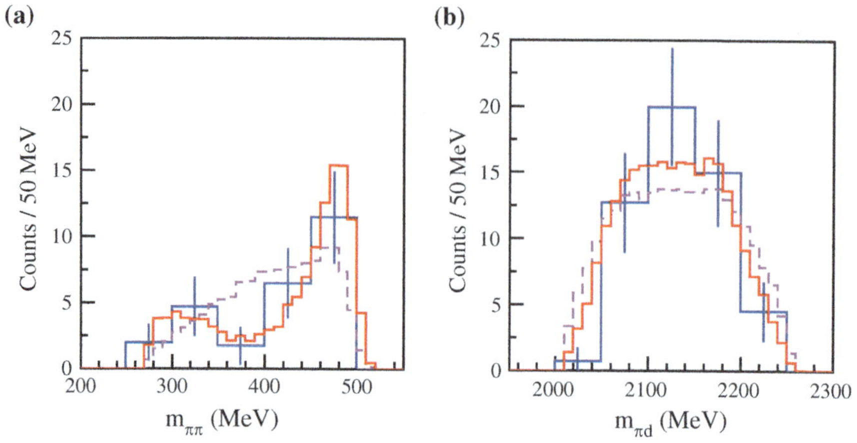

Fig. 6 Invariant mass distributions in ELPH experiment [45] $\gamma d \to d\pi^0\pi^0$ at $\sqrt{s} = 2.39$ GeV

width. Fortunately, our hadronic-model calculations [16, 17] offer resolution of this insufficiency by coupling to the tightly bound and compact $\Delta\Delta$ component of the d*(2380) dibaryon's wavefunction a $\pi N\Delta$ resonating component dominated asymptotically by a p-wave pion attached loosely to the near-threshold $N\Delta$ dibaryon $\mathscr{D}_{12}$ with size about 1.5–2 fm. Formally, one can recouple spins and isospins in this $\pi\mathscr{D}_{12}$ system, so as to assume an extended $\Delta\Delta$-like object. This explains why the preceding discussion of $\Gamma_{d^*\to NN\pi\pi}$ in terms of a $\Delta\Delta$ constituent model required a size larger than provided by quark-based RGM calculations [37] to reconcile with the reported value of $\Gamma_{d^*(2380)}$. We recall that the width calculated in our diffuse-structure $\pi N\Delta$ model [16, 17], as listed in Table 4, is in good agreement with the observed width of the d*(2380) dibaryon resonance.

Support for the role of the $\pi\mathscr{D}_{12}$ configuration in the decay of the d*(2380) dibaryon resonance is provided by a recent ELPH $\gamma d \to d\pi^0\pi^0$ experiment [45] looking for the d*(2380). While the data is not sufficiently conclusive, its fit in terms of a relativistic Breit-Wigner resonance shape with mass of 2370 MeV and width of 68 MeV is acceptable. Two invariant mass distributions at $\sqrt{s} = 2.39$ GeV from this experiment are shown in Fig. 6. The $\pi\pi$ mass distribution shown in (a) suggests a two-bump structure, fitted in solid red. The lower bump around 300 MeV is another manifestation of the ABC effect [46], also observed in $pn \to d\pi^0\pi^0$ by WASA-at-COSY [22, 42] and interpreted in Ref. [43] as due to a tightly bound $\Delta\Delta$ decay with reduced $\Delta \to N\pi$ phase space. The upper bump in (a) is apparently due to d*(2380) $\to \pi\mathscr{D}_{12}$ decay, where the corresponding $N\Delta$ dibaryon pionic decay width of order 100 MeV in (b) is not quenched owing to the $\mathscr{D}_{12}$ large size of order 2 fm.

Theoretical support for the relevance of the $\mathscr{D}_{12}(2150)$ $N\Delta$ dibaryon to the physics of the d*(2380) resonance is demonstrated in Fig. 7 by showing a $d\pi$ invariant-mass distribution peaking near the $N\Delta$ threshold as deduced from the $pn \to d\pi^0\pi^0$ reac-

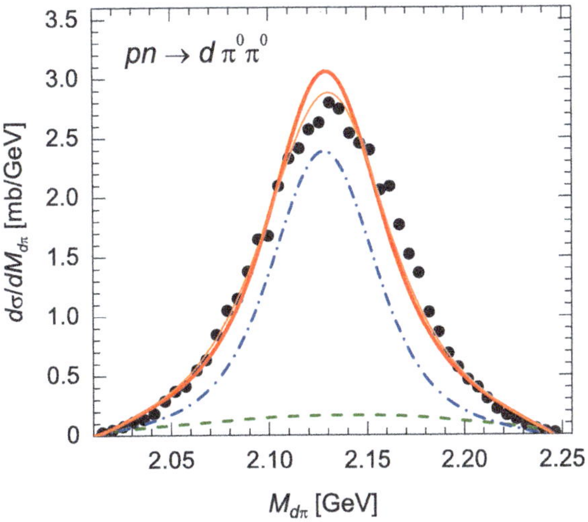

Fig. 7 The $pn \to d\pi^0\pi^0$ WASA-at-COSY $M_{d\pi}$ invariant-mass distribution [22] and, in solid lines, as calculated [47] for two input parametrizations of $\mathscr{D}_{12}(2150)$. The dot-dashed line gives the $\pi\mathscr{D}_{12}(2150)$ contribution to the two-body decay of the d*(2380) dibaryon, and the dashed line gives a σ-meson emission contribution. Figure adapted from Ref. [47]

Table 6 d*(2380) decay width branching ratios (BR in percents) from Ref. [43] for theory and from Refs. [48, 49] for experiment

Decay channel	$d\pi^0\pi^0$	$d\pi^+\pi^-$	$pn\pi^0\pi^0$	$pn\pi^+\pi^-$	$pp\pi^-\pi^0$	$nn\pi^+\pi^0$	$NN\pi$	NN	total
BR (th.)	11.2	20.4	11.6	25.8	4.7	4.7	8.3	13.3	100
BR (exp.)	14 ± 1	23 ± 2	12 ± 2	30 ± 5	6 ± 1	6 ± 1	≤ 9	12 ± 3	103

tion by which the d*(2380) was discovered [22]. This peaking, essentially at the $\mathscr{D}_{12}(2150)$ mass value, suggests that the $\pi\mathscr{D}_{12}$ two-body channel plays an important role in the decay modes of the d*(2380) dibaryon [47]. The width of this invariant-mass distribution, nevertheless, agrees roughly with $\Gamma_{d^*(2380)} = 80 \pm 10$ MeV irrespective of the underlying decay mechanism.

Recalling the $\Delta\Delta - \pi\mathscr{D}_{12}$ coupled channel nature of the d*(2380) in our hadronic model [16, 17], one may describe satisfactorily the d*(2380) total and partial decay widths in terms of an incoherent mixture of these relatively short-ranged ($\Delta\Delta$) and long-ranged ($\pi\mathscr{D}_{12}$) channels. This is demonstrated in Table 6 where the $NN\pi\pi$ calculated partial widths, totaling ≈60 MeV, are assigned a weight $\frac{5}{7}$ from $\Delta\Delta$ and a weight $\frac{2}{7}$ from $\pi\mathscr{D}_{12}$. This choice, ensuring that the partial decay width $\Gamma_{d^*\to NN\pi}$ does not exceed the upper limit of BR $\leq 9\%$ determined recently from *not* observing the single-pion decay branch [49], is by no means unique and the weights chosen here may be varied to some extent. For more details, see Ref. [43].

4 Conclusion

Substantiated by systematic production and decay studies in recent WASA-at-COSY experiments [14], the d*(2380) is the most spectacular dibaryon candidate at present. Following its early prediction in 1964 by Dyson and Xuong [13], it has been assigned in most theoretical works to a $\Delta\Delta$ quasibound state. Given the small width $\Gamma_{d^*(2380)} = 80 \pm 10\,\text{MeV}$ [27] with respect to twice the width of a free-space Δ, $\Gamma_\Delta \approx 115\,\text{MeV}$, its location far from thresholds makes it easier to discard a possible underlying threshold effect. However, as argued in this review following Ref. [43], the observed small width is much larger than what two *deeply bound* Δ baryons can yield upon decay. The d*(2380) therefore cannot be described in terms of a single compact $\Delta\Delta$ state as quark-based calculations derive it [36, 37]. A complementary quasi two-body component is provided within a $\pi N\Delta$ three-body hadronic model [16, 17] by the $\pi\mathcal{D}_{12}$ channel, in which the d*(2380) resonates. The $\mathcal{D}_{12}$ dibaryon stands here for the $I(J^P) = 1(2^+)$ $N\Delta$ near-threshold system that might possess a quasibound state S-matrix pole. It is a loose system of size 1.5–2 fm, as opposed to a compact $\Delta\Delta$ component of size 0.5–1 fm. It was also pointed out here, following Ref. [43], how the ABC low-mass enhancement in the $\pi^0\pi^0$ invariant mass distribution of the $pn \to d\pi^0\pi^0$ fusion reaction at $\sqrt{s} = 2.38\,\text{GeV}$ might be associated with a compact $\Delta\Delta$ component. The $\pi\mathcal{D}_{12}$ channel, in contrast, is responsible to the higher-mass structure of the $\pi^0\pi^0$ distribution and, furthermore, it gives rise to a non-negligible $d^* \to NN\pi$ single-pion decay branch, considerably higher than that obtained for a quark-based purely $\Delta\Delta$ configuration [50], but consistently with the upper limit of $\lesssim$9% determined recently by the WASA-at-COSY Collaboration [49]. A precise measurement of this decay width and BR will provide a valuable constraint on the $\pi\mathcal{D}_{12}$–$\Delta\Delta$ mixing parameter.

We end with a brief discussion of possible 6q admixtures in the essentially hadronic wavefunction of the d*(2380) dibaryon resonance. For this we refer to the recent 6q non-strange dibaryon variational calculation in Ref. [15] which depending on the assumed confinement potential generates a $^3S_1$ 6q dibaryon about 550–700 MeV above the deuteron, and a $^7S_3$ 6q dibaryon about 230–350 MeV above the d*(2380). Taking a typical 20 MeV potential matrix element from deuteron structure calculations and 600 MeV for the energy separation between the deuteron and the $^3S_1$ 6q dibaryon, one finds admixture amplitude of order 0.03 and hence 6q admixture probability of order 0.001 which is compatible with that discussed recently by Miller [51]. Using the same 20 MeV potential matrix element for the $\Delta\Delta$ dibaryon candidate and 300 MeV for the energy separation between the d*(2380) and the $^7S_3$ 6q dibaryon, one finds twice as large admixture amplitude and hence four times larger 6q admixture probability in the d*(2380), altogether smaller than 1%. These order-of-magnitude estimates demonstrate that long-range hadronic and short-range quark degrees of freedom hardly mix also for $\Delta\Delta$ configurations, and that the d*(2380) is extremely far from a pure 6q configuration. This conclusion is at odds with the conjecture made recently by Bashkanov, Brodsky and Clement [52] that 6q CC components dominate the wavefunctions of the $\Delta\Delta$ dibaryon candidates $\mathcal{D}_{03}$, identified with the observed d*(2380), and $\mathscr{D}_{30}$. Unfortunately, most of the quark-based calculations discussed in the present work combine quark-model input with hadronic-exchange model input in a loose way which discards their predictive power.

Acknowledgements I'm indebted to the organizers of the Frontiers of Science symposium in memory of Walter Greiner, held at FIAS, Frankfurt, June 2017, particularly to Horst Stöcker, for inviting me to participate in this special event and for supporting my trip. Special thanks are due to Humberto Garcilazo, together with whom the concept of pion assisted dibaryons was conceived, and also due to Heinz Clement for many stimulating exchanges on the physics of dibaryons and Jerry Miller for instructive discussions on 6q contributions to dibaryons.

References

1. E. Friedman, G. Soff, J. Phys. G **11**, L37 (1985)
2. H. Toki, T. Yamazaki, Phys. Lett. B **213**, 129 (1988)
3. T. Yamazaki, R.S. Hayano, K. Itahashi et al., Z. Phys. A **355**, 219 (1996)
4. T. Yamazaki, S. Hirenzaki, R.S. Hayano, H. Toki, Phys. Rep. **514**, 1 (2012)
5. E.E. Kolomeitsev, N. Kaiser, W. Weise, Phys. Rev. Lett. **90**, 092501 (2003). References to earlier work by W. Weise listed therein
6. E. Friedman, A. Gal, Nucl. Phys. A **928**, 128 (2014). References to earlier work listed therein
7. E. Friedman, A. Gal, Phys. Rep. **452**, 89 (2007)
8. J. Schaffner, C.B. Dover, A. Gal, C. Greiner, H. Stöcker, Phys. Rev. Lett. **71**, 1328 (1993)
9. J. Schaffner, C.B. Dover, A. Gal, C. Greiner, D.J. Millener, H. Stöcker, Ann. Phys. **235**, 35 (1994)
10. J. Schaffner-Bielich, A. Gal, Phys. Rev. C **62**, 034311 (2000)
11. J. Schaffner, A. Gal, I.N. Mishustin, H. Stöcker, W. Greiner, Phys. Lett. B **334**, 268 (1994)
12. A. Gal, Acta Phys. Pol. B **47**, 471 (2016)
13. F.J. Dyson, N.-H. Xuong, Phys. Rev. Lett. **13**, 815 (1964)
14. H. Clement, Prog. Part. Nucl. Phys. **93**, 195 (2017)
15. W. Park, A. Park, S.H. Lee, Phys. Rev. D **92**, 014037 (2015)
16. A. Gal, H. Garcilazo, Phys. Rev. Lett. **111**, 172301 (2013)
17. A. Gal, H. Garcilazo, Nucl. Phys. A **928**, 73 (2014)
18. R.A. Arndt, J.S. Hyslop III, L.D. Roper, Phys. Rev. D **35**, 128 (1987). References to earlier work listed therein
19. N. Hoshizaki, Phys. Rev. C **45**, R1424 (1992); Prog. Theor. Phys. **89**, 563 (1993). References to earlier work listed therein
20. T. Ueda, Phys. Lett. B **119**, 281 (1982)
21. P. Adlarson et al. (WASA-at-COSY Collaboration), Phys. Rev. Lett. **121**, 052001 (2018); **99**, 025201 (2019)
22. P. Adlarson et al. (WASA-at-COSY Collaboration), Phys. Rev. Lett. **106**, 242302 (2011)
23. P. Adlarson et al. (WASA-at-COSY Collaboration), Phys. Lett. B **721**, 229 (2013)
24. P. Adlarson et al. (WASA-at-COSY Collaboration), Phys. Lett. B **743**, 325 (2015)
25. P. Adlarson et al. (WASA-at-COSY Collaboration), Phys. Rev. C **88**, 055208 (2013)
26. G. Agakishiev et al., Phys. Lett. B **750**, 184 (2015)
27. P. Adlarson et al. (WASA-at-COSY Collaboration, SAID Data Analysis Center), Phys. Rev. C **90**, 035204 (2014); Phys. Rev. Lett. **112**, 202301 (2014)
28. P.J. Mulders, A.T. Aerts, J.J. de Swart, Phys. Rev. D **21**, 2653 (1980)
29. M. Oka, K. Yazaki, Phys. Lett. B **90**, 41 (1980)
30. M. Cvetič, B. Golli, N. Mankoč-Borštnik, M. Rosina, Phys. Lett. B **93**, 489 (1980)
31. P.J. Mulders, A.W. Thomas, J. Phys. G: Nucl. Part. Phys. **9**, 1159 (1983)

32. T. Goldman, K. Maltman, G.J. Stephenson Jr., K.E. Schmidt, F. Wang, Phys. Rev. C **39**, 1889 (1989)
33. X.Q. Zhang, Z.Y. Zhang, Y.W. Yu, P.N. Shen, Phys. Rev. C **60**, 045203 (1999)
34. R.D. Mota, A. Valcarce, F. Fernandez, D.R. Entem, H. Garcilazo, Phys. Rev. C **65**, 034006 (2002)
35. H. Clement et al. (CELSIUS-WASA Collaboration), Prog. Part. Nucl. Phys. **61**, 276 (2008)
36. H. Huang, J. Ping, F. Wang, Phys. Rev. C **89**, 034001 (2014). References to earlier work listed therein
37. Y. Dong, F. Huang, P. Shen, Z. Zhang, Phys. Rev. C **94**, 014003 (2016). References to earlier work listed therein
38. P. Adlarson et al. (WASA-at-COSY Collaboration), Phys. Lett. B **762**, 455 (2016)
39. F. Wang, J. Ping, H. Huang, arXiv:1711.01445v1. See also Y. Dong, P. Shen, Z. Zhang, Int. J. Mod. Phys. A **34** (18), 1950100 (2019)
40. R.A. Arndt, W.J. Briscoe, I.I. Strakovsky, R.L. Workman, Phys. Rev. C **76**, 025209 (2007)
41. A.V. Anisovich, R. Beck, E. Klempt, V.A. Nikonov, A.V. Sarantsev, U. Thoma, Eur. Phys. J. A **48**, 15 (2012)
42. M. Bashkanov, H. Clement, T. Skorodko, Nucl. Phys. A **958**, 129 (2017)
43. A. Gal, Phys. Lett. B **769**, 436 (2017)
44. F. Huang, Z.Y. Zhang, P.N. Shen, W.L. Wang, Chinese Phys. C **39**, 071001 (2015), arXiv:1408.0458v3 (nucl-th)
45. T. Ishikawa et al. (ELPH Experiment), Phys. Lett. B **772**, 398 (2017)
46. A. Abashian, N.E. Booth, K.M. Crowe, Phys. Rev. Lett. **5**, 258 (1960); Substantiated by N.E. Booth, A. Abashian, K.M. Crowe, Phys. Rev. Lett. **7**, 35 (1961)
47. M.N. Platonova, V.I. Kukulin, Nucl. Phys. A **946**, 117 (2016)
48. M. Bashkanov, H. Clement, T. Skorodko, Eur. Phys. J. A **51**, 87 (2015)
49. P. Adlarson et al. (WASA-at-COSY Collaboration), Phys. Lett. B **774**, 559 (2017)
50. Y. Dong, F. Huang, P. Shen, Z. Zhang, Phys. Lett. B **769**, 223 (2017)
51. G.A. Miller, Phys. Rev. C **89**, 045203 (2014)
52. M. Bashkanov, S.J. Brodsky, H. Clement, Phys. Lett. B **727**, 438 (2013)

Dense Matter in Neutron Star: Lessons from GW170817

Sarmistha Banik and Debades Bandyopadhyay

Abstract Neutron star merger event GW170817 sets an upper limit on the maximum mass of non-rotating neutron stars. Consequently, this event puts strong constraints on the dense matter equation of state (EoS). A comparative study of dense matter equations of state (EoSs) is presented here. It is found that the Λ hyperon EoS BHB$\Lambda\phi$ (Banik et al. in Astrophys J Suppl 24:22, 2014, [1]) constructed within the framework of the density dependent hadron field theory is favoured.

1 Introduction

S. Chandrasekhar predicted the mass limit for the first family of compact astrophysical objects known as White Dwarfs [2]. Next L. D. Landau should be credited for his idea about the second family of compact objects, as 'giant nucleus' [3]. After the discovery of neutrons, it was realised that the second family might be neutron stars [4]. First pulsar was discovered in 1967 [5]. We are celebrating 50 years of the discover of first pulsar in 2017. What could be a better celebration than finding the neutron star merger event GW170817 [6]. This stands out as a very important discovery in the history of mankind.

Neutron star merger event GW170817 was discovered both in gravitational waves and light. The gravitational wave signal was observed in LIGO detectors [6]. A short Gamma Ray Burst (sGRB) was recorded 1.7 s after the merger by the Fermi-GBM [7]. This, for the first time, established a link between a neutron star merger event

S. Banik (✉)
Department of Physics, Birla Institute of Technology and Science, Pilani, Hyderabad Campus, Jawahar Nagar, Kapra Mandal,
Medchal District, Hyderabad 500078, India
e-mail: sarmistha.banik@hyderabad.bits-pilani.ac.in

D. Bandyopadhyay
Astroparticle Physics and Cosmology Division, Centre for Astroparticle Physics, Saha Institute of Nuclear Physics, HBNI, 1/AF Bidhannagar, Kolkata 700064, India
e-mail: debades.bandyopadhyay@saha.ac.in

J. Kirsch et al. (eds.), *Discoveries at the Frontiers of Science*,
FIAS Interdisciplinary Science Series, https://doi.org/10.1007/978-3-030-34234-0_8

and sGRB. Later electromagnetic signals in visible, ultra-violet and infra-red bands were detected from the ejected matter which formed a 'kilonova'.

GW170817 is a boon to the nuclear astrophysics community because it allows to probe compositions and EoS in neutron star and r-process nucleosynthesis in the ejected neutron rich matter. The merger event provides crucial information about the remnant and neutron stars in the binary. The chirp mass is estimated to be $1.188^{+0.004}_{-0.002}$ $M_\odot$. Assuming low spins as found from observations of neutron stars in our Galaxy, individual neutron star mass in the binary ranges $1.17 - 1.60$ $M_\odot$. The massive remnant formed in the merger has a mass $2.74^{+0.04}_{-0.01}$ [6].

The outstanding question is what happened to the massive remnant formed in GW170817. The prompt collapse of it to a black hole is ruled out because large amount of matter was ejected. In this situation, either the remnant is a long lived massive neutron star or it collapsed to a black hole. Recent x-ray observation using the Chandra observatory indicates that the massive remnant might be a black hole [8].

It is possible to estimate the upper limit on the maximum mass (M^{TOV}_{max}) of the non-rotating neutron star if the remnant becomes a black hole through delayed collapse. Different groups have determined the upper limit on M^{TOV}_{max} from the multimessenger observation of GW170817 as well as from numerical relativity [9–11]. All these estimates converge to the same value of ~ 2.16 $M_\odot$ for the upper limit on M^{TOV}_{max}. It is already known from the observations of neutron stars that the most massive neutron star has a 2.01 $M_\odot$ which sets the lower limit on M^{TOV}_{max} [12]. All these information tell us that the maximum mass of non-rotating neutron stars should be in the range $2.01 < M^{TOV}_{max} < 2.16$. This constraint on M^{TOV}_{max} might severely restrict EoS models. This motivates us to carry out a comparative study of EoS models involving Banik, Hempel and Bandyopadhyay (BHB) EoS with hyperons in the density dependent relativistic hadron (DDRH) field theory [1, 13].

We organise the article in the following way. We introduce the density dependent hadron field theory and BHB$\Lambda\phi$ EoS in Sect. 2. Results are discussed in Sect. 3. We conclude in Sect. 4.

2 Equation of State for Neutron Star Matter

Equation of state is an important microphysical input for the study of core-collapse supernovae (CCSN), neutron stars and neutron star mergers [14, 15]. For CCSN and neutron star merger simulations, an EoS is a function of three parameters—density, temperature and proton fraction. These parameters vary over wide range of values. For example, density varies from 10^2 to 10^{15} g/cm^3, temperature from 0 to 150 MeV and proton fraction from 0 to 0.6. In this study, we focus on neutron star EoSs which are derived from the EoS constructed for CCSN and neutron star merger simulations. Particular, we describe here the BHB EoS and adopt the same for our calculation [1].

The compositions of matter in CCSN and neutron star changes with density, temperature and proton fraction. Below the saturation density (2.7×10^{14} g/cm^3) and low temperature, nuclei and nuclear clusters are present and make the matter inhomogeneous. In this case, non-uniform matter is made of light and heavy nuclei, nucleons and leptons in thermodynamic equilibrium. Matter above the saturation density is uniform. Several novel phases of matter such as hyperons, kaon condensate or quarks might appear at higher densities. We discuss both (non-)uniform matter in the following subsections.

2.1 Non-uniform Matter

Here the in-homogeneous matter is described by an extended version of the Nuclear Statistical Equilibrium (NSE) model that was developed by Hempel and Schaffner (HS) [16]. The extended NSE model takes into account interactions among nucleons, interaction of nuclei or nuclear clusters with the surrounding medium. Furthermore, the Coulomb interaction is considered.

Interactions among unbound nucleons are treated in the relativistic mean filed (RMF) approximation using a density dependent relativistic hadron field theory. Nuclei are considered as classical particles described by the Maxwell-Boltzmann statistics. Binding energies of thousands of nuclei entering into the calculation are obtained from the nuclear mass data table [17]. When experimental values are not available, theoretically calculated values are exploited [18]. Medium modifications of nuclei or nuclear clusters due to the screening of Coulomb energies of background electrons as well as corrections due to excited states and excluded volume effects are taken into account in this calculation.

The total canonical partition function of the in-homogeneous matter is given by,

$$Z(T, V, \{N_i\}) = Z_{nuc} \prod_{A,Z} Z_{A,Z}\, Z_{Coul}. \tag{1}$$

Here Z_{nuc}, $Z_{A,Z}$, Z_{Coul} represent partition functions corresponding to the contributions of unbound nucleons, nuclei and Coulomb interaction, respectively.

The free energy density is defined as

$$f = \sum_{A,Z} f^0_{A,Z}(T, n_{A,Z}) + f_{Coul}(n_e, n_{A,Z}) + \xi f^0_{nuc}(T, n'_n, n'_p) - T \sum_{A,Z} n_{A,Z} \ln\kappa\,, \tag{2}$$

where the first term gives the contribution of non-interacting nuclei, f_{Coul} corresponds to the Coulomb energy, the contribution of interacting nucleons f^0_{nuc} is multiplied by the available volume fraction of nucleons ξ, n'_n and n'_p are local neutron and proton number densities and the last term goes to infinity when available volume fraction of nuclei (κ) is zero near the saturation density. The number density of nuclei is given by the modified Saha equation [1, 16],

$$n_{A,Z} = \kappa\, g_{A,Z}(T) \left(\frac{M_{A,Z}T}{2\pi}\right)^{3/2} \exp\left(\frac{(A-Z)\mu^0_n + Z\mu^0_p - M_{A,Z} - E^{Coul}_{A,Z} - P^0_{nuc}V_{A,Z}}{T}\right), \tag{3}$$

where the meaning of different quantities in the equation can be found from Refs. [1, 16]. Finally, the pressure is calculated as mentioned in Refs. [1, 16].

2.2 Density Dependent Field Theory for Dense Matter

We calculate the EoS of uniform matter above the saturation density at finite temperature within the frame of a density dependent relativistic hadron field theory [1, 13]. In this case, the dense matter is made of neutrons, protons, hyperons and electrons. Being the lightest hyperons, Λ hyperons populate the dense matter first. Furthermore, heavier hyperons such as Σ and Ξ are excluded from this calculation because very little is known about their interaction in nuclear medium experimentally. The starting point here is the Lagrangian density for baryon-baryon interaction mediated by the exchange of σ, ω and ρ mesons. The interaction among Λ hyperons is taken into account by the exchange of ϕ mesons [1] as described by the Lagrangian density,

$$
\begin{aligned}
\mathcal{L}_B = &\sum_B \bar{\psi}_B \left(i\gamma_\mu \partial^\mu - m_B + g_{\sigma B}\sigma - g_{\omega B}\gamma_\mu \omega^\mu - g_{\phi B}\gamma_\mu \phi^\mu - g_{\rho B}\gamma_\mu \boldsymbol{\tau}_B \cdot \boldsymbol{\rho}^\mu\right)\psi_B \\
&+\frac{1}{2}\left(\partial_\mu \sigma \partial^\mu \sigma - m_\sigma^2 \sigma^2\right) - \frac{1}{4}\omega_{\mu\nu}\omega^{\mu\nu} \\
&+\frac{1}{2}m_\omega^2 \omega_\mu \omega^\mu - \frac{1}{4}\phi_{\mu\nu}\phi^{\mu\nu} + \frac{1}{2}m_\phi^2 \phi_\mu \phi^\mu \\
&-\frac{1}{4}\boldsymbol{\rho}_{\mu\nu} \cdot \boldsymbol{\rho}^{\mu\nu} + \frac{1}{2}m_\rho^2 \boldsymbol{\rho}_\mu \cdot \boldsymbol{\rho}^\mu .
\end{aligned} \tag{4}
$$

Here ψ_B denotes the baryon octets, $\boldsymbol{\tau}_B$ is the isospin operator and gs are density dependent meson-baryon couplings. It is to be noted that ϕ mesons are mediated among Λ hyperons only. The pressure is given by [1],

$$
\begin{aligned}
P = &-\frac{1}{2}m_\sigma^2\sigma^2 + \frac{1}{2}m_\omega^2\omega_0^2 + \frac{1}{2}m_\rho^2\rho_{03}^2 + \frac{1}{2}m_\phi^2\phi_0^2 + \Sigma^r \sum_{B=n,p,\Lambda} n_B \\
&+2T\sum_{i=n,p,\Lambda}\int \frac{d^3k}{(2\pi)^3}[ln(1+e^{-\beta(E^*-\nu_i)}) + ln(1+e^{-\beta(E^*+\nu_i)})] ,
\end{aligned} \tag{5}
$$

where the temperature is defined as $\beta = 1/T$ and $E^* = \sqrt{(k^2 + m_i^{*2})}$. This involves the rearrangement term Σ^r [1, 19] due to many-body correlations which is given by

$$
\Sigma^r = \sum_B [-g_{\sigma B}^{'}\sigma n_B^s + g_{\omega B}^{'}\omega_0 n_B + g_{\rho B}^{'}\tau_{3B}\rho_{03}n_B + g_{\phi B}^{'}\phi_0 n_B] , \tag{6}
$$

where $'$ denotes derivative with respect to baryon density of species B.

The energy density is

$$\epsilon = \frac{1}{2}m_\sigma^2\sigma^2 + \frac{1}{2}m_\omega^2\omega_0^2 + \frac{1}{2}m_\rho^2\rho_{03}^2 + \frac{1}{2}m_\phi^2\phi_0^2 \\ +2\sum_{i=n,p,\Lambda}\int\frac{d^3k}{(2\pi)^3}E^*\left(\frac{1}{e^{\beta(E^*-\nu_i)}+1}+\frac{1}{e^{\beta(E^*+\nu_i)}+1}\right). \quad (7)$$

Parameters of the Lagrangian density are computed using available experimental data at the saturation density. Meson-nucleon couplings are determined by fitting the properties of finite nuclei using some functional forms of density dependent couplings [13]. This parameter set is known as the DD2. For vector meson couplings of Λ hyperons, we exploit the SU(6) symmetry relations whereas the scalar coupling is obtained from Λ hypernuclei data with a potential depth of -30 MeV at the saturation density [20].

The descriptions of non-uniform and uniform matter are matched at the crust-core boundary in a thermodynamically consistent manner [1]. Charge neutrality and β-equilibrium conditions are imposed for neutron star matter.

3 Maximum Mass of Neutron Star

Here we discuss the results of our calculation. As discussed in the preceding section, we consider neutron star matter made of neutrons, protons, Λ hyperons and electrons in the DDRH model. The EoS corresponding to nucleons only matter is denoted as the DD2 whereas the EoS of dense matter involving Λ hyperons is known as the BHB$\Lambda\phi$. We also include the SFHo nuclear EoS of Steiner et al. in this discussion

Fig. 1 Pressure versus energy density (EoS) is shown for the DD2, BHB$\Lambda\phi$ and SFHo EoS models

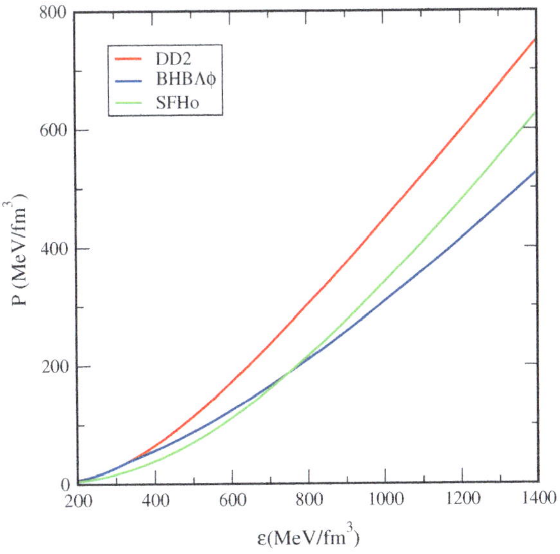

[21]. Figure 1 displays the EoSs (pressure versus energy density) corresponding to the DD2, BHBH$\Lambda\phi$ and SFHo models. It shows that the DD2 EoS is the stiffest among the three. Further we note that the SFHo EoS was softer over a certain region of energy density but becomes stiffer at higher densities than the BHB$\Lambda\phi$. However, it follows from the structure calculation using the Tolman-Oppenheimer-Volkoff (TOV) equation that the overall SFHo EoS is softer compared with the BHB$\Lambda\phi$ EoS. Maximum masses of non-rotating neutron stars are 2.42, 2.11 and 2.06 $M_\odot$ corresponding to the DD2, BHB$\Lambda\phi$ and SFHo EoS, respectively. All these EoSs are compatible with the observed 2 $M_\odot$ neutron star [12].

We also compute the structures of rotating neutron stars using the RNS code [22, 23]. Mass-radius relationships of (non)-rotating neutron stars are exhibited in Fig. 2. The sequences of non-rotating neutron stars (bottom curve) and uniformly rotating neutron stars (upper curve) at Keplerian frequencies are plotted for the DD2 EoS in the left panel and for the hyperon EoS BHB$\Lambda\phi$ in the right panel. Horizontal lines in both panels are fixed rest mass sequences. Those are denoted as normal and supramassive sequences. Rotating neutron stars evolve along those sequences keeping the total baryon mass conserved. The normal sequence finds its counterpart on the non-rotating star branch spinning down whereas neutron stars following the supramassive sequence would finally collapse into black holes. Any evolutionary sequence above the maximum mass rotating neutron star is known as the hypermassive sequence and a neutron star in this sequence would be stabilised

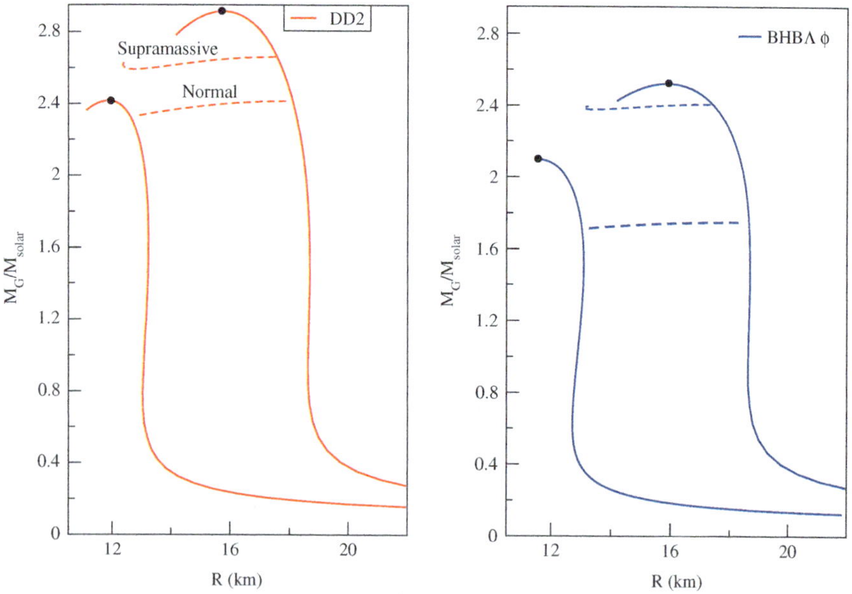

Fig. 2 Mass-radius relationship is shown for the DD2 EoS in left panel and BHB$\Lambda\phi$ in the right panel. In both panels, the bottom curve represents the non-rotating sequence and the upper curve corresponds to the sequence of neutron stars uniformly rotating at their Keplerian frequencies

only by differential rotation before collapsing into a black hole in a few tens of milli seconds. Recently, it was demonstrated that the relation between the maximum mass (M_{max}^{Rot}) of the rotating neutron star at the Keplerian frequency and that (M_{max}^{TOV}) of the non-rotating neutron star satisfied a universal relation [24, 25]. This relation is given by [24]

$$M_{max}^{Rot} = (1.203 \pm 0.022) M_{max}^{TOV} . \tag{8}$$

With this understanding of different evolutionary sequences respecting total baryon mass conservation, we discuss the fate of the massive remnant formed in merger event GW170817. The remnant could not be a hypermassive neutron star undergoing a prompt collapse to a black hole because a large amount of ejected matter was observed in the event [9]. This implies that the massive remnant existed for some duration. However, a long lived massive remnant is ruled out because of a sGRB sighted 1.7 s after the merger. It is inferred that the massive remnant collapsed to a black hole close to the maximum mass of a uniformly rotating sequence [10, 11]. This description might be intimately tied to the maximum mass of the non-rotating neutron star.

It is estimated from the observation of neutron star merger event GW170817 assuming low dimensionless spins for the neutron stars in the binary that the total binary mass was $\sim$2.74 $M_\odot$. The mass loss from the merged object due to emissions of gravitational waves and neutrinos and ejected neutron rich matter amounts to $\sim$0.15 $\pm$0.03 $M_\odot$ [26]. Consequently, the mass of the remnant reduced to $\sim$2.6 $M_\odot$. If we identify this mass of the remnant that might have collapsed into a black hole, with the maximum mass of the uniformly rotating neutron star at the Keplerian frequency i.e. M_{max}^{Rot} of Eq. (8), an upper limit on the maximum mass of non-rotating neutron stars might be obtained [10]. It follows from Eq. (8) that the upper limit is $\sim$2.16 $M_\odot$. It is already known from the observations of galactic pulsars that the lower limit on the maximum mass of non-rotating neutron stars is 2.01 $M_\odot$. All these information put together lead to

$$2.01 M_\odot \leq M_{max}^{TOV} < 2.16 M_\odot . \tag{9}$$

Different groups converged almost to the same value of the upper limit from different analyses of GW170817 [9–11, 26].

3.1 Constraint on EoS

We discuss the implications of the lower and upper limits of the maximum mass on EoSs. Mass-radius relationships corresponding to the DD2, BHB$\Lambda\phi$ and SFHo EoSs are plotted in Fig. 3. The lower and upper limits on the maximum mass are also indicated by two horizontal lines. It is evident from the figure that the BHB$\Lambda\phi$ and SFHo EoSs are consistent with both limits of the maximum mass. But this is not case with the DD2 EoS because it fails to satisfy the upper limit. It is to be noted that the

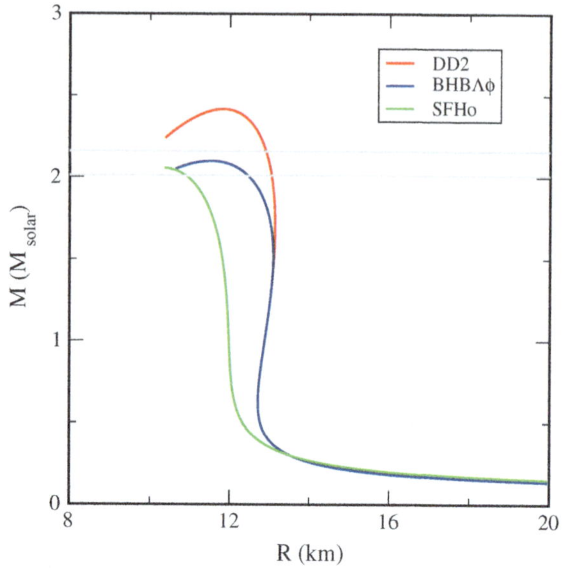

Fig. 3 Mass-radius relationships of non-rotating neutron stars are shown for the DD2, BHB$\Lambda\phi$ and SFHo EoSs. Horizontal lines denote the lower bound and upper bound on the maximum mass as given by Eq. (9)

DD2, BHB$\Lambda\phi$ and SFHo EoSs are being used for neutron star merger simulations by various groups [15, 26].

Next we perform a comparative study of different EoSs. Particularly, we look at the nuclear matter properties of all EoSs such as the saturation density (n_0), binding energy(E_0), incompressibility (K), symmetry energy (S) and its density slope (L). The nuclear matter properties of eight EoSs are recorded in Table 1. The last row of the table gives experimental values of nuclear matter properties [27]. First five of those EoSs for example, Lattimer-Swesty 200 (LS200) [28], Skyrme Lyon (SLy) [29], Müller-Serot 1 (MS1) [30], Akmal-Pandharipande-Ravenhall 4 (APR4) [31] and hyperon EoS H4 [32] were used in the analysis of GW170817 [6] because all of them satisfy the lower limit on the maximum mass. It is to be noted that all are nucleons only EoSs except the H4 EoS. A closer look at the nuclear matter properties at the saturation density of first five EoSs throw up important information about their behaviour at higher densities. It is evident from the table that one or more observables of nuclear matter in case of LS220, MS1, APR4 and H4 EoSs are not consistent with the experimental values. This leads to a very soft or stiff EoS in those cases. For example, high values of incompressibility (K) for the APR4 and H4 make them stiffer EoSs. The threshold for the appearance of hyperons is shifted to a lower density for a very stiff EoS leading to the large population of hyperons in dense matter and resulting in a lower maximum mass neutron star as it is happening in case of the H4 EoS. For the LS220 and MS1 EoSs, the density slope (L) of symmetry energy is much higher than the experimental value. As a result, the maximum mass for the MS1 EoS is higher than the upper limit of 2.16 $M_\odot$. However, the interplay between a lower value of K and higher value of L for the LS220 EoS determines the maximum mass which falls well within the limits of Eq. (9). Though the SLy EoS

Table 1 Nuclear matter properties of different EoSs used in the analysis of GW170817 [6] and in this article are recorded here. Experimental values of saturation density (n_0), binding energy (E_0), incompressibility (K), symmetry energy (S) and its density slope (L) are listed in the last row of this table [27]. Maximum mass of non-rotating neutron stars corresponding to each EoS is also shown here. Lower bound on the maximum mass of non-rotating neutron stars is mentioned in the last row [12]

EoS	n_0 (fm^{-3})	E$_0$ (MeV)	K (MeV)	S (MeV)	L (MeV)	M$_{max}$ (M$_\odot$)
LS220	0.1550	16.00	220	28.61	73.82	2.06
SLy	0.160	15.97	230	32.00	45.94	2.05
MS1	0.1484	15.75	250	35.00	110.00	2.77
APR4	0.160	16.00	266	32.59	58.46	2.19
H4	0.153	16.3	300	32.5	94.02	2.02
DD2	0.1491	16.02	243	31.67	55.04	2.42
SFHo	0.1583	16.19	245	31.57	47.10	2.06
BHB$\Lambda\phi$	0.1491	16.02	243	31.67	55.04	2.11
Exp.	~0.15	~16	240 ± 10	29.0–32.7	40.5–61.9	2.01 ± 0.04

is consistent with the experimental values and observational limits on the maximum mass, it is a non-relativistic EoS and superluminal behaviour could be a problem in this case at very high density (5–8 n_0) [31]. We have already discussed the last three EoSs of the Table. The nuclear matter properties of the DD2, SFHo and BHB$\Lambda\phi$ EoSs are in good agreement with the experimental values. However, it is concluded that the DD2 EoS is ruled out by Eq. (9). It is possible to further constrain EoSs using the measured tidal deformability from GW170817. Based on the tidal deformability of GW170817, the H4, APR4 and LS220 EoSs are excluded whereas the BHB$\Lambda\phi$ EoS is consistent with GW170817 data [33].

4 Summary and Conclusion

We have investigated the equations of state of dense matter within the framework of the density dependent relativistic hadron field theory. The nucleons only EoS is denoted as the DD2 whereas the Λ hyperon EoS is known as the BHB$\Lambda\phi$. The neutron star merger event GW170817 gives an upper limit on the maximum mass of non-rotating neutron stars whereas the lower limit is already known from the observations of pulsars. The upper and lower limits severely constraint the EoS as we have found through a comparative study of eight EoSs and their nuclear matter properties. It is found that the BHB$\Lambda\phi$ EoS is consistent with both limits of the maximum mass and the tidal deformability of GW170817.

Acknowledgements S.B. and D.B. gratefully remember the support and encouragement that they received always from Professor Walter Greiner.

References

1. S. Banik, M. Hempel, D. Bandyopadhyay, Astrophys. J. Suppl. **214**, 22 (2014)
2. S. Chandrasekhar, Astrophys. J. **74**, 81 (1931)
3. L.D. Landau, Phys. Zs. Sowjet. **1**, 285 (1932)
4. W. Baade, F. Zwicky, Phys. Rev. **45**, 138 (1934)
5. A. Hewish, S.J. Bell, J.D.H. Pilkington, P.F. Scott, R.A. Collins, Nature **217**, 709 (1968)
6. B.P. Abbott et al., Phys. Rev. Lett. **119**, 161101 (2017)
7. B.P. Abbott et al., Astrophys. J. Lett. **848**, L13 (2017)
8. D. Pooley, P. Kumar, J.C. Wheeler, B. Grossan, ApJL. **859**, 2 (2018)
9. B. Margalit, B.D. Metzger, ApJL. **850**, 2 (2017)
10. L. Rezzolla, E.R. Most, L.R. Weih, ApJL. **852**, L25 (2018)
11. M. Ruiz, S.L. Shapiro, A. Tsokaros, Phys. Rev. D **97**, 021501 (2018)
12. J. Antoniadis et al., Science **340**, 448 (2013)
13. S. Typel, G. Röpke, T. Klähn, D. Blaschke, H. Wolter, Phys. Rev. C **81**, 015803 (2010)
14. P. Char, S. Banik, D. Bandyopadhyay, Astrophys. J. **809**, 116 (2015)
15. D. Radice, S. Bernuzzi, W. Del Pozzo, L.F. Roberts, C. Ott, Astrophys. J. **842**, L10 (2017)
16. M. Hempel, J. Schaffner-Bielich, Nucl. Phys. A **837**, 210 (2010)
17. G. Audi, A.H. Wapstra, C. Thibault, Nucl. Phys. A **729**, 337 (2003)
18. P. Moller, J.R. Nix, W.D. Myers, W.J. Swiatecki, At. Data Nucl. Data Tables **59**, 185 (1995)
19. F. Hofmann, C.M. Keil, H. Lenske, Phys. Rev. C **64**, 025804 (2001)
20. J. Schaffner, I.N. Mishustin, Phys. Rev. C **53**, 1416 (1996)
21. A.W. Steiner, M. Hempel, T. Fischer, Astrophys. J. **774**, 17 (2013)
22. H. Komatsu, Y. Eriguchi, I. Hachisu, Mon. Not. R. Astron. Soc. **237**, 355 (1989); ibid Mon. Not. R. Astron. Soc. **239**, 153 (1989)
23. N. Stergioulas, J.L. Friedman, Astrophys. J. **444**, 306 (1995)
24. C. Breu, L. Rezzolla, MNRAS **459**, 646 (2016)
25. S.S. Lenka, P. Char, S. Banik, Int. J. Mod. Phys. **26**, 1750127 (2017)
26. M. Shibata et al., Phys. Rev. D **96**, 123012 (2017)
27. T. Fischer et al., Euro. Phys. J. A **50**, 46 (2014)
28. J.M. Lattimer, F.D. Swesty, Lattimer-Swesty EOS web site (1991–2012), http://www.astro.sunysb.edu/dswesty/lseos.html
29. F. Douchin, P. Haensel, Astron. Astrophys. **380**, 151 (2001)
30. H. Müller, B.D. Serot, Nucl. Phys. A **606**, 508 (1996)
31. A. Akmal, V.R. Pandharipande, D.G. Ravenhall, Phys. Rev. C **73**, 1804 (1998)
32. B.D. Lackey, M. Nayyar, B.J. Owen, Phys. Rev. D **73**, 024021 (2006)
33. D. Radice, A. Perego, F. Zappa, S. Bernuzzi, ApJL. **852**, L29 (2018)

Quark Matter in Neutron Stars

William M. Spinella, Fridolin Weber, Gustavo A. Contrera and Milva G. Orsaria

Abstract The nonlocal three-flavor Nambu-Jona-Lasinio model is used to study quark deconfinement in the cores of neutron stars (NSs). The quark-hadron phase transition is modeled using both the Maxwell construction and the Gibbs construction. For the Maxwell construction, we find that all NSs with core densities beyond the phase transition density are unstable. Therefore, no quark matter cores would exist inside such NSs. The situation is drastically different if the phase transition is treated as a Gibbs transition, resulting in stable NSs whose stellar cores are a mixture of hadronic matter and deconfined quarks. The largest fractions of quarks achieved in the quark-hadron mixed phase are around 50%. No choice of parametrization or composition leads to a pure quark matter core. The inclusion of repulsive vector interactions among the quarks is crucial since the equation of state (EoS) in the quark-hadron mixed phase is significantly softer than that of the pure hadronic phase.

W. M. Spinella (✉)
School of Physical Sciences and Technologies, Irvine Valley College, 5550 Irvine Center Drive, Irvine, CA 92618, USA
e-mail: wspinella@ivc.edu

F. Weber
Department of Physics, San Diego State University, 5500 Campanile Drive, San Diego, CA 92182, USA
e-mail: fweber@sdsu.edu

Center for Astrophysics and Space Sciences, University of California, San Diego, La Jolla, CA 92093, USA

G. A. Contrera
Facultad de Ciencias Exactas, IFLP, UNLP, CONICET, Diagonal 113 entre 63 y 64, La Plata 1900, Argentina
e-mail: contrera@fisica.unlp.edu.ar

G. A. Contrera · M. G. Orsaria
Grupo de Gravitación, Astrofísica y Cosmología, Facultad de Ciencias Astronómicas y Geofísicas, Universidad Nacional de La Plata, Paseo del Bosque S/N (1900), La Plata, Argentina
e-mail: morsaria@fcaglp.unlp.edu.ar

CONICET, Godoy Cruz 2290, C1425FQB Buenos Aires, Argentina

J. Kirsch et al. (eds.), *Discoveries at the Frontiers of Science*, FIAS Interdisciplinary Science Series, https://doi.org/10.1007/978-3-030-34234-0_9

1 Introduction

Neutron stars (NSs) are among the most dense astrophysical objects in the Universe [1]. They have radii of around 12 km and masses up to an observed 2.01 $M_{\odot}$ [2–4]. This suggests that these objects have extreme central densities on the order of $10^{15}\,\mathrm{g/cm^3}$, greater than the density of an atomic nucleus which is about $2.5 \times 10^{14}\,\mathrm{g/cm^3}$. At such extreme densities the primary constituents of NS matter are neutrons, protons, electrons and muons, and possibly also hyperons [5, 6], the delta isobar $\Delta(1232)$ [7], and/or deconfined quarks [8, 9]. More than that, if deconfined quarks should exist in the cores of neutron stars, they are likely to be in a color superconducting state (see, for instance, [10–12] and references therein). Investigating the possible existence of these building blocks of matter in the cores of neutron stars is a very active area of research [13] and the focus of this short review paper. A detailed discussion of the material presented here can be found in [14].

2 Description of Quark Matter

In this paper quark matter is described by an improved version of the nonlocal SU(3) Nambu-Jona-Lasinio (n3NJL) model [15, 16], which treats the quark-quark interaction as a one-gluon exchange [17, 18]. A unique scalar (σ_f) and vector (ω_f) field is assigned to each quark flavor and an additional mixing interaction is included. These interactions are parametrized by the coupling constants of the model: G_S is the strong coupling constant, G_V is the vector coupling constant, and H is the coupling constant associated with mixing. The quark masses $m_u = m_d$ and m_s, G_S, H, and the nonlocality parameter Λ_{nl} are determined by fitting the model to empirical values of the pion mass and decay constant (m_π, f_π), and the pseudoscalar η' meson mass ($m_{\eta'}$), while the vacuum pressure (Ω_0) is determined by imposing that the nonlocal thermodynamic potential equal zero at zero temperature and chemical potential, i.e., $\Omega_0 - \Omega^{\mathrm{NL}}(M, T = 0, \mu = 0) = 0$ [17–19]. The n3NJL parametrizations are provided in Table 1. In the mean-field approximation the thermodynamic potential of the model is given by [15, 16]

Table 1 Parametrizations of the n3NJL model [20]

Parameters	P119	P127
$m_u = m_d$ (MeV)	5.0	5.5
m_s (MeV)	119.3	127.8
Λ_{nl} (MeV)	843.0	780.6
$G_S \Lambda_{\mathrm{nl}}^2$ (MeV^2)	13.34	14.48
$H \Lambda_{\mathrm{nl}}^5$ (MeV^5)	−273.75	−267.24
Ω_0 (MeV/fm^3)	−239.51	−243.85

$$\Omega^{\rm NL}(M, T = 0, \mu) = -\frac{3}{\pi^3}\sum_f \int_0^\infty dk_0 \int_0^\infty dk \ \ln\left[\left[\widehat{p}_f^2 + M_f^2(p_f)\right]\frac{1}{p_f^2 + m_f^2}\right]$$
$$-\frac{3}{\pi^2}\sum_f \int_0^{\sqrt{\mu_f^2 - m_f^2}} dk\, k^2 \left[(\mu_f - E_f)\theta\left(\mu_f - m_f\right)\right]$$
$$-\frac{1}{2}\left[\sum_f \left(\bar{\sigma}_f \bar{S}_f + \frac{G_S}{2}\bar{S}_f^2\right) + \frac{H}{2}\bar{S}_u\,\bar{S}_d\,\bar{S}_s\right] - \sum_f \frac{\bar{\omega}_f^2}{4G_V}, \tag{1}$$

where $f \in \{u, d, s\}$ is the quark flavor, $\bar{\sigma}_f$ and $\bar{\omega}_f$, are the quark scalar and vector mean fields respectively, $E_f = \sqrt{k^2 + m_f^2}$ is the quark energy, $p_f^2 = (k_0 + i\mu_f)^2 + k^2$ is the quark four momentum, and $\bar{S}_f$ is the quark auxiliary field given by [15, 16]

$$\bar{S}_f = -\frac{6}{\pi^3}\int_0^\infty dk_0 \int_0^\infty dk\, g(p_f)\,\frac{M_f(p_f)}{\widehat{p}^2 + M_f^2(p_f)}. \tag{2}$$

The dynamical quark mass $M_f(p_f) = m_f + \bar{\sigma}_f g(p_f)$ depends on the form factor $g(p_f) = \exp(-p_f^2/\Lambda_{\rm nl}^2)$, which makes the model more consistent with results from lattice QCD calculations [21]. The scalar and vector mean field equations can be found by minimizing $\Omega^{\rm NL}$ with respect to the fields,

$$\frac{\partial \Omega^{\rm NL}}{\partial \bar{\sigma}_f} = \bar{\sigma}_f + G_S\bar{S}_f + \frac{1}{2}H\bar{S}_i\bar{S}_j = 0, \quad \frac{\partial \Omega^{\rm NL}}{\partial \bar{\omega}_f} = \bar{\omega}_f - 2G_V\frac{\partial \Omega^{\rm NL}}{\partial p_f} = 0, \tag{3}$$

for $f \in \{u, d, s\}$ and $i \neq j \neq f$. The simultaneous solution of this system is subject to the conditions of charge neutrality and baryon number conservation,

$$\sum_f n_f q_f + \sum_\lambda n_\lambda q_\lambda = 0, \quad n - \frac{1}{3}\sum_f n_f = 0, \tag{4}$$

respectively, determines the mean fields ($\bar{\sigma}_f, \bar{\omega}_f, f \in \{u, d, s\}$) and the baryon and electron chemical potentials (μ_n, μ_e). The individual quark chemical potentials are then specified by the chemical equilibrium condition

$$\mu_f = \frac{1}{3}\left(\mu_n - 3q_f\,\mu_e\right) \tag{5}$$

and the particle number densities follow from $n_f = \partial\Omega^{\rm NL}/\partial\mu_f$. The total pressure of the system comes from the quarks, leptons, and the vacuum and is given by

$$P_{\rm n3NJL} = \Omega_0 - \Omega^{\rm NL} - \Omega_l, \tag{6}$$

where Ω_0 is fixed by the condition that Eq. 1 vanishes at $T = \mu_f = 0$, and Ω_l denotes the lepton thermodynamic potential given by

$$\Omega_l = -\frac{1}{12\pi^2}\sum_\lambda \left[\mu_\lambda \sqrt{\mu_\lambda^2 - m_\lambda^2}\left(\mu_\lambda^2 - \frac{5}{2}m_\lambda^2\right) + \frac{3}{2}m_\lambda^4 \ln\left(\frac{\mu_\lambda + \sqrt{\mu_\lambda^2 + m_\lambda^2}}{m_\lambda}\right)\right]. \quad (7)$$

Finally, the total energy density of the system is given by

$$\mathscr{E}_{\mathrm{n3NJL}} = -P_{\mathrm{n3NJL}} + \sum_f n_f\, \mu_f + \sum_\lambda n_\lambda\, \mu_\lambda. \quad (8)$$

3 EoS of Quark-Hybrid Matter and Neutron Star Structure

Allowing for the possibility of quark deconfinement, a quark-hadron phase transition may commence when the pressure of the quark phase equals that of the hadronic phase. The nature of the phase transition depends on the surface tension, α, between the two phases that is still quite uncertain. Recent works have typically placed $\alpha \lesssim 30$ MeV/fm^2, though there are suggestions that values greater than 100 MeV/fm^2 could be possible too [22–25]. If $\alpha \gtrsim 70$ MeV/fm^2 the quark-hadron phase transition will be one of constant pressure with an equation of state that is discontinuous in energy density [26, 27]. The result is a sharp interface between phases of pure hadronic matter and pure quark matter at a given NS radius, as shown schematically in Fig. 1.

The phase transition is achieved by applying the Maxwell construction (at zero temperature, T)

$$P_H\,(\mu_n, T = 0) = P_Q\,(\mu_n, T = 0)\,, \quad (9)$$

where μ_n is the neutron chemical potential, and P_H and P_Q represent the pressure of the hadronic and the quark phase, respectively. Alternatively, if $\alpha \lesssim 70$ MeV/fm^2 the phase transition results in the formation of a stable coexistent (mixed) phase, gradually converting NS matter from hadronic matter to deconfined quark matter with increasing density, as shown in Fig. 2. This phase transition satisfies the Gibbs

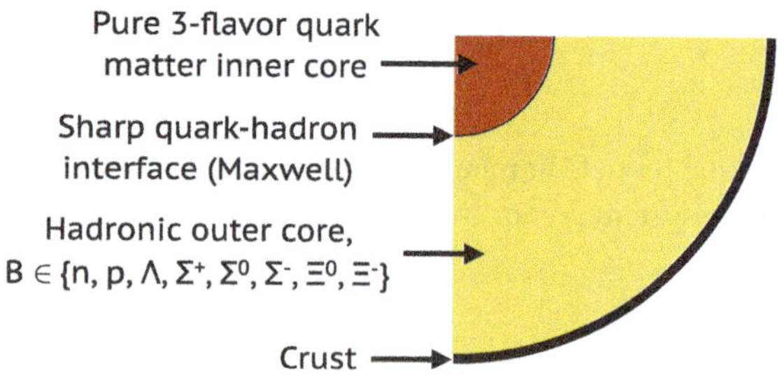

Fig. 1 Hypothetical NS cross section assuming a constant pressure (Maxwell construction) phase transition from hadronic matter to a pure quark matter core [14]

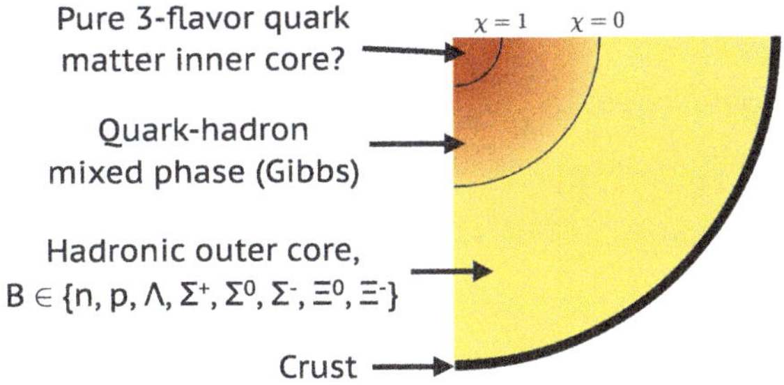

Fig. 2 Hypothetical NS cross section assuming a phase transition from hadronic matter to a quark-hadron mixed phase (Gibbs construction) and possibly a pure quark matter core [14]

condition for phase equilibrium,

$$P_H\left(\mu_n, \mu_e, T=0\right) = P_Q\left(\mu_n, \mu_e, T=0\right), \tag{10}$$

where μ_e is the electron chemical potential. The isospin restoring force favors a positively charged hadronic phase as an increased number of protons reduces isospin asymmetry. This necessarily results in a negatively charged quark phase, electric charge neutrality being achieved globally. Competition between the Coulomb and surface energies will cause the phases to arrange themselves into energy minimizing geometric configurations [28]. This rearrangement and its effect on neutrino emissivity were investigated in [29, 30].

A popular model used to determine P_H is the relativistic mean-field (RMF) theory [31, 32], which is based on a lagrangian of the following type [33–36],

$$\begin{aligned}
\mathscr{L} = & \sum_B \bar{\psi}_B \big[\gamma_\mu (i\partial^\mu - g_\omega \omega^\mu - \frac{1}{2} g_\rho \boldsymbol{\tau} \cdot \boldsymbol{\rho}^\mu) - (m_N - g_\sigma \sigma)\big]\psi_B \\
& + \frac{1}{2}(\partial_\mu \sigma \partial^\mu \sigma - m_\sigma^2 \sigma^2) - \frac{1}{3} b_\sigma m_N (g_\sigma \sigma)^3 - \frac{1}{4} c_\sigma (g_\sigma \sigma)^4 - \frac{1}{4}\omega_{\mu\nu}\,\omega^{\mu\nu} \\
& + \frac{1}{2} m_\omega^2\, \omega_\mu\, \omega^\mu + \frac{1}{2} m_\rho^2\, \boldsymbol{\rho}_\mu \cdot \boldsymbol{\rho}^\mu - \frac{1}{4}\boldsymbol{\rho}_{\mu\nu}\, \boldsymbol{\rho}^{\mu\nu} + \sum_{\lambda = e^-, \mu^-} \bar{\psi}_\lambda (i\gamma_\mu \partial^\mu - m_\lambda)\psi_\lambda \,.
\end{aligned} \tag{11}$$

This lagrangian describes baryons interacting via the exchange of scalar, vector, and isovector mesons (σ, ω, $\boldsymbol{\rho}$, respectively). The sum over B sums all baryon states that are present in neutron star matter at a given density. The quantities g_ρ, g_σ, and g_ω are the meson-baryon coupling constants. The coupling constants of cubic and quartic σ-meson self-interactions are denoted b_σ and c_σ. The quantities $\omega^{\mu\nu}$ $(= \partial^\mu \omega^\nu - \partial^\nu \omega^\mu)$ and $\boldsymbol{\rho}^{\mu\nu}$ $(= \partial^\mu \boldsymbol{\rho}^\nu - \partial^\nu \boldsymbol{\rho}^\mu)$ denote meson field tensors. The RMF approach is parametrized to reproduce the properties of symmetric nuclear matter at saturation density n_0 (see Table 2): the binding energy per nucleon (E_0), the nuclear incompressibility (K_0), the isospin asymmetry energy (J), and the effective mass (m^*/m_N). In addition, the RMF parametrizations used in this work employ a density-dependent isovector-meson-baryon coupling constant that can be fit to the slope of the asymmetry energy (L_0) at n_0, and the DD2 and ME2 parametrizations scalar-

Table 2 Properties of nuclear matter at saturation density for the hadronic parametrizations of this work

Saturation property	SWL [14]	GM1L [14, 37]	DD2 [38]	ME2 [39]
n_0 (fm^{-3})	0.150	0.153	0.149	0.152
E_0 (MeV)	−16.00	−16.30	−16.02	−16.14
K_0 (MeV)	260.0	300.0	242.7	250.9
m^*/m_N	0.70	0.70	0.56	0.57
J (MeV)	31.0	32.5	32.8	32.3
L_0 (MeV)	55.0	55.0	55.3	51.3

and vector-meson-baryon coupling constants are also density-dependent and are fit to properties of finite nuclei [14, 38, 39].

For the Gibbs transition (10), the nonlinear systems of equations following from (3) and (11) must be solved simultaneously along with the global charge neutrality condition

$$\sum_B n_B q_B + \sum_f n_f q_f + \sum_\lambda n_\lambda q_\lambda = 0\,, \tag{12}$$

where n_B denotes the number density of baryon B carrying an electric charge q_B. To parametrize the amount of quark matter in the mixed phase, the parameter χ is introduced, which represents the volume fraction of quark matter at a given density. This parameter enters into the nonlinear, coupled system of equations through the condition of baryon number conservation,

$$n - (1-\chi)\sum_B n_B - \frac{1}{3}\chi \sum_f n_f = 0\,. \tag{13}$$

The EoS can then be calculated from the relations

$$\varepsilon_{HQ} = (1-\chi)\varepsilon_H + \chi\varepsilon_Q\,, \quad P_{HQ} = \frac{1}{2}\left(P_H + P_Q\right)\,, \tag{14}$$

where ε_H and ε_Q are the respective energy densities of hadronic matter and quark matter in the mixed phase.

The system of Eqs. (3) and (11) is easier to solve for the Maxwell phase transition (9), where hadronic matter and quark matter are separately charge neutral and in chemical equilibrium so that Eqs. (3) and (11) are decoupled.

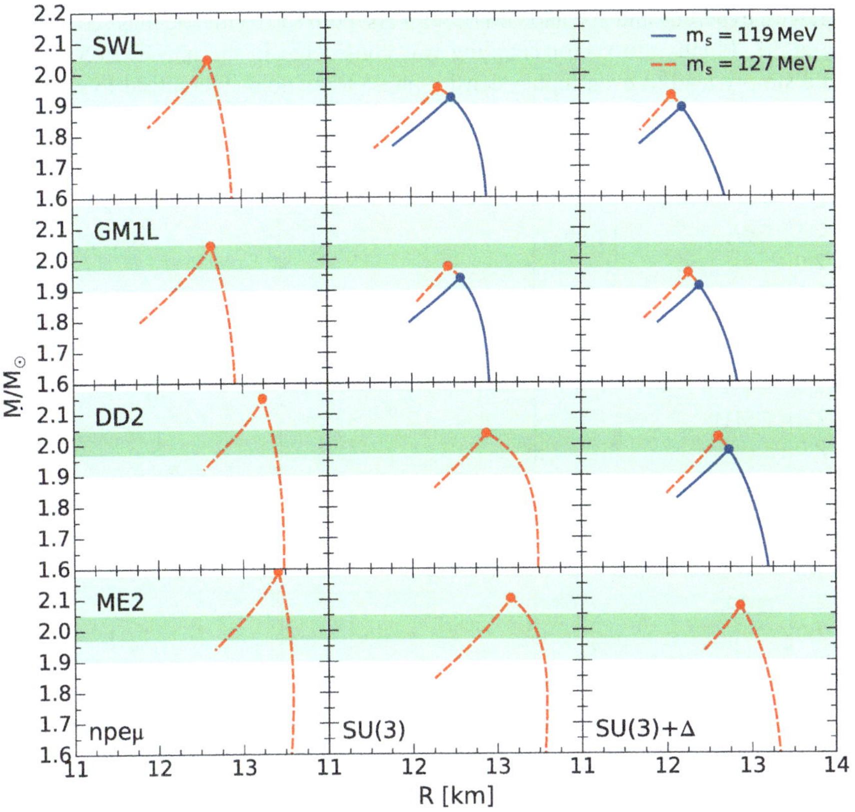

Fig. 3 Mass versus radius for NSs with a phase transition from hadronic matter to deconfined quark matter modeled using the Maxwell construction [14]. The dark (light) green shading indicates the 1σ (3σ) uncertainty range in the mass of PSR J0348 + 0432. The circle markers indicate the location of the phase transition which happens to coincide with the maximum mass for all parametrizations and compositions. (Left column) Only nucleons and leptons are included in the hadronic phase. (Center column) Hyperons are included with the vector meson-hyperon coupling constants given by the SU(3) ESC08 model. (Right column) Delta isobars are also included with the following couplings: $x_{\sigma\Delta} = x_{\omega\Delta} = 1.1$ and $x_{\rho\Delta} = 1.0$

3.1 The Maxwell Phase Transition

The mass-radius curves for quark-hadron hybrid EoSs constructed for the Maxwell transition with the P119 and P127 n3NJL parametrization are given in Fig. 3. EoSs that include pure deconfined quark matter are too soft to support stable NSs for all particle configurations and parametrizations. One interesting interpretation of this result is that the phase transition to deconfined quark matter is the limiting factor for the NS mass and density. Further, almost all configurations in Fig. 3 are consistent with the mass constraint set by PSR J0348+0432 at the 1σ level, with the exception of

SWL with hyperons and Δs, and GM1L with Δs, both satisfying the mass constraint only at 3σ. Finally, no vector coupling was considered in the n3NJL EoS as this would simply lead to even higher density phase transitions guaranteed to result in the same instability as observed for $G_V = 0$.

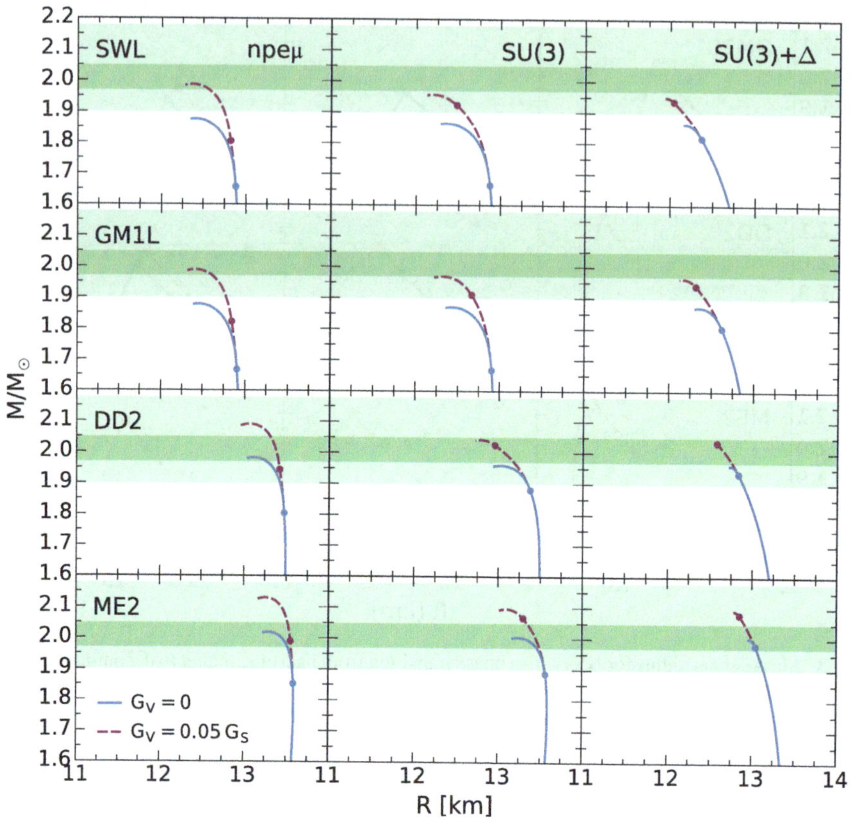

Fig. 4 The mass versus radius for NSs with a phase transition from hadronic matter to deconfined quark matter modeled using the Gibbs construction and the P119 n3NJL parametrization [14]. The dark (light) green shading indicates the 1σ (3σ) uncertainty range in the mass of PSR J0348+0432. The circle markers indicate the location of the phase transition. (Left column) Only nucleons and leptons are included in the hadronic phase. (Center column) Hyperons are included with the vector meson-hyperon coupling constants given by the SU(3) ESC08 model. (Right column) Delta isobars are also included with the following couplings: $x_{\sigma\Delta} = x_{\omega\Delta} = 1.1$ and $x_{\rho\Delta} = 1.0$

3.2 The Gibbs Phase Transition

The mass-radius curves for quark-hadron hybrid EoSs constructed using the Gibbs transition with the P119 n3NJL parametrization are given in Fig. 4. As an extreme we can first consider a quark-hadron hybrid EoS in the absence of hyperons or Δs ($npe\mu$ matter). In this case the SWL and GM1L parametrizations require the inclusion of the vector interaction to be consistent with the mass constraint, the maximum falling in the 1σ range. The DD2 and ME2 parametrizations satisfy the mass constraint at the 1σ level without requiring the inclusion of the vector interaction. All parametrizations have a fairly discoverable 0.15–0.2 $M_\odot$ range in the mass-radius curve in which a quark-hadron mixed phase appears in the core. The phase transition occurs in the fairly low density range 2.5–3 n_0, leading to quark matter fractions of 38–45% in the core. Free quarks are found to make up as much 4–6.3% of the total quarks in a NS, accounting for 6–8.8% of the total mass, both fractions increasing with the softness of the EoS.

Hyperons are included in all parametrizations with the vector meson-hyperon coupling constants given by the SU(3) ESC08 model [40]. The presence of hyperons lowers the maximum mass by less than 2% in almost all cases (exception DD2 with vector interaction), and moves the SWL and GM1L parametrizations outside the 1σ range of the mass constraint even with the inclusion of the vector interaction. The DD2 parametrization with hyperons requires the vector interaction to satisfy the mass constraint, while ME2 still does not. Without the vector interaction the quark-hadron mixed phase exists in about a 0.2 $M_\odot$ range for SWL and GM1L, but a smaller somewhat less discoverable $0.08 - 0.12\,M_\odot$ range for DD2 and ME2. Including the vector interaction reduces the overall mixed phase mass range to $\lesssim 0.06\,M_\odot$, with a range of only 0.01 $M_\odot$ for DD2, seriously limiting discoverability. Hyperons soften the EoS delaying the quark-hadron phase transition slightly without the vector interaction but drastically when the vector interaction is included. The quark matter fraction in the core reaches 35–42%, smaller than with the purely nucleonic hadronic EoSs, with a similar decrease in the total quark matter fraction and quark matter mass fraction.

Including Δs in addition to hyperons further delays the onset of the quark-hadron phase transition. However, the quark-hadron mixed phase EoS with Δs is very soft, with the result that the maximum masses are almost the same as with hyperons alone. The critical masses are within 0.02 $M_\odot$ of the maximum mass (excluding SWL for $G_V = 0$), leaving an extremely small range for discoverability. Surprisingly, in this small mass range the highest mixed phase quark matter fractions are achieved at almost 50%. However, even with these large χ values the high critical density indicates an extended hadronic phase and leads to a total quark matter fraction and quark matter mass fraction that are comparatively small.

Figure 5 shows the particle number densities for the DD2 parametrization for all particle configurations both with and without the vector interaction. Charge neutrality and the reduction of isospin asymmetry drive the NS composition in the quark-hadron mixed phase. At the onset the negatively charged d quark dominates the other quarks

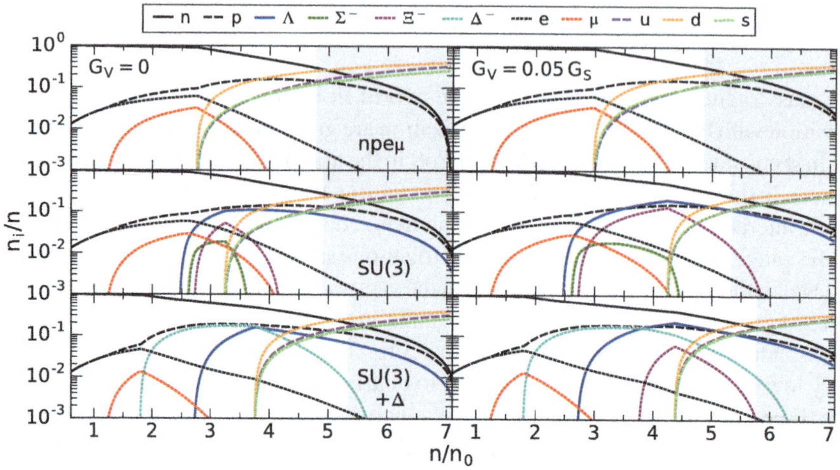

Fig. 5 The relative number density of particles as a function of baryon number density (in units of the saturation density) for the DD2 parametrization [14]. The gray shading indicates baryon number densities beyond the maximum. (Top row) Only nucleons and leptons are included in the hadronic phase. (Center row) Hyperons are included with the vector meson-hyperon coupling constants given by the SU(3) ESC08 model. (Bottom row) Delta isobars are also included with the following couplings: $x_{\sigma\Delta} = x_{\omega\Delta} = 1.1$ and $x_{\rho\Delta} = 1.0$

and replaces high energy leptons and negatively charged baryons. The s quark has a relatively high mass for a quark but it is still quite low compared to the baryons, so it also helps remove high energy leptons and negatively charged baryons from the system. The positively charged u quark replaces high energy protons and has a positive isospin ($I_u = +1/2$) which helps reduce isospin asymmetry. In $npe\mu$ and hyperonic matter the proton number density actually continues to increase at the onset of the mixed phase reducing isospin asymmetry. The negatively charged hyperons and Δs on the other hand all rapidly decline in number density, vanishing before the end of the mixed phase. Finally, the protons, neutrons, and Λs all exist up to the limit of the mixed phase.

4 Summary

We presented the nonlocal SU(3) Nambu-Jona-Lasinio model for the modeling of deconfined quark matter, and combined it with different quantum hadrodynamical models (RMFL, DDRMF) [14] to construct models for the quark-hadron hybrid EoS. The quark-hadron phase transition was modeled using the Maxwell construction given a high surface tension ($\gtrsim 70\,\mathrm{MeV/fm^2}$), and it was found that NSs with core densities beyond the phase transition density are unstable. Therefore, no pure quark matter core exists and the phase transition mass marks the endpoint of the NS mass-radius curve. Given a relatively low surface tension ($\lesssim 70\,\mathrm{MeV/fm^2}$) the quark-hadron phase transition was modeled using the Gibbs construction, resulting in a stable NS core where hadronic and deconfined quark matter coexist. The largest fractions of quark matter achieved in the quark-hadron mixed phase were around 50%; no choice of parametrization or composition led to a pure quark matter core. Finally, hybrid EoSs employing the SWL and GM1L hadronic parametrizations were shown to require the vector interaction to satisfy the mass constraint set by PSR J0348+0432.

Acknowledgements G.A.C. and M.G.O. thank CONICET and UNLP for financial support under Grants PIP 0714, G 140, G157, and X824. F.W. is supported by the National Science Foundation (USA) under Grants PHY-1411708 and PHY-1714068.

References

1. W. Becker (ed.), *Neutron Stars and Pulsars, Astrophysics and Space Science Library*, vol. 357 (Springer, 2009)
2. J. Antoniadis et al., Science **340**, 6131 (2013)
3. P. Demorest, T. Pennucci, S. Ransom, M. Roberts, J. Hessels, Nature **467**, 1018 (2010)
4. E. Fonseca et al., Astrophys. J. **832**, 167 (2016)
5. V.A. Ambartsumyan, G.S. Saakyan, Sov. Ast. **4**, 187 (1960)
6. N.K. Glendenning, Astrophys. J. **293**, 470 (1985)
7. A. Drago, A. Lavagno, G. Pagliara, Phys. Rev. D **89**, 043014 (2014); A. Drago, A. Lavagno, G. Pagliara, Daniele Pigato, Phys. Rev. D **90**, 065809 (2014)
8. D.D. Ivanenko, D.F. Kurdgelaidze, Astrophysics **1**, 251 (1965)
9. H. Fritzsch, M. Gell-Mann, H. Leutwyler, Phys. Lett. **47B**, 365 (1973)
10. K. Rajagopal, F. Wilczek, The condensed matter physics of QCD, in *At the Frontier of Particle Physics* (World Scientific, 2001), pp. 2061–2151
11. M.G. Alford, K. Rajagopal, T. Schaefer, A. Schmitt, Rev. Mod. Phys. **80**, 1455 (2008)
12. I.F. Ranea-Sandoval, M.G. Orsaria, S. Han, F. Weber, W.M. Spinella, Phys. Rev. D **96**, 065807 (2018)
13. M. Buballa et al., J. Phys. G: Nucl. Part. Phys. **41**, 123001 (2014)
14. W.M. Spinella, A Systematic Investigation of Exotic Matter in Neutron Stars. Ph.D. thesis, Claremont Graduate University & San Diego State University (2017)
15. M. Orsaria, H. Rodrigues, F. Weber, G.A. Contrera, Phys. Rev. C **87**, 023001 (2013)
16. M. Orsaria, H. Rodrigues, F. Weber, G.A. Contrera, Phys. Rev. C **89**, 015806 (2014)
17. G.A. Contrera, D. Gómez Dumm, N.N. Scoccola, Phys. Lett. B **661**, 113 (2008)
18. G.A. Contrera, D. Gómez Dumm, N.N. Scoccola, Phys. Rev. C **81**, 054005 (2010)
19. A. Scarpettini, D. Gómez Dumm, N.N. Scoccola, Phys. Rev. C **69**, 114018 (2004)
20. I.F. Ranea-Sandoval, S. Han, M.G. Orsaria, G.A. Contrera, F. Weber, M.G. Alford, Phys. Rev. C **93**, 045812 (2016)
21. M.B. Parappilly, P.O. Bownman, U.M. Heller, D.B. Leinweber, A.G. Williams, J.B. Zhang, Phys. Rev. C **73**, 054504 (2006)
22. L.F. Palhares, E.S. Fraga, Phys. Rev. C **82**, 125018 (2010)
23. M.B. Pinto, V. Koch, J. Randrup, Phys. Rev. C **86**, 025203 (2012)
24. B.W. Mintz, R. Stiele, R.O. Ramos, J. Schaffner-Bielich, Phys. Rev. C **87**, 036004 (2013)
25. G. Lugones, A.G. Grunfeld, M. Al Ajmi, Phys. Rev. C **88**, 045803 (2013)
26. D.N. Voskresensky, M. Yasuhira, T. Tatsumi, Nucl. Phys. A **723**, 291 (2003)
27. N. Yasutake, R. Lastowiecki, S. Benic, D. Blaschke, T. Maruyama, T. Tatsumi, Phys. Rev. C **89**, 065803 (2014)

28. N.K. Glendenning, Phys. Rep. **342**, 393 (2001)
29. X. Na, R. Xu, F. Weber, R. Negreiros, Phys. Rev. C **86**, 123016 (2012)
30. W.M. Spinella, F. Weber, G.A. Contrera, M.G. Orsaria, Eur. Phys. J. A **52**, 61 (2016)
31. J.D. Walecka, Ann. Phys. (NY) **83**, 491 (1974)
32. B.D. Serot, J.D. Walecka, Adv. Nucl. Phys. **16**, 1 (1986)
33. J. Boguta, A.R. Bodmer, Nucl. Phys. A **292**, 413 (1977)
34. J. Boguta, J. Rafelski, Phys. Lett. **71B**, 22 (1977)
35. J. Boguta, H. Stöcker, Phys. Lett. **120B**, 289 (1983)
36. R.D. Mellinger, F. Weber, W. Spinella, G.A. Contrera, M.G. Orsaria, Universe **3**, 5 (2017)
37. N.K. Glendenning, S.A. Moszkowski, Phys. Rev. Lett. **67**, 2414 (1991)
38. S. Typel, G. Ropke, T. Klahn, D. Blaschke, H.H. Wolter, Phys. Rev. C **81**, 015803 (2010)
39. G.A. Lalazissis, T. Niksic, D. Vretenar, P. Ring, Phys. Rev. C **71**, 024312 (2005)
40. T.A. Rijken, M.M. Nagels, Y. Yamamoto, Prog. Theor. Phys. Suppl. **185**, 14 (2010)

Binary Compact Star Mergers and the Phase Diagram of Quantum Chromodynamics

Matthias Hanauske and Horst Stöcker

1 Preface

This article is dedicated to one of my doctor fathers, Prof. Dr. Dr. h.c. mult. Walter Greiner, whom I first met during my undergraduate studies at the Goethe University Frankfurt in 1990. During this time Walter Greiner gave his legendary 5-semester course on theoretical physics and I had the fortune to be able to attend his first lecture on *Mechanic I*. Greiner's fashion and manner of presenting the philosophical and mathematical backgrounds of theoretical physics in a detailed way was outstanding. I also enjoyed his interposed personal stories on topics like e.g. scientific conferences and the wilderness of Africa's nature. Between 1998 and 2004 I worked as a doctoral student in his large international group in the Institute of Theoretical Physics (Campus Bockenheim). The present article summarizes some of the scientific results I have obtained during this period under his supervision. Prof. Greiner's works on the interplay of elementary particle physics and general relativity, contemplated from a present viewpoint, show that he was always ahead of time. Especially the first detection of a gravitational wave emanated from a binary neutron star merger, which had been discovered after his death on 17.08.2017 by the LIGO/VIRGO collaboration, demonstrates his outstanding scientific intuition.

M. Hanauske (✉) · H. Stöcker
Institut für Theoretische Physik, Max-von-Laue-Straße 1,
60438 Frankfurt, Germany

Frankfurt Institute for Advanced Studies, Ruth-Moufang-Straße 1,
60438 Frankfurt, Germany
e-mail: hanauske@fias.uni-frankfurt.de

H. Stöcker
GSI Helmholtzzentrum für Schwerionenforschung GmbH,
64291 Darmstadt, Germany

J. Kirsch et al. (eds.), *Discoveries at the Frontiers of Science*,
FIAS Interdisciplinary Science Series, https://doi.org/10.1007/978-3-030-34234-0_10

2 Introduction

Although I had attended Greiner's full course on theoretical physics, I had never a personal conversation with him during my undergraduate studies and I left Frankfurt in 1994 to begin my main diploma studies at the University of Konstanz under the supervision of Prof. Heinz Dehnen where I worked on gauge theories of gravitation.

All of the four interactions known till now can be described by gauge theories. Three of them are found to be Yang-Mills theories and are formulated within the Quantum electrodynamics (QED), the Weak Interaction and the Quantum chromodynamics (QCD) which describes the strong nuclear interaction. The unification of the weak and the electromagnetic interaction is known as the electroweak interaction and its unification with QCD is hoped to be eventually achieved within the Grand Unified theory (GUT). Gravity itself is also expected to be a gauge theory [1], and there are different ways to formulate such a theory. The gauge group is expected to be the Poincare group which—if localized—leads to an Einstein theory of gravity including torsion, the so called Riemann-Cartan theory [2, 3]. Other ways to construct a theory of gravity from fundamental principles are based on different gauge groups which acts on internal degrees of freedom of the particles and are therefore nearer to Yang-Mills theories (see e.g. the spin gauge theory of gravity [4, 5]), use the concept of a minimal length [6] or use the formalism of extended canonical transformations in the realm of covariant Hamiltonian field theory [7, 8]. These examples of alternative theories of gravity predict often differences between the classical Einstein theory of gravity and some of them can be ruled out using data from black hole shadows [9, 10] or timing of pulsar binaries [11].

I finished my diploma thesis in 1997 and returned to Frankfurt, where I had then finally the first personal contact with Walter Greiner. Some of my former student colleagues told me that Prof. Greiner was on the way of writing the last missing topic of his red book series on *General Relativity* (GR) and I was quite free-spirited and wrote an e-mail to him, where I asked him if he needs help in setting up the book. I wrote him that I was quite familiar with GR, explained the topic of my diploma and asked him for a temporary student job. He replied that he cannot offer me a temporary student job, but I should give a talk about my diploma in the "Palaver". Later, after my presentation, he told me that I can begin working in his group on a topic related to neutron stars, he helped me getting a PhD-scholarship and finally I stayed for seven years in his group working on my doctoral thesis.

The theory of neutron stars is in general a complicated interplay between all known forces, but if one restricts oneself to weakly magnetized neutron stars and equilibrated systems, only two forces dominate the system, namely the strongest (QCD) and the weakest force (gravity). Due to the different symmetry transformation groups of these two gauge theories, the covariant derivatives and as a result the underlying field equations are different, but the mathematical gauge theoretical structures are analogue

GR: Covariant derivative: $D_\nu v^\mu := v^\mu{}_{||\nu} = v^\mu{}_{|\nu} + \Gamma^\mu_{\beta\nu} v^\beta$

Field strength: $R^\alpha{}_{\sigma\nu\mu} v^\sigma = -\left[D_\mu, D_\nu\right] v^\alpha = \left(\Gamma^\alpha_{\sigma\nu|\mu} - \Gamma^\alpha_{\sigma\mu|\nu} + \Gamma^\alpha_{\beta\mu}\Gamma^\beta_{\sigma\nu} - \Gamma^\alpha_{\beta\nu}\Gamma^\beta_{\sigma\mu}\right) v^\sigma$

Lagrangian: $\mathcal{L}_{GR} = R$

QCD: Covariant derivative: $D_{\mu a}{}^b \psi_b = \left(\partial_\mu 1_a{}^b + i g_3 G_{\mu i} \tau^i{}_a{}^b\right) \psi_b = \psi_{a|\mu} + i g_3 G_{\mu a}{}^b \psi_b$

Field strength: $\mathcal{F}_{\mu\nu} = F_{\mu\nu a}{}^b = \frac{i}{g_3}\left[D_{\mu a}{}^c, D_{\nu c}{}^b\right] = F_{\mu\nu i} \tau^i{}_a{}^b$

Field strength: $F_{\mu\nu i} = G_{\nu i|\mu} G_{\nu i|\mu} - g_3 f^{jk}{}_i G_{\mu j} G_{\nu k}$

Lagrangian: $\mathcal{L}_{QCD} = \frac{1}{16\pi} F_{\mu\nu i} F^{\mu\nu i}$,

where the indices $_a{}^b$... stand for the color indices of QCD while the indices $_\mu{}^\nu$... are the spacetime coordinates of general relativity (GR). One main difference between the two theories is that its color-charges can not be seen by an outside observer (confinement effect of QCD).

In a binary neutron star merger the field contributions of QCD and GR are of the same order and we will see at the end of this article, that the liberation of the colourful quark phase (the deconfinement QCD phase transition) in the interior of a merger remnant and the general relativistic formation of a black hole (which macroscopically confines the star by forming the event horizon), happens in an hypermassive hybrid star after the post-merger phase, when the last collapsing phase of the mergers remnant begins. Unfortunately QCD is not solvable in the nonperturbative regime and, up to now, numerical solutions of QCD on a finite space-time lattice are still unable to describe neutron star matter or even finite nuclei or infinite nuclear matter. As a consequence several effective theoretical models of the hadronic interaction have been proposed (see e.g. [12, 13]), however, theses models are limited in there description to moderate temperature and densities and are therefore believed to be not applicable in a binary neutron star merger (BNS) scenario. If one wants to extends these theories to higher density and temperatures, it is believed that hadronic matter undergoes a phase transition to a deconfined state consisting of quarks and gluons, the so-called quark-gluon plasma.

The paper is structured as follows: After this brief introduction, Sect. 3 focuses on the construction of a hadron-quark phase transition (HQPT) in the interior of a compact star. In order to highlight the contributions of W. Greiner, the section accentuate the research results of Walter Greiner and various scientists of his group obtained more than 15 years ago (details can be found in my PhD-thesis *Properties of Compact Stars within QCD-motivated Models* [14] and in the underlying research articles [15–25]). In order to motivate, that the temperature and density values reached inside a BNS merger product requires an incorporation of a HQPT, Sect. 4 discusses a BNS merger in the context of the hadron-quark phase transition. The interior temperature and density structure of a neutron star merger product and the evolution of the hot and dense matter inside the hypermassive/supramassive neutron star (HMNS,SMNS) will be analysed and visualized in a (T-ρ/ρ_0) QCD phase diagram. These new results of are based on the articles [26–33]. A summary and outlook will be presented in Sect. 5.

3 Compact Stars in QCD-Motivated Models

The pressure inside a neutron star is primarily due to the Fermi pressure of the hadrons and their strong interaction with each other. The accepted underlying theory of strong interactions, QCD, is however not solvable in the nonperturbative regime. So far numerical solutions of QCD on a finite space-time lattice are unable to describe finite nuclei or infinite nuclear matter. As an alternative approach several effective models of hadronic interactions have been proposed and especially the Walecka model [34] and its nonlinear extensions have been quite successful and widely used for the description of hadronic matter and finite nuclei. These models are relativistic quantum field theories of baryons and mesons, but they do not consider essential features of QCD, like approximate $SU(3)_R \times SU(3)_L$ chiral symmetry or broken scale invariance. The Nambu–Jona-Lasinio (NJL) model [35–37] is an effective theory which has implemented these features of QCD but it lacks confinement and thereby fails to describe finite nuclei and nuclear matter.

This has lead W. Greiner and colleagues to construct a QCD-motivated chiral $SU(3)_L \times SU(3)_R$ model as an effective theory of strong interactions, which implements the main features of QCD. The model has been found to describe reasonably well the hadronic masses of the various SU(3) multiplets, finite nuclei, hypernuclei, excited nuclear matter [12, 13] and astrophysical observable data from neutron stars, like masses and radii [16].[1] The basic assumptions in the chiral model are (for details see [12–14]):

- The Lagrangian is constructed with respect to the nonlinear realization of chiral $SU(3)_L \times SU(3)_R$ symmetry.
- The masses of the heavy baryons and mesons are generated by spontaneous symmetry breaking using a Higgs mechanism.
- The masses of the pseudo-scalar mesons are generated by explicit symmetry breaking, since they are the Goldstone bosons of the model.
- A QCD-motivated field χ enters, which describes the condensed gluon field (dilaton).
- Baryons and mesons are grouped according to their quark structure.

The total Lagrangian of the chiral $SU(3)_L \times SU(3)_R$ model for neutron star matter can be written in the mean-field approximation as (for details see Refs. [12–14])

$$\mathcal{L} = \mathcal{L}_{\text{kin}} + \mathcal{L}_{\text{BM}} + \mathcal{L}_{\text{BV}} + \mathcal{L}_{\text{vec}} + \mathcal{L}_{0} + \mathcal{L}_{\text{SB}} + \mathcal{L}_{\text{lep}} \,, \tag{1}$$

where $\mathcal{L}_{\text{kin}}$ is the kinetic energy term of the baryons. The interaction Lagrangian of the different baryons with the various spin-0 and spin-1 mesons (the scalar (σ, ζ) and vector (ω, ϕ, ρ) mesons) are $\mathcal{L}_{\text{BM}}$ and $\mathcal{L}_{\text{BV}}$, respectively. It includes all the charge states of the baryon octet $(p, n, \Lambda, \Sigma^-, \Sigma^0, \Sigma^+, \Xi^-, \Xi^0)$. $\mathcal{L}_{\text{vec}}$ generates the masses

[1] We note that during the time of achieving these results the current observational constraint on the observed maximum mass in neutron stars, i.e., $2.01 \pm 0.04 M_\odot$ [38] were not known.

of the spin-1 mesons through the interactions with spin-0 mesons, and $\mathcal{L}_0$ gives the meson-meson interaction term which induce the spontaneous breaking of chiral symmetry. A salient feature of the model, the dilaton field χ, which can be identified with the gluon condensate, has been included. It accounts for the broken scale invariance of QCD at tree level through the logarithmic potential. $\mathcal{L}_{\mathrm{SB}}$ introduces an explicit symmetry breaking of the $\mathrm{U}(1)_A$, the $\mathrm{SU}(3)_V$, and the chiral symmetry. The last term $\mathcal{L}_{\mathrm{lep}}$ represents the free lepton Lagrangian.

In contrast to the environment in finite nuclei and heavy ion collisions, neutron star matter is charge neutral, in β-equilibrium and the strangeness is no more a conserved quantity. The particle composition of a neutron star depends strongly on its density profile and in the interior region hyperonic particle are expected to be present (see Fig. 1).

General relativistic astrophysics of neutron stars and nuclear/elementary particle physics are strongly connected and numerical simulations in both fields are strongly depend on the equation of state (EOS) of fundamental elementary matter [39]. Hot and dense matter created in high energy heavy ion collisions and mergers of a binary system of two neutron stars systems reach values where strongly interacting matter is expected to undergoing the deconfinement phase transition. Additionally, the density in the interior of a single, cold neutron star could reach values that a hadron-quark phase transition (HQPT) needs to be incorporated within the theoretical description of the underlying model. The primary concern of my doctoral thesis under the supervision of Walter Greiner was an comprehensive analysis of the astrophysical consequences of a HQPT in the interior of neutron stars. During that time the properties of hybrid stars containing both the hadrons and quarks had been only studied for some specific models (see [40, 41] for the pioneer works) and I started my research by implementing a HQPT within the chiral $SU(3)_L \times SU(3)_R$ model using a simple MIT-Bag model for the deconfined quark phase. The first paper on this topic appeared 2001 in a special issue of *General Relativity and Gravitation* dedicated to Prof. H.Dehnen on the occasion of his 65th birthday [17].[2]

The transition from hadronic matter to quark matter can occur either continuously in a mixed phase region using a Gibbs construction between the hadron and quark model or, assuming a large surface tension of the first deconfined quark bubbles, in a sharp discontinuous way using a Maxwell construction between the two separate models. In the following the construction of a first order HQPT in the interior of a hybrid star is briefly described and we assume that the surface tension of the deconfined quark phase has moderate values resulting in a mixed phase region between the pure hadronic phase and pure quark phase. At zero temperature this mixed phase should follow the Gibbs conditions

$$P^H(\mu_B, \mu_e) = P^Q(\mu_B, \mu_e) , \tag{2}$$

[2]During the time of my diploma-studies at the University Konstanz I was not aware that W. Greiner and H. Dehnen had been dear friends since the days of joint graduate time at the Freiburg University 60 years ago.

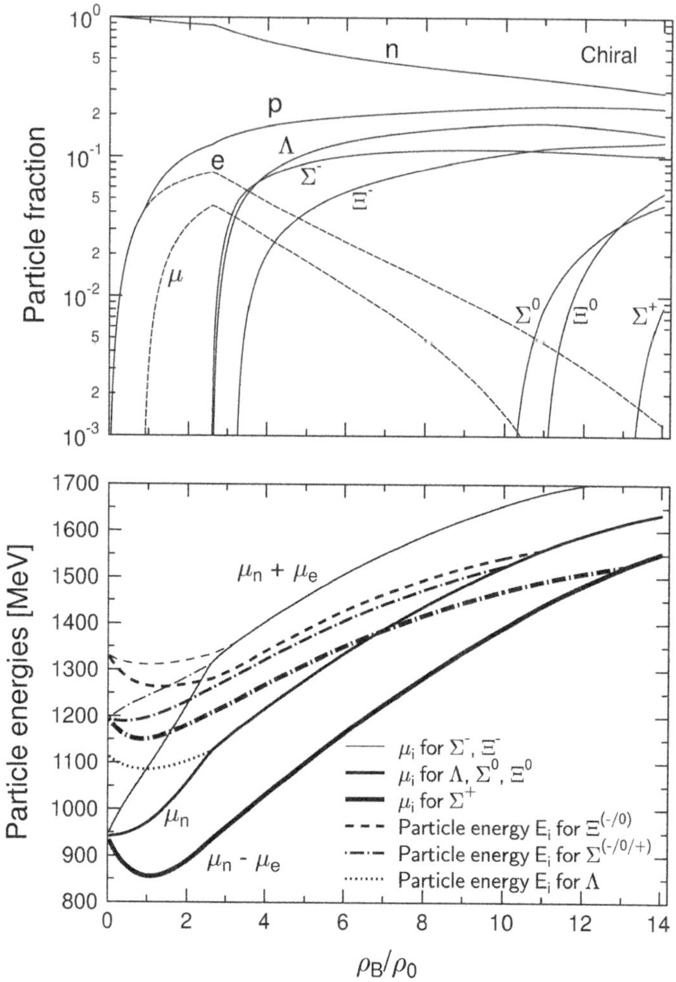

Fig. 1 Particle fraction of the chiral model (upper panel) and energy levels E_i of the baryonic particle species i and the chemical potentials μ_n, $\mu_n - \mu_e$ and $\mu_n + \mu_e$ (lower panel) as a function of the baryon density ρ_{B} in units of the nuclear-matter density ρ_0. Due to the high pressure inside the neutron star baryonic particles are able to transform into other particles, e.g. $n \Leftrightarrow p + e + \tilde{\nu}_e$ or $n + n \Rightarrow n + \Lambda + K$, where n, p, Λ and K stands for the neutron, proton, Λ hyperon and kaon. Whether and to which amount a particle species i decays into another one depends on the energy levels E_i and the chemical potential $\mu_i = b_i \mu_n - q_i \mu_e$, where b_i and q_i are the baryon number and the charge of the i'th baryon ($i = n, p, \Lambda, \Sigma^-, \Sigma^0, \Sigma^+, \Xi^-, \Xi^0$). The appearance of the charge-less Λ-hyperon takes place when its energy level $E_\Lambda \equiv \mu_n$ while the negatively charged Σ^--hyperon appears when $E_{\Sigma^-} \equiv \mu_n + \mu_e$

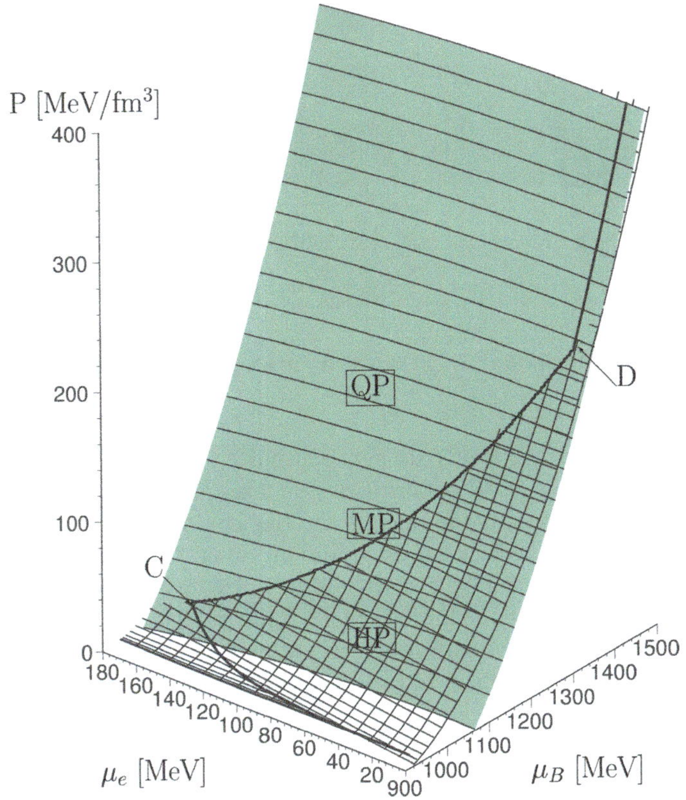

Fig. 2 Gibbs construction of hadron quark phase transition: pressure surface of the hadronic phase (HP) in CHM-model with hyperons (wired surface) and the quark phase (QP) in the MIT Bag model with $B^{1/4} = 180$ MeV (green solid surface) as a function of the two independent chemical potentials μ_B and μ_e. The solid black curve upon the hadronic surface describes charge-neutral neutron star matter at different densities. The beginning of the mixed phase (MP) takes place where the charge neutral star matter curve encounters with the intersection line of the two surfaces (marker "C"). The resulting volume fraction of quark matter increases during the mixed phase region until all of the matter is finally deconfined (marker "D")

where P^H and P^Q is the pressure in the hadronic and quark phase and $\mu_B = \mu_n$ and μ_e are the baryonic and charge chemical potentials. Figure 2 illustrates the construction of a HQPT using the method of a Gibbs construction. The figure shows the pressure surfaces for the pure hadronic (wired surface, CHM model) and pure quark phase (green solid surface, MIT Bag model) as functions of the chemical potentials μ_B and μ_e (for details see [14]). The mixed phase region connects the pure phases along the intersection line and the two markers "C" and "D" depicts the beginning and end point of the mixed phase region. The thick curve onto the surfaces shows the trajectory of charge neutral matter in β-equilibrium.

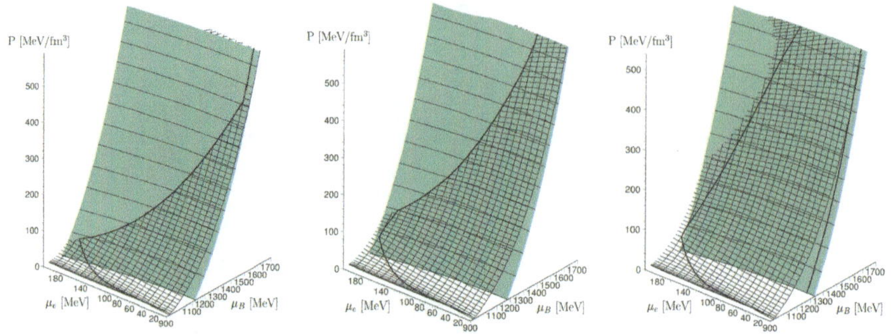

Fig. 3 Same as Fig. 2 but for $B^{1/4} = 190$ MeV (left), 200 MeV (middle) and 210 MeV (right)

The construction of the HQPT strongly depends on the Bag constant B and for $B^{1/4} = 180$ MeV the onset "C" of the transition takes place at rather low densities ($\rho_{\text{B}}/\rho_0 \approx 1.9$). By increasing the value of the Bag constant the green solid quark surface moves more to the back and as a result the onset "C" of the transition takes place at higher densities and the extension of the mixed phase region is more prolonged (see Fig. 3). For very high values of the Bag constant ($B^{1/4} \geq 205$ MeV) the intersection region of the two surfaces do not encounter with the charge neutral line of neutron star matter (see right picture in Fig. 3) and as a result no Gibbs construction from the hadronic CHM-model to the MIT-Bag model is possible any more. In such a scenario, the deconfinement transition would never take place, however, under the "strange matter hypothesis", the strange-quark phase could be the true ground state of matter and the whole neutron star would then transform into a pure quark star after exceeding a certain deconfinement barrier.

Figure 4 visualizes the particle fractions of the chiral-MIT-Bag model for $B^{1/4} = 190$ MeV (left picture) and $B^{1/4} = 200$ MeV (right picture) as a function of the baryon density ρ_{B}. For $B^{1/4} = 190$ MeV the HQPT starts before the onset of the first hyperonic species at $\rho_{\text{B}}/\rho_0 \approx 2.5$ and the mixed phase ends at $\rho_{\text{B}}/\rho_0 \approx 11.1$ and only small fractions of λ hyperons are present during the MP. For $B^{1/4} = 200$ MeV the MP starts after the appearance of the Λ, Σ^- and Ξ^- hyperon at $\rho_{\text{B}}/\rho_0 \approx 4.2$ and due to the shape of the two pressure surfaces (see middle picture in Fig. 3) it never reaches a pure quark phase region. The main impact of such a Gibbs construction on the EOS is a strong softening within the mixed phase region followed by a stiff pure quark matter region (see left picture in Fig. 5).

In order to calculate the properties of a neutron star, the gravitational force needs to be implemented in the model. Walter Greiner was always fascinated by Einstein's theory of general relativity, however, he did never believe in black holes and real singularities and especially in his last period of research he had several projects on alternative theories of gravity [42]. During the time of my PhD thesis we first started with classical Einstein's theory of general relativity and used the most simple model of a static, spherically symmetric compact star having an energy-momentum tensor of an ideal fluid. We used Einstein's theory of general relativity in connection with

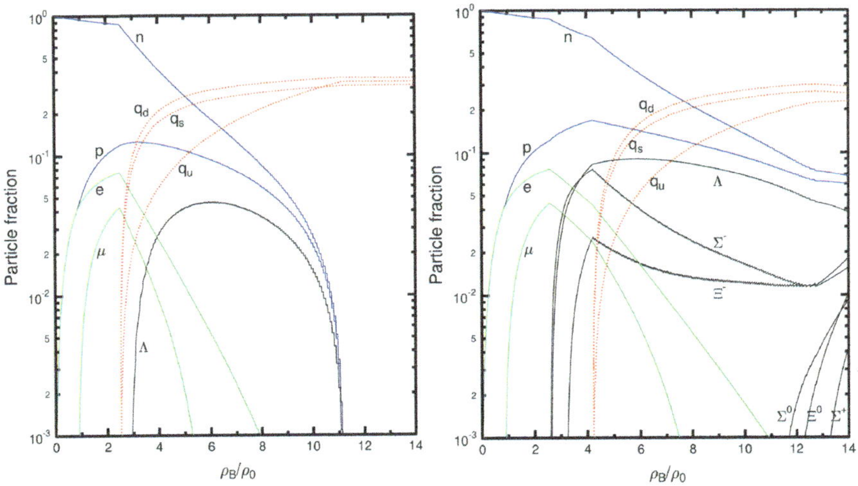

Fig. 4 Particle fraction of the chiral-MIT-Bag model for $B^{1/4} = 190$ MeV (left) and $B^{1/4} = 200$ MeV (right)

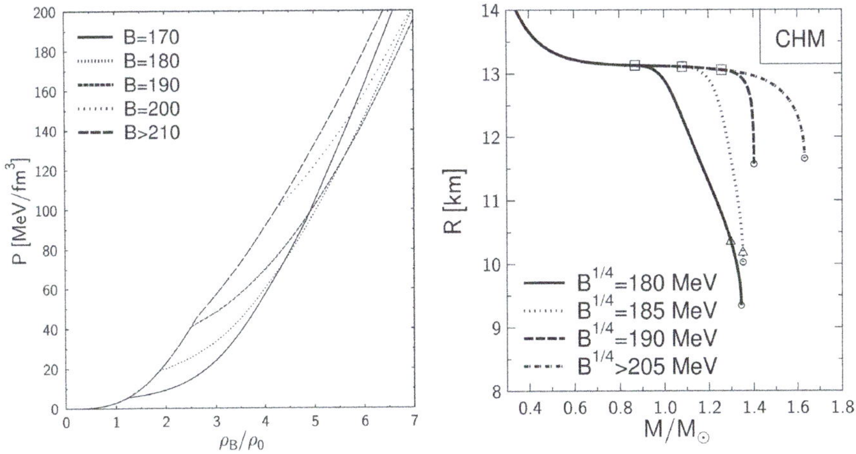

Fig. 5 EOS in the CHM-Bag model (left) and corresponding mass-radius relations (right). The rectangles mark the beginning and the triangles the end of the mixed phase region, while the open dots depicts the maximum mass star

the conservation laws for energy-momentum and rest mass as the groundings of the differential equations which we needed to solve. The Einstein equation and the conservation laws are summarized in the following set of highly non-linear differential equations:

$$R_{\mu\nu} - \frac{1}{2} g_{\mu\nu} R = 8\pi T_{\mu\nu} , \quad \nabla_\mu T^{\mu\nu} = 0 , \quad \nabla_\mu (\rho u^\mu) = 0 , \tag{3}$$

where $T_{\mu\nu}$ is the energy-momentum tensor, $R_{\mu\nu}$ is the Ricci tensor, which contains first and second derivatives of the space-time metric $g_{\mu\nu}$, ∇_μ is the covariant derivative and u^μ is the four velocity of the star's fluid. The Einstein equation (first equation in (3)) describes in which way the space-time structure need to bend (left hand side of the equation) if energy-momentum is present (right hand side of the equation).

In a static, spherical symmetrical condition, the metric $g_{\mu\nu}$ and the infinitesimal line element ds can be written as

$$ds^2 = g_{\mu\nu} dx^\mu dx^\nu = -e^{2\nu(r)} dt^2 + \left(1 - \frac{2m(r)}{r}\right)^{-1} dr^2 + r^2 d\theta^2 + r^2 \sin^2\theta \, d\phi^2 . \tag{4}$$

In the following we will use the expression of an ideal fluid energy-momentum tensor to describe the neutron star matter

$$T_{\mu\nu} = (e + P) u_\mu u_\nu + P g_{\mu\nu} , \quad e = \rho (1 + \epsilon) , \tag{5}$$

where e is the energy density, P the pressure, ρ the rest mass density and ϵ the internal energy density of the neutron star fluid.

By inserting the Ansatz of the metric (4) together with the expression for the energy momentum tensor (5) into the equations (3) one derives the Tollman-Openheimer-Volkov (TOV) equations (see e.g. [43]):

$$\frac{dm}{dr} = 4\pi r^2 e , \quad \frac{d\nu}{dr} = \frac{m + 4\pi r^3 P}{r(2m - r)} , \quad \frac{dP}{dr} = -(e + P) \frac{d\nu}{dr} . \tag{6}$$

For a given equation of state (EoS: a function $P(e)$) the TOV equation can be solved numerically by fixing the central pressure P_c and integrating outwards to the star's surface where the pressure is zero. By changing the value of the central pressure P_c a sequence of neutron stars can be calculated, where each star has a different mass and radius (mass-radius relation, see right picture in Fig. 5). From this mass-radius relation we see that there exists a maximum value of the total mass of the star (M_{max}), which cannot be exceeded. One can show (for details see [44]) that if the energy density is higher than the energy density of the maximum mass star e_{max}, the neutron stars become unstable and either need to collapse to a black hole or migrate back to a stable neutron star configuration.

Although the main consequences of a HQPT have been calculated and published by us already in 2001 [17], I stayed four years longer in Greiner's group working

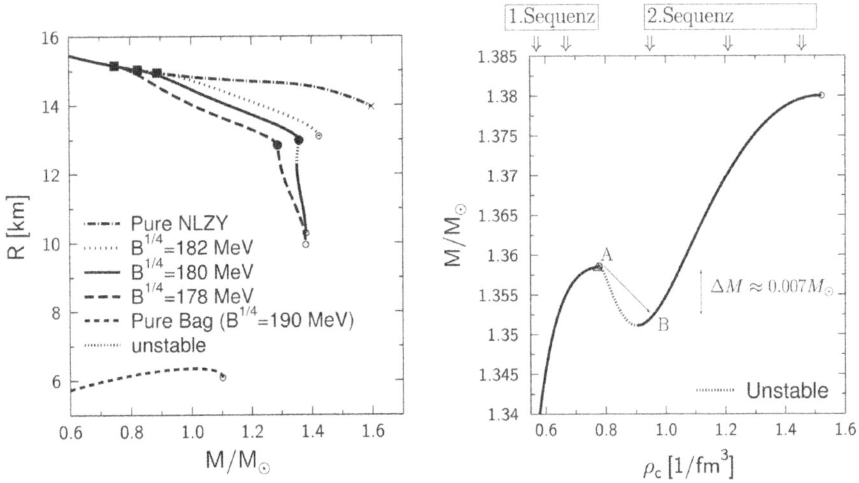

Fig. 6 Left: Radius R as a function of the total mass M of hybrid stars in NLZYY-Bag model using the Gibbs's construction of a HQPT. The picture shows that within a specific region of the Bag-constant B twin stars are possible. The curve at the bottom of the picture are pure quark stars in the MIT-bag model ($B^{1/4} = 190$ MeV). Right: The twin star collapse; total gravitational mass of hybrid stars of the first and second sequence as a function of the central baryonic density ρ_c in the NLZYY-Bag180 model

on my doctoral thesis. Together with various scientist of Greiner's group we calculated the properties of neutron stars, pure quark/strange stars and hybrid stars using different model for the hadronic (CHM, TM1, TM1YY, NLZ, NLZYY) and quark phase (MIT-Bag, NJL, MQM), which we had combined using a Gibbs- or Maxwell construction. In a detailed analysis we focused on the astrophysical consequences of a HQPT in the interior of a compact star and grouped the astrophysical observables of the quark-gluon plasma in the following way (see Chap. 3.5 in [14]): Evidence of the HQPT based on mass and radius properties, rotational behaviours, twin star properties and by means of a future gravitational wave detection. The effects of a strong HQPT have been investigated in the context of static [17, 20, 22] and uniformly rotating hybrid stars [23, 24] and the results showed that tremendous changes in the star properties might occur including the existence of a third family of compact stars—the so called "twin stars" [45] (see Fig. 6). More than 10 years before the first detection of a gravitational wave [46], we speculated about the astrophysical observables of a HQPT in a single compact star system (e.g. a gravitational wave bust during a twin star collapse) and estimated properties of the emitted gravitational waves in a binary hybrid star merger scenario (see pp. 75–78 and p. 190 in [14] and [47]). We found that the GW-frequency at the moment of collision in a neutron star

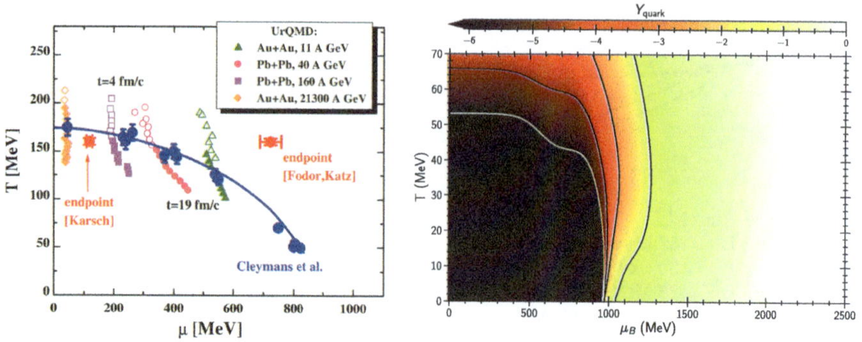

Fig. 7 The QCD phase diagram. Left: Quasi-quantitative results of the QCD phase diagram obtained by numerical calculations (the picture has been created by Elena Bratkovskaya in 2004). Right: Quark fraction Y_q of hot and dense matter within a new version of the former CHM-model, called the quark-hadron chiral parity-doublet model (QχP model). The picture has been created by Ayon Mukherjee in 2018

merger (f_{peak}) is lower than in a hybrid or quark star merger and estimated various other GW-properties (see Tables 2.1 and 2.2 in [14]).[3]

So far, we have only summarized the "old works" we have been calculated closely together with Prof. Greiner. After 2004 several observations in the field of astro- and heavy ion physics had been made, including the current observational constraint on the observed maximum mass in neutron stars, i.e., $2.01 \pm 0.04 M_{\odot}$ [38], which caused the hyperon puzzle and constrained the parameters in the EOSs further. Additionally, the implementation of the HQPT within the CHM model have been extended to finite temperatures and the former CHM model had been separated into two branches (the quark-hadron chiral parity-doublet model (QχP model, see Fig. 7) and the chiral mean field (CMF) model). Last but not least, simulations of BNS merger within full general relativity had become possible and where finally detected in 2017.

4 Numerical Relativity of Binary Compact Star Mergers

Not even two years after the first detection of a gravitational wave (GW) emanated from the inward spiral and merger of pairs of a black holes by LIGO, GWs from a binary neutron star merger has been discovered on 17.08.2017 by the LIGO/VIRGO collaboration (GW170817 [49]). This long-awaited event in conjunction with the independently detected gamma-ray burst (GRB170817A [50]) and further electromagnetic radiation [51] marked the beginning of the combined new field of gravitational-wave and multi-messenger astronomy. With the use of the extracted

[3]It needed almost 15 years that these estimates were confirmed by the numerical simulations of Bauswein et al. [48].

tidal deformation of the two neutron stars in the late inspiral phase of GW170817 it is possible to constrain the EOS of neutron star matter—however the most interesting part of the GW, the post-merger GW emission, has so far not been observed but will possibly been detected within the next observing run [52].

The neutron star merger scenario of GW170817 is in good agreement with numerical simulations of binary neutron star mergers performed in full general relativistic hydrodynamics. During the last decades a large number of numerical-relativity simulations of merging neutron star binaries have been performed and the emitted GWs and the interior structure of the generated HMNS in the post-merger phase have been analysed in detail (see e.g. [26]). The extracted constraints on the mass of the total system of GW170817, in combination with the limitations on the EOS, which were estimated using the extracted tidal deformation of the two neutron stars right before merger [49], results in a neutron star merger scenario which is remarkably similar to the `APR4-M135` simulation discussed in [26]. As a result of the binary merger, a fast, differentially rotating compact object is produced, dubbed a HMNS. Matter in the interior of this object reaches densities of up to several times normal nuclear matter, and temperatures could reach $T \sim 50$–100 MeV. The numerical simulations show that—after the violent transient post-merger phase—the HMNS stabilizes, after $\approx$10 ms, resulting in a quasi-stable configuration with a specific rotation profile. The observation of the gamma-ray burst GRB 170817A [50], which was detected with a time delay of $\simeq$1.7 s with respect to the merger time, indicates the collapse of the HMNS at a post-merger time $\simeq$1 s. During the late post-merger simulation, the value of central rest-mass density is increasing to $\rho_c \simeq 5\,\rho_{\rm nuc}$ for the `APR4-M135` run. However, for such high densities, the EOS is still poorly constrained as the frequency spectrogram of the post merger phase has not been detected in the GW170817 event. High energy heavy ion collision data are compatible with a hadron-quark phase transition (HQPT), which then shall also be present in the interior of the HMNS produced [39]. By analysing the power spectral density profile of the post-merger emission of a future event of the next observing run of the LIGO/VIRGO collaboration, the GW signal can set tight constraints on the high density regime of the EOS of elementary matter. The modification of the EOS due to a potential influence of a HQPT and the impact of strange quark matter on the EOS, which is currently solely probed in relativistic heavy ion collisions, might be imprinted in the post-merger phase of the emitted GW of a merging compact star binary. Hybrid star mergers represent optimal astrophysical laboratories to investigate the phase structure of QCD and in addition with the observations from heavy ion collisions will possibly provide a conclusive picture on the QCD phase structure at high density and temperature [39].

In order to justify that a HQPT is expected to happen during the post-merger phase of a BNS merger Fig. 8 shows a comparison of a neutron star merger simulation with two typical simulation of a heavy ion collision. The density and temperature evolution of a BNS merger has similarities and differences to an heavy ion collision. Figure 8 shows the time evolution of the maximum value of the temperature in MeV (triangles) and rest mass density in units of the nuclear matter density (diamonds) of the BNS merger simulation in a (T-ρ/ρ_0) QCD phase diagram. The simulation is based on the LS220 EOS with a total gravitational mass of $2 \times 1.350\,M_\odot$ evolved without the

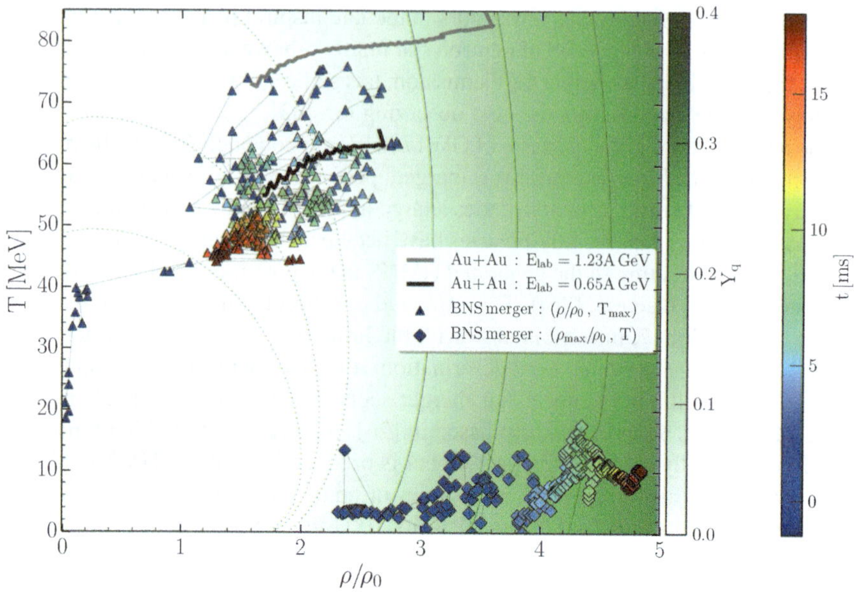

Fig. 8 Evolution of the maximum value of the temperature (triangles) and rest mass density (diamonds) at the equatorial plane in the interior of a HMNS using the simulation results of the `LS220-M135` run (for details see [26, 53]). The color coding of triangles/diamonds indicate the time of the simulation after merger in milliseconds (see colorbar at the right border of the figure). The grey and black curve shows the trajectories of two heavy ion collision simulations with energies $E_{lab} = 1.23A$ GeV and $E_{lab} = 0.65A$ GeV within the quark-hadron chiral parity-doublet model (QχP model). The colorbar right next to the picture displays the quark fraction Y_q of the corresponding hot and dense matter within the QχP model. Contour lines of the quark fraction were taken at $Y_q = [0.0001, 0.001, 0.01]$ (green dotted lines) and $Y_q = [0.1, 0.2, 0.3]$ (green solid lines)

assumption of π-symmetry (for details see [26]). Figure 8 illustrates the deficit of the present neutron star merger simulations. On the one hand it is evident from the numerical results that the density values reached during the post merger phase exceed the threshold where a purely hadronic description of the matter is meaningful. On the other hand, almost all of the used models for the EOSs do not contain a transition to a quark phase and so far no simulation of a binary compact star merger containing a strong HQPT has been performed. However, the effects of a strong HQPT have been investigated in the context of static [17, 20, 22] and uniformly rotating hybrid stars [23, 24] and the results show that tremendous changes in the star properties might occur including the existence of a third family of compact stars—the so called "twin stars" [45].

The consequences of a rearrangement of a compact star due to the quark core formation has been already discussed in [20] and it was found that a rapid transition between the twin stars will release an amount of energy of about 10^{52} ergs and that such a transition should be accompanied by the prompt neutrino burst and the delayed gamma-ray burst. A time dependent numerical simulation of such a twin star collapse (TSC) have however only been performed recently [28]. A detailed analysis of the time dependence of a TSC, followed by a twin star oscillation (TSO) will be presented within a forthcoming article.

4.1 Theoretical Background of a BNS Merger Simulation

Einstein's theory of general relativity and the resulting general relativistic conservation laws for energy-momentum in connection with the rest mass conservation (see Eq. 3) are the theoretical groundings of a BNS merger simulation. In order to solve the evolution of a merging neutron star binary system numerically, Eq. (3) needs to be rewritten, because its structure is not well posed. To reformulate Eq. (3), the so called (3 + 1)-split is used, which starts by slicing the 4-dimensional manifold $\mathcal{M}$ into 3-dimensional space-like hypersurfaces Σ_t. The space-time metric $g_{\mu\nu}$ is then sub-classified into a purely spatial metric γ_{ij}, a lapse function α and a shift vector β_i ($\mu, \nu = 0, 1, 2, 3$ and $i, j = 1, 2, 3$):

$$g_{\mu\nu} = \begin{pmatrix} -\alpha^2 + \beta_i\beta^i & \beta_i \\ \beta_i & \gamma_{ij} \end{pmatrix} . \tag{7}$$

The lapse function α describes the difference between the coordinate time t and the proper time τ of a fluid particle ($d\tau = \alpha\ dt$). The shift vector β^i measures how the coordinates are shifted on the spatial slice if the fluid particle moves an infinitesimal time step further. The ϕ-component of the shift vector β^ϕ describes the dragging of local inertial frames within the highly differentially rotating BNS merger product [26].

By inserting the metric (Eq. 7) into the Einstein equation (Eq. 3) one can restate the equations into a system of first order differential equations, the so called Arnowitt-Deser-Misner (ADM) equations. As the ADM equations are still not'well posed' they need to be further transformed using a conformal traceless formulation (for details see [43]). The numerical simulations have been performed in full general relativity using the `Einstein Toolkit` [54] where the `BSSNOK` conformal traceless formulation of the Einstein equations [55–57] using a "1 + log" slicing condition and a "Gamma-driver" shift condition [58, 59] were used. The covariant conservation of energy, momentum and rest mass, formulated within the general-relativistic hydrodynamics equations (see second and third equation in Eq. (3)), are cast in the conservative Valencia formulation [43, 60]. The evolution of these hydrodynamics equations was done using the `WhiskyTHC` code [61, 62]. A numerical grid with an adaptive mesh refinement approach based on the `Carpet` mesh-refinement driver [63] was used to both increase resolution and extend the spatial domain.

In the next section the results of a BNS merger simulation will be presented which bases on a "hot", i.e., temperature dependent, Lattimer-Swesty (LS220) EOS [64].

The total gravitational mass of the binary is $2 \times 1.350\, M_{\odot}$ and the run was evolved without the assumption of π-symmetry (for details see [26, 30, 53, 65]).

4.2 Neutron Star Mergers and the Phase Diagram of QCD

In the BNS merger simulation of Fig. 8 we had neglected the structure of the density and temperature profiles of the HMNS and have only indicated the evolution of the maximum values of the temperature and density reached inside the neutron star merger product. The maximum values of the density (diamonds) and temperature (triangles) do not coincide spatially and the distributions of the rest-mass density and temperature profiles in the interior of the HMNS have a strongly spatial and time dependent structure.

A BNS merger simulation is like a great love story and can be separated in different phases. It begins with the first face-to-face encounter of the two lovers (neutron stars), which is probable the greatest serendipity. During this inspiral phase (Viennese waltz phase) the two stars are separated by a certain distance and orbit around each other. Due to the emission of GWs their separating distance decreases with time. Among the currently known 2900 neutron stars, there are some which are in binary systems where the companion of the neutron star is either a normal star, a planet, a white dwarf or again a neutron star. One of the most impressive binary neutron star system is the so called *Double Pulsar: PSR J0737-3039A/B*, which has been discovered in 2003 [66–68]. The two neutron stars in this binary system, which are only separated by 800,000 km, orbit around each other with an orbital period of 147 min and a mean velocities of one million km/h. Additionally, each neutron star rotates around its rotational axis and the whole movement looks similar to a Viennese waltz dance, where each neutron star corresponds to an individual dancer who dances with an invisible companion. The distance between the two neutron stars decreases with time and finally the two objects need to merge and the two lovers touch each other and become a couple (see Fig. 9). In this phase, the temperature effects can be neglected in good approximation and only at the star surfaces the temperature reaches non-neglectable values. During the last orbits the stars become tidally deformed and the temperature in the low density regime, near to the region where the two NSs touch each other, increases rapidly (see Fig. 9).

At merger time ($t = 0$ ms), where the emitted gravitational wave of the newly born remnant reaches its maximum value, the temperature hot spot of the newly born remnant reaches values up to $T \approx 75$ MeV (see blue triangles in Fig. 8). The density maxima are almost at the center and the high temperature regions are placed between them (for details see [30]). The merger and the following violent, early post-merger phase (Disco-fox phase) is characterized by a pronounced density double-core structure and hot temperature regions which are smeared out in areas between the double-core density maxima (see Fig. 10). The movement of advanced Disco-fox dancers consist of two separate motions: A coming closer and a subsequently removing of the two bodies and a shared rotation with respect to the static dance

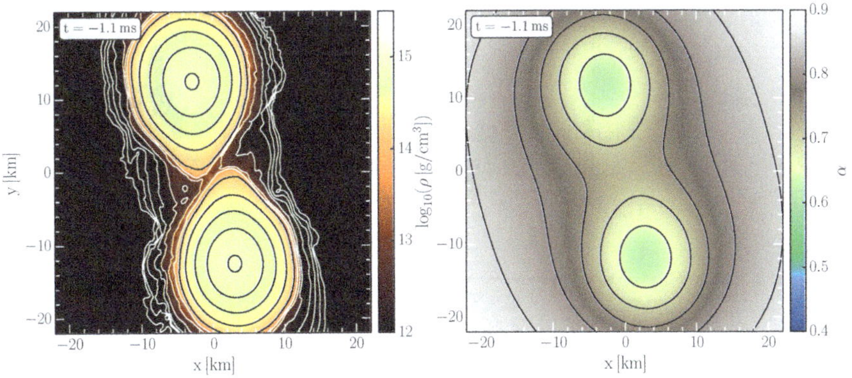

Fig. 9 Logarithm of the rest mass density $\log(\rho\ [\mathrm{g/cm^3}])$ and lapse function α in the xy-plane in the late inspiral phase ($t = -1.1$ ms) within the LS220-M135 simulation (for details see [26, 30, 53]). The black density contour lines were taken at $\rho = [0.5, 1, 1.5, ..]\,\rho_0$ while the white density contours have a logarithmic distance to indicate the low density crust matter ($\log(\rho) \in [10, 14]$ $\log([\mathrm{g/cm^3}])$). The black contour lines of the lapse function were taken at $\alpha = [0.55, 0.6, 0.65, ..]$ while the white density contour indicate $\alpha = 0.5$

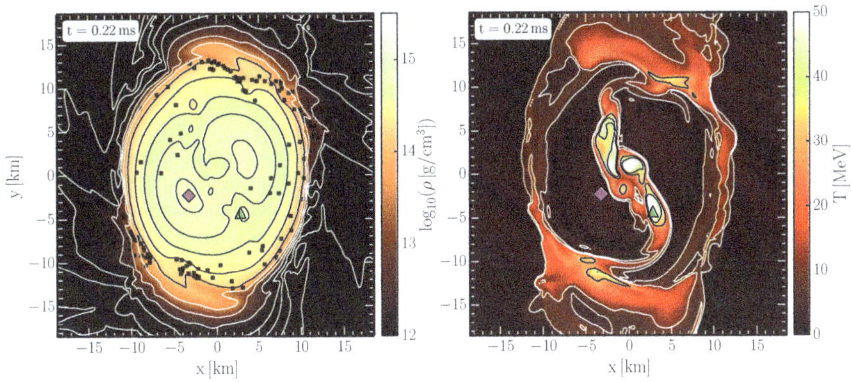

Fig. 10 Distributions of the rest-mass density (left picture) and temperature (right picture) within the Disco-fox phase at time $t = 0.22$ ms. The green triangle marks the maximum value of the temperature while the magenta diamond indicates the maximum of the density. Additionally several tracer particles that remain close to the equatorial plane are visualized with black dots

floor. The two dancers correspond now to the two, still separately visible double-core density maxima of the HMNS (see left picture in Fig. 10) and the movement of the double-core structure within the first ≈2 ms after the merger looks quite similar to a Disco-fox.

Figure 11 illustrate the spatial allocation of the density and temperature values reached in a (T-ρ/ρ_0) plot during late inspiral phase (upper row of pictures in Fig. 11) and early post-merger phase (lower picture). The color of a density/temperature point ((ρ-T)-fluid element) indicates now its radial position r measured from the

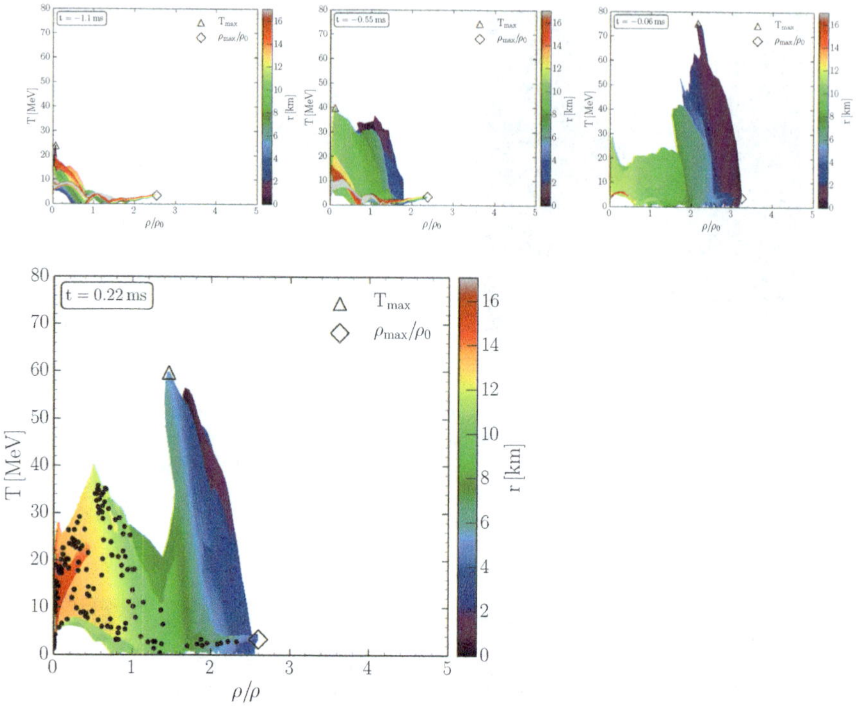

Fig. 11 Lower picture: Density-temperature profiles inside the inner region of the HMNS of the `LS220-M135` simulation in the style of a (T-ρ/ρ_0) QCD phase diagram plot at time $t = 0.22$ ms. The color-coding indicate the radial position r of the corresponding fluid element inside the HMNS. The open triangle/diamond marks the maximum value of the temperature/density. Additionally several tracer particles that remain close to the equatorial plane are visualized with black dots. The upper pictures display the same property but at three time-snapshots within the late inspiral phase ($t = -1.1, 0.55, 0.06$ ms)

origin of the simulation $(x, y) = (0, 0)$ on the equatorial plane ($z = 0$). In order to track the motion of individual fluid cells, tracer particles had been used within the `LS220-M135` simulation (for details see [53, 65]). Initially placed at $t = 0$ near to the surface of the newly born remnant the tracer particles diffuse both spatially and in the (ρ-T)-plane, however, in the following only these tracers will be visualized that stays in the inner region of the HMNS. The flowlines of these tracer particles can be visualized using the method of a "corotating frame". In a corotating frame, each grid point is rotating at a frequency that is half the angular frequency of the instantaneous emitted gravitational waves, $\Omega_{\rm GW}$ and it corresponds to the collective rotation of the whole HMNS. In a "corotating frame" the observer is transformed within the rotating frame of the dancing couple and the movement of the two density double-cores (dancers) takes place on a straight line.[4]

After the violent and transforming, early post merger phase a new phase begins (the Merengue phase). In the time interval 2 ms $< t <$ 4 ms the double-core structure disappears and the maximum density value shifts to the central region of the HMNS. The high temperature regions transform to two temperature hot spots and they move further out. The density distribution at this post merger time has been named'peanut' shape but the highest value of the density ρ is located in the center of the HMNS. In a Merengue dance the dancers are so close together that distinguishing the individuals is difficult (no double-core structure). In close embrace, the united couple quickly rotate around each other and this movement describes in a clear way the motion of the HMNS at $t >$ 4 ms. The high temperature values ($T >$ 40 MeV) are reached now in regions where the density is in a range of 1–2.5 ρ_0, while the maximum density values are always at moderate temperatures $T <$ 20 MeV. The temperature hot spots have moved further out and the interior of the HMNS, where the maximum of the density is located, has become denser. The tracer particles have diffused over the entire inner region of the HMNS and populate almost the whole area of the (T-ρ/ρ_0) plane. Some of these tracers circulate around the high temperature hot spots, others populate the low temperature high dense inner region and some are moving in the outer surface of the HMNS within the low density regime (see Fig. 12).

At later post merger times (see Figs. 13 and 14) the temperature hot spots have smeared out to become a ring like structure, the'peanut' shape has been dissolved and the area populated in the (T-ρ/ρ_0) plane has been constricted to a small quasi stable region. The central region of the HMNS consists of high dense matter ($\rho/\rho_0 \approx 5$) at moderate temperature values $T \approx$ 10 MeV while the maximum of the temperature is reached at the top of the temperature ring like structure at $r \approx$ 6 km at moderate density values ($\rho/\rho_0 \approx 2$).

The results of the `LS220-M135` simulation presented so far show that within the first 20 ms the matter inside the HMNS populate areas in the QCD phase diagram where an inclusion of the quark degrees of freedom in the EOS is necessary. Especially in the interior region of the HMNS for $t >$ 10 ms the density reaches values where a non neglectable amount of deconfined quark matter is expected to be present. Numerical-relativity simulations of merging neutron star binaries show that the emitted GW and the interior structure of the generated hypermassive neutron stars depends strongly on the equation of state and the total mass of the system. Figure 15 shows the impact of the used EOS and total mass on the populate areas in the QCD phase diagram at a post merger time of $t \approx$ 25 ms. While the maximum value of the density reached within the `LS220-M135` run exceeds values $\rho/\rho_0 = 3.5$, the density of the HMNS within the `DD2-M135` simulation never reaches $\rho/\rho_0 = 3$. By increasing the total mass of the system, the density and temperature values reached during the post merger phase also increases (see Fig. 15: `DD2-M145`).

---

[4]In a corotating frame the Diskofox looks more like a West Cost Swing dance and the merger snapshot can be regarded as the first "sugar push" of the West Cost Swing couple.

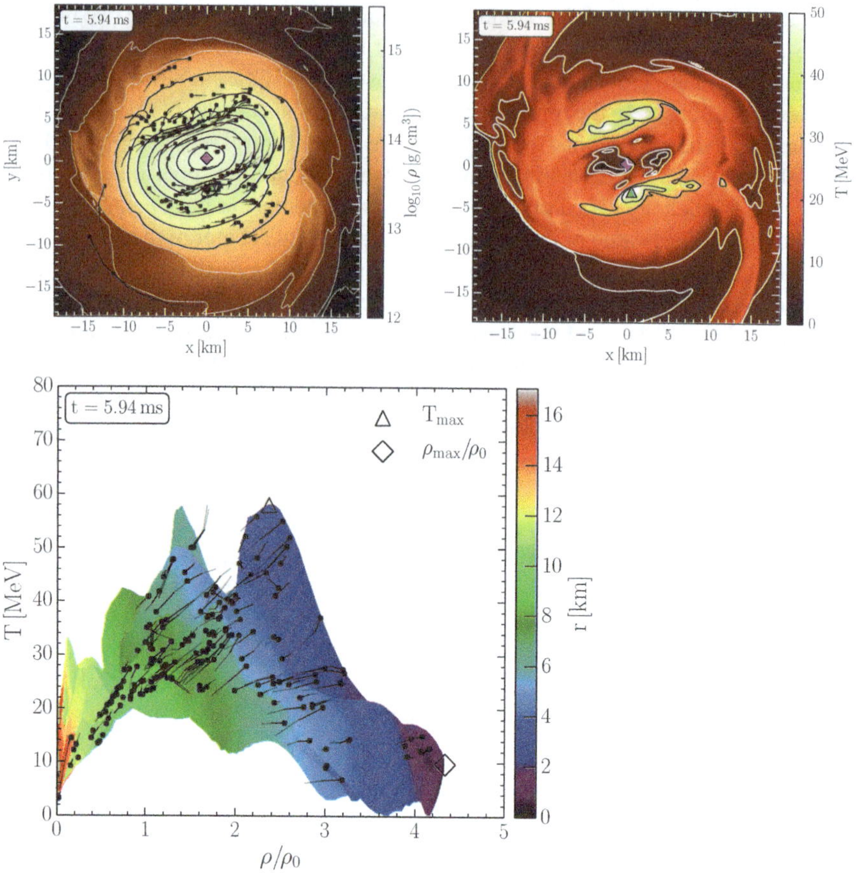

Fig. 12 Same as Fig. 10 (upper pictures) and Fig. 11 (lower picture) but at but time $t = 5.94$ ms. The final part of the tracer flowlines for the last $\Delta t = \simeq 0.19$ ms are shown and the small black dots are used to indicate the tracer position at the time indicated in the frame. The initial parts of the trajectories have increasing transparency so as to highlight the final part of the trajectories

5 Summary and Outlook

The results of the `LS220-M135` and `DD2-M135/145` simulation have based on temperature dependent, but purely hadronic EOSs. So far only two recent simulations have been performed which includes both, a temperature dependence in the EOS and an implementation of a strong HQPT [48, 69]. However, BNS merger simulation which uses for the cold part a QCD-motivated EOS with a HQPT and a separate ideal-fluid component to account for the thermal contributions have been investigated earlier (see e.g. [26]) and the implementation of a strong HQPT, including the possibility of a twin star behaviour, in a binary hybrid star merger scenario is currently under construction and the preliminary results show that the appearance

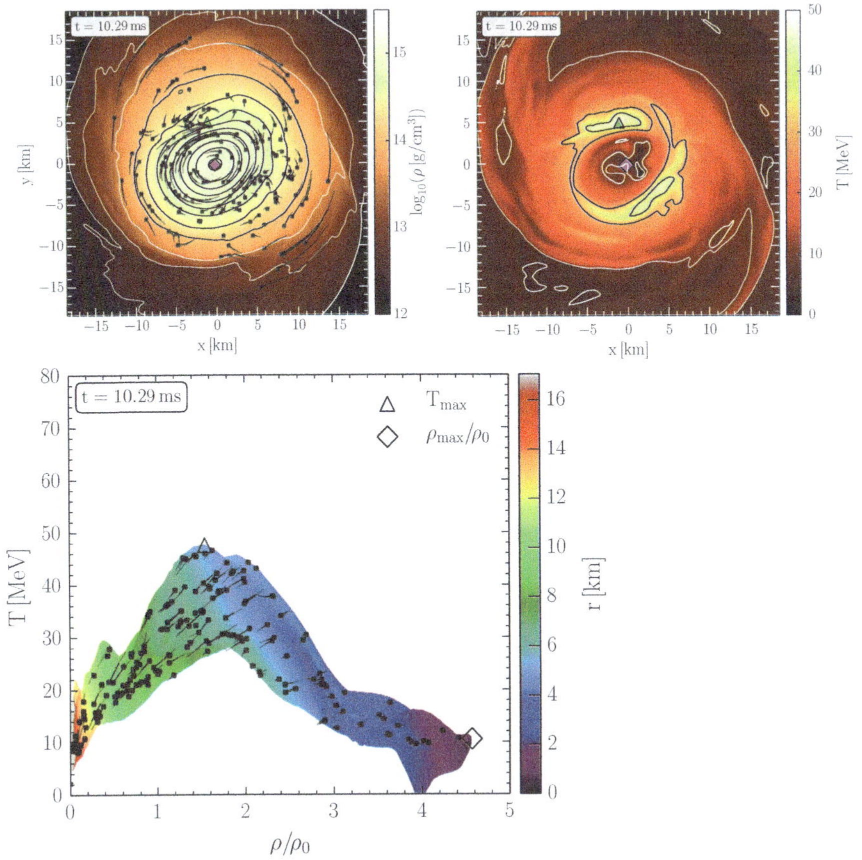

Fig. 13 Same as Fig. 12 but at but time $t = 10.29$ ms. The neutron star matter at this post-merger time populates a smaller area $\mathcal{A}$ in the $(T\text{-}\rho/\rho_0)$ plane compared to $t = 5.94$ ms. The motions of the tracer particles are restricted within the surface of $\mathcal{A}$ and at its outer borders they are reflected. The two pronounces temperature maxima at $t = 5.94$ ms have almost merged to one maximum at this post-merger time

of the hadron-quark phase transition in the interior region of the HMNS will change the spectral properties of the emitted GW if it is strong enough. If the unstable twin star region is reached during the "post-transient" phase, the f_2-frequency peak of the GW signal will change due to the speed up of the differentially rotating HMNS.

In this article, the research results I have obtained together with Walter Greiner and various scientists (calculated 15 years ago) had been combined with current simulations of BNS mergers. In the period 1998–2004 we had only dreamed of compact star merger simulations and both the sophisticated computer programs, which solves the Einstein and hydrodynamic equations in full general relativity, and the highly developed supercomputer clusters were not available at that time. I finally finished and defended my doctoral thesis in 2004 and the PhD committee (Prof. W.Greiner,

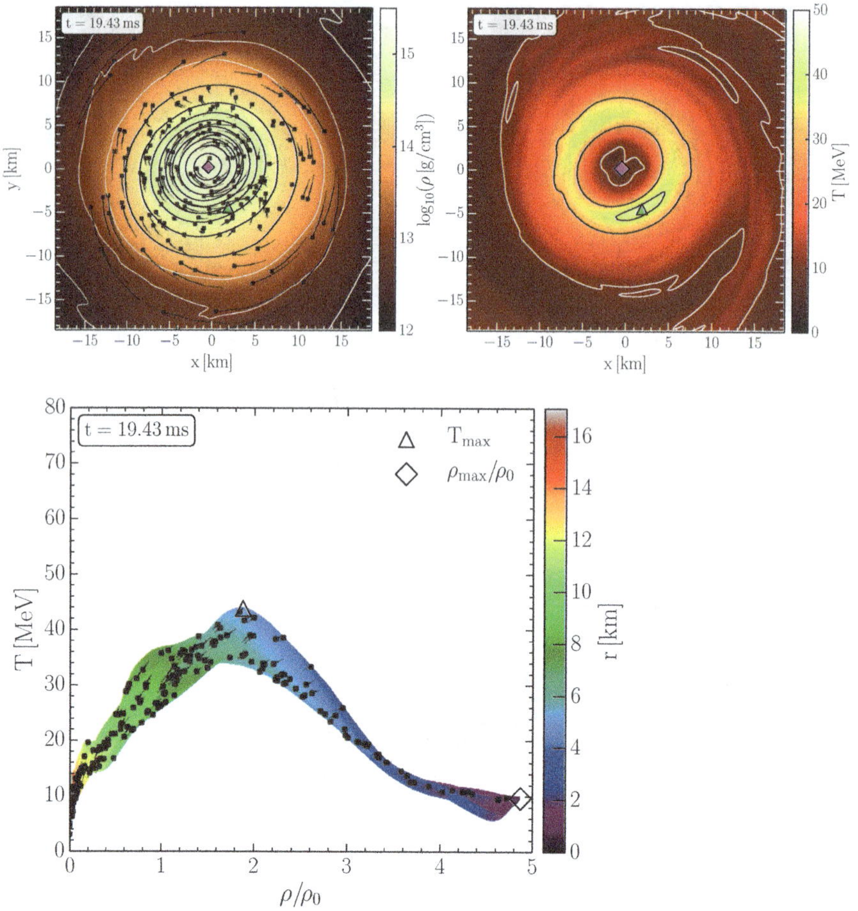

Fig. 14 Same as Fig. 12 but at but time $t = 19.43$ ms. For post-merger times $t > 15$ ms the HMNS/SMNS reaches a quasi stable configuration. The unusual temperature ring-like structure is closely related with the rotation profile Ω of the differentially rotating HMNS ($\Omega = \alpha\, v^{\phi}\ -\ \beta^{\phi}$, where v^{ϕ} and β^{ϕ} describes the ϕ-component of the three-velocity and shift vector, see [26] for details)

Prof. H. Stöcker, Prof. R. Stock and Prof. W. Martienssen) graded it with "very good". In 2004 the physics department and the newly founded Frankfurt Institute for Advanced Studies (FIAS) moved from campus Bockenheim to the campus Riedberg and I decided to turn my back on relativistic astrophysics and stayed at the Bockenheim campus working as a research assistant at the department of economics in the "Information Systems" group of Prof. W. König. However in 2006 I met Walter Greiner again on the *Eleventh Marcel Grossmann Meeting on General Relativity* in Berlin where I gave an unorthodox presentation on "Black Holes and the German Reichstag" [70] in which I illustrated that the modern glass dome of the newly con-

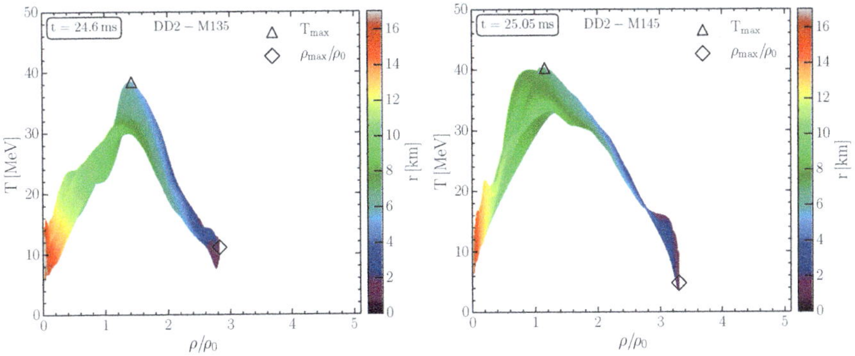

Fig. 15 (T-ρ/ρ_0) density-temperature profiles of the DD2-M135 and DD2-M140 simulation

structed core of the Reichstag building visualizes in a good way the main properties of a black hole. The event horizon of a black hole marks a certain threshold and the collapse of a neutron star to a black hole looks for an outside observer like a frozen picture of the collapsing star.[5] A HMNS will also collapse to a black hole and this last dance of the two lovers can be described by the end phase of a Tango, where the motion of the dancers is suddenly freezing and the emission of gravitational wave stops. Figure 16 summarizes the different dances of a BNS merger simulation and connects them with the emitted gravitational-wave amplitude of a BNS merger simulation (ALF2-M135 run, for details see [26]). During the collapse of the HMNS to a black hole (Tango phase) the density in the inner region increases rapidly, the quarks get free and the color charge of the deconfined pure quark phase becomes visible—however, the formation of the event horizon prevents that this new degree of color can be observed from the outside and the whole deconfined quark phase gets macroscopically confined by general relativity (see pp. 186–188 in [14]). Last but not least it should be mentioned, that all of the results presented so far where based on Einstein's theory of general relativity and alternative theories of gravity might change the whole picture of a BNS merger simulation.

In 2011 I finished my second doctoral thesis on evolutionary quantum game theory and network science of socio-economic systems [71] and since 2014 I am again employed at the ITP/FIAS working together with Prof. L. Rezzolla and Prof. H. Stöcker on the HQPT in BNS mergers. Additionally, I am a lecturer at the Goethe University (see lecture on "General relativity with the computer" [http://fias.uni-frankfurt.de/~hanauske/VARTC/index.html] and "Physics of socio-economic systems" [http://fias.uni-frankfurt.de/~hanauske/VPSOC/index.html]).

Acknowledgements We thank L. Bovard, J. Steinheimer, A. Motornenko, E. Most, J. Papenfort, S. Schramm, E. Bratkovskaya and L. Rezzolla for their scientific contributions.

[5] However, the color of this frozen picture will be infinitely red-shifted quite rapidly.

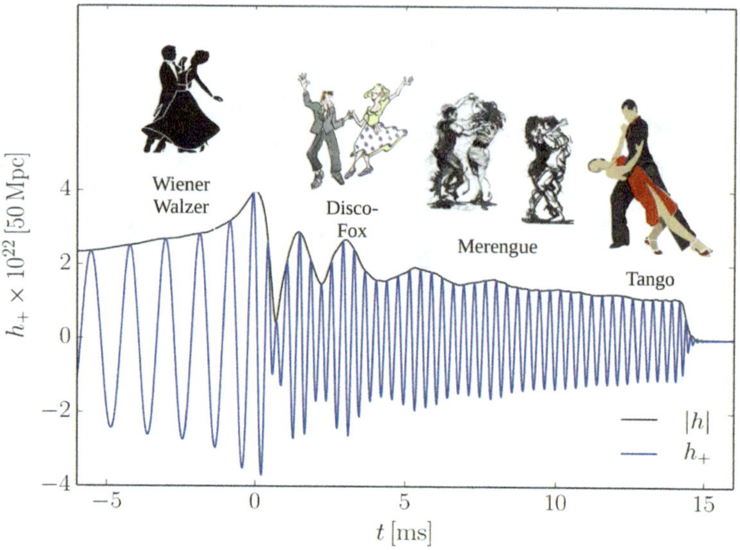

Fig. 16 Illustration of the different dances within a BNS merger simulation: gravitational-wave amplitude $|h|$ (black line) and strain amplitude in the + polarisation h_+ (blue line) for the `ALF2-M135` binary at a distance of 50 Mpc

References

1. J. Baez, J.P. Muniain, *Gauge Fields, Knots and Gravity*, vol. 4 (World Scientific Publishing Company, 1994)
2. F.W. Hehl, P. von der Heyde, G.D. Kerlick, J.M. Nester, Rev. Mod. Phys. **48**(3), 393–416 (1976), https://link.aps.org/doi/10.1103/RevModPhys.48.393
3. F.W. Hehl, Gauge theory of gravity and spacetime, in *Towards a Theory of Spacetime Theories* (Springer, 2017), pp. 145–169
4. H. Dehnen, E. Hitzer, Int. J. Theor. Phys. **33**, 575–592 (1994)
5. A. Geitner, M. Hanauske, E. Hitzer (1998), arXiv:gr-qc/9801048
6. P.O. Hess, T. Boller, *Walter Greiner Memorial Volume*, vol. 199 (2018)
7. J. Struckmeier, J. Muench, D. Vasak, J. Kirsch, M. Hanauske, H. Stoecker, Phys. Rev. D **95**, 124048 (2017)
8. J. Struckmeier, P. Liebrich, J. Muench, M. Hanauske, J. Kirsch, D. Vasak, L. Satarov, H. Stoecker (2017), arXiv:1711.10333
9. Y. Mizuno, Z. Younsi, C.M. Fromm, O. Porth, M. De Laurentis, H. Olivares, H. Falcke, M. Kramer, L. Rezzolla, Nat. Astron. **1** (2018)
10. T. Boller, P.O. Hess, *Walter Greiner Memorial Volume*, vol. 209 (2018)
11. W. Zhu, I. Stairs, P. Demorest, D.J. Nice, J. Ellis, S. Ransom, Z. Arzoumanian, K. Crowter, T. Dolch, R. Ferdman et al., Astrophys. J. **809**, 41 (2015)
12. P. Papazoglou, S. Schramm, J. Schaffner-Bielich, H. Stöcker, W. Greiner, Phys. Rev. C **57**, 2576 (1998)
13. P. Papazoglou, D. Zschiesche, S. Schramm, J. Schaffner-Bielich, H. Stöcker, W. Greiner, Phys. Rev. C **59**, 411 (1999)
14. M. Hanauske, *Properties of Compact Stars Within QCD-Motivated Models*. Ph.D. thesis J.W. Goethe-University Frankfurt, Germany (2004)

15. S. Pal, M. Hanauske, I. Zakout, H. Stöcker, W. Greiner, Phys. Rev. C **60**, 015802 (1999)
16. M. Hanauske, D. Zschiesche, S. Pal, S. Schramm, H. Stöcker, W. Greiner, Astrophys. J. **537**, 958–963 (2000)
17. M. Hanauske, W. Greiner, Gen. Relativ. Gravit. **33**, 739–755 (2001)
18. M. Hanauske, L.M. Satarov, I.N. Mishustin, H. Stöcker, W. Greiner, Phys. Rev. D **64**, 043005 (2001)
19. J. Schaffner-Bielich, M. Hanauske, H. Stöcker, W. Greiner, Phys. Rev. Lett. **89**, 171101 (2002)
20. I.N. Mishustin, M. Hanauske, A. Bhattacharyya, L.M. Satarov, H. Stöcker, W. Greiner, Phys. Lett. B **552** 1–8 (2003), arXiv:hep-ph/0210422
21. M. Hanauske, D. Zschiesche, U. Eichmann, L.M. Satarov, I.N. Mishustin, J. Schaffner-Bielich, H. Stoecker, W. Greiner, Theory of compact stars, in *XEUS—Studying the Evolution of the Hot Universe*, held at MPE Garching, ed. by G. Hasinger, Th. Boller, A.N. Parmer. MPE Report 281 (2003), p. 277, 11–13 Mar 2002
22. I. Shovkovy, M. Hanauske, M. Huang, Phys. Rev. D **67**, 103004 (2003)
23. S. Banik, M. Hanauske, D. Bandyopadhyay, W. Greiner, Phys. Rev. D **70**, 123004 (2004)
24. A. Bhattacharyya, S.K. Ghosh, M. Hanauske, S. Raha, Phys. Rev. C **71**, 048801 (2005)
25. S. Banik, M. Hanauske, D. Bandyopadhyay, J. Phys. G Nucl. Phys. **31**, S841–S848 (2005)
26. M. Hanauske, K. Takami, L. Bovard, L. Rezzolla, J.A. Font, F. Galeazzi, H. Stöcker, Phys. Rev. D **96** 043004 (2017), arXiv:1611.07152
27. M. Hanauske, J. Steinheimer, L. Bovard, A. Mukherjee, S. Schramm, K. Takami, J. Papenfort, N. Wechselberger, L. Rezzolla, H. Stöcker, Concluding remarks: connecting relativistic heavy ion collisions and neutron star mergers by the equation of state of dense Hadron- and Quark matter as signalled by gravitational waves. J. Phys. Conf. Ser. **878**(1) (article id. 012031) (2017)
28. M. Hanauske, Z.S. Yilmaz, C. Mitropoulos, L. Rezzolla, H. Stöcker, Gravitational waves from binary compact star mergers in the context of strange matter, in *17th International Conference on Strangeness in Quark Matter (SQM 2017)*, Utrecht, The Netherlands. EPJ Web of Conferences, ed. by A. Mischke, P. Kuijer, id. 20004 (2018), p. 20004
29. M.G. Alford, L. Bovard, M. Hanauske, L. Rezzolla, K. Schwenzer, Phys. Rev. Lett. **120**, 041101 (2018)
30. M. Hanauske, L. Bovard, J. Astrophys. Astron. **39**, 45 (2018)
31. M. Hanauske, L. Bovard, E. Most, J. Papenfort, J. Steinheimer, A. Motornenko, V. Vovchenko, V. Dexheimer, S. Schramm, H. Stöcker, Universe **5** (2019). ISSN 2218-1997, https://www.mdpi.com/2218-1997/5/6/156
32. M. Hanauske, J. Steinheimer, A. Motornenko, V. Vovchenko, L. Bovard, E.R. Most, L.J. Papenfort, S. Schramm, H. Stöcker, Particles **2**, 44–56 (2019). ISSN 2571-712X, https://www.mdpi.com/2571-712X/2/1/4
33. M. Hanauske, L. Bovard, J. Steinheimer, A. Motornenko, V. Vovchenko, S. Schramm, V. Dexheimer, J. Papenfort, E.R. Most, H. Stöcker, J. Phys. Conf. Ser. **1271**, 012023 (2019)
34. J. Walecka, Oxf. Stud. Nucl. Phys. **16**, 1–610 (1995)
35. Y. Nambu et al., Phys. Rev. **122**, 345 (1961)
36. I.N. Mishustin, L.M. Satarov, H. Stöcker, W. Greiner, Phys. Rev. C **59**, 3343 (1999)
37. I.N. Mishustin, L.M. Satarov, H. Stöcker, W. Greiner, Phys. Rev. C **62**, 034901 (2000)
38. J. Antoniadis, P.C.C. Freire, N. Wex, T.M. Tauris, R.S. Lynch, M.H. van Kerkwijk, M. Kramer, C. Bassa, V.S. Dhillon, T. Driebe, J.W.T. Hessels, V.M. Kaspi, V.I. Kondratiev, N. Langer, T.R. Marsh, M.A. McLaughlin, T.T. Pennucci, S.M. Ransom, I.H. Stairs, J. van Leeuwen, J.P.W Verbiest, D.G. Whelan, Science **340**, 448 (2013), arXiv:1304.6875
39. M. Hanauske, J. Steinheimer, L. Bovard, A. Mukherjee, S. Schramm, K. Takami, J. Papenfort, N. Wechselberger, L. Rezzolla, H. Stöcker, Concluding remarks: connecting relativistic heavy ion collisions and neutron star mergers by the equation of state of dense hadron-and quark matter as signalled by gravitational waves. J. Phys. Conf. Ser. **878**, 012031 (2017) (IOP Publishing)
40. N. Glendenning, *Compact Stars: Nuclear Physics, Particle Physics, and General Relativity*. Astronomy and Astrophysics Library (Springer, New York, 2000). ISBN 9780387989778, https://books.google.de/books?id=BBSOgxe2S1AC

41. F. Weber, *Pulsars as Astrophysical Laboratories for Nuclear and Particle Physics* (IoP, Bristol, 1999)
42. T. Schönenbach, G. Caspar, P.O. Hess, T. Boller, A. Müller, M. Schäfer, W. Greiner, *Experimental Tests of Pseudo-Complex General Relativity* (2016), p. 111
43. L. Rezzolla, O. Zanotti, *Relativistic Hydrodynamics* (Oxford University Press, Oxford, UK, 2013). ISBN 9780198528906
44. J.M. Bardeen, K.S. Thorne, D.W. Meltzer, Astrophys. J. **145**, 505 (1966)
45. N.K. Glendenning, C. Kettner, Astron. Astrophys. **353**, L9 (2000), arXiv:astro-ph/9807155
46. The LIGO Scientific Collaboration and the Virgo Collaboration, Phys. Rev. Lett. **116**, 061102 (2016), arXiv:1602.03837
47. M. Hanauske, *GSI Annual Report*, vol. 96 (2003)
48. A. Bauswein, N.U.F. Bastian, D.B. Blaschke, K. Chatziioannou, J.A. Clark, T. Fischer, M. Oertel (2018), arXiv:1809.01116
49. B.P. Abbott et al. (LIGO Scientific Collaboration and Virgo Collaboration), Phys. Rev. Lett. **119**(16), 161101 (2017), https://link.aps.org/doi/10.1103/PhysRevLett.119.161101
50. LIGO Scientific Collaboration, Virgo Collaboration, Gamma-Ray Burst Monitor F and INTEGRAL, Astrophys. J. Lett. **848**, L13 (2017), arXiv:1710.05834, http://stacks.iop.org/2041-8205/848/i=2/a=L13
51. The LIGO Scientific Collaboration, the Virgo Collaboration, B.P. Abbott, R. Abbott, T.D. Abbott, F. Acernese, K. Ackley, C. Adams, T. Adams, P. Addesso et al. (LIGO Scientific Collaboration and Virgo Collaboration), Astrophys. J. Lett. **848**, L12 (2017), http://stacks.iop.org/2041-8205/848/i=2/a=L12
52. The LIGO Scientific Collaboration, the Virgo Collaboration, B.P. Abbott, R. Abbott, T.D. Abbott, F. Acernese, K. Ackley, C. Adams, T. Adams, P. Addesso et al. (2017), arXiv:1710.09320
53. L. Bovard, D. Martin, F. Guercilena, A. Arcones, L. Rezzolla, O. Korobkin, Phys. Rev. D **96**, 124005 (2017), arXiv:1709.09630
54. The Einstein Toolkit Consortium, http://seinsteintoolkit.org
55. T. Nakamura, K. Oohara, Y. Kojima, Prog. Theor. Phys. Suppl. **90**, 1–218 (1987)
56. M. Shibata, T. Nakamura, Phys. Rev. D **52**, 5428–5444 (1995)
57. T.W. Baumgarte, S.L. Shapiro, Phys. Rev. D **59**, 024007 (1999), arXiv:gr-qc/9810065
58. M. Alcubierre, B. Brügmann, P. Diener, M. Koppitz, D. Pollney, E. Seidel, R. Takahashi, Phys. Rev. D **67**, 084023 (2003), arXiv:gr-qc/0206072
59. D. Pollney, C. Reisswig, L. Rezzolla, B. Szilágyi, M. Ansorg, B. Deris, P. Diener, E.N. Dorband, M. Koppitz, A. Nagar, E. Schnetter, Phys. Rev. D **76**, 124002 (2007), arXiv:0707.2559
60. J.A. Font, Living Rev. Relativ. **6**, 4 (2008), arXiv:0704.2608, http://www.livingreviews.org/lrr-2008-7
61. D. Radice, L. Rezzolla, F. Galeazzi, Mon. Not. R. Astron. Soc. L **437**, L46–L50 (2014). arXiv:1306.6052
62. D. Radice, L. Rezzolla, F. Galeazzi, Class. Quantum Gravity **31**, 075012 (2014), arXiv:1312.5004
63. E. Schnetter, S.H. Hawley, I. Hawke, Class. Quantum Gravity **21**, 1465–1488 (2004), arXiv:gr-qc/0310042
64. J.M. Lattimer, F.D. Swesty, Nucl. Phys. A **535**, 331–376 (1991)
65. L. Bovard, L. Rezzolla, Class. Quantum Gravity **34**, 215005 (2017), arXiv:1705.07882, http://stacks.iop.org/0264-9381/34/i=21/a=215005
66. A. Lyne, M. Burgay, M. Kramer, A. Possenti, R. Manchester, F. Camilo, M. McLaughlin, D. Lorimer, N. D'Amico, B. Joshi et al., Science **303**, 1153–1157 (2004)
67. M. Kramer, I. Stairs, Annu. Rev. Astron. Astrophys. **46**, 541–572 (2008)
68. M. Burgay, M. Kramer, M. McLaughlin, Bull. Astron. Soc. India **42**, 101–119 (2014)
69. E.R. Most, L.J. Papenfort, V. Dexheimer, M. Hanauske, S. Schramm, H. Stöcker, L. Rezzolla (2018), arXiv:1807.03684
70. M. Hanauske, Phys. World **18**, 64 (2005)
71. M. Hanauske, *Evolutionäre Quanten-Spieltheorie im Kontext sozio-ökonomischer Systeme*. Ph.D. thesis J.W. Goethe-University Frankfurt, Germany (2011)

Simulations of Accretions Disks at the Frequency Used of the Event Horizon Telescope

Peter O. Hess

Abstract Within the pseudo-complex General Relativity, simulations of accretion disks for SgrA* and M87 are presented at the frequency 250 GHz as used in the Event Horizon Telescope. Differences to the standard theory of General Relativity are pointed out, as the presence of a dark ring followed by a bright one near the position of the black hole.

1 Introduction

In this contribution I will resume on one of the last interests of Prof. Walter Greiner, namely on how to avoid the event horizon around a so-called black hole and what the consequences are. I consider it as a privilege having known Prof. Greiner since 1976, when he accepted me as a student with a topic on geometric models in nuclear physics. Since then he was my teacher and friend. The collaboration never stopped, though the topics changed. Here, I will talk on one of the last topics he was interested in. This was not the only one, because his interests were very broad. Unfortunately, Prof. Greiner passed away too soon!

The *General Relativity* (GR) is one of the most successful theories developed so far, it has been confirmed by many observations [1], though, only for weak gravitational fields. Recently, gravitational waves have been observed [2], whose source must be the fusion of two black holes, i.e., a system with very strong gravitational fields. However, as we will see further below, the interpretation of the source depends on the assumed theory applied. That the GR is also valid in its present form for very strong gravitational fields have to be validated yet.

In the past, several attempts have been undertaken to extend the GR algebraically. For example, in [3, 4] the metric has been changed to a complex one, where the

P. O. Hess (✉)
Instituto de Ciencias Nucleares, UNAM, Circuito Exterior, C.U., A.P. 70-543, 04510 Mexico City, Mexico
e-mail: hess@nucleares.unam.mx

Frankfurt Institute for Advanced Studies, Johann Wolfgang Goethe Universität, Ruth-Moufang-Str. 1, 60438 Frankfurt am Main, Germany

J. Kirsch et al. (eds.), *Discoveries at the Frontiers of Science*,
FIAS Interdisciplinary Science Series, https://doi.org/10.1007/978-3-030-34234-0_11

real part corresponds to the standard metric and the imaginary one is associated to the electromagnetic field tensor. M. Born proposed another extension of the length element squared [5, 6], including the contribution of the linear momentum in order to achieve a symmetry between coordinates and momenta. The momentum part, however, depends on the mass of the particle under consideration. In [7] this was resolved, substituting the momentum part by a dependence in the four velocity components. Due to dimensional reasons, a minimal length appears. This length element can be recovered, using a complex extension of the coordinates, as proposed in [8, 9].

In [10] all possible algebraic extensions were studied and the properties of the theory for weak gravitational fields were analyzed. The important result is that only two kinds of algebraic versions are allowed, namely real and pseudo-complex coordinates (in [10] another name was used). Only in these cases no ghost solutions and/or tachyons appear.

In [11] a first attempt was published, using pseudo-complex coordinates. The final version was discussed in [12, 13]. While in [11] a modified variational principle was used, in [11] it is shown that the same can be achieved with a standard variational principle plus a constraint, namely that the length element squared has to be real. For more details, please consult the book [14] on the pseudo-complex General Relativity (pcGR).

In what follows, I will resume the main properties of the pseudo-complex theory and afterward some applications will be discussed.

2 The Pseudo-complex General Relativity

In an algebraic extension of GR the real coordinates are extended to a different type, namely to be complex, pseudo-complex (pc) (other names for that also exist in literature, as hyperbolic, hypercomplex, para-complex, and more) quaternions, etc. In [10] it was shown that only the pc-coordinates make sense, they do not lead to ghost and/or tachyon solutions in the limit of weak gravitational fields.

Pc-variables have the structure

$$X^\mu = x^\mu + I y^\mu \quad , \tag{1}$$

with $I^2 = +1$ and where x^μ is the standard coordinate in space-time and y^μ the pseudo-complex component.

In what follows, some elementary properties of pseudo-complex variables will be resumed:

- Instead of the division in a pseudo-real and a pseudo-complex component, there is an alternative form

$$X^\mu = X^\mu_+ \sigma_+ + X^\mu_- \sigma_-$$
$$\sigma_\pm = \frac{1}{2}(1 \pm I) \quad . \tag{2}$$

- The $\sigma_\pm$ satisfy the relations

$$\sigma_\pm^2 = \sigma_\pm \; , \; \sigma_+ \sigma_- = 0 \quad , \tag{3}$$

  which is called the *zero-divisor basis*.
- The last relation in (3), states that an element proportional to σ_+ multiplied by one proportional to σ_- gives zero, i.e., there is a *zero-divisor* and the pc-variables do not form a field but rather a ring.
- In both zero-divisor component ($\sigma_\pm$) the analysis is very similar to the standard complex analysis.

The metric is also pseudo-complex, namely

$$g_{\mu\nu} = g^+_{\mu\nu} \sigma_+ + g^-_{\mu\nu} \sigma_- \quad . \tag{4}$$

Because of the existence of the zero-divisor, a GR can be constructed independently in the component of σ_+ and σ_-.

This results in the problem on how to connect both components and get real answers.

Here, we will resume the latest formulation, published in [13, 14], using a standard variational principle through the implementation of the constrained of a real line element. We will not use the modified variational principle as in [11, 12, 14], though in [14] the standard variational principle was partially already discussed as an alternative.

The infinitesimal pc length element squared is given by

$$d\omega^2 = g_{\mu\nu} dX^\mu dX^\nu \; = \; g^+_{\mu\nu} dX^\mu_+ dX^\nu_+ \sigma_+ + g^-_{\mu\nu} dX^\mu_- dX^\nu_- \sigma_- \tag{5}$$

written in the zero-divisor components.

The connection of the two zero-divisor components is achieved, demanding that the infinitesimal length element squared in *is real*, i.e., in the zero divisor components

$$(\sigma_+ - \sigma_-) \left(g^+_{\mu\nu} dX^\mu_+ dX^\nu_+ - g^-_{\mu\nu} dX^\mu_- dX^\nu_- \right) = 0 \quad . \tag{6}$$

The $X^\mu_\pm$ denote the local pc-position of a fluid element, which are fixed but arbitrary. The 4-velocity components are defined by $u^\mu = \frac{dx^\mu}{ds}$.

The action proposed is

$$S = \int dx^4 \sqrt{-g} \, (\mathcal{R} + 2\alpha) \quad , \tag{7}$$

where $\mathscr{R}$ is the pc-Riemann scalar. The last term in the action integral allows to introduce the cosmological constant in cosmological models, where α has to be constant in order not to violate the Lorentz symmetry. This, however changes, when a system with a uniquely defined center is considered, which has spherical (Schwarzschild) or axial (Kerr) symmetry. In these cases, the α is allowed to be a function in r, for the Schwarzschild solution, and a function in r and ϑ, for the Kerr solution.

The variation of the action with respect to the metric is $\frac{\delta S}{\delta g_{\mu\nu}}\delta g_{\mu\nu} = \frac{\delta S_+}{\delta g^+_{\mu\nu}}\delta g^+_{\mu\nu}\sigma_+ + \frac{\delta S_-}{\delta g^-_{\mu\nu}}\delta g^-_{\mu\nu}\sigma_- = 0$, which leads to the equations of motion

$$
\begin{aligned}
\mathscr{R}^{\pm}_{\mu\nu} - \frac{1}{2}g^{\pm}_{\mu\nu}\mathscr{R}_{\pm} &= \lambda\left(\dot{X}^{\pm}_{\mu}\dot{X}^{\pm}_{\nu}\right) + \alpha g^{\pm}_{\mu\nu} \\
&= \lambda u_\mu u_\nu + \lambda\left(\dot{y}_\mu\dot{y}_\nu \pm u_\mu\dot{y}_\nu \pm u_\nu\dot{y}_\mu\right) + \alpha g^{\pm}_{\mu\nu} \\
&= 8\pi T^{\Lambda}_{\pm\,\mu\nu} \quad ,
\end{aligned} \tag{8}
$$

where the derivative of the coordinate ($X^{\mu}_{\pm}$) with respect to time (or the length parameter S) was expressed in terms of the pseudo-real and pseudo-imaginary component of the $X^{\mu}_{\pm}$. The right hand side of the Einstein equations is identified with an energy-momentum tensor, using the appropriate expression of λ [14] (also, see further below).

When only the pseudo-real part is taken, the right hand side of (8) can be rewritten in terms of an energy-momentum tensor of an ideal anisotropic fluid: Renaming $\lambda = 8\pi\tilde{\lambda}$, $\alpha = 8\pi\tilde{\alpha}$, $\tilde{\lambda} = \left(p^{\Lambda}_{\vartheta} + \rho_{\Lambda}\right)$, $\tilde{\alpha} = p^{\Lambda}_{\vartheta}$ and $\tilde{\lambda}y_\mu y_\nu = \left(p^{\Lambda}_r - p^{\Lambda}_{\vartheta}\right)k_\mu k_\nu$, the real part of the energy momentum tensor acquires the structure

$$
T^{\Lambda}_{\mu\nu,R} = (\rho_\Lambda + p^{\Lambda}_{\vartheta})u_\mu u_\nu + p^{\Lambda}_{\vartheta}g_{\mu\nu} + \left(p^{\Lambda}_r - p^{\Lambda}_{\vartheta}\right)k_\mu k_\nu \quad , \tag{9}
$$

where p^{Λ}_{ϑ} and p^{Λ}_r are the tangential and radial pressure respectively. Within pcGR, the fluid is anisotropic due to the presence of y_μ.

Up to now, the path followed mainly mathematical arguments. The theory, as it is, cannot tell the origin of the energy-momentum tensor, only that it has to be there and that it has the property af a dark energy. A fully quantized GR should provide the answer. In the absence of a such theory, a phenomenological approach is an alternative. The pc-GR is much more general because it contains the information on a minimal length [14], but for practical reasons we restrict to the pseudo-real part of this theory.

A hint to the possible origin of the dark energy is given in [15], where semi-classical quantum mechanical calculations were performed [16] (Casimir effect). Due to the presence of a curved back-ground, vacuum fluctuations appear which for the central mass increase toward the center. In [15] These fluctuations explode at the Schwarzschild radius. However, at the Schwarzschild radius the method is not valid any more, because the gravitational field strength is too large. This result provides evidence that the density of the dark energy has to increase significantly toward the center

A general conjecture emerges: *mass not only curves the space but also changes the space- (vacuum-) properties, which in turn influences the metric*. As a consequence *the event horizon may disappears*!

In our phenomenological approach, we treat according to the above finding the vacuum fluctuations as a classical ideal anisotropic fluid, thus we are free to propose a different fall-off of the negative energy density, which is finite at the Schwarzschild radius.

For the dark energy density we use

$$\rho_\Lambda = \frac{B}{8\pi r^5} \quad , \tag{10}$$

which is strong enough in order not to contribute to the known observations within the solar system and others with not too strong gravitational fields [1].

With the assumed density, the metric for the Kerr solution changes to [17]

$$\begin{aligned}
g_{00} &= -\frac{r^2 - 2mr + a^2\cos^2\vartheta + \frac{B}{2r}}{r^2 + a^2\cos^2\vartheta} \quad , \\
g_{11} &= \frac{r^2 + a^2\cos^2\vartheta}{r^2 - 2mr + a^2 + \frac{B}{2r}} \quad , \\
g_{22} &= r^2 + a^2\cos^2\vartheta \quad , \\
g_{33} &= (r^2 + a^2)\sin^2\vartheta + \frac{a^2\sin^4\vartheta\left(2mr - \frac{B}{2r}\right)}{r^2 + a^2\cos^2\vartheta} \quad , \\
g_{03} &= \frac{-a\sin^2\vartheta\, 2mr + a\frac{B}{2r}\sin^2\vartheta}{r^2 + a^2\cos^2\vartheta} \quad ,
\end{aligned} \tag{11}$$

where $0m \leq a \leq 1m$ is the spin parameter of the Kerr solution, in units of m. The Schwarzschild solution is obtained, setting $a = 0$. The parameter $B = bm^3$ measures the coupling of the dark energy to the central mass, where the value of b has to be fixed and is considered as a universal parameter.

3 Simulation of Accretions Disks: A Very Simple Model

First, some basic properties and differences of pcGR to GR are pointed out, for the case of a central mass:

- The orbital frequency in pcGR is always lower to the one in GR [12].
- The orbital frequency in pcGR shows a maximum at $r = 1.72\,\mathrm{m}$ (independent on the rotational parameter a), after which it falls off to zero at two-thirds of the Schwarzschild radius, where the surface is estimated to be. This feature is independent on a. Because orbits near that maximum have very similar orbital frequencies, friction will be low and the emission of light should show a minimum,

expresses as a dark ring. Further in, the differences in the orbital frequency are large again, increasing the friction and a bright ring should occur.

- Also in pcGR last stable orbits exists, though the structure is more complicated: For $a < 0.5$ m (approximately) the last stable orbit in pcGR follows the one of GR, but further in. Thus, the particles will release more energy, which is distributed in the disk and therefore the disk will appear brighter. For the values of a mentioned, the orbits do not reach the maximum of the orbital frequency, thus no dark ring followed by a bright one can occur. For $a > 0.5$ m (approximately) all orbits are stable and the maximum in the orbital frequency are reached. As a consequence, pictures of an accretion disk will show a dark ring followed by a bright one further in. This is a robust feature and should show up in any disk model, more involved than the one we use.
- The dark disk is smaller in pcGR than in GR.

For the accretion disk we use the model from [18]. It assumes a thin disk which is optically thick at the equator ($\theta = \frac{\pi}{2}$). The loss of energy is through the emission of light. The beauty of this model is in the use of conservation laws, as the conservation of mass, energy and angular momentum. Details, as the viscosity and accretion rate are hidden within the parameters of the model. The formulation of the theory is independent on the metric and, thus, can also be applied to pcGR. For the simulation we use the GYOTO routine [19], which applies the raytracing method and allows the use an arbitrary metric, once provided by the user (see [12]). Though the model is quite simple, it will already expose the difference between GR and pcGR, which is similar for all more sophisticated models,

There exist many other more sophisticated models for accretion disks (see as an example [20]) but in all these models, the basic differences between GR and pcGR, which we will discuss in this contribution, are equivalent.

Under the hypothesis of a minimal value for B, simulations of accretion disks have been performed in [12]. We did calculations at the frequency of 250 GHz [21]. Also the field of view was chosen arbitrarily and the temperature of the disk near the inner edge has not been evaluated before.

The result of the simulations depends on several not well determined parameters, as the accretion rate, the rotational parameter a, the field of view and the mass of the black hole. Thus, the specific intensities contain many uncertainties. Nevertheless, the important part is the *relative comparison* of pcGR to GR. In pcGR the intensities are much larger than in GR and this alone is a robust result and should be observable.

A further unknown parameter is the inclination angle of an accretion disk with respect to the observer. In order to cover a wide range, we performed calculation from 10° to 80° in steps of ten degrees. We present only a very restricted selection. The corresponding fits-files for all orientations, within GR and pc-GR, can be send on request.

The main parameters for Sgr A* are its mass and the accretion rate. The mass of the black hole is known to be about four million solar masses [22]. and an upper limit of the accretion rate is estimated in [23] with the value of 8×10^{-5} solar masses per year. Under the supposition that an accretion disk forms around SgrA*, the same value for the accretion rate will be used. Especially, for SgrA* there is no proof for the existence of an accretion disk [24], where individual flares were reported, consistent with time-separated events.

We assume a spatial resolution of 0.1 microarcsec over a field of 1000 pixels, resulting into a field of view of 100 μ as.

The simulations provide figures for the *specific intensity* at the fixed, above mentioned frequency. The unit for the specific intensity is erg $cm^{-2}Hz^{-1}sr^{-1}$ [25].

The version of the raytracing routine, provided by Vincent et al. [19], developed numerical instabilities for very large distances. Because the specific intensity is independent of the distance, the simulations were performed at a distance of 1000 m, with m in units of cm. The field of view was correspondingly adjusted, which for SgrA* at a distance of 1000 m is about 0.022 rad.

Figures 1 and 2 show the simulation for a disk at 80° inclination angle for GR and pcGR, respectively. As a rotational parameter we use $a = 0.9$ m, taking into account that according to [26] this parameter has to be larger than 0.5 m. This value is above 0.4 m, i.e., stable circular orbits exist up to the surface and, thus, the appearance of the accretion disk will be very similar for all cases of $a > 0.5$ m to what is seen in Figs. 1 and 2. Note, that the intensity in pcGR is significantly larger than in GR. Also, at a distance (about 1.72 m) a dark ring occurs, which is the consequence of the maximum of the orbital frequency of a particle in a circular orbit [12]. Near this distance, friction is low (and therefore viscosity) and the emission of light reaches a minimum (see also Sect. 2). This is also noted in Fig. 3, where the intensity is plotted versus the pixel number in y (in each direction of the detector we have 1000 pixels and the curve is taken along the y-column at x = 500 pixels). The intensity reaches

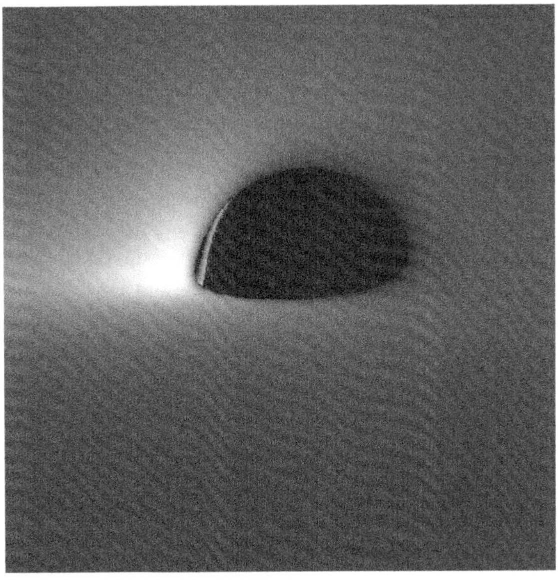

Fig. 1 Infinite, counter clockwise rotating geometrically thin accretion disk around the static and rotating compact object SgrA* viewed from an inclination of 80°. The figure shows the original disk model by [18], for GR and at $a = 0.9$ m. The specific intensity is calculated at 250 GHz and normalized to the brightest spot in pcGR for $a = 0.9$ m (see Fig. 2). A remnant of the first Einstein ring is seen in the left part of the dark area

Fig. 2 Infinite, counter clockwise rotating geometrically thin accretion disk around the static and rotating compact object SgrA* viewed from an inclination of 80°. The figure shows the original disk model by [18], for the pcGR and at $a = 0.9\,\mathrm{m}$. The specific intensity is calculated at 250 GHz. Note the appearance of a dark ring, followed further in by a bright ring

a minimum at the distance in question and for lower radial distances the curve raises again, which corresponds to the bright ring shown in Fig. 2. In GR this behavior is not seen. Also, the dark area is larger in GR compared to pcGR which is a consequence that the surface of the star in GR is at smaller radial distances than the event horizon in GR.

In case no accretion disk exists, the model can still be applied but using cuts as in Fig. 3. An approaching cloud should follow this emission profile as a function in the radial distance to the massive object. The left panel depicts the intensity profile when the cloud approaches the observer, while the right panel is for receding from the observer.

We also estimated the temperature in the disk, performing calculations in steps of 0.5 in $log(\nu)$, where ν is the frequency in Hz. The starting value is at $log(\nu) = 0.5$, which corresponds to approximately 3.16 Hz, the next one is $log(\nu) = 1$, corresponding to 10 Hz, etc. In total we take 40 values such that the last frequency considered is 10^{20} Hz, i.e., $log(\nu) = 20$. Taking the brightest spot near the inner edge of the accretion disk (before the dark ring) and using Wien's displacement law ($\mathrm{T} = 1.702 \times 10^{-11}\nu_{\mathrm{max}}$, where ν_{max} is the frequency at the maximum in units of Hz), the temperature can be estimated to be $\mathrm{T} = 1.702 \times 10^{-11}\nu_{\mathrm{max}}$. For $a = 0\,\mathrm{m}$ we obtained in GR and pcGR an estimated temperature of approximately 50,000 °K

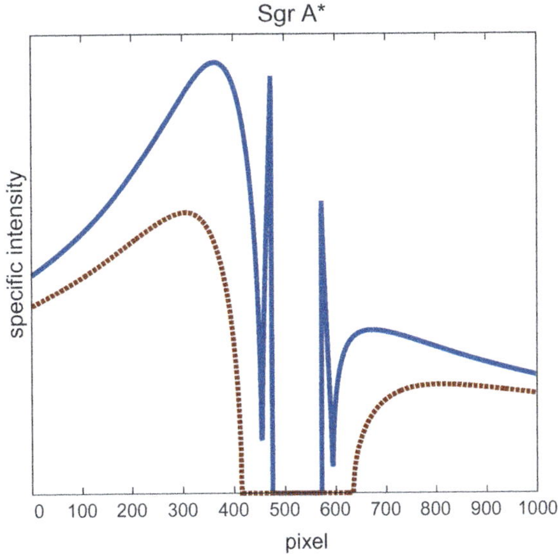

Fig. 3 Comparison between GR (color red (online) and dashed) and pcGR (color blue (online) and solid) of the intensities along the column at 500 pixels, for SgrA*. Scales are not shown but the upper line corresponds to the numerical value 3.3×10^{-12} (liner scale). The figure shows the result for the rotation parameter is $a = 0.9\,\mathrm{m}$. The inclination angle is 80°. A dark ring is seen as zero intensity, while the bright ring is presented by a large peak further in. For $a = 0.9\,\mathrm{m}$ the GR does not show peaks because a possible Einstein ring is hidden behind the disk. Note also the size of the central black disk, which is smaller in pcGR. The line for the inner bright ring goes sharply to zero, approaching the surface of the star

($\log(\nu) \approx 4.7$), while for $a = 0.9\,\mathrm{m}$ is was around $1.7 \times 10^5\,\mathrm{K}$ ($\log(\nu) \approx 5.2$). This is in agreement with a study published in [27], where the temperature for the Schwarzschild case ($a = 0\,\mathrm{m}$) ranges from $log(\nu)$ of 4.4–5.2 and in the case of a Kerr black from the range is 5–5.7.

4 Conclusions

In this contribution we resumed the main properties of the *pseudo-complex General Relativity* (pcGR). As an application the theory was applied to a disk model where robust differences between GR and pcGR should be observable, as the appearance of a dirk finge followed by a bright ring further in and the overall brightness in pcGR is significantly larger than in GR.

There are more results, presented in the talk, as the interpretation of the source of the gravitational waves observed end of 2015. While in GR the source is situated at a distance of approximately 1.3 billion light years and includes two black holes of the size of 30 solar masses each, in pcGR the source is at the remote past of the universe and correspond to the merger of two black holes of several thousand masses [28]. This dispute can only be resolved when in coincidence a bright is observed, which allows to deduce the distance.

Acknowledgements P.O.H. acknowledges financial support from DGAPA (IN100418) and CONA-CyT (No. 251817).

References

1. C.M. Will, Living Rev. Relativ. **9**, 3 (2006)
2. B.P. Abbott et al., LIGO Scientific Collaboration and Virgo Collaboration. Phys. Rev. Lett. **116**, 061102 (2016)
3. A. Einstein, Ann. Math. **46**, 578 (1945)
4. A. Einstein, Rev. Mod. Phys. **20**, 35 (1948)
5. M. Born, Proc. R. Soc. A **165**, 291 (1938)
6. M. Born, Rev. Mod. Phys. **21**, 463 (1949)
7. E.R. Caianiello, Nuovo Cim. Lett. **32**, 65 (1981)
8. C. Mantz, T. Prokopec (2008), arXiv:gr-qc—0804.0213v1
9. C. Mantz, T. Prokopec, Found. Phys. **41**, 1597 (2011)
10. P.F. Kelly, R.B. Mann, Class. Quantum Gravity **3**, 705 (1986)
11. P.O. Hess, W. Greiner, Int. J. Mod. Phys. E **18**, 51 (2009)
12. T. Schönenbach, G. Caspar, P.O. Hess, T. Boller, A. Müller, M. Schäfer, W. Greiner, MNRAS **442**, 121 (2014)
13. P.O. Hess, W. Greiner, *Memorial 100 years of GR* (World Scientific, 2017)
14. P.O. Hess, M. Schäfer, W. Greiner, *Pseudo-Complex General Rleativity* (Springer, Heidelberg, 2015)
15. M. Visser, Phys. Rev. D **54**, 5116 (1996)
16. N.D. Birrell, P.C.W. Davies, *Quantum Fields in Curved Space* (Cambridge University Press, Cambridge, 1994)
17. T. Schönenbach, G. Caspar, P.O. Hess, T. Boller, A. Müller, M. Schäfer, W. Greiner, MNRAS **430**, 2999 (2013)
18. D.N. Page, K.S. Thorne, ApJ **191**, 499 (1974)
19. F.H. Vincent, T. Paumard, E. Gourgoulhon, G. Perrin, Class. Quantum Gravity **28**, 225011 (2011)
20. W. Kluzniak, S. Rappaport, ApJ **671**, 1990 (2007)
21. Event horizon telescope (2016), http://www.eventhorizontelescope.org
22. R. Schödel, T. Ott, R. Genzel et al., Nature **419**, 694 (2002)
23. E. Quataert, R. Narayan, M.J. Reid, Astrophys. J. **517**, L101 (1999)
24. G. Ponti, E. George, S. Scaringi et al. (2017). Sent to MNRAS
25. GYOTO manual (2015), http://www.gyoto.obspm.fr/GyotoManual.pdf
26. R. Genzel, R. Schödel, T. Ott et al., Nature **425**, 934G (2003)
27. E.W. Bonning, L. Cheng, G.A.S. Shields, S. Salvander, K. Gebhard, ApJ **659**, 211 (2007)
28. P.O. Hess, MNRAS **462**, 3026 (2016)

Generic Theory of Geometrodynamics from Noether's Theorem for the Diff(M) Symmetry Group

Jürgen Struckmeier, David Vasak and Johannes Kirsch

Abstract We work out the most general theory for the interaction of spacetime geometry and matter fields—commonly referred to as geometrodynamics—for spin-0 and spin-1 particles. Actually, we present a Hamilton–Lagrange–Noether formulation of the gauge theory of gravitation. It is based on the minimum set of postulates to be introduced, namely (i) the action principle and (ii) the form-invariance of the action under the (local) diffeomorphism group. The second postulate thus implements the *Principle of General Relativity*, also referred to as the *Principle of General Covariance*. According to Noether's theorem, this *physical symmetry* gives rise to a conserved Noether current, from which the complete set of theories compatible with both postulates can be deduced. This finally results in a new generic Einstein-type equation, which can be interpreted as an energy-momentum balance equation emerging from the Lagrangian $\mathscr{L}_R$ for the source-free dynamics of gravitation and the energy-momentum tensor of the source system $\mathscr{L}_0$. Provided that the system has no other symmetries—such as SU(N)—the *canonical* energy-momentum tensor turns out to be the correct source term of gravitation. For the case of massive spin-1 particles, this entails an increased weighting of the kinetic energy over the mass

Dedicated to the memory of Prof. Dr. Walter Greiner, our teacher, mentor, and friend.
In this contribution, we present the canonical transformation formalism in the realm of classical field theory, where spacetime is treated as a dynamical quantity, and apply it to formulate the gauge theory of gravitation. In this respect, it generalizes the Extended Hamilton–Lagrange–Jacobi formalism of relativistic point dynamics. Walter was very much interested in this formalism and therefore added several chapters on the matter to the second edition of his textbook "Classical Mechanics" [1]. To quote Walter from the Preface to the Second Edition: "It may come as a surprise that even for the time-honored subject such as Classical Mechanics in the formulation of Lagrange and Hamilton, new aspects may emerge." We are sure, Walter would have loved the following elaboration.

J. Struckmeier (✉) · D. Vasak · J. Kirsch
Frankfurt Institute for Advanced Studies (FIAS), Ruth-Moufang-Strasse 1, 60438 Frankfurt am Main, Germany
e-mail: struckmeier@fias.uni-frankfurt.de

J. Struckmeier
Goethe-Universität, Max-von-Laue-Strasse 1, 60438 Frankfurt am Main, Germany

GSI Helmholtzzentrum für Schwerionenforschung GmbH, Planckstrasse 1, 64291 Darmstadt, Germany

J. Kirsch et al. (eds.), *Discoveries at the Frontiers of Science*,
FIAS Interdisciplinary Science Series, https://doi.org/10.1007/978-3-030-34234-0_12

as the source of gravity, compared to the *metric* energy momentum tensor, which constitutes the source of gravity in Einstein's General Relativity. We furthermore confirm that a massive vector field necessarily acts as a source for torsion of spacetime. From the viewpoint of our generic Einstein-type equation, Einstein's General Relativity constitutes the particular case for scalar and massless vector particle fields, and the Hilbert Lagrangian $\mathscr{L}_{R,\mathrm{H}}$ as the model for the source-free dynamics of gravitation.

1 Introduction

The covariant Hamiltonian formalism in the realm of classical field theories was recently shown to allow generalized canonical transformations which also include arbitrary active diffeomorphisms within the underlying spacetime manifold M [2]. With this framework at hand, it is now possible to isolate the complete set of theories which are based on the action principle and the Principle of General Relativity, i.e., the condition for a system to be form-invariant under the diffeomorphism group. This approach naturally leads to a Palatini formulation, where the metric and the affine connection *a priori* represent independent dynamical quantities. Systems complying with these principles are thus required to have the $\mathrm{Diff}(M)$ group as an intrinsic symmetry group. As known from Noether's theorem, each symmetry is associated with a pertaining conserved Noether current j_{N}^{μ}. The main favor of the Noether approach is that the obtained condition for a conserved Noether current directly leads to the respective field equations, which describe the *coupling* of metric and connection to the given source fields of gravitation.

With our actual Noether approach, it is possible to derive the most general form of an Einstein-type equation for a closed system of scalar and vector fields in dynamical spacetime satisfying the Principle of General Relativity—including a spacetime with torsion and without the restriction to a covariantly conserved metric. We thereby isolate the complete set of possible theories of geometrodynamics for scalar and vector matter and derive a new form of a generic Einstein-type equation.

Similar to all gauge theories, the Noether approach to geometrodynamics provides the *coupling* of the fields of the given system to the spacetime geometry, but does *not* fix the Hamiltonian resp. the Lagrangian $\mathscr{L}_R$ describing the dynamics of the "free" (uncoupled) gravitational field, hence the gravitational field dynamics in classical vacuum. The Hilbert Lagrangian $\mathscr{L}_{R,\mathrm{H}}$—which entails the Einstein tensor of standard General Relativity—is the simplest example. Based on analogy with other classical field theories, Einstein himself already proposed a Lagrangian $\mathscr{L}_R$ quadratic in the Riemann tensor [3], which will be discussed here as an amendment to the conventional Hilbert Lagrangian $\mathscr{L}_{R,\mathrm{H}}$. Remarkably, the field equation emerging from this Lagrangian is equally satisfied by the Schwarzschild and even the Kerr metric in the case of classical vacuum [4].

The source term of gravity is shown to be given by the *canonical* energy-momentum tensor, provided that the given system has no additional symmetries—

such as a $SU(N)$ symmetry—besides the $\mathrm{Diff}(M)$ symmetry. This entails an increased weight of the kinetic energy over the mass in their roles as sources of gravity. Also, a massive vector field is shown to *necessarily* induce a torsion of spacetime—which is in perfect agreement with previous works of Hehl et al. (see, for instance, [5]).

We review in Sect. 2 the formalism of canonical transformations in the covariant Hamiltonian description of classical field theories. After having formulated Noether's theorem [6] in the realm of covariant Hamiltonian field theory in Sect. 3, we proceed in Sect. 4 with the canonical transformation representation of finite $\mathrm{Diff}(M)$ symmetry transformations.

In order to work out the conserved Noether current for the $\mathrm{Diff}(M)$ symmetry transformation, the finite transformation is reformulated in Sect. 5 as the pertaining infinitesimal transformation. The detailed discussion of the conserved Noether current then follows in Sect. 8. It will be shown that the *zero-energy principle*—hence a vanishing energy-momentum tensor of the total system of source fields and dynamic spacetime—emerges as a direct consequence.

The most general Einstein-type equation of geometrodynamics is presented in Sect. 9. This equation is shown to be equivalent to the "consistency equation" of Ref. [2] by means of an *identity*, which holds for scalar-valued functions of arbitrary tensors and the metric. This identity also provides the correlation of the metric (Hilbert) and the canonical energy-momentum tensors of a given Lagrangian system. We discuss in Sect. 10 possible Lagrangians $\mathscr{L}_R$ for the dynamics of the free gravitational field and set up a generalized field equation quadratic and linear in the Riemann–Cartan tensor. Finally, we discuss the correlation of the spin part of the energy-momentum tensor of the source system with the torsion of spacetime, described by the then skew-symmetric part of the Ricci tensor.

2 Canonical Transformations Under a Dynamic Spacetime

The requirement of form-invariance of the action functional for a real scalar field ϕ and a vector field a_μ—in conjunction with their respective canonical conjugate fields π^μ and $p^{\nu\mu}$—in a local inertial frame, hence under a *static* spacetime, is formulated as

$$\delta \int_{\Omega_y} \left(\pi^\alpha \frac{\partial \phi}{\partial y^\alpha} + p^{\beta\alpha} \frac{\partial a_\beta}{\partial y^\alpha} - \mathscr{H} - \frac{\partial \mathscr{F}_1^\alpha}{\partial y^\alpha} \right) \mathrm{d}^4 y \stackrel{!}{=} \delta \int_{\Omega_y} \left(\Pi^\alpha \frac{\partial \Phi}{\partial y^\alpha} + P^{\beta\alpha} \frac{\partial A_\beta}{\partial y^\alpha} - \mathscr{H}' \right) \mathrm{d}^4 y. \tag{1}$$

For the transition to non-inertial frames, the volume form $\mathrm{d}^4 y$ is expressed in terms of $\mathrm{d}^4 x$ and $\mathrm{d}^4 X$ by means of the respective Jacobians

$$\mathrm{d}^4 y = \left| \frac{\partial y}{\partial x} \right| \mathrm{d}^4 x, \qquad \mathrm{d}^4 y = \left| \frac{\partial y}{\partial X} \right| \mathrm{d}^4 X.$$

We assume the local inertial system $\left(y^0, \ldots, y^3\right)$ to be endowed by the Minkowski metric $\eta_{jk} = \operatorname{diag}(1, -1, -1, -1)$. The metrics $g_{\mu\nu}(x)$ and $G_{\mu\nu}(X)$ in the respective non-inertial frames $\left(x^0, \ldots, x^3\right)$ and $\left(X^0, \ldots, X^3\right)$ are then

$$g_{\mu\nu} = \eta_{jk} \frac{\partial y^j}{\partial x^\mu} \frac{\partial y^k}{\partial x^\mu}, \qquad G_{\mu\nu} = \eta_{jk} \frac{\partial y^j}{\partial X^\mu} \frac{\partial y^k}{\partial X^\mu},$$

and hence their determinants, owing to $\det \eta_{jk} = -1$, are

$$g \equiv \det g_{\mu\nu} = -\left|\frac{\partial y}{\partial x}\right|^2, \qquad G \equiv \det G_{\mu\nu} = -\left|\frac{\partial y}{\partial X}\right|^2.$$

The volume form $\mathrm{d}^4 y$ thus writes in terms of the determinants of the respective metric

$$\mathrm{d}^4 y = \sqrt{-g}\, \mathrm{d}^4 x, \qquad \mathrm{d}^4 y = \sqrt{-G}\, \mathrm{d}^4 X. \tag{2}$$

The requirement of form-invariance of the action functional for real scalar and vector fields under transitions $x^\mu \mapsto X^\mu$ is formulated as

$$\delta \int_{\Omega_x} \left(\tilde{\pi}^\alpha \frac{\partial \phi}{\partial x^\alpha} + \tilde{p}^{\beta\alpha} \frac{\partial a_\beta}{\partial x^\alpha} - \tilde{\mathscr{H}} - \frac{\partial \tilde{\mathscr{F}}_1^\alpha}{\partial x^\alpha}\right) \mathrm{d}^4 x \stackrel{!}{=} \delta \int_{\Omega_X} \left(\tilde{\Pi}^\alpha \frac{\partial \Phi}{\partial X^\alpha} + \tilde{P}^{\beta\alpha} \frac{\partial A_\beta}{\partial X^\alpha} - \tilde{\mathscr{H}}'\right) \mathrm{d}^4 X, \tag{3}$$

where the factors $\sqrt{-g}$ and $\sqrt{-G}$ are absorbed into the respective momentum fields, thereby converting them into relative tensors of weight $w = 1$, i.e., into *tensor densities*:

$$\tilde{\pi}^\beta = \pi^\beta \sqrt{-g}, \qquad \tilde{\Pi}^\beta = \Pi^\beta \sqrt{-G},$$

and similarly for all other momenta. The Hamiltonians are thus converted into scalar densities.

As the action integral is to be varied, Eq. (3) implies that the *integrands* may differ by the divergence of a vector density $\tilde{\mathscr{F}}_1^\mu(x)$ whose variation vanishes on the boundary $\partial\Omega_x$ of the integration region Ω_x within spacetime

$$\delta \int_{\Omega_x} \frac{\partial \tilde{\mathscr{F}}_1^\alpha}{\partial x^\alpha} \mathrm{d}^4 x = \delta \oint_{\partial\Omega_x} \tilde{\mathscr{F}}_1^\alpha \, \mathrm{d}S_\alpha \stackrel{!}{=} 0. \tag{4}$$

The addition of a term $\partial \tilde{\mathscr{F}}_1^\alpha / \partial x^\alpha$ to the integrand which can be converted into a surface integral—commonly referred to briefly as a *surface term*—thus does not modify the variation of the action integral. This means that the integrand is only determined up to the divergence of the functions $\tilde{\mathscr{F}}_1^\mu(\Phi, \phi, A, a, x)$. With the transformation rule of the volume form from Eq. (2), and $\tilde{\mathscr{F}}_1^\mu$ to be taken at x, the integrand condition obtained from Eq. (3) for an extended canonical transformation thus writes

$$\tilde{\pi}^{\beta}\frac{\partial\phi}{\partial x^{\beta}}+\tilde{p}^{\alpha\beta}\frac{\partial a_{\alpha}}{\partial x^{\beta}}-\tilde{\mathscr{H}}-\left(\tilde{\Pi}^{\beta}\frac{\partial\Phi}{\partial X^{\beta}}+\tilde{P}^{\alpha\beta}\frac{\partial A_{\alpha}}{\partial X^{\beta}}-\tilde{\mathscr{H}}'\right)\left|\frac{\partial X}{\partial x}\right| \tag{5}$$
$$=\frac{\partial\tilde{\mathscr{F}}_1^{\beta}}{\partial\phi}\frac{\partial\phi}{\partial x^{\beta}}+\frac{\partial\tilde{\mathscr{F}}_1^{\alpha}}{\partial\Phi}\frac{\partial X^{\beta}}{\partial x^{\alpha}}\frac{\partial\Phi}{\partial X^{\beta}}+\frac{\partial\tilde{\mathscr{F}}_1^{\beta}}{\partial a_{\alpha}}\frac{\partial a_{\alpha}}{\partial x^{\beta}}+\frac{\partial\tilde{\mathscr{F}}_1^{\xi}}{\partial A_{\alpha}}\frac{\partial X^{\beta}}{\partial x^{\xi}}\frac{\partial A_{\alpha}}{\partial X^{\beta}}+\left.\frac{\partial\tilde{\mathscr{F}}_1^{\alpha}}{\partial x^{\alpha}}\right|_{\mathrm{expl}}.$$

Comparing the coefficients yields the transformation rules

$$\tilde{\pi}^{\mu}(x)=\frac{\partial\tilde{\mathscr{F}}_1^{\mu}}{\partial\phi}\qquad \tilde{\Pi}^{\mu}(X)=-\frac{\partial\tilde{\mathscr{F}}_1^{\beta}}{\partial\Phi}\frac{\partial X^{\mu}}{\partial x^{\beta}}\left|\frac{\partial x}{\partial X}\right| \tag{6a}$$

$$\tilde{p}^{\nu\mu}(x)=\frac{\partial\tilde{\mathscr{F}}_1^{\mu}}{\partial a_{\nu}}\qquad \tilde{P}^{\nu\mu}(X)=-\frac{\partial\tilde{\mathscr{F}}_1^{\beta}}{\partial A_{\nu}}\frac{\partial X^{\mu}}{\partial x^{\beta}}\left|\frac{\partial x}{\partial X}\right| \tag{6b}$$

$$\left.\tilde{\mathscr{H}}'\right|_X=\left(\left.\tilde{\mathscr{H}}\right|_x+\left.\frac{\partial\tilde{\mathscr{F}}_1^{\alpha}}{\partial x^{\alpha}}\right|_{\mathrm{expl}}\right)\left|\frac{\partial x}{\partial X}\right|\Rightarrow\left.\tilde{\mathscr{H}}'\right|_x-\left.\tilde{\mathscr{H}}\right|_x=\left.\frac{\partial\tilde{\mathscr{F}}_1^{\alpha}}{\partial x^{\alpha}}\right|_{\mathrm{expl}}. \tag{6c}$$

The generating function $\tilde{\mathscr{F}}_1^{\mu}(\Phi,\phi,A,a,x)$ may be Legendre-transformed into an equivalent generating function $\tilde{\mathscr{F}}_2^{\mu}(\tilde{\Pi},\phi,\tilde{P},a,x)$ according to

$$\tilde{\mathscr{F}}_2^{\mu}=\tilde{\mathscr{F}}_1^{\mu}-\Phi\frac{\partial\tilde{\mathscr{F}}_1^{\mu}}{\partial\Phi}-A_{\alpha}\frac{\partial\tilde{\mathscr{F}}_1^{\mu}}{\partial A_{\alpha}}=\tilde{\mathscr{F}}_1^{\mu}+\left(\Phi\,\tilde{\Pi}^{\beta}+A_{\alpha}\,\tilde{P}^{\alpha\beta}\right)\frac{\partial x^{\mu}}{\partial X^{\beta}}\left|\frac{\partial X}{\partial x}\right|. \tag{7}$$

In order to derive the divergence of $\tilde{\mathscr{F}}_2^{\mu}(x)$, we make use of the identity [2] for the right-hand side factor. Thus

$$\frac{\partial\tilde{\mathscr{F}}_1^{\alpha}}{\partial x^{\alpha}}=\frac{\partial\tilde{\mathscr{F}}_2^{\alpha}}{\partial x^{\alpha}}-\left|\frac{\partial X}{\partial x}\right|\frac{\partial}{\partial X^{\beta}}\left(\Phi\,\tilde{\Pi}^{\beta}+A_{\alpha}\,\tilde{P}^{\alpha\beta}\right). \tag{8}$$

Inserting Eq. (8) into the integrand condition (5), we encounter the modified integrand condition for a generating function of type $\tilde{\mathscr{F}}_2^{\mu}$, to be taken at the spacetime event x

$$\tilde{\pi}^{\beta}\frac{\partial\phi}{\partial x^{\beta}}+\tilde{p}^{\alpha\beta}\frac{\partial a_{\alpha}}{\partial x^{\beta}}-\tilde{\mathscr{H}}+\left(\Phi\delta_{\xi}^{\beta}\frac{\partial\tilde{\Pi}^{\xi}}{\partial X^{\beta}}+A_{\alpha}\delta_{\xi}^{\beta}\frac{\partial\tilde{P}^{\alpha\xi}}{\partial X^{\beta}}+\tilde{\mathscr{H}}'\right)\left|\frac{\partial X}{\partial x}\right|$$
$$=\frac{\partial\tilde{\mathscr{F}}_2^{\beta}}{\partial\phi}\frac{\partial\phi}{\partial x^{\beta}}+\frac{\partial\tilde{\mathscr{F}}_2^{\beta}}{\partial a_{\alpha}}\frac{\partial a_{\alpha}}{\partial x^{\beta}}+\frac{\partial\tilde{\mathscr{F}}_2^{\eta}}{\partial\tilde{\Pi}^{\xi}}\frac{\partial X^{\beta}}{\partial x^{\eta}}\frac{\partial\tilde{\Pi}^{\xi}}{\partial X^{\beta}}+\frac{\partial\tilde{\mathscr{F}}_2^{\eta}}{\partial\tilde{P}^{\alpha\xi}}\frac{\partial X^{\beta}}{\partial x^{\eta}}\frac{\partial\tilde{P}^{\alpha\xi}}{\partial X^{\beta}}+\left.\frac{\partial\tilde{\mathscr{F}}_2^{\alpha}}{\partial x^{\alpha}}\right|_{\mathrm{expl}}, \tag{9}$$

and hence the transformation rules

$$\tilde{\pi}^{\mu}(x) = \frac{\partial \tilde{\mathscr{F}}_2^{\mu}}{\partial \phi} \qquad \delta_{\nu}^{\mu} \Phi(X) = \frac{\partial \tilde{\mathscr{F}}_2^{\eta}}{\partial \tilde{\Pi}^{\nu}} \frac{\partial X^{\mu}}{\partial x^{\eta}} \left| \frac{\partial x}{\partial X} \right| \tag{10a}$$

$$\tilde{p}^{\,\nu\mu}(x) = \frac{\partial \tilde{\mathscr{F}}_2^{\mu}}{\partial a_{\nu}} \qquad \delta_{\nu}^{\mu} A_{\alpha}(X) = \frac{\partial \tilde{\mathscr{F}}_2^{\eta}}{\partial \tilde{P}^{\alpha\nu}} \frac{\partial X^{\mu}}{\partial x^{\eta}} \left| \frac{\partial x}{\partial X} \right| \tag{10b}$$

$$\tilde{\mathscr{H}}'\Big|_X = \left(\tilde{\mathscr{H}}\Big|_x + \left.\frac{\partial \tilde{\mathscr{F}}_2^{\alpha}}{\partial x^{\alpha}}\right|_{\text{expl}} \right) \left| \frac{\partial x}{\partial X} \right| \Rightarrow \tilde{\mathscr{H}}'\Big|_x - \tilde{\mathscr{H}}\Big|_x = \left.\frac{\partial \tilde{\mathscr{F}}_2^{\alpha}}{\partial x^{\alpha}}\right|_{\text{expl}} . \tag{10c}$$

In any case, the integrands of the action integrals (3) must be world scalars in order to keep their form under general spacetime transformations. This finally ensures that the canonical field equations emerge as *tensor equations*. Furthermore, the generating function $\tilde{\mathscr{F}}_1^{\mu}(\Phi, \phi, A, a, x)$ may also be Legendre-transformed into an equivalent generating function of type $\tilde{\mathscr{F}}_3^{\mu}(\tilde{\pi}, \Phi, \tilde{p}, A, x)$ according to

$$\tilde{\mathscr{F}}_3^{\mu} = \tilde{\mathscr{F}}_1^{\mu} - \phi \frac{\partial \tilde{\mathscr{F}}_1^{\mu}}{\partial \phi} - a_{\alpha} \frac{\partial \tilde{\mathscr{F}}_1^{\mu}}{\partial a_{\alpha}} = \tilde{\mathscr{F}}_1^{\mu} - \phi\, \tilde{\pi}^{\mu} - a_{\alpha}\, \tilde{p}^{\alpha\mu}, \tag{11}$$

hence

$$\frac{\partial \tilde{\mathscr{F}}_1^{\alpha}}{\partial x^{\alpha}} = \frac{\partial \tilde{\mathscr{F}}_3^{\alpha}}{\partial x^{\alpha}} + \frac{\partial}{\partial x^{\beta}} \left(\phi\, \tilde{\pi}^{\beta} + a_{\alpha}\, \tilde{p}^{\alpha\beta} \right). \tag{12}$$

Inserting Eq. (12) into the integrand condition (5), we encounter the modified integrand condition for a generating function of type $\tilde{\mathscr{F}}_3^{\mu}(x)$, to be taken at the spacetime event x

$$\begin{aligned}
&- \phi\, \delta_{\xi}^{\beta} \frac{\partial \tilde{\pi}^{\xi}}{\partial x^{\beta}} - a_{\alpha} \delta_{\xi}^{\beta} \frac{\partial \tilde{p}^{\,\alpha\xi}}{\partial x^{\beta}} - \tilde{\mathscr{H}} - \left(\tilde{\Pi}^{\beta} \frac{\partial \Phi}{\partial X^{\beta}} + \tilde{P}^{\,\alpha\beta} \frac{\partial A_{\alpha}}{\partial X^{\beta}} - \tilde{\mathscr{H}}' \right) \left| \frac{\partial X}{\partial x} \right| \\
&= \frac{\partial \tilde{\mathscr{F}}_3^{\eta}}{\partial \Phi} \frac{\partial X^{\beta}}{\partial x^{\eta}} \frac{\partial \Phi}{\partial X^{\beta}} + \frac{\partial \tilde{\mathscr{F}}_3^{\eta}}{\partial A_{\alpha}} \frac{\partial X^{\beta}}{\partial x^{\eta}} \frac{\partial A_{\alpha}}{\partial X^{\beta}} + \frac{\partial \tilde{\mathscr{F}}_3^{\beta}}{\partial \tilde{\pi}^{\xi}} \frac{\partial \tilde{\pi}^{\xi}}{\partial x^{\beta}} + \frac{\partial \tilde{\mathscr{F}}_3^{\beta}}{\partial \tilde{p}^{\alpha\xi}} \frac{\partial \tilde{p}^{\alpha\xi}}{\partial x^{\beta}} + \left.\frac{\partial \tilde{\mathscr{F}}_3^{\alpha}}{\partial x^{\alpha}}\right|_{\text{expl}} ,
\end{aligned} \tag{13}$$

and hence the transformation rules by comparing the coefficients

$$\tilde{\Pi}^{\mu}(X) = -\frac{\partial \tilde{\mathscr{F}}_3^{\kappa}}{\partial \Phi} \frac{\partial X^{\mu}}{\partial x^{\kappa}} \left| \frac{\partial x}{\partial X} \right| \qquad \delta_{\nu}^{\mu}\, \phi(x) = -\frac{\partial \tilde{\mathscr{F}}_3^{\mu}}{\partial \tilde{\pi}^{\nu}} \tag{14a}$$

$$\tilde{P}^{\,\nu\mu}(X) = -\frac{\partial \tilde{\mathscr{F}}_3^{\kappa}}{\partial A_{\nu}} \frac{\partial X^{\mu}}{\partial x^{\kappa}} \left| \frac{\partial x}{\partial X} \right| \qquad \delta_{\nu}^{\mu}\, a_{\alpha}(x) = -\frac{\partial \tilde{\mathscr{F}}_3^{\mu}}{\partial \tilde{p}^{\alpha\nu}} \tag{14b}$$

$$\tilde{\mathscr{H}}'\Big|_X = \left(\tilde{\mathscr{H}}\Big|_x + \left.\frac{\partial \tilde{\mathscr{F}}_3^{\alpha}}{\partial x^{\alpha}}\right|_{\text{expl}} \right) \left| \frac{\partial x}{\partial X} \right| \Rightarrow \tilde{\mathscr{H}}'\Big|_x - \tilde{\mathscr{H}}\Big|_x = \left.\frac{\partial \tilde{\mathscr{F}}_3^{\alpha}}{\partial x^{\alpha}}\right|_{\text{expl}} . \tag{14c}$$

In order to allow the description of a *dynamic* spacetime, the Hamiltonians are presumed to depend in addition on the metric $g_{\alpha\lambda}$ and the—in general—non-symmetric connection $\gamma^{\alpha}{}_{\xi\eta}$, in conjunction with their respective conjugates, $\tilde{k}^{\alpha\lambda\beta}$ and $\tilde{q}_{\alpha}{}^{\xi\eta\beta}$. Compared to Eq. (13), we thus encounter an extended integrand condition for a generating function of type $\tilde{\mathscr{F}}_3^{\mu}(x)$,

$$\begin{aligned}
&-\phi\,\delta^{\beta}_{\xi}\frac{\partial\tilde{\pi}^{\xi}}{\partial x^{\beta}}-a_{\alpha}\delta^{\beta}_{\xi}\frac{\partial\tilde{p}^{\alpha\xi}}{\partial x^{\beta}}-g_{\alpha\lambda}\delta^{\beta}_{\xi}\frac{\partial\tilde{k}^{\alpha\lambda\xi}}{\partial x^{\beta}}-\gamma^{\alpha}{}_{\lambda\eta}\,\delta^{\beta}_{\xi}\frac{\partial\tilde{q}_{\alpha}{}^{\lambda\eta\xi}}{\partial x^{\beta}}-\tilde{\mathscr{H}}\\
&-\left(\tilde{\Pi}^{\beta}\frac{\partial\Phi}{\partial X^{\beta}}+\tilde{P}^{\alpha\beta}\frac{\partial A_{\alpha}}{\partial X^{\beta}}+\tilde{K}^{\alpha\lambda\beta}\frac{\partial G_{\alpha\lambda}}{\partial X^{\beta}}+\tilde{Q}_{\alpha}{}^{\xi\eta\beta}\frac{\partial\Gamma^{\alpha}{}_{\xi\eta}}{\partial X^{\beta}}-\tilde{\mathscr{H}}'\right)\left|\frac{\partial X}{\partial x}\right|\\
&=\frac{\partial\tilde{\mathscr{F}}_3^{\eta}}{\partial\Phi}\frac{\partial X^{\beta}}{\partial x^{\eta}}\frac{\partial\Phi}{\partial X^{\beta}}+\frac{\partial\tilde{\mathscr{F}}_3^{\eta}}{\partial A_{\alpha}}\frac{\partial X^{\beta}}{\partial x^{\eta}}\frac{\partial A_{\alpha}}{\partial X^{\beta}}+\frac{\partial\tilde{\mathscr{F}}_3^{\eta}}{\partial G_{\alpha\lambda}}\frac{\partial X^{\beta}}{\partial x^{\eta}}\frac{\partial G_{\alpha\lambda}}{\partial X^{\beta}}+\frac{\partial\tilde{\mathscr{F}}_3^{\kappa}}{\partial\Gamma^{\alpha}{}_{\xi\eta}}\frac{\partial X^{\beta}}{\partial x^{\kappa}}\frac{\partial\Gamma^{\alpha}{}_{\xi\eta}}{\partial X^{\beta}}\\
&+\frac{\partial\tilde{\mathscr{F}}_3^{\beta}}{\partial\tilde{\pi}^{\xi}}\frac{\partial\tilde{\pi}^{\xi}}{\partial x^{\beta}}+\frac{\partial\tilde{\mathscr{F}}_3^{\beta}}{\partial\tilde{p}^{\alpha\xi}}\frac{\partial\tilde{p}^{\alpha\xi}}{\partial x^{\beta}}+\frac{\partial\tilde{\mathscr{F}}_3^{\beta}}{\partial\tilde{k}^{\alpha\lambda\xi}}\frac{\partial\tilde{k}^{\alpha\lambda\xi}}{\partial x^{\beta}}+\frac{\partial\tilde{\mathscr{F}}_3^{\beta}}{\partial\tilde{q}_{\alpha}{}^{\lambda\eta\xi}}\frac{\partial\tilde{q}_{\alpha}{}^{\lambda\eta\xi}}{\partial x^{\beta}}+\left.\frac{\partial\tilde{\mathscr{F}}_3^{\beta}}{\partial x^{\beta}}\right|_{\text{expl}},
\end{aligned} \tag{14d}$$

and hence the additional transformation rules for the metric tensor $g_{\alpha\beta}$ and for the connection $\gamma^{\eta}{}_{\alpha\beta}$, together with their respective conjugates, $\tilde{k}^{\alpha\beta\mu}$ and $\tilde{q}_{\eta}{}^{\alpha\beta\mu}$:

$$\tilde{K}^{\alpha\beta\mu}(X)=-\frac{\partial\tilde{\mathscr{F}}_3^{\kappa}}{\partial G_{\alpha\beta}}\frac{\partial X^{\mu}}{\partial x^{\kappa}}\left|\frac{\partial x}{\partial X}\right| \qquad \delta^{\mu}_{\nu}\,g_{\alpha\beta}(x)=-\frac{\partial\tilde{\mathscr{F}}_3^{\mu}}{\partial\tilde{k}^{\alpha\beta\nu}} \tag{14e}$$

$$\tilde{Q}_{\eta}{}^{\alpha\beta\mu}(X)=-\frac{\partial\tilde{\mathscr{F}}_3^{\kappa}}{\partial\Gamma^{\eta}{}_{\alpha\beta}}\frac{\partial X^{\mu}}{\partial x^{\kappa}}\left|\frac{\partial x}{\partial X}\right| \qquad \delta^{\mu}_{\nu}\,\gamma^{\eta}{}_{\alpha\beta}(x)=-\frac{\partial\tilde{\mathscr{F}}_3^{\mu}}{\partial\tilde{q}_{\eta}{}^{\alpha\beta\nu}}. \tag{14f}$$

3 Generalized Noether Theorem

In this section, we set up a generating function of type $\tilde{\mathscr{F}}_3^{\mu}$ to define a general *infinitesimal* canonical transformation. Specifically, for a sample system of a scalar field ϕ and a vector field a_{μ} in a metric-affine space, the infinitesimal transformation rules are derived from the generating function

$$\left.\tilde{\mathscr{F}}_3^{\mu}\right|_x=\left.\left(-\tilde{\pi}^{\mu}\Phi-\tilde{p}^{\alpha\mu}A_{\alpha}-\tilde{k}^{\xi\lambda\mu}G_{\xi\lambda}-\tilde{q}_{\eta}{}^{\alpha\beta\mu}\Gamma^{\eta}{}_{\alpha\beta}-\varepsilon\,\tilde{j}^{\mu}_{\mathrm{N}}\right)\right|_x, \tag{15}$$

wherein $\tilde{j}^{\mu}_{\mathrm{N}}=\tilde{j}^{\mu}_{\mathrm{N}}\left(\tilde{\pi},\phi,\tilde{p},a,\tilde{k},g,\tilde{q},\gamma,x\right)$. For $\varepsilon=0$, Eq. (15) thus generates the identity transformation for all dynamical quantities the autonomous Hamiltonian $\tilde{\mathscr{H}}=\tilde{\mathscr{H}}\left(\tilde{\pi},\phi,\tilde{p},a,\tilde{k},g,\tilde{q},\gamma\right)$ depends on. All contributions of the general transformation rules (14) that are associated with a non-identical mapping of fields and spacetime are encoded in the particular expression for $\tilde{j}^{\mu}_{\mathrm{N}}(x)$. The transformation rules (14) now read:

$$\delta^{\mu}_{\nu}\,\delta\phi = -\varepsilon\frac{\partial\tilde{j}_{\mathrm{N}}^{\mu}}{\partial\tilde{\pi}^{\nu}},\qquad \delta\tilde{\pi}^{\mu} = \varepsilon\frac{\partial\tilde{j}_{\mathrm{N}}^{\mu}}{\partial\phi},\qquad \delta^{\mu}_{\nu}\,\delta a_{\alpha} = -\varepsilon\frac{\partial\tilde{j}_{\mathrm{N}}^{\mu}}{\partial\tilde{p}^{\alpha\nu}},\qquad \delta\tilde{p}^{\alpha\mu} = \varepsilon\frac{\partial\tilde{j}_{\mathrm{N}}^{\mu}}{\partial a_{\alpha}} \tag{16a}$$

$$\delta^{\mu}_{\nu}\,\delta g_{\lambda\xi} = -\varepsilon\frac{\partial\tilde{j}_{\mathrm{N}}^{\mu}}{\partial\tilde{k}^{\lambda\xi\nu}},\quad \delta\tilde{k}^{\lambda\xi\mu} = \varepsilon\frac{\partial\tilde{j}_{\mathrm{N}}^{\mu}}{\partial g_{\lambda\xi}},\quad \delta^{\mu}_{\nu}\,\delta\gamma^{\eta}{}_{\alpha\beta} = -\varepsilon\frac{\partial\tilde{j}_{\mathrm{N}}^{\mu}}{\partial\tilde{q}_{\eta}{}^{\alpha\beta\nu}},\quad \delta\tilde{q}_{\eta}{}^{\alpha\beta\mu} = \varepsilon\frac{\partial\tilde{j}_{\mathrm{N}}^{\mu}}{\partial\gamma^{\eta}{}_{\alpha\beta}} \tag{16b}$$

and

$$\delta\tilde{\mathscr{H}}\Big|_{\mathrm{CT}} \equiv \tilde{\mathscr{H}}'\Big|_{x} - \tilde{\mathscr{H}}\Big|_{x} = \frac{\partial\tilde{\mathscr{F}}_{3}^{\alpha}}{\partial x^{\alpha}}\bigg|_{\mathrm{expl}} = -\varepsilon\frac{\partial\tilde{j}_{\mathrm{N}}^{\alpha}}{\partial x^{\alpha}}\bigg|_{\mathrm{expl}}. \tag{16c}$$

The explicit representation of the infinitesimal transformation rules (16) for the generating function $\tilde{\mathscr{F}}_{3}^{\mu}$ of a diffeomorphism will be presented below.

On the other hand, for a closed system, where the Hamiltonian does not explicitly depend on x, the variation of the Hamiltonian emerging from the variation of the fields follows as

$$\begin{aligned}\delta\tilde{\mathscr{H}} = {}& \frac{\partial\tilde{\mathscr{H}}}{\partial\phi}\delta\phi + \frac{\partial\tilde{\mathscr{H}}}{\partial\tilde{\pi}^{\alpha}}\delta\tilde{\pi}^{\alpha} + \frac{\partial\tilde{\mathscr{H}}}{\partial a_{\alpha}}\delta a_{\alpha} + \frac{\partial\tilde{\mathscr{H}}}{\partial\tilde{p}^{\alpha\beta}}\delta\tilde{p}^{\alpha\beta} + \frac{\partial\tilde{\mathscr{H}}}{\partial g_{\lambda\xi}}\delta g_{\lambda\xi} + \frac{\partial\tilde{\mathscr{H}}}{\partial\tilde{k}^{\lambda\xi\beta}}\delta\tilde{k}^{\lambda\xi\beta}\\ &+ \frac{\partial\tilde{\mathscr{H}}}{\partial\gamma^{\eta}{}_{\lambda\xi}}\delta\gamma^{\eta}{}_{\lambda\xi} + \frac{\partial\tilde{\mathscr{H}}}{\partial\tilde{q}_{\eta}{}^{\lambda\xi\beta}}\delta\tilde{q}_{\eta}{}^{\lambda\xi\beta}.\end{aligned}$$

Inserting the covariant canonical field equations [7],

$$-\frac{\partial\tilde{\pi}^{\beta}}{\partial x^{\beta}} = \frac{\partial\tilde{\mathscr{H}}}{\partial\phi}\quad \frac{\partial\phi}{\partial x^{\alpha}} = \frac{\partial\tilde{\mathscr{H}}}{\partial\tilde{\pi}^{\alpha}}\quad -\frac{\partial\tilde{p}^{\alpha\beta}}{\partial x^{\beta}} = \frac{\partial\tilde{\mathscr{H}}}{\partial a_{\alpha}}\quad \frac{\partial a_{\alpha}}{\partial x^{\beta}} = \frac{\partial\tilde{\mathscr{H}}}{\partial\tilde{p}^{\alpha\beta}} \tag{17a}$$

$$-\frac{\partial\tilde{k}^{\lambda\xi\beta}}{\partial x^{\beta}} = \frac{\partial\tilde{\mathscr{H}}}{\partial g_{\lambda\xi}}\quad \frac{\partial g_{\lambda\xi}}{\partial x^{\beta}} = \frac{\partial\tilde{\mathscr{H}}}{\partial\tilde{k}^{\lambda\xi\beta}}\quad -\frac{\partial\tilde{q}_{\eta}{}^{\lambda\xi\beta}}{\partial x^{\beta}} = \frac{\partial\tilde{\mathscr{H}}}{\partial\gamma^{\eta}{}_{\lambda\xi}}\quad \frac{\partial\gamma^{\eta}{}_{\lambda\xi}}{\partial x^{\beta}} = \frac{\partial\tilde{\mathscr{H}}}{\partial\tilde{q}_{\eta}{}^{\lambda\xi\beta}}, \tag{17b}$$

the variation of $\tilde{\mathscr{H}}$ is expressed along the system's spacetime evolution

$$\begin{aligned}\delta\tilde{\mathscr{H}} = {}& -\frac{\partial\tilde{\pi}^{\beta}}{\partial x^{\alpha}}\delta^{\alpha}_{\beta}\,\delta\phi + \frac{\partial\phi}{\partial x^{\alpha}}\delta\tilde{\pi}^{\alpha} - \frac{\partial\tilde{p}^{\alpha\beta}}{\partial x^{\xi}}\delta^{\xi}_{\beta}\,\delta a_{\alpha} + \frac{\partial a_{\alpha}}{\partial x^{\beta}}\delta\tilde{p}^{\alpha\beta}\\ &- \frac{\partial\tilde{k}^{\lambda\xi\beta}}{\partial x^{\alpha}}\delta^{\alpha}_{\beta}\,\delta g_{\lambda\xi} + \frac{\partial g_{\lambda\xi}}{\partial x^{\beta}}\delta\tilde{k}^{\lambda\xi\beta} - \frac{\partial\tilde{q}_{\eta}{}^{\lambda\xi\beta}}{\partial x^{\alpha}}\delta^{\alpha}_{\beta}\,\delta\gamma^{\eta}{}_{\lambda\xi} + \frac{\partial\gamma^{\eta}{}_{\lambda\xi}}{\partial x^{\beta}}\delta\tilde{q}_{\eta}{}^{\lambda\xi\beta}.\end{aligned} \tag{18}$$

With the transformation rules (16), this writes in terms of the derivatives of $\tilde{j}_{\mathrm{N}}^{\mu}$

$$\begin{aligned}
\delta\tilde{\mathscr{H}} &= \varepsilon\frac{\partial\tilde{\pi}^{\beta}}{\partial x^{\alpha}}\frac{\partial\tilde{j}_{\mathrm{N}}^{\alpha}}{\partial\tilde{\pi}^{\beta}} + \varepsilon\frac{\partial\phi}{\partial x^{\alpha}}\frac{\partial\tilde{j}_{\mathrm{N}}^{\alpha}}{\partial\phi} + \varepsilon\frac{\partial\tilde{p}^{\alpha\beta}}{\partial x^{\xi}}\frac{\partial\tilde{j}_{\mathrm{N}}^{\xi}}{\partial\tilde{p}^{\alpha\beta}} + \varepsilon\frac{\partial a_{\alpha}}{\partial x^{\beta}}\frac{\partial\tilde{j}_{\mathrm{N}}^{\beta}}{\partial a_{\alpha}} \\
&\quad + \varepsilon\frac{\partial\tilde{k}^{\lambda\xi\beta}}{\partial x^{\alpha}}\frac{\partial\tilde{j}_{\mathrm{N}}^{\alpha}}{\partial\tilde{k}^{\lambda\xi\beta}} + \varepsilon\frac{\partial g_{\lambda\xi}}{\partial x^{\beta}}\frac{\partial\tilde{j}_{\mathrm{N}}^{\beta}}{\partial g_{\lambda\xi}} + \varepsilon\frac{\partial\tilde{q}_{\eta}{}^{\lambda\xi\beta}}{\partial x^{\alpha}}\frac{\partial\tilde{j}_{\mathrm{N}}^{\alpha}}{\partial\tilde{q}_{\eta}{}^{\lambda\xi\beta}} + \varepsilon\frac{\partial\gamma^{\eta}{}_{\lambda\xi}}{\partial x^{\beta}}\frac{\partial\tilde{j}_{\mathrm{N}}^{\beta}}{\partial\gamma^{\eta}{}_{\lambda\xi}} \\
&= \varepsilon\frac{\partial\tilde{j}_{\mathrm{N}}^{\alpha}}{\partial x^{\alpha}} - \varepsilon\left.\frac{\partial\tilde{j}_{\mathrm{N}}^{\alpha}}{\partial x^{\alpha}}\right|_{\mathrm{expl}} \\
&= \varepsilon\frac{\partial\tilde{j}_{\mathrm{N}}^{\alpha}}{\partial x^{\alpha}} + \delta\tilde{\mathscr{H}}\Big|_{\mathrm{CT}}.
\end{aligned}$$

The requirement that both variations, $\delta\tilde{\mathscr{H}}$ and $\delta\tilde{\mathscr{H}}\big|_{\mathrm{CT}}$, weakly coincide ensures that the canonical transformation defines a symmetry transformation. Then, the vector $\tilde{j}_{\mathrm{N}}^{\mu}$ in the generating function (15) defines the (weakly) conserved Noether current

$$\delta\tilde{\mathscr{H}} - \delta\tilde{\mathscr{H}}\Big|_{\mathrm{CT}} \stackrel{\not\equiv}{=} 0 \quad\Leftrightarrow\quad \frac{\partial\tilde{j}_{\mathrm{N}}^{\alpha}}{\partial x^{\alpha}} \stackrel{\not\equiv}{=} 0. \tag{19}$$

4 Finite Diff(M) Transformation

For a closed system of a scalar field ϕ and a vector field a_{μ} in a metric-affine space, the generating function of type $\tilde{\mathscr{F}}_3^{\mu}$ for the canonical transformation of the (active) diffeomorphisms that build the Diff(M) symmetry group is set up as follows in order to ensure the proper transformation behavior of the fields:

$$\begin{aligned}
\tilde{\mathscr{F}}_3^{\mu}\Big|_{x} &= -\tilde{\pi}^{\mu}\,\Phi - \tilde{p}^{\alpha\mu}\,A_{\beta}\frac{\partial X^{\beta}}{\partial x^{\alpha}} - \tilde{k}^{\alpha\beta\mu}G_{\xi\lambda}\frac{\partial X^{\xi}}{\partial x^{\alpha}}\frac{\partial X^{\lambda}}{\partial x^{\beta}} \\
&\quad - \tilde{q}_{\eta}{}^{\alpha\beta\mu}\left(\Gamma^{\tau}{}_{\xi\lambda}\frac{\partial x^{\eta}}{\partial X^{\tau}}\frac{\partial X^{\xi}}{\partial x^{\alpha}}\frac{\partial X^{\lambda}}{\partial x^{\beta}} + \frac{\partial x^{\eta}}{\partial X^{\tau}}\frac{\partial^2 X^{\tau}}{\partial x^{\alpha}\partial x^{\beta}}\right).
\end{aligned} \tag{20}$$

We remind that a tilde denotes that the respective quantity represents a tensor density, i.e., a relative tensor of weight $w = 1$. Notice that the transformed metric tensor $G_{\xi\lambda}(X)$ is *symmetric*, which induces the tensor $\tilde{k}^{\alpha\beta\mu}$ to be symmetric in its first index pair, α, β.

From the general rules (14), the specific generating function (20) yields the following particular transformation rules for involved fields and their conjugates:

$$\phi(x) = \Phi(X), \qquad a_{\alpha}(x) = A_{\beta}(X)\frac{\partial X^{\beta}}{\partial x^{\alpha}}, \tag{21a}$$

$$\tilde{\Pi}^{\mu}(X) = \tilde{\pi}^{\kappa}(x)\frac{\partial X^{\mu}}{\partial x^{\kappa}}\left|\frac{\partial x}{\partial X}\right|, \qquad \tilde{P}^{\nu\mu}(X) = \tilde{p}^{\alpha\kappa}(x)\frac{\partial X^{\nu}}{\partial x^{\alpha}}\frac{\partial X^{\mu}}{\partial x^{\kappa}}\left|\frac{\partial x}{\partial X}\right|, \tag{21b}$$

and for the "metric and connection fields" and their conjugates,

$$g_{\alpha\beta}(x) = G_{\xi\lambda}(X)\frac{\partial X^\xi}{\partial x^\alpha}\frac{\partial X^\lambda}{\partial x^\beta}, \qquad \gamma^{\eta}_{\ \alpha\beta}(x) = \Gamma^{\tau}_{\ \xi\lambda}(X)\frac{\partial x^\eta}{\partial X^\tau}\frac{\partial X^\xi}{\partial x^\alpha}\frac{\partial X^\lambda}{\partial x^\beta} + \frac{\partial x^\eta}{\partial X^\tau}\frac{\partial^2 X^\tau}{\partial x^\alpha \partial x^\beta}, \tag{21c}$$

$$\tilde{K}^{\xi\lambda\mu}(X) = \tilde{k}^{\alpha\beta\kappa}(x)\frac{\partial X^\xi}{\partial x^\alpha}\frac{\partial X^\lambda}{\partial x^\beta}\frac{\partial X^\mu}{\partial x^\kappa}\left|\frac{\partial x}{\partial X}\right|, \quad \tilde{Q}_{\tau}^{\ \xi\lambda\mu}(X) = \tilde{q}_{\eta}^{\ \alpha\beta\kappa}(x)\frac{\partial x^\eta}{\partial X^\tau}\frac{\partial X^\xi}{\partial x^\alpha}\frac{\partial X^\lambda}{\partial x^\beta}\frac{\partial X^\mu}{\partial x^\kappa}\left|\frac{\partial x}{\partial X}\right|. \tag{21d}$$

Note that the connection field $\gamma^{\eta}_{\ \alpha\beta}(x)$ is not assumed to be symmetric in α and β. Finally, the particular transformation rule for the covariant Hamiltonian follows from the general rule (14c) as:

$$\begin{aligned}
\delta\tilde{\mathcal{H}}\Big|_{\mathrm{CT}} = \tilde{\mathcal{H}}'\Big|_x - \tilde{\mathcal{H}}\Big|_x &= -\tilde{p}^{\alpha\mu}a_\sigma\frac{\partial x^\sigma}{\partial X^\xi}\frac{\partial^2 X^\xi}{\partial x^\alpha \partial x^\mu} \\
&- \tilde{k}^{\alpha\beta\mu}\left(g_{\sigma\beta}\frac{\partial x^\sigma}{\partial X^\xi}\frac{\partial^2 X^\xi}{\partial x^\alpha \partial x^\mu} + g_{\alpha\sigma}\frac{\partial x^\sigma}{\partial X^\xi}\frac{\partial^2 X^\xi}{\partial x^\beta \partial x^\mu}\right) \\
&+ \tilde{q}_{\eta}^{\ \alpha\beta\mu}\left(\gamma^{\sigma}_{\ \alpha\beta}\frac{\partial x^\eta}{\partial X^\xi}\frac{\partial^2 X^\xi}{\partial x^\sigma \partial x^\mu} - \gamma^{\eta}_{\ \sigma\beta}\frac{\partial x^\sigma}{\partial X^\xi}\frac{\partial^2 X^\xi}{\partial x^\alpha \partial x^\mu} - \gamma^{\eta}_{\ \alpha\sigma}\frac{\partial x^\sigma}{\partial X^\xi}\frac{\partial^2 X^\xi}{\partial x^\beta \partial x^\mu}\right. \\
&\quad + \frac{\partial x^\eta}{\partial X^\tau}\frac{\partial^2 X^\tau}{\partial x^\sigma \partial x^\beta}\frac{\partial x^\sigma}{\partial X^\xi}\frac{\partial^2 X^\xi}{\partial x^\alpha \partial x^\mu} \\
&\quad \left. + \frac{\partial x^\eta}{\partial X^\tau}\frac{\partial^2 X^\tau}{\partial x^\sigma \partial x^\alpha}\frac{\partial x^\sigma}{\partial X^\xi}\frac{\partial^2 X^\xi}{\partial x^\beta \partial x^\mu} - \frac{\partial x^\eta}{\partial X^\xi}\frac{\partial^3 X^\xi}{\partial x^\alpha \partial x^\beta \partial x^\mu}\right).
\end{aligned} \tag{21e}$$

We will show that the transformation following from the generating function (20) defines a symmetry transformation, in the sense that any action integral in one point of the manifold M is mapped to an action integral *of the same form* at another point of M [8]. In other words, the *Principle of General Relativity* is implemented here via the generating function $\tilde{\mathcal{F}}_3^\mu$.

5 Infinitesimal Diff(M) Transformation

In order to work out the particular form of the Noether current $\tilde{j}_{\mathrm{N}}^\mu$ which is associated with an invariance of a given field theory under *active* diffeomorphisms, the generating function for the corresponding *infinitesimal* canonical transformation needs to be set up first. To this end, a local parameter vector $h^\mu(x)$ is introduced, which defines the (active) *infinitesimal local diffeomorphism* $x^\mu \mapsto X^\mu$ on a manifold M:

$$X^\mu = x^\mu + \varepsilon\, h^\mu(x), \qquad L_h = h^\alpha(x)\frac{\partial}{\partial x^\alpha}, \tag{22}$$

with $\varepsilon \ll 1$ and L_h the Lie algebra generators of the diffeomorphism group of M. To first order in ε, the spacetime dependent coefficients of (20) are then expressed as

$$\frac{\partial X^\mu}{\partial x^\nu} = \delta^\mu_\nu + \varepsilon\frac{\partial h^\mu}{\partial x^\nu}, \qquad \frac{\partial x^\mu}{\partial X^\nu} = \delta^\mu_\nu - \varepsilon\frac{\partial h^\mu}{\partial x^\nu}, \qquad \frac{\partial^2 X^\tau}{\partial x^\alpha \partial x^\beta} = \varepsilon\frac{\partial^2 h^\tau}{\partial x^\alpha \partial x^\beta},$$
$$\left|\frac{\partial x}{\partial X}\right| = 1 - \varepsilon\frac{\partial h^\tau}{\partial x^\tau}. \tag{23}$$

The finite transformation rules (21a)–(21d) of the fields now follow up to first order in ε as

$$\delta\phi(x) = \Phi(x) - \phi(x) = -\varepsilon h^\beta \frac{\partial\phi}{\partial x^\beta} = -\varepsilon\mathscr{L}_h\phi(x) \tag{24a}$$

$$\delta a_\alpha(x) = A_\alpha(x) - a_\alpha(x) = -\varepsilon\left(h^\beta\frac{\partial a_\alpha}{\partial x^\beta} + a_\beta\frac{\partial h^\beta}{\partial x^\alpha}\right) = -\varepsilon\mathscr{L}_h a_\alpha(x) \tag{24b}$$

$$\delta g_{\xi\lambda}(x) = G_{\xi\lambda}(x) - g_{\xi\lambda}(x) = -\varepsilon\left(h^\beta\frac{\partial g_{\xi\lambda}}{\partial x^\beta} + g_{\beta\lambda}\frac{\partial h^\beta}{\partial x^\xi} + g_{\beta\xi}\frac{\partial h^\beta}{\partial x^\lambda}\right) = -\varepsilon\mathscr{L}_h g_{\xi\lambda}(x) \tag{24c}$$

$$\begin{aligned}\delta\gamma^{\eta}{}_{\lambda\tau}(x) = \Gamma^{\eta}{}_{\lambda\tau}(x) - \gamma^{\eta}{}_{\lambda\tau}(x) &= -\varepsilon\left(h^\beta\frac{\partial\gamma^{\eta}{}_{\lambda\tau}}{\partial x^\beta} - \gamma^{\beta}{}_{\lambda\tau}\frac{\partial h^\eta}{\partial x^\beta} + \gamma^{\eta}{}_{\beta\tau}\frac{\partial h^\beta}{\partial x^\lambda}\right.\\ &\qquad\left. + \gamma^{\eta}{}_{\lambda\beta}\frac{\partial h^\beta}{\partial x^\tau} + \frac{\partial^2 h^\eta}{\partial x^\lambda \partial x^\tau}\right)\\ &= -\varepsilon\mathscr{L}_h\gamma^{\eta}{}_{\lambda\tau}(x).\end{aligned} \tag{24d}$$

The differences of the scalar, vector, and tensor fields are actually proportional to conventional (non-covariant) *Lie derivatives* along the vector field $h^\beta(x)$, denoted by $\mathscr{L}_h$ [9, 10]. This reflects the general fact that a finite-dimensional continuous group of transformations which lie infinitesimally close to the identity define the Lie algebra of said group. The difference of the non-tensorial connection field $\gamma^{\eta}{}_{\lambda\tau}(x)$ is equally proportional to its Lie derivative by virtue of its transformation property (21d) (see Schouten [11, Eq. (II 10.34)]).

The corresponding differences of the conjugate momentum tensors are proportional to conventional Lie derivatives of the respective tensor *densities*

$$\delta\tilde{\pi}^\mu(x) = \tilde{\Pi}^\mu(x) - \tilde{\pi}^\mu(x) = \varepsilon\left[-h^\beta\frac{\partial\tilde{\pi}^\mu}{\partial x^\beta} + \tilde{\pi}^\beta\left(\frac{\partial h^\mu}{\partial x^\beta} - \delta^\mu_\beta\frac{\partial h^\tau}{\partial x^\tau}\right)\right] = -\varepsilon\mathscr{L}_h\tilde{\pi}^\mu(x) \tag{25a}$$

$$\begin{aligned}\delta\tilde{p}^{\alpha\mu}(x) = \tilde{P}^{\alpha\mu}(x) - \tilde{p}^{\alpha\mu}(x) &= \varepsilon\left[-h^\beta\frac{\partial\tilde{p}^{\alpha\mu}}{\partial x^\beta} + \tilde{p}^{\beta\mu}\frac{\partial h^\alpha}{\partial x^\beta} + \tilde{p}^{\alpha\beta}\left(\frac{\partial h^\mu}{\partial x^\beta} - \delta^\mu_\beta\frac{\partial h^\tau}{\partial x^\tau}\right)\right]\\ &= -\varepsilon\mathscr{L}_h\tilde{p}^{\alpha\mu}(x)\end{aligned} \tag{25b}$$

$$\delta\tilde{k}^{\lambda\tau\mu}(x) = \tilde{K}^{\lambda\tau\mu}(x) - \tilde{k}^{\lambda\tau\mu}(x) = \varepsilon\left[-h^\beta\frac{\partial\tilde{k}^{\lambda\tau\mu}}{\partial x^\beta} + \tilde{k}^{\beta\tau\mu}\frac{\partial h^\lambda}{\partial x^\beta} + \tilde{k}^{\lambda\beta\mu}\frac{\partial h^\tau}{\partial x^\beta}\right.$$

$$
\begin{aligned}
&+\tilde{k}^{\lambda\tau\beta}\left(\frac{\partial h^{\mu}}{\partial x^{\beta}}-\delta^{\mu}_{\beta}\frac{\partial h^{\tau}}{\partial x^{\tau}}\right)\Bigg]\\
&=-\varepsilon\mathscr{L}_{h}\tilde{k}^{\lambda\tau\mu}(x)
\end{aligned}
\tag{25c}
$$

and

$$
\begin{aligned}
\delta\tilde{q}_{\eta}{}^{\lambda\tau\mu}(x)=\tilde{Q}_{\eta}{}^{\lambda\tau\mu}(x)-\tilde{q}_{\eta}{}^{\lambda\tau\mu}(x)=\varepsilon\Bigg[&-h^{\beta}\frac{\partial\tilde{q}_{\eta}{}^{\lambda\tau\mu}}{\partial x^{\beta}}-\tilde{q}_{\beta}{}^{\lambda\tau\mu}\frac{\partial h^{\beta}}{\partial x^{\eta}}+\tilde{q}_{\eta}{}^{\beta\tau\mu}\frac{\partial h^{\lambda}}{\partial x^{\beta}}+\tilde{q}_{\eta}{}^{\lambda\beta\mu}\frac{\partial h^{\tau}}{\partial x^{\beta}}\\
&+\tilde{q}_{\eta}{}^{\lambda\tau\beta}\left(\frac{\partial h^{\mu}}{\partial x^{\beta}}-\delta^{\mu}_{\beta}\frac{\partial h^{\tau}}{\partial x^{\tau}}\right)\Bigg]=-\varepsilon\mathscr{L}_{h}\tilde{q}_{\eta}{}^{\lambda\tau\mu}(x).
\end{aligned}
\tag{25d}
$$

As a particular feature of the covariant Hamiltonian formalism of field theories, merely the *divergences* of momentum fields are determined by the system's Hamiltonian and not the individual components of the respective canonical momentum tensor. The canonical momentum tensors are thus determined only up to additional divergence-free tensors. The transformation rule for the *divergence* of the momenta $\tilde{\pi}^{\mu}$ is set up on the basis of Eq. (25a)

$$
\begin{aligned}
\frac{\partial\left(\delta\tilde{\pi}^{\mu}(x)\right)}{\partial x^{\mu}}=\frac{\partial\tilde{\Pi}^{\mu}}{\partial x^{\mu}}-\frac{\partial\tilde{\pi}^{\mu}}{\partial x^{\mu}}&=\varepsilon\left[\cancel{-\frac{\partial h^{\beta}}{\partial x^{\mu}}\frac{\partial\tilde{\pi}^{\mu}}{\partial x^{\beta}}}-h^{\beta}\frac{\partial^{2}\tilde{\pi}^{\mu}}{\partial x^{\beta}\partial x^{\mu}}+\frac{\partial\tilde{\pi}^{\beta}}{\partial x^{\mu}}\left(\cancel{\frac{\partial h^{\mu}}{\partial x^{\beta}}}-\delta^{\mu}_{\beta}\frac{\partial h^{\tau}}{\partial x^{\tau}}\right)\right]\\
&=-\varepsilon\left(h^{\mu}\frac{\partial^{2}\tilde{\pi}^{\beta}}{\partial x^{\mu}\partial x^{\beta}}+\frac{\partial h^{\mu}}{\partial x^{\mu}}\frac{\partial\tilde{\pi}^{\beta}}{\partial x^{\beta}}\right)\\
&=-\varepsilon\frac{\partial}{\partial x^{\mu}}\left(h^{\mu}\frac{\partial\tilde{\pi}^{\beta}}{\partial x^{\beta}}\right).
\end{aligned}
$$

We are thus allowed to replace the transformation rule (25a) by

$$
\delta\tilde{\pi}^{\mu}=-\varepsilon h^{\mu}\frac{\partial\tilde{\pi}^{\beta}}{\partial x^{\beta}}=\varepsilon h^{\mu}\frac{\partial\tilde{\mathscr{H}}}{\partial\phi},
\tag{26a}
$$

which amounts to replacing the momentum tensor components $\tilde{\pi}^{\mu}$ by modified components of an equivalent momentum tensor with the same divergence. Similarly, the divergences of the momenta $\tilde{p}^{\alpha\mu}$ transform as

$$
\begin{aligned}
\frac{\partial\tilde{P}^{\alpha\mu}}{\partial x^{\mu}}-\frac{\partial\tilde{p}^{\alpha\mu}}{\partial x^{\mu}}=\varepsilon\Bigg[&\cancel{-\frac{\partial h^{\beta}}{\partial x^{\mu}}\frac{\partial\tilde{p}^{\alpha\mu}}{\partial x^{\beta}}}-h^{\beta}\frac{\partial^{2}\tilde{p}^{\alpha\mu}}{\partial x^{\beta}\partial x^{\mu}}+\frac{\partial\tilde{p}^{\beta\mu}}{\partial x^{\mu}}\frac{\partial h^{\alpha}}{\partial x^{\beta}}+\tilde{p}^{\beta\mu}\frac{\partial^{2}h^{\alpha}}{\partial x^{\beta}\partial x^{\mu}}\\
&+\frac{\partial\tilde{p}^{\alpha\beta}}{\partial x^{\mu}}\left(\cancel{\frac{\partial h^{\mu}}{\partial x^{\beta}}}-\delta^{\mu}_{\beta}\frac{\partial h^{\tau}}{\partial x^{\tau}}\right)\Bigg]\\
=-\varepsilon\Bigg(&h^{\mu}\frac{\partial^{2}\tilde{p}^{\alpha\beta}}{\partial x^{\mu}\partial x^{\beta}}-\frac{\partial\tilde{p}^{\beta\mu}}{\partial x^{\mu}}\frac{\partial h^{\alpha}}{\partial x^{\beta}}-\tilde{p}^{\beta\mu}\frac{\partial^{2}h^{\alpha}}{\partial x^{\beta}\partial x^{\mu}}+\frac{\partial\tilde{p}^{\alpha\beta}}{\partial x^{\beta}}\frac{\partial h^{\mu}}{\partial x^{\mu}}\Bigg)
\end{aligned}
$$

$$= -\varepsilon \frac{\partial}{\partial x^{\mu}} \left(h^{\mu} \frac{\partial \tilde{p}^{\alpha\beta}}{\partial x^{\beta}} - \tilde{p}^{\beta\mu} \frac{\partial h^{\alpha}}{\partial x^{\beta}} \right).$$

The transformation rule for the momenta $\tilde{p}^{\alpha\mu}$ from Eq. (25b) can thus equivalently be expressed as

$$\delta \tilde{p}^{\alpha\mu} = -\varepsilon \left(h^{\mu} \frac{\partial \tilde{p}^{\alpha\beta}}{\partial x^{\beta}} - \tilde{p}^{\beta\mu} \frac{\partial h^{\alpha}}{\partial x^{\beta}} \right) = \varepsilon \left(h^{\mu} \frac{\partial \tilde{\mathscr{H}}}{\partial a_{\alpha}} + \tilde{p}^{\beta\mu} \frac{\partial h^{\alpha}}{\partial x^{\beta}} \right). \tag{26b}$$

The corresponding modified transformation rules apply for the momenta $\tilde{k}^{\lambda\xi\mu}$ and $\tilde{q}_{\eta}^{\ \lambda\tau\mu}$:

$$\delta \tilde{k}^{\lambda\xi\mu} = \varepsilon \left(h^{\mu} \frac{\partial \tilde{\mathscr{H}}}{\partial g_{\lambda\xi}} + \tilde{k}^{\beta\xi\mu} \frac{\partial h^{\lambda}}{\partial x^{\beta}} + \tilde{k}^{\lambda\beta\mu} \frac{\partial h^{\xi}}{\partial x^{\beta}} \right) \tag{26c}$$

$$\delta \tilde{q}_{\eta}^{\ \lambda\tau\mu} = \varepsilon \left(h^{\mu} \frac{\partial \tilde{\mathscr{H}}}{\partial \gamma^{\eta}_{\ \lambda\tau}} - \tilde{q}_{\beta}^{\ \lambda\tau\mu} \frac{\partial h^{\beta}}{\partial x^{\eta}} + \tilde{q}_{\eta}^{\ \beta\tau\mu} \frac{\partial h^{\lambda}}{\partial x^{\beta}} + \tilde{q}_{\eta}^{\ \lambda\beta\mu} \frac{\partial h^{\tau}}{\partial x^{\beta}} \right). \tag{26d}$$

6 The Noether Theorem in the Form of Eq. (19)

Finally, the finite canonical transformation rule for the Hamiltonian density from Eq. (21e) has the infinitesimal representation according to Eqs. (23):

$$\delta \tilde{\mathscr{H}}\Big|_{\mathrm{CT}} = -\varepsilon \left[\frac{\partial^2 h^{\beta}}{\partial x^{\alpha} \partial x^{\mu}} \left(\tilde{p}^{\alpha\mu} a_{\beta} + \tilde{k}^{\lambda\alpha\mu} g_{\beta\lambda} + \tilde{k}^{\alpha\lambda\mu} g_{\lambda\beta} + \tilde{q}_{\eta}^{\ \lambda\alpha\mu} \gamma^{\eta}_{\ \lambda\beta} + \tilde{q}_{\eta}^{\ \alpha\lambda\mu} \gamma^{\eta}_{\ \beta\lambda} \right.\right.$$
$$\left.\left. -\tilde{q}_{\beta}^{\ \eta\lambda\mu} \gamma^{\alpha}_{\ \eta\lambda} \right) + \frac{\partial^3 h^{\beta}}{\partial x^{\alpha} \partial x^{\eta} \partial x^{\mu}} \tilde{q}_{\beta}^{\ \alpha\eta\mu} \right].$$

In contrast, the variation of $\delta \tilde{\mathscr{H}}$ emerging from the transformations of the fields, as stated by Eq. (18), has the following infinitesimal representation according to Eqs. (24) and (26):

$$\delta \tilde{\mathscr{H}} = \varepsilon \left[\frac{\partial h^{\beta}}{\partial x^{\alpha}} \frac{\partial}{\partial x^{\mu}} \left(\tilde{p}^{\alpha\mu} a_{\beta} + \tilde{k}^{\lambda\alpha\mu} g_{\beta\lambda} + \tilde{k}^{\alpha\lambda\mu} g_{\lambda\beta} + \tilde{q}_{\eta}^{\ \lambda\alpha\mu} \gamma^{\eta}_{\ \lambda\beta} + \tilde{q}_{\eta}^{\ \alpha\lambda\mu} \gamma^{\eta}_{\ \beta\lambda} - \tilde{q}_{\beta}^{\ \eta\lambda\mu} \gamma^{\alpha}_{\ \eta\lambda} \right) \right.$$
$$\left. + \frac{\partial^2 h^{\beta}}{\partial x^{\alpha} \partial x^{\eta}} \frac{\partial \tilde{q}_{\beta}^{\ \alpha\eta\mu}}{\partial x^{\mu}} \right].$$

The difference of the transformations of the Hamiltonian yields the divergence:

$$\delta\tilde{\mathscr{H}} - \delta\tilde{\mathscr{H}}\Big|_{\mathrm{CT}} = \varepsilon\frac{\partial}{\partial x^{\mu}}\left[\frac{\partial h^{\beta}}{\partial x^{\alpha}}\left(\tilde{p}^{\alpha\mu}\, a_{\beta} + \tilde{k}^{\lambda\alpha\mu} g_{\beta\lambda} + \tilde{k}^{\alpha\lambda\mu} g_{\lambda\beta} + \tilde{q}_{\eta}{}^{\lambda\alpha\mu}\gamma^{\eta}{}_{\lambda\beta} + \tilde{q}_{\eta}{}^{\alpha\lambda\mu}\gamma^{\eta}{}_{\beta\lambda} - \tilde{q}_{\beta}{}^{\eta\lambda\mu}\gamma^{\alpha}{}_{\eta\lambda}\right) + \frac{\partial^2 h^{\beta}}{\partial x^{\alpha}\partial x^{\eta}}\tilde{q}_{\beta}{}^{\alpha\eta\mu}\right]. \tag{27}$$

According to Noether's theorem from Eq. (19), the divergence of the Noether current $\tilde{j}^{\mu}_{\mathrm{N}}$ vanishes exactly if $\delta\tilde{\mathscr{H}} - \delta\tilde{\mathscr{H}}\Big|_{\mathrm{CT}} = 0$. Then, the finite canonical transformation defined by the generating function (20) establishes a *symmetry transformation* which leaves the form of the given system invariant under the Diff(M) symmetry group—and thereby establishes the *General Principle of Relativity*. As we will see in the next section, requiring a conserved Noether current, namely a (weakly) vanishing of the right-hand side of Eq. (27), provides us with the generic theory of geometrodynamics.

7 The Noether Current $\tilde{j}^{\mu}_{\mathrm{N}}$ Associated with Diffeomorphism

The total set of canonical transformation rules (16) for the infinitesimal local translation (22) is:

$$-\frac{1}{\varepsilon}\delta^{\mu}_{\nu}\,\delta\phi = \frac{\partial\tilde{j}^{\mu}_{\mathrm{N}}}{\partial\tilde{\pi}^{\nu}} \stackrel{!}{=} \delta^{\mu}_{\nu} h^{\beta}\frac{\partial\phi}{\partial x^{\beta}} \tag{28a}$$

$$\frac{1}{\varepsilon}\delta\tilde{\pi}^{\mu} = \frac{\partial\tilde{j}^{\mu}_{\mathrm{N}}}{\partial\phi} \stackrel{!}{=} h^{\mu}\frac{\partial\tilde{\mathscr{H}}}{\partial\phi} \tag{28b}$$

$$-\frac{1}{\varepsilon}\delta^{\mu}_{\nu}\,\delta a_{\alpha} = \frac{\partial\tilde{j}^{\mu}_{\mathrm{N}}}{\partial\tilde{p}^{\alpha\nu}} \stackrel{!}{=} \delta^{\mu}_{\nu}\left(h^{\beta}\frac{\partial a_{\alpha}}{\partial x^{\beta}} + a_{\beta}\frac{\partial h^{\beta}}{\partial x^{\alpha}}\right) \tag{28c}$$

$$\frac{1}{\varepsilon}\delta\tilde{p}^{\alpha\mu} = \frac{\partial\tilde{j}^{\mu}_{\mathrm{N}}}{\partial a_{\alpha}} \stackrel{!}{=} h^{\mu}\frac{\partial\tilde{\mathscr{H}}}{\partial a_{\alpha}} + \tilde{p}^{\beta\mu}\frac{\partial h^{\alpha}}{\partial x^{\beta}} \tag{28d}$$

$$-\frac{1}{\varepsilon}\delta^{\mu}_{\nu}\,\delta g_{\xi\lambda} = \frac{\partial\tilde{j}^{\mu}_{\mathrm{N}}}{\partial\tilde{k}^{\lambda\xi\nu}} \stackrel{!}{=} \delta^{\mu}_{\nu}\left(h^{\beta}\frac{\partial g_{\xi\lambda}}{\partial x^{\beta}} + g_{\beta\lambda}\frac{\partial h^{\beta}}{\partial x^{\xi}} + g_{\beta\xi}\frac{\partial h^{\beta}}{\partial x^{\lambda}}\right) \tag{28e}$$

$$\frac{1}{\varepsilon}\delta\tilde{k}^{\lambda\xi\mu} = \frac{\partial\tilde{j}^{\mu}_{\mathrm{N}}}{\partial g_{\lambda\xi}} \stackrel{!}{=} h^{\mu}\frac{\partial\tilde{\mathscr{H}}}{\partial g_{\lambda\xi}} + \tilde{k}^{\beta\xi\mu}\frac{\partial h^{\lambda}}{\partial x^{\beta}} + \tilde{k}^{\lambda\beta\mu}\frac{\partial h^{\xi}}{\partial x^{\beta}} \tag{28f}$$

and

$$-\frac{1}{\varepsilon}\delta^{\mu}_{\nu}\,\delta\gamma^{\eta}{}_{\lambda\tau} = \frac{\partial\tilde{j}^{\mu}_{\mathrm{N}}}{\partial\tilde{q}_{\eta}{}^{\lambda\tau\nu}} \stackrel{!}{=} \delta^{\mu}_{\nu}\left(h^{\beta}\frac{\partial\gamma^{\eta}{}_{\lambda\tau}}{\partial x^{\beta}} - \gamma^{\beta}{}_{\lambda\tau}\frac{\partial h^{\eta}}{\partial x^{\beta}} + \gamma^{\eta}{}_{\beta\tau}\frac{\partial h^{\beta}}{\partial x^{\lambda}} + \gamma^{\eta}{}_{\lambda\beta}\frac{\partial h^{\beta}}{\partial x^{\tau}} + \frac{\partial^2 h^{\eta}}{\partial x^{\lambda}\partial x^{\tau}}\right) \tag{28g}$$

$$\frac{1}{\varepsilon}\delta\tilde{q}_{\eta}^{\ \lambda\tau\mu} = \frac{\partial \tilde{j}_{\mathrm{N}}^{\mu}}{\partial \gamma^{\eta}_{\ \lambda\tau}} \stackrel{!}{=} h^{\mu}\frac{\partial \tilde{\mathscr{H}}}{\partial \gamma^{\eta}_{\ \lambda\tau}} - \tilde{q}_{\beta}^{\ \lambda\tau\mu}\frac{\partial h^{\beta}}{\partial x^{\eta}} + \tilde{q}_{\eta}^{\ \beta\tau\mu}\frac{\partial h^{\lambda}}{\partial x^{\beta}} + \tilde{q}_{\eta}^{\ \lambda\beta\mu}\frac{\partial h^{\tau}}{\partial x^{\beta}}. \tag{28h}$$

The Noether current $\tilde{j}_{\mathrm{N}}^{\mu}$, which defines the particular infinitesimal transformation rules (28), follows as

$$\boxed{\tilde{j}_{\mathrm{N}}^{\mu} = h^{\beta}\tilde{B}_{\beta}^{\ \mu} + \frac{\partial h^{\beta}}{\partial x^{\alpha}}\tilde{C}_{\beta}^{\ \alpha\mu} + \frac{\partial^{2} h^{\beta}}{\partial x^{\alpha}\partial x^{\eta}}\,\tilde{q}_{\beta}^{\ \alpha\eta\mu},} \tag{29}$$

wherein $\tilde{B}_{\beta}^{\ \mu}$ abbreviates the sum of all terms emerging from Eqs. (28) proportional to h^{β}

$$\begin{aligned}\tilde{B}_{\beta}^{\ \mu} &= \tilde{\pi}^{\mu}\frac{\partial\phi}{\partial x^{\beta}} + \tilde{p}^{\alpha\mu}\frac{\partial a_{\alpha}}{\partial x^{\beta}} + \tilde{k}^{\alpha\lambda\mu}\frac{\partial g_{\alpha\lambda}}{\partial x^{\beta}} + \tilde{q}_{\eta}^{\ \alpha\lambda\mu}\frac{\partial \gamma^{\eta}_{\ \alpha\lambda}}{\partial x^{\beta}} \\ &\quad - \delta_{\beta}^{\mu}\left(\tilde{\pi}^{\tau}\frac{\partial\phi}{\partial x^{\tau}} + \tilde{p}^{\alpha\tau}\frac{\partial a_{\alpha}}{\partial x^{\tau}} + \tilde{k}^{\alpha\lambda\tau}\frac{\partial g_{\alpha\lambda}}{\partial x^{\tau}} + \tilde{q}_{\eta}^{\ \alpha\lambda\tau}\frac{\partial \gamma^{\eta}_{\ \alpha\lambda}}{\partial x^{\tau}} - \tilde{\mathscr{H}}\right),\end{aligned} \tag{30a}$$

whereas $\tilde{C}_{\beta}^{\ \alpha\mu}$ stands for the collection of the terms proportional to $\partial h^{\beta}/\partial x^{\alpha}$:

$$\tilde{C}_{\beta}^{\ \alpha\mu} = \tilde{p}^{\alpha\mu}\,a_{\beta} + \tilde{k}^{\lambda\alpha\mu}g_{\beta\lambda} + \tilde{k}^{\alpha\lambda\mu}g_{\lambda\beta} + \tilde{q}_{\eta}^{\ \lambda\alpha\mu}\gamma^{\eta}_{\ \lambda\beta} + \tilde{q}_{\eta}^{\ \alpha\lambda\mu}\gamma^{\eta}_{\ \beta\lambda} - \tilde{q}_{\beta}^{\ \eta\lambda\mu}\gamma^{\alpha}_{\ \eta\lambda}\,. \tag{30b}$$

According to the Hamiltonian form of Noether's theorem, the canonical transformation rules emerging from the generating function $\tilde{\mathscr{F}}_{3}^{\mu}$ (Eq. (15)) represent a symmetry transformation of the given system exactly if the divergence of the function $\tilde{j}_{\mathrm{N}}^{\mu}(x)$ contained therein vanishes:

$$\frac{\partial \tilde{j}_{\mathrm{N}}^{\mu}}{\partial x^{\mu}} = 0 \qquad \Leftrightarrow \qquad \tilde{j}_{\mathrm{N}}^{\mu} \text{ in } \tilde{\mathscr{F}}_{3}^{\mu} \text{ defines an (infinitesimal) symmetry transformation.}$$

Then, $\tilde{j}_{\mathrm{N}}^{\mu}$ represents the conserved Noether current. Here, $\tilde{B}_{\beta}^{\ \mu}$ is actually the *local* Hamiltonian representation of the canonical energy-momentum tensor density of the total dynamical system consisting of scalar and vector fields in conjunction with a dynamic metric and connection. As will be derived in the following section, the terms emerging from $\tilde{C}_{\beta}^{\ \alpha\mu}$ in the condition for a *conserved* Noether current convert the *partial* derivatives in $\tilde{B}_{\beta}^{\ \mu}$ into *covariant* derivatives—and hence into the *global* Hamiltonian representation of the canonical energy-momentum tensor of the total system.

For the particular case of $h^{\beta}(x) = \bar{h}^{\beta} + \tilde{h}^{\beta}_{\ \alpha}x^{\alpha}$ a *linear* function of x, Eq. (22) defines a generalized global Poincaré transformation,

$$X^{\beta}=x^{\beta}+\varepsilon\left(\bar{h}^{\beta}+\tilde{h}^{\beta}{}_{\alpha}x^{\alpha}\right),\qquad \bar{h}^{\beta},\tilde{h}^{\beta}{}_{\alpha}=\text{const.},$$

which reduces to a proper Poincaré transformation if the covariant representation of the matrix $\tilde{h}^{\beta}{}_{\alpha}$ is skew-symmetric. The condition for a conserved Noether current is then

$$\frac{\partial \tilde{j}_{\mathrm{N}}^{\mu}}{\partial x^{\mu}}=h^{\beta}\frac{\partial \tilde{B}_{\beta}{}^{\mu}}{\partial x^{\mu}}+\frac{\partial h^{\beta}}{\partial x^{\alpha}}\left(\tilde{B}_{\beta}{}^{\alpha}+\frac{\partial \tilde{C}_{\beta}{}^{\alpha\mu}}{\partial x^{\mu}}\right)\stackrel{!}{=}0,$$

which directly yields the conservation equations [12, 13]

$$\frac{\partial \tilde{B}_{\beta}{}^{\alpha}}{\partial x^{\alpha}}=0,\qquad \tilde{B}^{[\beta\alpha]}+\frac{\partial \tilde{C}^{[\beta\alpha]\mu}}{\partial x^{\mu}}=0$$

for both the energy-momentum and the angular momentum. The skew-symmetric part of $\tilde{p}^{\alpha\mu}a^{\beta}$ in α and β in the sum $\tilde{C}^{\beta\alpha\mu}$ then defines the canonical spin tensor of the source vector field. The Noether current (29) thus generalizes this case to a dynamic spacetime geometry if $h^{\beta}(x)$ stands for a differentiable vector function of spacetime with non-vanishing second and third derivatives.

8 Discussion of the Conserved Noether Current

For the general case, the divergence of the Noether current (29) is obtained as

$$\begin{aligned}\frac{\partial \tilde{j}_{\mathrm{N}}^{\mu}}{\partial x^{\mu}}&=h^{\beta}\frac{\partial \tilde{B}_{\beta}{}^{\mu}}{\partial x^{\mu}}+\frac{\partial h^{\beta}}{\partial x^{\alpha}}\left(\tilde{B}_{\beta}{}^{\alpha}+\frac{\partial \tilde{C}_{\beta}{}^{\alpha\mu}}{\partial x^{\mu}}\right)+\frac{\partial^{2} h^{\beta}}{\partial x^{\alpha}\partial x^{\eta}}\left(\tilde{C}_{\beta}{}^{\alpha\eta}+\frac{\partial \tilde{q}_{\beta}{}^{\alpha\eta\mu}}{\partial x^{\mu}}\right)\\&\quad+\frac{\partial^{3} h^{\beta}}{\partial x^{\alpha}\partial x^{\eta}\partial x^{\mu}}\,\tilde{q}_{\beta}{}^{\alpha\eta\mu}\stackrel{!}{=}0.\end{aligned}\qquad(31)$$

With this equation involving a vanishing *partial* derivative of the Noether current $\tilde{j}_{\mathrm{N}}^{\mu}$, it establishes a proper (local) conservation law. Yet, the field equations emerging from Eq. (31) will turn out to be tensor equations and thus hold invariantly in any reference frame. As $h^{\beta}(x)$ is supposed to be an arbitrary function of x, Eq. (31) has $4+16+40+80=140$ independent coefficients and thus defines a finite-dimensional subgroup of the infinite-dimensional diffeomorphism group. We note that in this description the torsion degrees of freedom do not emerge from separate dynamical quantities but are implicitly contained in the additional freedom of the Riemann–Cartan tensor, which is constructed on the basis of the 64 non-symmetric connection coefficients $\gamma^{\eta}{}_{\alpha\beta}$. In contrast, the Riemann tensor of Einstein's general relativity emerges from the 40 connection coefficients—referred to as Christoffel symbols—that are symmetric in their lower index pair.

The four separate conditions for each order of derivatives of $h^{\beta}(x)$ will be worked out in the following sections.

8.1 Condition 1: Term Proportional to the Third Partial Derivatives of h^{β}

The only term proportional to the third derivative of h^{β} is the canonical momentum $\tilde{q}_{\beta}^{\ \alpha\eta\mu}$, hence the dual of the partial x^{μ}-derivative of the connection $\gamma^{\beta}_{\ \alpha\eta}$. A necessary and sufficient condition for this term to vanish is that the (generally non-zero) momentum $\tilde{q}_{\beta}^{\ \alpha\eta\mu}$ is skew-symmetric in one of the index pairs formed out of α, η, and μ. We choose here the last index pair, namely η and μ, and define

$$\boxed{\tilde{q}_{\beta}^{\ \alpha\eta\mu} = -\tilde{q}_{\beta}^{\ \alpha\mu\eta}} \tag{32}$$

which implies that $\tilde{q}_{\beta}^{\ \alpha\eta\mu}$ need not in addition be skew-symmetric in α and η. Equation (31) then simplifies to

$$\frac{\partial \tilde{j}_{\mathrm{N}}^{\mu}}{\partial x^{\mu}} = h^{\beta}\frac{\partial \tilde{B}_{\beta}^{\ \mu}}{\partial x^{\mu}} + \frac{\partial h^{\beta}}{\partial x^{\alpha}}\left(\tilde{B}_{\beta}^{\ \alpha} + \frac{\partial \tilde{C}_{\beta}^{\ \alpha\mu}}{\partial x^{\mu}}\right) + \frac{\partial^2 h^{\beta}}{\partial x^{\alpha}\partial x^{\eta}}\left(\tilde{C}_{\beta}^{\ \alpha\eta} + \frac{\partial \tilde{q}_{\beta}^{\ \alpha\eta\mu}}{\partial x^{\mu}}\right) \stackrel{!}{=} 0. \tag{33}$$

8.2 Condition 2: Terms Proportional to the Second Partial Derivatives of h^{β}

A zero divergence of the Noether current for any symmetry transformation (22)—hence for arbitrary functions $h^{\beta}(x)$—requires in particular that the sum of terms related to the *second* derivatives of $h^{\beta}(x)$ in Eq. (33) vanishes. This means with $\tilde{C}_{\beta}^{\ \alpha\eta}$ from Eq. (30b) inserted into the last term of Eq. (33):

$$\frac{\partial^2 h^{\beta}}{\partial x^{\alpha}\partial x^{\eta}}\left(\frac{\partial \tilde{q}_{\beta}^{\ \alpha\eta\mu}}{\partial x^{\mu}} + \tilde{p}^{\alpha\eta}\,a_{\beta} + \tilde{k}^{\lambda\alpha\eta}g_{\beta\lambda} + \tilde{k}^{\alpha\lambda\eta}g_{\lambda\beta} + \cancel{\tilde{q}_{\tau}^{\ \lambda\alpha\eta}\gamma^{\tau}_{\ \lambda\beta}} + \tilde{q}_{\tau}^{\ \alpha\lambda\eta}\gamma^{\tau}_{\ \beta\lambda} - \tilde{q}_{\beta}^{\ \tau\lambda\eta}\gamma^{\alpha}_{\ \tau\lambda}\right) \stackrel{!}{=} 0.$$

Due to the symmetry of the second partial derivatives of h^{β} in α and η, the term $\tilde{q}_{\tau}^{\ \lambda\alpha\eta}\gamma^{\tau}_{\ \lambda\beta}$ drops out by virtue of the skew-symmetry condition (32). As no symmetries in α and η are implied in the remaining terms, one encounters the sufficient condition

$$\boxed{\frac{\partial \tilde{q}_{\beta}^{\ \alpha\eta\mu}}{\partial x^{\mu}} + \tilde{p}^{\alpha\eta}\,a_{\beta} + \tilde{k}^{\lambda\alpha\eta}g_{\beta\lambda} + \tilde{k}^{\alpha\lambda\eta}g_{\lambda\beta} - \tilde{q}_{\xi}^{\ \alpha\eta\lambda}\gamma^{\xi}_{\ \beta\lambda} - \tilde{q}_{\beta}^{\ \xi\lambda\eta}\gamma^{\alpha}_{\ \xi\lambda} = 0.} \tag{34}$$

We can express Eq. (34) equivalently as the tensor equation

$$\tilde{q}_{\beta}{}^{\alpha\eta\mu}{}_{;\mu} + \tilde{p}^{\alpha\eta}\,a_{\beta} + \tilde{k}^{\lambda\alpha\eta} g_{\beta\lambda} + \tilde{k}^{\alpha\lambda\eta} g_{\lambda\beta} - \tilde{q}_{\beta}{}^{\alpha\tau\mu} s^{\eta}{}_{\tau\mu} - 2\tilde{q}_{\beta}{}^{\alpha\eta\mu} s^{\xi}{}_{\mu\xi} = 0, \qquad (35)$$

with $s^{\eta}{}_{\tau\mu} \equiv \gamma^{\eta}{}_{[\tau\mu]}$ the Cartan torsion tensor. It agrees with the corresponding field equation (56) of Ref. [2]. Its implications will be discussed in Sects. 9.2 and 9.4.

With Eq. (34), the condition (33) for the divergence of the Noether current now further simplifies to

$$\frac{\partial \tilde{j}_{\mathrm{N}}^{\mu}}{\partial x^{\mu}} = h^{\beta} \frac{\partial \tilde{B}_{\beta}{}^{\mu}}{\partial x^{\mu}} + \frac{\partial h^{\beta}}{\partial x^{\alpha}} \left(\tilde{B}_{\beta}{}^{\alpha} + \frac{\partial \tilde{C}_{\beta}{}^{\alpha\mu}}{\partial x^{\mu}} \right) \stackrel{!}{=} 0. \qquad (36)$$

8.3 Condition 3: Terms Proportional to the First Partial Derivatives of h^{β}

For a generally conserved Noether current, the coefficient proportional to the first derivative of h^{β} in Eq. (36) must vanish as well, hence

$$\tilde{B}_{\beta}{}^{\alpha} + \frac{\partial \tilde{C}_{\beta}{}^{\alpha\mu}}{\partial x^{\mu}} \stackrel{!}{=} 0. \qquad (37)$$

Equation (37) writes in expanded form with $\tilde{C}_{\beta}{}^{\alpha\mu}$ from Eq. (30b)

$$\tilde{B}_{\beta}{}^{\alpha} + \frac{\partial}{\partial x^{\mu}} \left(\tilde{p}^{\alpha\mu}\,a_{\beta} + \tilde{k}^{\lambda\alpha\mu} g_{\beta\lambda} + \tilde{k}^{\alpha\lambda\mu} g_{\lambda\beta} + \tilde{q}_{\eta}{}^{\lambda\alpha\mu} \gamma^{\eta}{}_{\lambda\beta} + \tilde{q}_{\eta}{}^{\alpha\lambda\mu} \gamma^{\eta}{}_{\beta\lambda} - \tilde{q}_{\beta}{}^{\eta\lambda\mu} \gamma^{\alpha}{}_{\eta\lambda} \right) = 0,$$

which is expressed equivalently inserting Eq. (34)

$$\tilde{B}_{\beta}{}^{\alpha} + \frac{\partial}{\partial x^{\mu}} \left(-\frac{\partial \tilde{q}_{\beta}{}^{\alpha\mu\eta}}{\partial x^{\eta}} + \tilde{q}_{\eta}{}^{\lambda\alpha\mu} \gamma^{\eta}{}_{\lambda\beta} \right) = 0.$$

This equation reduces due to the skew-symmetry of $\tilde{q}_{\beta}{}^{\alpha\mu\eta}$ in its last index pair to

$$\boxed{\tilde{B}_{\beta}{}^{\alpha} + \frac{\partial}{\partial x^{\mu}} \left(\tilde{q}_{\eta}{}^{\lambda\alpha\mu} \gamma^{\eta}{}_{\lambda\beta} \right) = 0.} \qquad (38)$$

As $\tilde{B}_{\beta}{}^{\alpha}$—defined by Eq. (30a)—is the local representation of the canonical energy-momentum tensor of the *total* system of source fields and dynamic spacetime, Eq. (38) establishes a correlation of this (pseudo-)tensor with the dynamic spacetime. The explicit form of this equation will be discussed in Sect. 8.5.

8.4 Condition 4: Term Proportional to h^β

Finally, the term proportional to h^β in Eq. (36) must separately vanish

$$\boxed{\frac{\partial \tilde{B}_\beta{}^\mu}{\partial x^\mu} = 0.} \tag{39}$$

Equation (39) thus establishes a *local* energy and momentum conservation law of the *total* system of scalar and vector source fields on the one hand, and the dynamic spacetime, described by the metric and the connection on the other hand. It turns out to coincide with the divergence of Eq. (38) by virtue of the skew-symmetry of $\tilde{q}_\eta{}^{\lambda\alpha\mu}$ in its last index pair

$$\begin{aligned}
0 &= \frac{\partial}{\partial x^\mu}\left(\tilde{B}_\beta{}^\mu + \frac{\partial \tilde{q}_\eta{}^{\lambda\mu\alpha}}{\partial x^\alpha}\gamma^\eta{}_{\lambda\beta} + \tilde{q}_\eta{}^{\lambda\mu\alpha}\frac{\partial \gamma^\eta{}_{\lambda\beta}}{\partial x^\alpha}\right)\\
&= \frac{\partial \tilde{B}_\beta{}^\mu}{\partial x^\mu} + \frac{\partial^2 \tilde{q}_\eta{}^{\lambda\mu\alpha}}{\partial x^\mu \partial x^\alpha}\gamma^\eta{}_{\lambda\beta} + \frac{\partial \tilde{q}_\eta{}^{\lambda\mu\alpha}}{\partial x^\alpha}\frac{\partial \gamma^\eta{}_{\lambda\beta}}{\partial x^\mu} + \frac{\partial \tilde{q}_\eta{}^{\lambda\mu\alpha}}{\partial x^\mu}\frac{\partial \gamma^\eta{}_{\lambda\beta}}{\partial x^\alpha} + \tilde{q}_\eta{}^{\lambda\mu\alpha}\frac{\partial^2 \gamma^\eta{}_{\lambda\beta}}{\partial x^\mu \partial x^\alpha}\\
&= \frac{\partial \tilde{B}_\beta{}^\mu}{\partial x^\mu}.
\end{aligned}$$

Equation (39) is thus equivalent to a vanishing divergence of Eq. (38) as the divergence of its last term vanishes identically. From Eqs. (39) and (37), one concludes that

$$0 = \frac{\partial \tilde{B}_\beta{}^\alpha}{\partial x^\alpha} = -\frac{\partial^2 \tilde{C}_\beta{}^{\alpha\mu}}{\partial x^\alpha \partial x^\mu} = -\frac{\partial^2}{\partial x^\alpha \partial x^\mu}\left(\tilde{C}_\beta{}^{\alpha\mu} + \frac{\partial \tilde{q}_\beta{}^{\alpha\mu\lambda}}{\partial x^\lambda}\right),$$

which is indeed satisfied owing to Eq. (34). This demonstrates the consistency of the set of equations (32), (34), (38), and (39), which were obtained from the Noether condition (31).

8.5 Amended Canonical Energy-Momentum $\tilde{\Theta}_\nu{}^\mu$

We now express Eq. (38) in expanded form by inserting the local representation of the canonical energy-momentum tensor $\tilde{B}_\beta{}^\alpha$ from Eq. (30a) and the field equation for the partial divergence of $\tilde{q}_\eta{}^{\lambda\alpha\mu}$ from Eq. (34):

$$\begin{aligned}
\tilde{\Theta}_\beta{}^\alpha &\equiv \tilde{B}_\beta{}^\alpha + \frac{\partial}{\partial x^\mu}\left(\tilde{q}_\eta{}^{\lambda\alpha\mu}\gamma^\eta{}_{\lambda\beta}\right)\\
&= \tilde{\pi}^\alpha\frac{\partial \phi}{\partial x^\beta} + \tilde{p}^{\xi\alpha}\frac{\partial a_\xi}{\partial x^\beta} - \tilde{p}^{\xi\alpha}a_\eta\gamma^\eta{}_{\xi\beta} + \tilde{k}^{\xi\lambda\alpha}\frac{\partial g_{\xi\lambda}}{\partial x^\beta} - \tilde{k}^{\lambda\xi\alpha}g_{\eta\lambda}\gamma^\eta{}_{\xi\beta} - \tilde{k}^{\xi\lambda\alpha}g_{\lambda\eta}\gamma^\eta{}_{\xi\beta}
\end{aligned}$$

$$+\tilde{q}_{\eta}^{\ \xi\lambda\alpha}\left(\frac{\partial\gamma^{\eta}_{\ \xi\lambda}}{\partial x^{\beta}}-\frac{\partial\gamma^{\eta}_{\ \xi\beta}}{\partial x^{\lambda}}+\gamma^{\eta}_{\ \tau\beta}\,\gamma^{\tau}_{\ \xi\lambda}-\gamma^{\eta}_{\ \tau\lambda}\,\gamma^{\tau}_{\ \xi\beta}\right)$$
$$-\,\delta^{\alpha}_{\beta}\left(\tilde{\pi}^{\tau}\frac{\partial\phi}{\partial x^{\tau}}+\tilde{p}^{\,\xi\tau}\frac{\partial a_{\xi}}{\partial x^{\tau}}+\tilde{k}^{\xi\lambda\tau}\frac{\partial g_{\xi\lambda}}{\partial x^{\tau}}+\tilde{q}_{\eta}^{\ \xi\lambda\tau}\frac{\partial\gamma^{\eta}_{\ \xi\lambda}}{\partial x^{\tau}}-\tilde{\mathcal{H}}\right)=0. \quad (40)$$

The term proportional to $\tilde{q}_{\eta}^{\ \xi\lambda\alpha}$ is exactly the Riemann tensor, defined in the convention of Misner et al. [14] by

$$R^{\eta}_{\ \xi\beta\lambda}=\frac{\partial\gamma^{\eta}_{\ \xi\lambda}}{\partial x^{\beta}}-\frac{\partial\gamma^{\eta}_{\ \xi\beta}}{\partial x^{\lambda}}+\gamma^{\eta}_{\ \tau\beta}\,\gamma^{\tau}_{\ \xi\lambda}-\gamma^{\eta}_{\ \tau\lambda}\,\gamma^{\tau}_{\ \xi\beta}\,. \quad (41)$$

We remark that the tensor (41) is actually the *Riemann–Cartan tensor*, as it is defined here from a non-symmetric connection, $\gamma^{\eta}_{\ [\xi\lambda]}\not\equiv 0$. Moreover, the torsion—hence the addressed skew-symmetric part of the connection—does *not* emerge as a separate dynamic quantity in our description as all terms containing the connection are absorbed into the covariant derivatives and into the Riemann–Cartan tensor.

Equation (40) now writes equivalently after merging the partial derivatives with the γ-dependent terms into covariant derivatives:

$$\tilde{\Theta}_{\beta}^{\ \alpha}=\tilde{\pi}^{\alpha}\frac{\partial\phi}{\partial x^{\beta}}+\tilde{p}^{\,\xi\alpha}\,a_{\xi;\beta}+\tilde{k}^{\xi\lambda\alpha}\,g_{\xi\lambda;\beta}-\tilde{q}_{\eta}^{\ \xi\lambda\alpha}\,R^{\eta}_{\ \xi\lambda\beta}$$
$$-\,\delta^{\alpha}_{\beta}\left(\tilde{\pi}^{\tau}\frac{\partial\phi}{\partial x^{\tau}}+\tilde{p}^{\,\xi\tau}\frac{\partial a_{\xi}}{\partial x^{\tau}}+\tilde{k}^{\xi\lambda\tau}\frac{\partial g_{\xi\lambda}}{\partial x^{\tau}}+\tilde{q}_{\eta}^{\ \xi\lambda\tau}\frac{\partial\gamma^{\eta}_{\ \xi\lambda}}{\partial x^{\tau}}-\tilde{\mathcal{H}}\right)=0. \quad (42)$$

The remaining partial derivatives can similarly be rewritten as tensors if we subtract the corresponding "gauge Hamiltonian" terms from the total Hamiltonian $\tilde{\mathcal{H}}(\tilde{\pi},\phi,\tilde{p},a,\tilde{k},g,\tilde{q},\gamma)$:

$$\tilde{\mathcal{H}}_{0}=\tilde{\mathcal{H}}-\tilde{\mathcal{H}}_{\mathrm{G}}, \quad (43)$$

with $\tilde{\mathcal{H}}_{\mathrm{G}}(\tilde{p},a,\tilde{k},g,\tilde{q},\gamma)$ given by

$$\tilde{\mathcal{H}}_{\mathrm{G}}=\tilde{p}^{\,\xi\tau}a_{\eta}\gamma^{\eta}_{\ \xi\tau}+\tilde{k}^{\xi\lambda\tau}\,g_{\eta\xi}\gamma^{\eta}_{\ \lambda\tau}+\tilde{k}^{\lambda\xi\tau}\,g_{\xi\eta}\gamma^{\eta}_{\ \lambda\tau}+\tilde{q}_{\eta}^{\ \xi\lambda\tau}\gamma^{\eta}_{\ \alpha\lambda}\,\gamma^{\alpha}_{\ \xi\tau}\,, \quad (44)$$

which agrees with the gauge Hamiltonian derived in Eq. (32) of Ref. [2]. The partial derivatives of the fields in Eq. (42) are thus converted into covariant derivatives, whereas the partial derivative of the connection reemerges as one-half the Riemann tensor:

$$\tilde{\Theta}_{\nu}^{\ \mu}=\tilde{\pi}^{\mu}\frac{\partial\phi}{\partial x^{\nu}}+\tilde{p}^{\,\alpha\mu}a_{\alpha;\nu}+\tilde{k}^{\alpha\beta\mu}g_{\alpha\beta;\nu}-\tilde{q}_{\eta}^{\ \alpha\beta\mu}R^{\eta}_{\ \alpha\beta\nu}$$
$$-\,\delta^{\mu}_{\nu}\left(\tilde{\pi}^{\tau}\frac{\partial\phi}{\partial x^{\tau}}+\tilde{p}^{\,\alpha\tau}a_{\alpha;\tau}+\tilde{k}^{\alpha\beta\tau}g_{\alpha\beta;\tau}-\tfrac{1}{2}\tilde{q}_{\eta}^{\ \alpha\beta\tau}R^{\eta}_{\ \alpha\beta\tau}-\tilde{\mathcal{H}}_{0}\right)=0. \quad (45)$$

$\tilde{\Theta}_{\nu}{}^{\mu}$ obviously represents the canonical energy-momentum tensor of the closed total system of dynamical fields and spacetime. The non-existing covariant x^{τ}-derivative of the connection $\gamma^{\eta}{}_{\alpha\beta}$ happens to be replaced by $-\frac{1}{2}R^{\eta}{}_{\alpha\beta\tau}$.

9 Discussion of the Field Equations in the Lagrangian Description

9.1 Lagrangian Representation of the Noether Condition (45)

The sum in parentheses of Eq. (42) represents the Lagrangian $\tilde{\mathscr{L}}$ of the total dynamical system established by the scalar field, the vector field, the metric, and the connection. This Lagrangian must be a world scalar density in order for Eq. (42) to be a tensor equation, hence to be form-invariant under the $\mathrm{Diff}(M)$ symmetry group. The Lagrangian $\tilde{\mathscr{L}}$ is thus equivalently expressed as

$$\tilde{\mathscr{L}} = \tilde{\pi}^{\tau}\frac{\partial\phi}{\partial x^{\tau}} + \tilde{p}^{\alpha\tau}a_{\alpha;\tau} + \tilde{k}^{\alpha\beta\tau}g_{\alpha\beta;\tau} - \tfrac{1}{2}\tilde{q}_{\eta}{}^{\alpha\beta\tau}R^{\eta}{}_{\alpha\beta\tau} - \tilde{\mathscr{H}}_0. \tag{46}$$

The field equations (35) and (45) will be rewritten in the following on the basis of this Lagrangian. The canonical momenta are obtained from the Lagrangian (46) as

$$\tilde{\pi}^{\mu} = \frac{\partial\tilde{\mathscr{L}}}{\partial\left(\frac{\partial\phi}{\partial x^{\mu}}\right)}, \qquad \tilde{p}^{\alpha\mu} = \frac{\partial\tilde{\mathscr{L}}}{\partial a_{\alpha;\mu}} \tag{47a}$$

$$\tilde{k}^{\alpha\beta\mu} = \frac{\partial\tilde{\mathscr{L}}}{\partial g_{\alpha\beta;\mu}}, \qquad -\tfrac{1}{2}\tilde{q}_{\eta}{}^{\alpha\beta\mu} = \frac{\partial\tilde{\mathscr{L}}}{\partial R^{\eta}{}_{\alpha\beta\mu}}. \tag{47b}$$

The Noether condition (45)—hence the assertion of a vanishing canonical energy-momentum tensor $\tilde{\Theta}_{\nu}{}^{\mu}=0$ of the total system of dynamical fields and spacetime described by $\tilde{\mathscr{L}}$—is now encountered in the equivalent form:

$$\tilde{\Theta}_{\nu}{}^{\mu} = \frac{\partial\tilde{\mathscr{L}}}{\partial\left(\frac{\partial\phi}{\partial x^{\mu}}\right)}\frac{\partial\phi}{\partial x^{\nu}} + \frac{\partial\tilde{\mathscr{L}}}{\partial a_{\alpha;\mu}}a_{\alpha;\nu} + \frac{\partial\tilde{\mathscr{L}}}{\partial g_{\alpha\beta;\mu}}g_{\alpha\beta;\nu} + 2\frac{\partial\tilde{\mathscr{L}}}{\partial R^{\eta}{}_{\alpha\beta\mu}}R^{\eta}{}_{\alpha\beta\nu} - \delta^{\mu}_{\nu}\tilde{\mathscr{L}} = 0. \tag{48}$$

We may now split the Lagrangian $\tilde{\mathscr{L}}$ of the total system into a Lagrangian $\tilde{\mathscr{L}}_0$ for the dynamics of the base fields ϕ and a_{μ}, a Lagrangian $\tilde{\mathscr{L}}_g$ for the dynamics of the metric $g_{\mu\nu}$, and a Lagrangian $\tilde{\mathscr{L}}_R$ for the dynamics of the free gravitational field $R^{\eta}{}_{\alpha\beta\nu}$ according to

$$\tilde{\mathscr{L}} = \tilde{\mathscr{L}}_0 + \tilde{\mathscr{L}}_R + \tilde{\mathscr{L}}_g,\quad \tilde{\mathscr{L}}_0 = \tilde{\mathscr{L}}_0(\phi,\partial\phi,a,\partial a,g,\gamma),\quad \tilde{\mathscr{L}}_g = \tilde{\mathscr{L}}_g(g,\partial g,\gamma),$$
$$\tilde{\mathscr{L}}_R = \tilde{\mathscr{L}}_R(\gamma,\partial\gamma,g),$$

where each Lagrangian represents separately a world scalar density. As no derivative with respect to the metric appears in Eq. (48), we are allowed to divide all terms by $\sqrt{-g}$, whereby the field equation acquires the form of the *generic Einstein-type equation*:

$$2\frac{\partial\mathscr{L}_R}{\partial R^{\eta}{}_{\alpha\beta\mu}}R^{\eta}{}_{\alpha\beta\nu}+\frac{\partial\mathscr{L}_g}{\partial g_{\alpha\beta;\mu}}g_{\alpha\beta;\nu}-\delta^{\mu}_{\nu}\left(\mathscr{L}_R+\mathscr{L}_g\right)=-\frac{\partial\mathscr{L}_0}{\partial\left(\frac{\partial\phi}{\partial x^{\mu}}\right)}\frac{\partial\phi}{\partial x^{\nu}}-\frac{\partial\mathscr{L}_0}{\partial a_{\alpha;\mu}}a_{\alpha;\nu}+\delta^{\mu}_{\nu}\mathscr{L}_0. \tag{49}$$

The left-hand side of Eq. (49), pertaining to $\mathscr{L}_R+\mathscr{L}_g$, can be regarded as the *canonical* energy-momentum tensor of dynamical spacetime. With the right-hand side the negative canonical energy-momentum tensor of the system $\mathscr{L}_0$,

$$\theta_{\nu}{}^{\mu}\equiv\frac{\partial\mathscr{L}_0}{\partial\left(\frac{\partial\phi}{\partial x^{\mu}}\right)}\frac{\partial\phi}{\partial x^{\nu}}+\frac{\partial\mathscr{L}_0}{\partial a_{\alpha;\mu}}a_{\alpha;\nu}-\delta^{\mu}_{\nu}\mathscr{L}_0, \tag{50}$$

Eq. (49) thus establishes a *generally covariant energy-momentum balance relation*, with the coupling of spacetime and source fields induced by both, the metric $g_{\alpha\beta}$ and the connection $\gamma^{\alpha}{}_{\xi\tau}$. Hence, the energy-momentum tensor of the closed total system $\mathscr{L}=\mathscr{L}_0+\mathscr{L}_R+\mathscr{L}_g$ is equal to zero—a result also found by Jordan [15] and, independently in each case, by Sciama [16], Feynman [17, p. 10], and Hawking [18], based on different physical reasoning. A vanishing total energy-momentum tensor of the universe—taken, by definition, as a closed system—is commonly referred to as the *zero-energy principle*. As we see now, the zero-energy principle follows directly from the requirement that the action integral (3) be diffeomorphism invariant, which has the physical content that the laws of physics should take the same form in all reference frames—which is exactly the gist of the *Principle of General Relativity*.

For the particular case of a covariantly conserved metric, $g_{\alpha\beta;\nu}\equiv 0$, the respective terms in Eq. (49) drop out. This yields the generic Einstein-type equation for the case of *metric compatibility*:

$$\boxed{2\frac{\partial\mathscr{L}_R}{\partial R^{\eta}{}_{\alpha\beta\mu}}R^{\eta}{}_{\alpha\beta\nu}-\delta^{\mu}_{\nu}\mathscr{L}_R=-\theta_{\nu}{}^{\mu}.} \tag{51}$$

It applies for all Lagrangians $\mathscr{L}_R$ which (i) describe the observed dynamics of the "free" (uncoupled) gravitational field and (ii) entail a consistent field equation with regard to its trace, its symmetries, and its covariant derivatives. Obviously, the expression

$$\vartheta_{\nu}{}^{\mu}\equiv 2\frac{\partial\mathscr{L}_R}{\partial R^{\eta}{}_{\alpha\beta\mu}}R^{\eta}{}_{\alpha\beta\nu}-\delta^{\mu}_{\nu}\mathscr{L}_R$$

can be interpreted as the *energy-momentum tensor of spacetime*. The zero-energy principle from Eq. (49) then reduces to $\vartheta_{\nu}{}^{\mu}+\theta_{\nu}{}^{\mu}=0$.

In general, the covariant and contravariant representations of Eq. (51) are not necessarily symmetric in ν and μ and thus include a possible spin of the source field and the then emerging torsion of spacetime. We thus obtain the following relation of spin and torsion:

$$\boxed{\frac{\partial \mathscr{L}_R}{\partial R^{\eta}{}_{\alpha\beta\nu}} R^{\eta}{}_{\alpha\beta}{}^{\mu} - \frac{\partial \mathscr{L}_R}{\partial R^{\eta}{}_{\alpha\beta\mu}} R^{\eta}{}_{\alpha\beta}{}^{\nu} = \theta^{[\nu\mu]}.} \tag{52}$$

For the Hilbert Lagrangian—and even for Lagrangians with additional quadratic terms in $R^{\eta}{}_{\alpha\beta}{}^{\mu}$— the left-hand side of this equation simplifies to the skew-symmetric part of the Ricci tensor

$$R^{[\nu\mu]} = 8\pi G\, \theta^{[\nu\mu]}. \tag{53}$$

For the particular case of the Hilbert Lagrangian $\mathscr{L}_R = \mathscr{L}_{R,H} = -R/16\pi G$, and $\theta^{\nu\mu}$ being the canonical energy-momentum tensor of the Klein–Gordon system—which is symmetric and coincides with the metric one—Eq. (51) yields the proper Einstein equation, in conjunction with a vanishing source term for the torsion of spacetime. We will discuss these issues in detail in Sect. 10.

9.2 Consistency Relation

As the second partial derivative of $\tilde{q}_{\beta}{}^{\alpha\eta\mu}$ vanishes identically by virtue of Eq. (32), we conclude from the field equation Eq. (34):

$$\frac{\partial^2 \tilde{q}_{\beta}{}^{\alpha\eta\mu}}{\partial x^{\eta}\partial x^{\mu}} \equiv 0 \quad \Rightarrow \quad \frac{\partial}{\partial x^{\eta}}\left(\tilde{p}^{\alpha\eta}\, a_{\beta} + \tilde{k}^{\lambda\alpha\eta} g_{\beta\lambda} + \tilde{k}^{\alpha\lambda\eta} g_{\lambda\beta} + \tilde{q}_{\xi}{}^{\alpha\lambda\eta}\gamma^{\xi}{}_{\beta\lambda} - \tilde{q}_{\beta}{}^{\xi\lambda\eta}\gamma^{\alpha}{}_{\xi\lambda}\right) = 0.$$

Two terms cancel after inserting the divergence of $\tilde{q}_{\beta}{}^{\alpha\eta\mu}$ from Eq. (34)

$$\begin{aligned}
&\frac{\partial \tilde{p}^{\alpha\eta}}{\partial x^{\eta}} a_{\beta} + \tilde{p}^{\alpha\eta}\frac{\partial a_{\beta}}{\partial x^{\eta}} + 2\frac{\partial \tilde{k}^{\lambda\alpha\eta}}{\partial x^{\eta}} g_{\lambda\beta} + 2\tilde{k}^{\lambda\alpha\eta}\frac{\partial g_{\lambda\beta}}{\partial x^{\eta}}\\
&-\left(\tilde{p}^{\alpha\lambda}\, a_{\xi} + 2\tilde{k}^{\tau\alpha\lambda} g_{\tau\xi} - \tilde{q}_{\eta}{}^{\alpha\lambda\tau}\gamma^{\eta}{}_{\xi\tau} - \cancel{\tilde{q}_{\xi}{}^{\tau\eta\lambda}\gamma^{\alpha}{}_{\tau\eta}}\right)\gamma^{\xi}{}_{\beta\lambda} + \tilde{q}_{\xi}{}^{\alpha\lambda\eta}\frac{\partial \gamma^{\xi}{}_{\beta\lambda}}{\partial x^{\eta}}\\
&+\left(\tilde{p}^{\xi\lambda}\, a_{\beta} + 2\tilde{k}^{\tau\xi\lambda} g_{\tau\beta} - \cancel{\tilde{q}_{\eta}{}^{\xi\lambda\tau}\gamma^{\eta}{}_{\beta\tau}} - \tilde{q}_{\beta}{}^{\eta\tau\lambda}\gamma^{\xi}{}_{\eta\tau}\right)\gamma^{\alpha}{}_{\xi\lambda} - \tilde{q}_{\beta}{}^{\xi\lambda\eta}\frac{\partial \gamma^{\alpha}{}_{\xi\lambda}}{\partial x^{\eta}} = 0.
\end{aligned} \tag{54}$$

The terms in Eq. (54) can now be rearranged such that the partial derivatives are converted into covariant derivatives:

$$\begin{aligned}
&\left(\frac{\partial \tilde{p}^{\alpha\eta}}{\partial x^{\eta}} + \tilde{p}^{\xi\eta}\gamma^{\alpha}{}_{\xi\eta}\right) a_{\beta} + \tilde{p}^{\alpha\eta}\left(\frac{\partial a_{\beta}}{\partial x^{\eta}} - a_{\xi}\gamma^{\xi}{}_{\beta\eta}\right)\\
&+2\left(\frac{\partial \tilde{k}^{\lambda\alpha\eta}}{\partial x^{\eta}} + \tilde{k}^{\xi\alpha\eta}\gamma^{\lambda}{}_{\xi\eta} + \tilde{k}^{\lambda\xi\eta}\gamma^{\alpha}{}_{\xi\eta}\right) g_{\lambda\beta} + 2\tilde{k}^{\lambda\alpha\eta}\left(\frac{\partial g_{\lambda\beta}}{\partial x^{\eta}} - g_{\xi\beta}\gamma^{\xi}{}_{\lambda\eta} - g_{\lambda\xi}\gamma^{\xi}{}_{\beta\eta}\right)
\end{aligned}$$

$$+\tilde{q}_{\xi}^{\ \alpha\lambda\eta}\left(\frac{\partial\gamma^{\xi}_{\ \beta\lambda}}{\partial x^{\eta}}+\gamma^{\xi}_{\ \tau\eta}\gamma^{\tau}_{\ \beta\lambda}\right)-\tilde{q}_{\beta}^{\ \xi\lambda\eta}\left(\frac{\partial\gamma^{\alpha}_{\ \xi\lambda}}{\partial x^{\eta}}+\gamma^{\alpha}_{\ \tau\eta}\gamma^{\tau}_{\ \xi\lambda}\right)=0. \tag{55}$$

Due to the skew-symmetry of $\tilde{q}_{\xi}^{\ \alpha\lambda\eta}$ in its last index pair, the last two terms are one-half the Riemann curvature tensor (41), respectively. The sums proportional to a_β and $g_{\lambda\beta}$ in Eq. (55) can be expressed as canonical field equations [2] of the Hamiltonian $\tilde{\mathscr{H}}_0=\tilde{\mathscr{H}}-\tilde{\mathscr{H}}_{\mathrm{G}}$:

$$\begin{aligned}
\frac{\partial\tilde{p}^{\alpha\eta}}{\partial x^{\eta}}+\tilde{p}^{\xi\eta}\gamma^{\alpha}_{\ \xi\eta}&=\tilde{p}^{\alpha\eta}_{\ \ ;\eta}-2\tilde{p}^{\alpha\xi}s^{\eta}_{\ \xi\eta} &&=-\frac{\partial\tilde{\mathscr{H}}_0}{\partial a_\alpha}=\frac{\partial\tilde{\mathscr{L}}}{\partial a_\alpha}\\
\frac{\partial\tilde{k}^{\lambda\alpha\eta}}{\partial x^{\eta}}+\tilde{k}^{\xi\alpha\eta}\gamma^{\lambda}_{\ \xi\eta}+\tilde{k}^{\lambda\xi\eta}\gamma^{\alpha}_{\ \xi\eta}&=\tilde{k}^{\lambda\alpha\eta}_{\ \ \ ;\eta}-2\tilde{k}^{\lambda\alpha\xi}s^{\eta}_{\ \xi\eta} &&=-\frac{\partial\tilde{\mathscr{H}}_0}{\partial g_{\lambda\alpha}}=\frac{\partial\tilde{\mathscr{L}}}{\partial g_{\lambda\alpha}}.
\end{aligned}$$

With Eqs. (47), the consistency relation (55) follows as

$$\frac{\partial\tilde{\mathscr{L}}}{\partial a_\alpha}a_\beta+\frac{\partial\tilde{\mathscr{L}}}{\partial a_{\alpha;\eta}}a_{\beta;\eta}+2\frac{\partial\tilde{\mathscr{L}}}{\partial g_{\lambda\alpha}}g_{\lambda\beta}+2\frac{\partial\tilde{\mathscr{L}}}{\partial g_{\lambda\alpha;\eta}}g_{\lambda\beta;\eta}+\frac{\partial\tilde{\mathscr{L}}}{\partial R^{\xi}_{\ \alpha\lambda\eta}}R^{\xi}_{\ \beta\lambda\eta}-\frac{\partial\tilde{\mathscr{L}}}{\partial R^{\beta}_{\ \xi\lambda\eta}}R^{\alpha}_{\ \xi\lambda\eta}=0, \tag{56}$$

which is exactly the Lagrangian representation of the consistency relation of Ref. [2].

For metric compatibility ($g_{\lambda\beta;\eta}\equiv 0$), Eq. (56) can be split into two groups, namely the terms depending on the Lagrangian $\mathscr{L}_R$ for the "free" gravitational field on the left-hand side and the Lagrangian $\mathscr{L}_0$ of the scalar and vector source fields on the right-hand side. This yields, after dividing by $\sqrt{-g}$:

$$\begin{aligned}
2\frac{\partial\mathscr{L}_R}{\partial g^{\nu\beta}}g^{\mu\beta}-\frac{\partial\mathscr{L}_R}{\partial R^{\tau}_{\ \mu\beta\lambda}}R^{\tau}_{\ \nu\beta\lambda}+\frac{\partial\mathscr{L}_R}{\partial R^{\nu}_{\ \tau\beta\lambda}}R^{\mu}_{\ \tau\beta\lambda}-\delta^{\mu}_{\nu}\,\mathscr{L}_R&=-2\frac{\partial\mathscr{L}_0}{\partial g^{\nu\beta}}g^{\mu\beta}+\frac{\partial\mathscr{L}_0}{\partial a_{\mu;\beta}}a_{\nu;\beta}\\
&\quad+\frac{\partial\mathscr{L}_0}{\partial a_\mu}a_\nu+\delta^{\mu}_{\nu}\mathscr{L}_0,
\end{aligned} \tag{57}$$

where the derivatives of the Lagrangian densities $\tilde{\mathscr{L}}=\mathscr{L}\sqrt{-g}$ with respect to the metric are replaced by the corresponding derivatives of the Lagrangians $\mathscr{L}$ as

$$\frac{2}{\sqrt{-g}}\frac{\partial\tilde{\mathscr{L}}}{\partial g^{\nu\beta}}g^{\mu\beta}=2\frac{\partial\mathscr{L}}{\partial g^{\nu\beta}}g^{\mu\beta}-\delta^{\mu}_{\nu}\mathscr{L},\qquad \frac{2}{\sqrt{-g}}\frac{\partial\tilde{\mathscr{L}}}{\partial g_{\mu\beta}}g_{\nu\beta}=2\frac{\partial\mathscr{L}}{\partial g_{\mu\beta}}g_{\nu\beta}+\delta^{\mu}_{\nu}\mathscr{L}. \tag{58}$$

9.3 *Equivalence of the Noether Condition (48) and the Consistency Relation (56)*

On first sight, the consistency relation (56) appears to emerge in addition to the Noether condition (48). Yet, as it turns out, both conditions are *equivalent*. In order to prove this, we sum up both equations:

$$\frac{\partial\tilde{\mathscr{L}}}{\partial\left(\frac{\partial\phi}{\partial x^\alpha}\right)}\frac{\partial\phi}{\partial x^\beta}+\frac{\partial\tilde{\mathscr{L}}}{\partial a_{\alpha;\eta}}a_{\beta;\eta}+\frac{\partial\tilde{\mathscr{L}}}{\partial a_{\eta;\alpha}}a_{\eta;\beta}+\frac{\partial\tilde{\mathscr{L}}}{\partial a_\alpha}a_\beta+\frac{\partial\tilde{\mathscr{L}}}{\partial g_{\lambda\eta;\alpha}}g_{\lambda\eta;\beta}+2\frac{\partial\tilde{\mathscr{L}}}{\partial g_{\lambda\alpha;\eta}}g_{\lambda\beta;\eta}+2\frac{\partial\tilde{\mathscr{L}}}{\partial g_{\lambda\alpha}}g_{\lambda\beta}$$
$$+2\frac{\partial\tilde{\mathscr{L}}}{\partial R^{\xi}{}_{\lambda\eta\alpha}}R^{\xi}{}_{\lambda\eta\beta}+\frac{\partial\tilde{\mathscr{L}}}{\partial R^{\xi}{}_{\alpha\lambda\eta}}R^{\xi}{}_{\beta\lambda\eta}-\frac{\partial\tilde{\mathscr{L}}}{\partial R^{\beta}{}_{\xi\lambda\eta}}R^{\alpha}{}_{\xi\lambda\eta}\equiv\delta^\alpha_\beta\,\tilde{\mathscr{L}}. \tag{59}$$

According to Eq. (91) of Corollary 2, the resulting equation (59) represents an *identity*—and thus does not constitute a dynamical law, i.e., an equation of motion. Hence, Eq. (56) holds if Eq. (48) is satisfied and vice versa.

9.4 *Correlation of Spin and Torsion*

The field equation (35) has the Lagrangian representation

$$\left(\frac{\partial\tilde{\mathscr{L}}}{\partial R^{\eta}{}_{\alpha\beta\mu}}\right)_{;\mu}=\frac{1}{2}\frac{\partial\tilde{\mathscr{L}}}{\partial a_{\alpha;\beta}}a_\eta+\frac{\partial\tilde{\mathscr{L}}}{\partial g_{\tau\alpha;\beta}}g_{\tau\eta}+\frac{\partial\tilde{\mathscr{L}}}{\partial R^{\eta}{}_{\alpha\tau\mu}}s^{\beta}{}_{\tau\mu}+2\frac{\partial\tilde{\mathscr{L}}}{\partial R^{\eta}{}_{\alpha\beta\mu}}s^{\tau}{}_{\mu\tau}. \tag{60}$$

Splitting again the total Lagrangian $\tilde{\mathscr{L}}$ into the sum of the Lagrangian $\tilde{\mathscr{L}}_R$ for the free gravitational field, the Lagrangian $\tilde{\mathscr{L}}_g$ for nonmetricity, and the Lagrangian $\tilde{\mathscr{L}}_0$ describing the dynamics of the scalar and vector field, this yields

$$\left(\frac{\partial\tilde{\mathscr{L}}_R}{\partial R^{\eta}{}_{\alpha\beta\mu}}\right)_{;\mu}=\frac{1}{2}\frac{\partial\tilde{\mathscr{L}}_0}{\partial a_{\alpha;\beta}}a_\eta+\frac{\partial\tilde{\mathscr{L}}_g}{\partial g_{\tau\alpha;\beta}}g_{\tau\eta}+\frac{\partial\tilde{\mathscr{L}}_R}{\partial R^{\eta}{}_{\alpha\tau\mu}}s^{\beta}{}_{\tau\mu}+2\frac{\partial\tilde{\mathscr{L}}_R}{\partial R^{\eta}{}_{\alpha\beta\mu}}s^{\tau}{}_{\mu\tau}. \tag{61}$$

For the usual case of a covariantly conserved metric, hence for metric compatibility, we can divide by $\sqrt{-g}$

$$\left(\frac{\partial\mathscr{L}_R}{\partial R_{\eta\alpha\beta\mu}}\right)_{;\mu}-\frac{\partial\mathscr{L}_R}{\partial R_{\eta\alpha\tau\mu}}s^{\beta}{}_{\tau\mu}-2\frac{\partial\mathscr{L}_R}{\partial R_{\eta\alpha\beta\mu}}s^{\tau}{}_{\mu\tau}=\frac{\partial\tilde{\mathscr{L}}_g}{\partial g_{\eta\alpha;\beta}}+\frac{1}{2}\frac{\partial\mathscr{L}_0}{\partial a_{\alpha;\beta}}a^\eta. \tag{62}$$

Equation (62) can now be split into a symmetric part in η and α

$$\frac{\partial\tilde{\mathscr{L}}_g}{\partial g_{\eta\alpha;\beta}}+\frac{1}{4}\left(\frac{\partial\mathscr{L}_0}{\partial a_{\alpha;\beta}}a^\eta+\frac{\partial\mathscr{L}_0}{\partial a_{\eta;\beta}}a^\alpha\right)=0, \tag{63}$$

and a skew-symmetric part in η and α

$$\boxed{\left(\frac{\partial \mathscr{L}_R}{\partial R_{\eta\alpha\beta\mu}}\right)_{;\mu} - \frac{\partial \mathscr{L}_R}{\partial R_{\eta\alpha\tau\mu}} s^{\beta}{}_{\tau\mu} - 2\frac{\partial \mathscr{L}_R}{\partial R_{\eta\alpha\beta\mu}} s^{\tau}{}_{\mu\tau} = \frac{1}{4}\left(\frac{\partial \mathscr{L}_0}{\partial a_{\alpha;\beta}} a^{\eta} - \frac{\partial \mathscr{L}_0}{\partial a_{\eta;\beta}} a^{\alpha}\right).} \tag{64}$$

We will show in the next section that the leftmost term, hence the divergence associated with $\mathscr{L}_R$, vanishes for the Hilbert Lagrangian $\mathscr{L}_{R,H}$. For that case, Eq. (64) yields an algebraic equation for the torsion emerging from the vector field a_μ. All other choices of $\mathscr{L}_R$, yield a non-algebraic equation for the correlation of spin and torsion. So, the question whether spacetime torsion either propagates or is merely tied to spinning matter depends on the model for the dynamics of the gravitational field in classical vacuum. In any case, for a non-vanishing right-hand side of Eq. (64), the vector field necessarily acts as a source of torsion of spacetime.

10 Sample Lagrangians

10.1 Lagrangian $\mathscr{L}_R$ of the "Free" Gravitational Field

With Eq. (49) and the equivalent equation (56), we have derived energy-momentum balance equations for the *interaction* of given source fields—whose free dynamics is described by $\mathscr{L}_0$—with the gravitational field, whose free dynamics is described by $\mathscr{L}_R$. The Lagrangians $\mathscr{L}_0$ and $\mathscr{L}_R$ for the dynamics in the absence of any coupling must be set up on the basis of physical reasoning. From analogies to other field theories, we chose a Lagrangian $\mathscr{L}_R(R, g)$ as a sum of a constant and of linear and quadratic terms in the Riemann curvature tensor [2]

$$\mathscr{L}_R = \frac{1}{4} R^{\eta}{}_{\alpha\beta\tau}\left[g_1 R^{\alpha}{}_{\eta\xi\lambda} g^{\beta\xi} g^{\tau\lambda} + \frac{1}{8\pi G}\left(\delta^{\tau}_{\eta}\, g^{\alpha\beta} - \delta^{\beta}_{\eta}\, g^{\alpha\tau}\right)\right] + \frac{\Lambda}{8\pi G}. \tag{65}$$

The terms in parentheses can be regarded as the Riemann tensor of the *maximally symmetric* 4-dimensional manifold

$$\hat{R}^{\alpha}{}_{\eta}{}^{\beta\tau} = \frac{1}{8\pi G}\left(\delta^{\tau}_{\eta}\, g^{\alpha\beta} - \delta^{\beta}_{\eta}\, g^{\alpha\tau}\right),$$

which can be interpreted as the "ground state" of spacetime [19]. The coupling constant g_1 is dimensionless, whereas G has the natural dimension of Length^2. The Lagrangian (65) is thus the sum of the Hilbert Lagrangian $\mathscr{L}_{R,\mathrm{H}}$

$$\mathscr{L}_{R,\mathrm{H}} = -\frac{1}{16\pi G}(R - 2\Lambda) \tag{66}$$

plus a Lagrangian quadratic in the Riemann tensor. The latter was proposed earlier by A. Einstein in a personal letter to H. Weyl [3].

To set up the energy-momentum balance equation (51) for the Lagrangian $\mathscr{L}_R$, we first calculate

$$\frac{\partial \mathscr{L}_R}{\partial R^{\eta}{}_{\alpha\beta\mu}} = \frac{g_1}{2} R^{\alpha}{}_{\eta\xi\lambda} g^{\beta\xi} g^{\mu\lambda} + \frac{1}{32\pi G}\left(\delta^{\mu}_{\eta}\, g^{\alpha\beta} - \delta^{\beta}_{\eta}\, g^{\alpha\mu}\right) = \frac{g_1}{2} R^{\alpha}{}_{\eta}{}^{\beta\mu} + \frac{1}{4}\hat{R}^{\alpha}{}_{\eta}{}^{\beta\mu}, \tag{67}$$

and hence

$$\begin{aligned} 2\frac{\partial \mathscr{L}_R}{\partial R^{\eta}{}_{\alpha\beta\mu}} R^{\eta}{}_{\alpha\beta\nu} &= g_1 R^{\alpha}{}_{\eta\xi\lambda} g^{\beta\xi} g^{\mu\lambda} R^{\eta}{}_{\alpha\beta\nu} + \frac{1}{16\pi G}\left(\delta^{\mu}_{\eta}\, g^{\alpha\beta} - \delta^{\beta}_{\eta}\, g^{\alpha\mu}\right) R^{\eta}{}_{\alpha\beta\nu} \\ &= -g_1 R_{\eta}{}^{\alpha\beta\mu} R^{\eta}{}_{\alpha\beta\nu} - \frac{1}{8\pi G} R^{\mu}{}_{\nu}. \end{aligned}$$

The energy-momentum balance equation (51) now follows as

$$g_1 R_{\eta}{}^{\alpha\beta\mu} R^{\eta}{}_{\alpha\beta\nu} + \frac{1}{8\pi G} R^{\mu}{}_{\nu} + \delta^{\mu}_{\nu}\, \mathscr{L}_R = \theta_{\nu}{}^{\mu},$$

which finally yields the Einstein-type equation after inserting the Lagrangian $\mathscr{L}_R$ from Eq. (65):

$$g_1\left(R_{\eta}{}^{\alpha\beta\mu} R^{\eta}{}_{\alpha\beta\nu} - \tfrac{1}{4}\delta^{\mu}_{\nu} R_{\eta}{}^{\alpha\beta\tau} R^{\eta}{}_{\alpha\beta\tau}\right) + \frac{1}{8\pi G}\left(R^{\mu}{}_{\nu} - \tfrac{1}{2}\delta^{\mu}_{\nu} R + \Lambda\delta^{\mu}_{\nu}\right) = \theta_{\nu}{}^{\mu}. \tag{68}$$

Note that here the contravariant representations of the energy-momentum tensor $\theta^{\nu\mu}$ as well as that of the Ricci tensor $R^{\mu\nu}$ are not necessarily symmetric. Thus, Eq. (68) can be split into a symmetric and a skew-symmetric equation in μ and ν (see also Kibble [20] and Sciama [21]):

$$g_1\left(R_{\eta}{}^{\alpha\beta\mu} R^{\eta}{}_{\alpha\beta}{}^{\nu} - \tfrac{1}{4} g^{\mu\nu} R_{\eta}{}^{\alpha\beta\tau} R^{\eta}{}_{\alpha\beta\tau}\right) + \frac{1}{8\pi G}\left(R^{(\mu\nu)} - \tfrac{1}{2} g^{\mu\nu} R + \Lambda g^{\mu\nu}\right) = \theta^{(\mu\nu)} \tag{69a}$$

$$\frac{1}{8\pi G} R^{[\mu\nu]} = \theta^{[\nu\mu]}. \tag{69b}$$

Setting $g_1 = 0$ in Eq. (69a), hence neglecting the term proportional to quadratic in the Riemann tensor, reduces Eq. (68) to the form of the Einstein equation proper. Equation (69b) is the representation of the general Eq. (52) for the particular Lagrangian $\mathscr{L}_R$ from Eq. (65).

The field equation (64), which describes the coupling of spin and torsion, emerges for the Lagrangian (65) as the differential equation:

$$\frac{g_1}{2} R^{\alpha\eta\beta\mu}{}_{;\mu} - \left(\tfrac{1}{2} g_1 R^{\alpha\eta\tau\mu} + \tfrac{1}{4}\hat{R}^{\alpha\eta\tau\mu}\right) s^{\beta}{}_{\tau\mu} + \left(g_1 R^{\alpha\eta\beta\mu} + \tfrac{1}{2}\hat{R}^{\alpha\eta\beta\mu}\right) s^{\tau}{}_{\tau\mu}$$
$$= \tfrac{1}{4}\left(\frac{\partial \mathscr{L}_0}{\partial a_{\alpha;\beta}} a^{\eta} - \frac{\partial \mathscr{L}_0}{\partial a_{\eta;\beta}} a^{\alpha}\right). \tag{70}$$

For the Hilbert Lagrangian (66), this reduces to the algebraic equation:

$$2\hat{R}^{\alpha\eta\beta\mu} s^{\tau}{}_{\tau\mu} - \hat{R}^{\alpha\eta\tau\mu} s^{\beta}{}_{\tau\mu} = \frac{\partial \mathscr{L}_0}{\partial a_{\alpha;\beta}} a^{\eta} - \frac{\partial \mathscr{L}_0}{\partial a_{\eta;\beta}} a^{\alpha}. \tag{71}$$

Thus, a massive spin-1 particle field a_{μ} always acts as a source of torsion of spacetime. The right-hand side will be specified for the Proca system in Sect. 10.3. Obviously, the spin-0 particle field ϕ, i.e., the Klein–Gordon system, to be discussed in the following section, does not act as a source of torsion as Eq. (71) is identically satisfied for $s^{\beta}{}_{\tau\mu} \equiv 0$.

10.2 Klein–Gordon Lagrangian $\mathscr{L}_0$

The Klein–Gordon Lagrangian density $\tilde{\mathscr{L}}_0(\phi, \partial_{\nu}\phi, g^{\mu\nu})$ for a system of a real scalar field ϕ in a dynamic spacetime is given by

$$\tilde{\mathscr{L}}_0 = \tfrac{1}{2}\left(\frac{\partial\phi}{\partial x^{\alpha}}\frac{\partial\phi}{\partial x^{\beta}} g^{\alpha\beta} - m^2\phi^2\right)\sqrt{-g}. \tag{72}$$

For this Lagrangian, the identity (91) takes on the particular form for a symmetric metric $g^{\alpha\beta}$:

$$2\frac{\partial\tilde{\mathscr{L}}_0}{\partial g^{\nu\lambda}} g^{\mu\lambda} - \frac{\partial\tilde{\mathscr{L}}_0}{\partial\left(\frac{\partial\phi}{\partial x^{\mu}}\right)}\frac{\partial\phi}{\partial x^{\nu}} \equiv -\delta^{\mu}_{\nu}\tilde{\mathscr{L}}_0,$$

hence, dividing by $\sqrt{-g}$:

$$\frac{2}{\sqrt{-g}}\frac{\partial\tilde{\mathscr{L}}_0}{\partial g^{\nu\lambda}} g^{\mu\lambda} \equiv \frac{\partial\mathscr{L}_0}{\partial\left(\frac{\partial\phi}{\partial x^{\mu}}\right)}\frac{\partial\phi}{\partial x^{\nu}} - \delta^{\mu}_{\nu}\mathscr{L}_0, \tag{73}$$

As $\tilde{\mathscr{L}}_0$ does not depend on the partial derivative of the metric, the left-hand side of Eq. (73) defines the mixed tensor representation of the *metric* energy-momentum tensor $T_{\nu}{}^{\mu}$ of the Klein–Gordon system (72), whereas the right-hand side represents its *canonical* energy-momentum tensor $\theta_{\nu}{}^{\mu}$:

$$T_{\nu}{}^{\mu} = \frac{2}{\sqrt{-g}}\frac{\partial\tilde{\mathscr{L}}_0}{\partial g^{\nu\lambda}} g^{\mu\lambda}, \qquad \theta_{\nu}{}^{\mu} = \frac{\partial\mathscr{L}_0}{\partial\left(\frac{\partial\phi}{\partial x^{\mu}}\right)}\frac{\partial\phi}{\partial x^{\nu}} - \delta^{\mu}_{\nu}\mathscr{L}_0. \tag{74}$$

Both tensors thus coincide for this system and have the explicit symmetric contravariant form:

$$T^{\nu\mu} = \frac{\partial\phi}{\partial x^\alpha}\frac{\partial\phi}{\partial x^\beta} g^{\alpha\nu} g^{\beta\mu} - \tfrac{1}{2} g^{\nu\mu}\left(\frac{\partial\phi}{\partial x^\alpha}\frac{\partial\phi}{\partial x^\beta}\, g^{\alpha\beta} - m^2\,\phi^2\right) = \theta^{\nu\mu} = \theta^{(\nu\mu)}. \quad (75)$$

The Euler–Lagrange equation for the Lagrangian (72) follows for a covariantly conserved metric as

$$g^{\alpha\beta}\left(\frac{\partial^2\phi}{\partial x^\alpha \partial x^\beta} - \gamma^{\xi}_{\ \alpha\beta}\frac{\partial\phi}{\partial x^\xi} - 2s^{\xi}_{\ \alpha\xi}\frac{\partial\phi}{\partial x^\beta}\right) + m^2\phi = 0.$$

The second derivative of ϕ as well as the term proportional to $\gamma^{\xi}_{\ \alpha\beta}$ are no tensors. Yet their sum is just the covariant x^β-derivative of the covector $\partial\phi/\partial x^\alpha$,

$$g^{\alpha\beta}\left[\left(\frac{\partial\phi}{\partial x^\alpha}\right)_{;\beta} - 2s^{\xi}_{\ \alpha\xi}\frac{\partial\phi}{\partial x^\beta}\right] + m^2\phi = 0 \quad (76)$$

and thus holds as a tensor equation in any reference system. The term related to the torsion vector $s^{\xi}_{\ \alpha\xi}$ states that the dynamics of the scalar field ϕ also couples to the torsion of spacetime, but does not act as a source of torsion as the canonical energy-momentum tensor (75) is symmetric.

For non-zero torsion neither the covariant divergence of the Einstein tensor $G^{\mu\nu}$ nor the covariant divergence of the energy-momentum tensor vanishes. One thus encounters from the conventional Einstein equation

$$G^{\alpha}_{\ \ \mu;\alpha} \equiv \left(R^{\alpha}_{\ \ \mu} - \tfrac{1}{2}\delta^{\alpha}_{\mu}R\right)_{;\alpha} = 8\pi G\, T^{\alpha}_{\ \ \mu;\alpha}\,, \quad (77)$$

the following explicit form for a covariantly conserved metric

$$\tfrac{1}{2}R^{\beta\xi}_{\ \ \ \mu\tau}s^{\tau}_{\ \beta\xi} + R^{\tau\beta}_{\ \ \ \tau\xi}s^{\xi}_{\ \beta\mu} = 8\pi G\,\frac{\partial\phi}{\partial x^\alpha} g^{\alpha\beta}\left(\frac{\partial\phi}{\partial x^\mu}s^{\xi}_{\ \beta\xi} + \frac{\partial\phi}{\partial x^\xi}s^{\xi}_{\ \beta\mu}\right). \quad (78)$$

Equation (78) is obviously satisfied for an identically vanishing torsion tensor $s^{\xi}_{\ \beta\mu} \equiv 0$ and thus shows that the spacetime dynamics of the Klein–Gordon system (72) is compatible with an identically vanishing torsion of spacetime.

10.3 Proca Lagrangian

The Proca Lagrangian density $\tilde{\mathscr{L}}_0\left(a_\mu, \partial_\nu a_\mu, g^{\mu\nu}, \gamma^{\xi}_{\ \mu\nu}\right)$ writes

$$\tilde{\mathscr{L}}_0 = \left(-\tfrac{1}{4}f_{\alpha\beta}\,f_{\xi\eta}\,g^{\alpha\xi}\,g^{\beta\eta} + \tfrac{1}{2}m^2 a_\alpha\,a_\xi\,g^{\alpha\xi}\right)\sqrt{-g}, \qquad f_{\alpha\beta} = a_{\beta;\alpha} - a_{\alpha;\beta} = -f_{\beta\alpha}, \quad (79)$$

with $f_{\alpha\beta}$ denoting the skew-symmetric field tensor. With the particular Lagrangian (79), the identity (91) takes on the form:

$$2\frac{\partial\tilde{\mathscr{L}}_0}{\partial g^{\nu\lambda}}g^{\mu\lambda}-\frac{\partial\tilde{\mathscr{L}}_0}{\partial a_{\mu;\lambda}}a_{\nu;\lambda}-\frac{\partial\tilde{\mathscr{L}}_0}{\partial a_{\lambda;\mu}}a_{\lambda;\nu}-\frac{\partial\tilde{\mathscr{L}}_0}{\partial a_{\mu}}a_{\nu}\equiv-\delta^{\mu}_{\nu}\tilde{\mathscr{L}}_0. \tag{80}$$

Dividing Eq. (80) by $\sqrt{-g}$, its leftmost term represents the *metric* energy-momentum tensor $T_{\nu}^{\;\mu}$ from Eq. (74), as $\tilde{\mathscr{L}}_0$ does not depend on the derivative of the metric. By virtue of the identity (80), $T_{\nu}^{\;\mu}$ can equivalently be obtained from the derivatives with respect to the fields:

$$\begin{aligned}T_{\nu}^{\;\mu}&=\frac{\partial\mathscr{L}_0}{\partial a_{\mu;\lambda}}a_{\nu;\lambda}+\frac{\partial\mathscr{L}_0}{\partial a_{\lambda;\mu}}a_{\lambda;\nu}+\frac{\partial\mathscr{L}_0}{\partial a_{\mu}}a_{\nu}-\delta^{\mu}_{\nu}\mathscr{L}_0\\&=\theta_{\nu}^{\;\mu}+\frac{\partial\mathscr{L}_0}{\partial a_{\mu;\lambda}}a_{\nu;\lambda}+\frac{\partial\mathscr{L}_0}{\partial a_{\mu}}a_{\nu},\end{aligned}$$

wherein $\theta_{\nu}^{\;\mu}$ denotes the *canonical* energy-momentum tensor

$$\theta_{\nu}^{\;\mu}=\frac{\partial\mathscr{L}_0}{\partial a_{\lambda;\mu}}a_{\lambda;\nu}-\delta^{\mu}_{\nu}\mathscr{L}_0, \tag{81}$$

and thus

$$\theta^{\nu\mu}=T^{\nu\mu}-\frac{\partial\mathscr{L}_0}{\partial a_{\mu;\lambda}}a^{\nu}_{\;;\lambda}-\frac{\partial\mathscr{L}_0}{\partial a_{\mu}}a^{\nu}. \tag{82}$$

With

$$\frac{\partial\mathscr{L}_0}{\partial a_{\mu;\nu}}=f^{\mu\nu},$$

the energy-momentum tensors follow as:

$$T_{\nu}^{\;\mu}=f^{\lambda\mu}\left(a_{\lambda;\nu}-a_{\nu;\lambda}\right)+m^2\,a_{\nu}\,a^{\mu}\qquad+\tfrac{1}{4}\delta^{\mu}_{\nu}\left(f_{\alpha\beta}\,f^{\alpha\beta}-2m^2a_{\alpha}\,a^{\alpha}\right) \tag{83}$$

$$\theta_{\nu}^{\;\mu}=f^{\lambda\mu}\,a_{\lambda;\nu}\qquad+\tfrac{1}{4}\delta^{\mu}_{\nu}\left(f_{\alpha\beta}\,f^{\alpha\beta}-2m^2a_{\alpha}\,a^{\alpha}\right). \tag{84}$$

Our conclusion is that for a Proca system the asymmetric *canonical* energy-momentum tensor $\theta^{\,\nu\mu}$ represents the correct source term for gravitation. Remarkably, the tensor $\theta^{\,\nu\mu}$ thus entails an increased weighting of the kinetic energy over the mass as compared to the *metric* energy momentum tensor $T^{\nu\mu}$ in their roles as the source of gravity. This holds independently of the particular model for the "free" (uncoupled) gravitational field, whose dynamics is encoded in the Lagrangian $\mathscr{L}_R$ of Eq. (51).

From the Euler–Lagrange field equation for the vector field a_{μ},

$$f^{\mu\alpha}_{\;\;\;;\alpha}-2f^{\mu\beta}s^{\alpha}_{\;\;\beta\alpha}-m^2a^{\mu}=0,$$

the covariant divergence of the canonical energy-momentum tensor is obtained as

$$\theta_{\nu}{}^{\mu}{}_{;\mu} = f^{\alpha\beta}\left[-a_{\xi}\,R^{\xi}{}_{\alpha\beta\nu} + 2\left(a_{\alpha;\xi}s^{\xi}{}_{\beta\nu} + a_{\alpha;\nu}s^{\xi}{}_{\beta\xi}\right)\right].$$

This tensor does generally not vanish in a curved spacetime, even if we neglect torsion. Hence, from

$$G^{\alpha}{}_{\nu;\alpha} \equiv \left(R^{\alpha}{}_{\nu} - \tfrac{1}{2}\delta^{\alpha}_{\nu}R\right)_{;\alpha} = 8\pi G\,\theta_{\nu}{}^{\alpha}{}_{;\alpha}, \tag{85}$$

which has the following explicit form for the Proca system,

$$\tfrac{1}{2}R^{\beta\xi}{}_{\nu\tau}s^{\tau}{}_{\beta\xi} + R^{\tau\beta}{}_{\tau\xi}s^{\xi}{}_{\beta\nu} = 8\pi G\,f^{\alpha\beta}\left[-a_{\xi}\,R^{\xi}{}_{\alpha\beta\nu} + 2\left(a_{\alpha;\xi}s^{\xi}{}_{\beta\nu} + a_{\alpha;\nu}s^{\xi}{}_{\beta\xi}\right)\right], \tag{86}$$

we conclude that the Einstein equation is consistent only if torsion is included. In contrast to the corresponding equation for the Klein–Gordon system from Eq. (78), this equation has no solution for a vanishing torsion.

The skew-symmetric part of the canonical energy-momentum tensor (84) of the Proca system follows as

$$\theta_{[\nu\mu]} = \tfrac{1}{2}\left(f_{\nu\beta}\,a^{\beta}{}_{;\mu} - f_{\mu\beta}\,a^{\beta}{}_{;\nu}\right) = \tfrac{1}{2}\left(a_{\mu;\beta}\,a^{\beta}{}_{;\nu} - a_{\nu;\beta}\,a^{\beta}{}_{;\mu}\right), \tag{87}$$

which yields, according to Eq. (69b), the skew-symmetric part of the Ricci tensor from the contraction of the generalized first Bianchi identity [22]:

$$R_{[\nu\mu]} \equiv s^{\alpha}{}_{\alpha\nu;\mu} - s^{\alpha}{}_{\alpha\mu;\nu} + s^{\alpha}{}_{\nu\mu;\alpha} - 2s^{\alpha}{}_{\beta\alpha}s^{\beta}{}_{\nu\mu} = 4\pi G\left(a_{\mu;\beta}\,a^{\beta}{}_{;\nu} - a_{\nu;\beta}\,a^{\beta}{}_{;\mu}\right). \tag{88}$$

Again, this equation satisfied only with a non-vanishing torsion.

11 Summary and Conclusions

The minimum set of postulates for a theory of spacetime geometry interacting with matter is that (i) the theory should be derived from an action principle and that (ii) the *Principle of General Relativity* should hold, i.e., the theory should be diffeomorphism-invariant. An appropriate basis for the formulation of such a theory is given by Noether's theorem—which directly follows from the action principle: it provides for any symmetry of the given action a pertaining conserved Noether current. Noether's theorem is most efficiently formulated in the Hamiltonian formalism by means of the generating function of the respective infinitesimal symmetry transformation as this function directly yields the weakly conserved Noether current. From the latter, one can then set up the most general field equation of geometrodynamics for systems with Diff(M) symmetry. The covariant field-theoretical version of the canonical transformation formalism is applied to work out the particular form of the action integral that is maintained under this symmetry.

The general recipe to set up this equation is as follows:

1. Establish the covariant representation of the canonical energy-momentum tensor for the given matter Lagrangian $\mathscr{L}_0$—which must be a *world scalar*. This means for a system of a scalar field ϕ and a massive vector field a_α

$$\theta_\nu^{\ \mu} = \frac{\partial \mathscr{L}_0}{\partial\left(\frac{\partial\phi}{\partial x^\mu}\right)}\frac{\partial\phi}{\partial x^\nu} + \frac{\partial \mathscr{L}_0}{\partial a_{\alpha;\mu}} a_{\alpha;\nu} - \delta_\nu^\mu \mathscr{L}_0.$$

   The covariant form of the canonical energy-momentum tensor thus contains *direct* coupling terms of the vector field a_α and the connection $\gamma^\alpha_{\ \mu\nu}$, and thereby also causes a coupling to a torsion of spacetime. For vector fields which represent the classical limit of massive spin particles, the correct source of gravitation is constituted by the canonical energy-momentum tensor and *not* by the conventionally used metric (Hilbert) energy-momentum tensor—in agreement with Hehl [12]. The source term changes for systems with additional symmetries—such as a system with additional U(1) symmetry, in which case the metric (Hilbert) energy-momentum tensor turns out to be the appropriate source term [23].
2. With the source term and the *postulated* Lagrangians for both, the dynamics of the "free" (uncoupled) connection, $\mathscr{L}_R$, and of the metric, $\mathscr{L}_g$, the new and most general equation of geometrodynamics is given by

$$2\frac{\partial \mathscr{L}_R}{\partial R^\eta_{\ \alpha\beta\mu}} R^\eta_{\ \alpha\beta\nu} + \frac{\partial \mathscr{L}_g}{\partial g_{\alpha\beta;\mu}} g_{\alpha\beta;\nu} - \delta_\nu^\mu\left(\mathscr{L}_R + \mathscr{L}_g\right) = -\theta_\nu^{\ \mu}.$$

   On first sight, $\mathscr{L}_R$ may be any world scalar formed out of the Riemann tensor $R^\mu_{\ \tau\alpha\beta}$ and the metric $g^{\mu\nu}$. Correspondingly, $\mathscr{L}_g$ may be any world scalar formed out of the covariant derivative of the metric and the metric itself. Yet, as the Noether current is merely *weakly* conserved, the choices of $\mathscr{L}_R$ and $\mathscr{L}_g$ are actually restricted by the requirement that the subsequent field equation is consistent with regard to its trace and its covariant divergence.
3. In the particular case of metric compatibility, hence for a covariantly conserved metric, the connection and the metric are correlated according to

$$g_{\alpha\beta;\nu} \equiv \frac{\partial g_{\alpha\beta}}{\partial x^\nu} - g_{\tau\beta}\gamma^\tau_{\ \alpha\nu} - g_{\alpha\tau}\gamma^\tau_{\ \beta\nu} = 0,$$

   and the correlation of the Riemann tensor $R^\eta_{\ \alpha\beta\mu}$ to the source simplifies to the following form of a generic Einstein-type equation:

$$\vartheta_\nu^{\ \mu} \equiv 2\frac{\partial \mathscr{L}_R}{\partial R^\eta_{\ \alpha\beta\mu}} R^\eta_{\ \alpha\beta\nu} - \delta_\nu^\mu \mathscr{L}_R = -\theta_\nu^{\ \mu}.$$

The left-hand side associated with the Lagrangian $\mathscr{L}_R$ describing the dynamics of the gravitational field in classical vacuum, can be interpreted as the covariant canonical energy-momentum tensor of spacetime, which balances the canonical energy-momentum tensor of matter on the right-hand side. $\vartheta_\nu{}^\mu + \theta_\nu{}^\mu = 0$ represents the zero energy principle.

The simplest case for zero torsion is given by the Hilbert Lagrangian (66), which directly yields the Einstein tensor on the left-hand side. This requires the energy-momentum tensor to be symmetric as well as its covariant divergence to be zero in order for the resulting field equation to be consistent.

Summarizing, our generic theory of geometrodynamics generalizes Einstein's General Relativity as follows:

1. The description of the dynamics of the "free" gravitational fields is not restricted to the Hilbert Lagrangian. As was shown by Hayashi and Shirafuji [24], the most general Lagrangian compatible with also parity invariance can be at most quadratic in the Riemann–Cartan curvature tensor.
In the case of a quadratic and linear dependence of $\mathscr{L}_R(R, g)$ on the Riemann tensor, the field equation
$$g_1 \left(R_\eta{}^{\alpha\beta\mu} R^\eta{}_{\alpha\beta\nu} - \tfrac{1}{4}\delta_\nu^\mu R_\eta{}^{\alpha\beta\tau} R^\eta{}_{\alpha\beta\tau} \right) + \frac{1}{8\pi G} \left(R^\mu{}_\nu - \tfrac{1}{2}\delta_\nu^\mu R + \Lambda \delta_\nu^\mu \right) = \theta_\nu{}^\mu,$$
is encountered [2]. It equally complies with the Principle of General Relativity. The additional term proportional to the *dimensionless* coupling constant g_1 is equally satisfied by the Schwarzschild and the Kerr metric [4] in the absence of torsion. Yet, it entails a different description of the dynamics of spacetime in the case of a non-vanishing source term $\theta_\nu{}^\mu$ as compared to the solution based on only the Einstein tensor—which follows setting $g_1 = 0$.
2. The generalized theory is not restricted to a covariantly conserved metric, hence to *metric compatibility*.
3. The spacetime is not assumed to be generally torsion-free. Based on the Riemann–Cartan curvature tensor, the generalized theory allows for sources of gravity which generate and couple to a torsion of spacetime. This applies in particular to those vector fields, which represent the classical limit of massive spin-1 particles. For this case, the *canonical* energy-momentum tensor is the appropriate source term. Its skew-symmetric part is then related to the skew-symmetric part of the then non-symmetric Ricci tensor according to
$$R^{[\nu\mu]} = 8\pi G\, \theta^{[\nu\mu]}.$$
This equation states that a skew-symmetric part of the canonical energy-momentum tensor is necessarily associated with a non-vanishing torsion of spacetime. The corresponding additional degrees of freedom are encoded in a Riemann tensor that is *not symmetric* under exchange of its first and second index pair, which gives rise to a non-symmetric Ricci tensor.

Moreover, for the case of the Hilbert Lagrangian, hence for $g_1 = 0$, one encounters the following field equation from the covariant derivatives of the Einstein- and the energy-momentum tensors of the Proca system:

$$\tfrac{1}{2}R^{\beta\xi}{}_{\nu\tau}\,s^{\tau}{}_{\beta\xi} + R^{\tau\beta}{}_{\tau\xi}\,s^{\xi}{}_{\beta\nu} = 8\pi G\, f^{\alpha\beta}\left[-a_{\xi}\,R^{\xi}{}_{\alpha\beta\nu} + 2\left(a_{\alpha;\xi}s^{\xi}{}_{\beta\nu} + a_{\alpha;\nu}s^{\xi}{}_{\beta\xi}\right)\right],$$

which necessarily gives rise to a non-vanishing torsion.

Acknowledgements First of all, we want to remember our revered academic teacher Walter Greiner, whose charisma and passion for physics inspired us to stay engaged in physics for all of our lives.

The authors thank Patrick Liebrich and Julia Lienert (Goethe University Frankfurt am Main and FIAS), and Horst Stoecker (FIAS, GSI, and Goethe University Frankfurt am Main) for valuable discussions. D.V. and J.K. thank the Fueck Foundation for its support.

Identity for a Scalar-Valued Function S of an (n, m)-Tensor T and the Metric

Proposition 1 *Let $S = S(g, T) \in \mathbb{R}$ be a scalar-valued function constructed from the metric tensor $g_{\mu\nu}$ and an (n, m)-tensor $T^{\xi_1\ldots\xi_n}{}_{\eta_1\ldots\eta_m}$, where $(m - n)/2 \in \mathbb{Z}$. Then the following identity holds:*

$$\begin{aligned}\frac{\partial S}{\partial g_{\mu\beta}}g_{\nu\beta} + \frac{\partial S}{\partial g_{\beta\mu}}g_{\beta\nu} - \frac{\partial S}{\partial T^{\nu\,\xi_2\ldots\xi_n}{}_{\eta_1\ldots\eta_m}}T^{\mu\,\xi_2\ldots\xi_n}{}_{\eta_1\ldots\eta_m} - \cdots - \frac{\partial S}{\partial T^{\xi_1\ldots\xi_{n-1}\,\nu}{}_{\eta_1\ldots\eta_m}}T^{\xi_1\ldots\xi_{n-1}\,\mu}{}_{\eta_1\ldots\eta_m}\\ + \frac{\partial S}{\partial T^{\xi_1\ldots\xi_n}{}_{\mu\,\eta_2\ldots\eta_m}}T^{\xi_1\ldots\xi_n}{}_{\nu\,\eta_2\ldots\eta_m} + \cdots + \frac{\partial S}{\partial T^{\xi_1\ldots\xi_n}{}_{\eta_1\ldots\eta_{m-1}\,\mu}}T^{\xi_1\ldots\xi_n}{}_{\eta_1\ldots\eta_{m-1}\,\nu} \equiv 0\,\delta^{\mu}_{\nu}.\end{aligned} \tag{89}$$

Proof The induction hypothesis is immediately verified for scalars constructed from second rank tensors and, if necessary, the metric, hence, $S = T^{\alpha}{}_{\alpha}$, $S = T^{\alpha\beta}\,g_{\alpha\beta}$, and $S = T_{\alpha\beta}\,g^{\alpha\beta}$. Let Eq. (89) hold for an (n, m)-tensor $T^{\xi_1\ldots\xi_n}{}_{\eta_1\ldots\eta_m}$. We first consider an $(n+1, m+1)$-tensor $\bar{T}^{\xi_1\ldots\xi_n\alpha}{}_{\eta_1\ldots\eta_m\alpha}$ with the last indices contracted in order to again make up a scalar. Setting up S according to (89) with the tensor $\bar{T}$, one encounters the two additional terms

$$\begin{aligned}&-\frac{\partial S}{\partial \bar{T}^{\xi_1\ldots\xi_n\nu}{}_{\eta_1\ldots\eta_m\alpha}}\bar{T}^{\xi_1\ldots\xi_n\mu}{}_{\eta_1\ldots\eta_m\alpha} + \frac{\partial S}{\partial \bar{T}^{\xi_1\ldots\xi_n\,\alpha}{}_{\eta_1\ldots\eta_m\,\mu}}\bar{T}^{\xi_1\ldots\xi_n\,\alpha}{}_{\eta_1\ldots\eta_m\,\nu}\\ &= -\delta^{\alpha}_{\nu}\,\bar{T}^{\xi_1\ldots\xi_n\mu}{}_{\eta_1\ldots\eta_m\alpha} + \delta^{\mu}_{\alpha}\,\bar{T}^{\xi_1\ldots\xi_n\,\alpha}{}_{\eta_1\ldots\eta_m\,\nu} = 0.\end{aligned}$$

Equation (89) thus also holds for the scalar S formed from the $(n+1, m+1)$-tensor $\bar{T}$.

For the case that $\bar{T}$ represents an $(n+2, m)$-tensor $\bar{T}^{\xi_1\ldots\xi_n\alpha\beta}{}_{\eta_1\ldots\eta_m}$, the scalar S must have one additional factor $g_{\alpha\beta}$. One thus encounters four additional terms:

$$\begin{aligned}
& -\frac{\partial S}{\partial \bar{T}^{\xi_1\ldots\xi_n\nu\beta}{}_{\eta_1\ldots\eta_m}} \bar{T}^{\xi_1\ldots\xi_n\mu\beta}{}_{\eta_1\ldots\eta_m} - \frac{\partial S}{\partial \bar{T}^{\xi_1\ldots\xi_n\,\alpha\nu}{}_{\eta_1\ldots\eta_m}} \bar{T}^{\xi_1\ldots\xi_n\,\alpha\mu}{}_{\eta_1\ldots\eta_m} + \frac{\partial S}{\partial g_{\mu\beta}} g_{\nu\beta} + \frac{\partial S}{\partial g_{\alpha\mu}} g_{\alpha\nu} \\
&= -\delta^{\alpha}_{\nu}\,\bar{T}^{\xi_1\ldots\xi_n\mu\beta}{}_{\eta_1\ldots\eta_m}\, g_{\alpha\beta} - \delta^{\beta}_{\nu}\,\bar{T}^{\xi_1\ldots\xi_n\,\alpha\mu}{}_{\eta_1\ldots\eta_m}\, g_{\alpha\beta} + \delta^{\mu}_{\alpha}\, g_{\nu\beta}\bar{T}^{\xi_1\ldots\xi_n\alpha\beta}{}_{\eta_1\ldots\eta_m} + \delta^{\mu}_{\beta}\, g_{\alpha\nu}\bar{T}^{\xi_1\ldots\xi_n\alpha\beta}{}_{\eta_1\ldots\eta_m} \\
&= -g_{\nu\beta}\bar{T}^{\xi_1\ldots\xi_n\mu\beta}{}_{\eta_1\ldots\eta_m} - g_{\alpha\nu}\bar{T}^{\xi_1\ldots\xi_n\,\alpha\mu}{}_{\eta_1\ldots\eta_m} + g_{\nu\beta}\bar{T}^{\xi_1\ldots\xi_n\mu\beta}{}_{\eta_1\ldots\eta_m} + g_{\alpha\nu}\bar{T}^{\xi_1\ldots\xi_n\alpha\mu}{}_{\eta_1\ldots\eta_m} \\
&= 0.
\end{aligned}$$

For the case that $\bar{T}$ represents an $(n, m+2)$-tensor $\bar{T}^{\xi_1\ldots\xi_n}{}_{\eta_1\ldots\eta_m\alpha\beta}$, the scalar S must have one additional factor $g^{\alpha\beta}$. Owing to

$$\frac{\partial S}{\partial g_{\mu\beta}} g_{\nu\beta} = -\frac{\partial S}{\partial g^{\alpha\nu}} g^{\alpha\mu}, \qquad \frac{\partial S}{\partial g_{\alpha\mu}} g_{\alpha\nu} = -\frac{\partial S}{\partial g^{\nu\beta}} g^{\mu\beta},$$

one thus encounters the four additional terms:

$$\begin{aligned}
& \frac{\partial S}{\partial \bar{T}^{\xi_1\ldots\xi_n}{}_{\eta_1\ldots\eta_m\mu\beta}} \bar{T}^{\xi_1\ldots\xi_n}{}_{\eta_1\ldots\eta_m\nu\beta} + \frac{\partial S}{\partial \bar{T}^{\xi_1\ldots\xi_n}{}_{\eta_1\ldots\eta_m\,\alpha\mu}} \bar{T}^{\xi_1\ldots\xi_n}{}_{\eta_1\ldots\eta_m\,\alpha\nu} - \frac{\partial S}{\partial g^{\alpha\nu}} g^{\alpha\mu} - \frac{\partial S}{\partial g^{\nu\beta}} g^{\mu\beta} \\
&= \delta^{\mu}_{\alpha}\,\bar{T}^{\xi_1\ldots\xi_n}{}_{\eta_1\ldots\eta_m\nu\beta}\, g^{\alpha\beta} + \delta^{\mu}_{\beta}\,\bar{T}^{\xi_1\ldots\xi_n}{}_{\eta_1\ldots\eta_m\,\alpha\nu}\, g^{\alpha\beta} - \delta^{\beta}_{\nu}\, g^{\alpha\mu}\bar{T}^{\xi_1\ldots\xi_n}{}_{\eta_1\ldots\eta_m\alpha\beta} - \delta^{\alpha}_{\nu}\, g^{\mu\beta}\bar{T}^{\xi_1\ldots\xi_n}{}_{\eta_1\ldots\eta_m\alpha\beta} \\
&= g^{\mu\beta}\bar{T}^{\xi_1\ldots\xi_n}{}_{\eta_1\ldots\eta_m\nu\beta} + g^{\alpha\mu}\bar{T}^{\xi_1\ldots\xi_n}{}_{\eta_1\ldots\eta_m\,\alpha\nu} - g^{\alpha\mu}\bar{T}^{\xi_1\ldots\xi_n}{}_{\eta_1\ldots\eta_m\alpha\nu} - g^{\mu\beta}\bar{T}^{\xi_1\ldots\xi_n}{}_{\eta_1\ldots\eta_m\nu\beta} \\
&= 0.
\end{aligned}$$

The derivative of a Lagrangian $\mathscr{L}$ with respect to the metric $g_{\mu\nu}$ can thus always be replaced by the derivatives with respect to the appertaining tensors T that are made into a scalar by means of the metric. The identity thus provides the correlation of the *metric* and the *canonical* energy-momentum tensors of a given system.

Corollary 1 *The contraction of Eq. (89) then yields a condition for the scalar S:*

$$\frac{\partial S}{\partial g_{\alpha\beta}} g_{\alpha\beta} \equiv \frac{n-m}{2} \frac{\partial S}{\partial T^{\xi_1\ldots\xi_n}{}_{\eta_1\ldots\eta_m}} T^{\xi_1\ldots\xi_n}{}_{\eta_1\ldots\eta_m}. \tag{90}$$

Proof Contracting Eq. (89) directly yields Eq. (90).

Corollary 2 *Let $\tilde{\mathscr{L}} = \tilde{\mathscr{L}}(g, T_k) \in \mathbb{R}$ be a* scalar density *(i.e. a relative scalar of weight one) valued function of the (symmetric) metric $g^{\mu\nu}$ and a sum of k tensors $T_k{}^{\xi_1\ldots\xi_{n_k}}{}_{\eta_1\ldots\eta_{m_k}}$ of respective rank (n_k, m_k), where $(m_k - n_k)/2 \in \mathbb{Z}$. Then the following identity holds:*

$$\begin{aligned}
& 2\frac{\partial \tilde{\mathscr{L}}}{\partial g_{\mu\beta}} g_{\nu\beta} - \frac{\partial \tilde{\mathscr{L}}}{\partial T_k{}^{\nu\xi_2\ldots\xi_{n_k}}{}_{\eta_1\ldots\eta_{m_k}}} T_k{}^{\mu\xi_2\ldots\xi_{n_k}}{}_{\eta_1\ldots\eta_{m_k}} - \cdots - \frac{\partial \tilde{\mathscr{L}}}{\partial T_k{}^{\xi_1\ldots\xi_{n_k-1}\nu}{}_{\eta_1\ldots\eta_{m_k}}} T_k{}^{\xi_1\ldots\xi_{n_k-1}\mu}{}_{\eta_1\ldots\eta_{m_k}} \\
& + \frac{\partial \tilde{\mathscr{L}}}{\partial T_k{}^{\xi_1\ldots\xi_{n_k}}{}_{\mu\eta_2\ldots\eta_{m_k}}} T_k{}^{\xi_1\ldots\xi_{n_k}}{}_{\nu\eta_2\ldots\eta_{m_k}} + \cdots + \frac{\partial \tilde{\mathscr{L}}}{\partial T_k{}^{\xi_1\ldots\xi_{n_k}}{}_{\eta_1\ldots\eta_{m_k-1}\mu}} T_k{}^{\xi_1\ldots\xi_{n_k}}{}_{\eta_1\ldots\eta_{m_k-1}\nu} \equiv \delta^{\mu}_{\nu}\tilde{\mathscr{L}}.
\end{aligned} \tag{91}$$

Proof Combine Eq. (89) with (58).

Equation (91) is obviously a representation of Euler's theorem on homogeneous functions in the realm of tensor calculus.

Examples for Identities (89) Involving the Riemann Tensor

Riemann Tensor Squared

As a scalar, any Lagrangian $\mathscr{L}_R(R, g)$ built from the Riemann–Cartan tensor (41) and the metric satisfies the identity (89)

$$2\frac{\partial \mathscr{L}_R}{\partial g^{\nu\beta}} g^{\mu\beta} \equiv \frac{\partial \mathscr{L}_R}{\partial R^{\tau}{}_{\mu\beta\lambda}} R^{\tau}{}_{\nu\beta\lambda} - \frac{\partial \mathscr{L}_R}{\partial R^{\nu}{}_{\tau\beta\lambda}} R^{\mu}{}_{\tau\beta\lambda} + 2\frac{\partial \mathscr{L}_R}{\partial R^{\eta}{}_{\alpha\beta\mu}} R^{\eta}{}_{\alpha\beta\nu}. \tag{92}$$

The factors "2" emerge from the symmetry of the metric and the skew-symmetry of the Riemann–Cartan tensor in its last index pair. The identity is easily verified for a Lagrangian linear and quadratic in the Riemann tensor:

$$\mathscr{L}_R = R^{\eta}{}_{\alpha\beta\tau} \left[\tfrac{1}{4} R^{\alpha}{}_{\eta\xi\lambda} g^{\beta\xi} g^{\tau\lambda} - g_2 \left(\delta^{\beta}_{\eta}\, g^{\alpha\tau} - \delta^{\tau}_{\eta}\, g^{\alpha\beta} \right) \right].$$

The left-hand side of Eq. (92) evaluates to

$$2\frac{\partial \mathscr{L}_R}{\partial g^{\nu\beta}} g^{\mu\beta} = -R^{\eta\alpha\beta\mu}\, R_{\eta\alpha\beta\nu} + g_2 \left(R_{\nu}{}^{\mu} + R^{\mu}{}_{\nu} \right),$$

which indeed agrees with the terms obtained from the right-hand side:

$$\frac{\partial \mathscr{L}_R}{\partial R^{\tau}{}_{\mu\alpha\beta}} R^{\tau}{}_{\nu\alpha\beta} - \frac{\partial \mathscr{L}_R}{\partial R^{\nu}{}_{\tau\alpha\beta}} R^{\mu}{}_{\tau\alpha\beta} + 2\frac{\partial \mathscr{L}_R}{\partial R^{\tau}{}_{\alpha\beta\mu}} R^{\tau}{}_{\alpha\beta\nu} = -R^{\eta\alpha\beta\mu} R_{\eta\alpha\beta\nu} + g_2 \left(R_{\nu}{}^{\mu} + R^{\mu}{}_{\nu} \right).$$

Ricci Scalar

The Ricci scalar R is defined as the following contraction of the Riemann tensor

$$R = R_{\eta\alpha\xi\lambda}\, g^{\eta\xi}\, g^{\alpha\lambda}. \tag{93}$$

With the scalar $\mathscr{L}_R = R$ and the tensor T the Riemann tensor, the general Eq. (89) takes on the particular form

$$\frac{\partial R}{\partial g^{\nu\beta}} g^{\mu\beta} + \frac{\partial R}{\partial g^{\beta\nu}} g^{\beta\mu} - \frac{\partial R}{\partial R_{\mu\alpha\xi\lambda}} R_{\nu\alpha\xi\lambda} - \frac{\partial R}{\partial R_{\eta\mu\xi\lambda}} R_{\eta\nu\xi\lambda} - \frac{\partial R}{\partial R_{\eta\alpha\mu\lambda}} R_{\eta\alpha\nu\lambda} - \frac{\partial R}{\partial R_{\eta\alpha\xi\mu}} R_{\eta\alpha\xi\nu} \equiv 0. \tag{94}$$

Without making use of the symmetries of the Riemann tensor and the metric, this identity is actually fulfilled as

$$\begin{aligned}\frac{\partial R}{\partial g^{\nu\beta}} g^{\mu\beta} &= R_{\eta\alpha\xi\lambda} \left(\delta^{\eta}_{\nu}\, \delta^{\xi}_{\beta}\, g^{\alpha\lambda} + g^{\eta\xi}\, \delta^{\alpha}_{\nu}\, \delta^{\lambda}_{\beta}\right) g^{\beta\mu} \\ &= \left(R_{\nu\alpha\beta\lambda}\, g^{\alpha\lambda} + g^{\eta\xi}\, R_{\eta\nu\xi\beta}\right) g^{\beta\mu} \\ &= R_{\nu\alpha}{}^{\mu\alpha} + R^{\xi}{}_{\nu\xi}{}^{\mu}.\end{aligned}$$

Similarly

$$\frac{\partial R}{\partial g^{\beta\nu}} g^{\beta\mu} = R^{\mu}{}_{\alpha\nu}{}^{\alpha} + R^{\xi\mu}{}_{\xi\nu}.$$

The derivative terms of the Riemann tensor are

$$\frac{\partial R}{\partial R_{\mu\alpha\xi\lambda}} R_{\nu\alpha\xi\lambda} = \delta^{\mu}_{\eta}\, g^{\eta\xi}\, g^{\alpha\lambda} R_{\nu\alpha\xi\lambda} = R_{\nu\alpha}{}^{\mu\alpha}$$

and

$$\begin{aligned}\frac{\partial R}{\partial R_{\eta\mu\xi\lambda}} R_{\eta\nu\xi\lambda} &= \delta^{\mu}_{\alpha}\, g^{\eta\xi}\, g^{\alpha\lambda} R_{\eta\nu\xi\lambda} = R^{\xi}{}_{\nu\xi}{}^{\mu} \\ \frac{\partial R}{\partial R_{\eta\alpha\mu\lambda}} R_{\eta\alpha\nu\lambda} &= \delta^{\mu}_{\xi}\, g^{\eta\xi}\, g^{\alpha\lambda} R_{\eta\alpha\nu\lambda} = R^{\mu}{}_{\alpha\nu}{}^{\alpha} \\ \frac{\partial R}{\partial R_{\eta\alpha\xi\mu}} R_{\eta\alpha\xi\nu} &= \delta^{\mu}_{\lambda}\, g^{\eta\xi}\, g^{\alpha\lambda} R_{\eta\alpha\xi\nu} = R^{\xi\mu}{}_{\xi\nu},\end{aligned}$$

which obviously cancel the four terms emerging from the derivatives with respect to the metric.

Making now use of the skew-symmetries of the Riemann tensor in its first and second index pair and of the symmetry of the metric, Eq. (94) simplifies to

$$\frac{\partial R}{\partial g^{\nu\beta}} g^{\mu\beta} \equiv \frac{\partial R}{\partial R_{\mu\alpha\xi\lambda}} R_{\nu\alpha\xi\lambda} + \frac{\partial R}{\partial R_{\eta\alpha\xi\mu}} R_{\eta\alpha\xi\nu}. \tag{95}$$

For zero torsion, the Riemann tensor has the additional symmetry on exchange of both index pairs. Then

$$\frac{\partial R}{\partial g^{\nu\beta}} g^{\mu\beta} \equiv 2 \frac{\partial R}{\partial R_{\mu\alpha\xi\lambda}} R_{\nu\alpha\xi\lambda} \quad \Leftrightarrow \quad \frac{\partial R}{\partial g^{\nu\mu}} \equiv 2 \frac{\partial R}{\partial R^{\mu}{}_{\alpha\xi\lambda}} R_{\nu\alpha\xi\lambda}. \tag{96}$$

Ricci Tensor Squared

The scalar made of the (not necessarily symmetric) Ricci tensor $R_{\eta\alpha}$ is defined by the following contraction with the metric

$$\mathscr{L}_R = R_{\eta\alpha} R_{\xi\lambda}\, g^{\eta\xi}\, g^{\alpha\lambda}. \tag{97}$$

With Eq. (97), the general Eq. (89) now takes on the particular form

$$\frac{\partial \mathscr{L}_R}{\partial g^{\nu\beta}} g^{\mu\beta} + \frac{\partial \mathscr{L}_R}{\partial g^{\beta\nu}} g^{\beta\mu} - \frac{\partial \mathscr{L}_R}{\partial R_{\mu\beta}} R_{\nu\beta} - \frac{\partial \mathscr{L}_R}{\partial R_{\beta\mu}} R_{\beta\nu} \equiv 0. \tag{98}$$

Without making use of the symmetries of the Ricci tensor and the metric, this identity is actually fulfilled as

$$\begin{aligned}\frac{\partial \mathscr{L}_R}{\partial g^{\nu\beta}} g^{\mu\beta} &= R_{\eta\alpha} R_{\xi\lambda} \left(\delta^{\eta}_{\nu}\, \delta^{\xi}_{\beta}\, g^{\alpha\lambda} + g^{\eta\xi}\, \delta^{\alpha}_{\nu}\, \delta^{\lambda}_{\beta}\right) g^{\beta\mu} \\ &= \left(R_{\nu\alpha} R_{\beta\lambda}\, g^{\alpha\lambda} + g^{\eta\xi}\, R_{\eta\nu} R_{\xi\beta}\right) g^{\beta\mu} \\ &= R_{\nu\beta} R^{\mu\beta} + R_{\beta\nu} R^{\beta\mu}.\end{aligned}$$

Similarly

$$\frac{\partial \mathscr{L}_R}{\partial g^{\beta\nu}} g^{\beta\mu} = R_{\nu\beta} R^{\mu\beta} + R_{\beta\nu} R^{\beta\mu}.$$

The derivative terms of the Ricci tensor are

$$\frac{\partial \mathscr{L}_R}{\partial R_{\mu\beta}} R_{\nu\beta} = \left(\delta^{\mu}_{\eta}\delta^{\beta}_{\alpha} R_{\xi\lambda} + R_{\eta\alpha}\delta^{\mu}_{\xi}\delta^{\beta}_{\lambda}\right) g^{\eta\xi}\, g^{\alpha\lambda}\, R_{\nu\beta} = 2R_{\nu\beta}\, R^{\mu\beta}$$

and

$$\frac{\partial \mathscr{L}_R}{\partial R_{\beta\mu}} R_{\beta\nu} = \left(\delta^{\beta}_{\eta}\delta^{\mu}_{\alpha} R_{\xi\lambda} + R_{\eta\alpha}\delta^{\beta}_{\xi}\delta^{\mu}_{\lambda}\right) g^{\eta\xi}\, g^{\alpha\lambda}\, R_{\beta\nu} = 2R_{\beta\nu}\, R^{\beta\mu},$$

which obviously cancel the four terms emerging from the derivatives with respect to the metric.

For zero torsion, the Ricci tensor is symmetric. Then

$$\frac{\partial \mathscr{L}_R}{\partial g^{\nu\beta}} g^{\mu\beta} \equiv \frac{\partial \mathscr{L}_R}{\partial R_{\mu\beta}} R_{\nu\beta} \qquad \Leftrightarrow \qquad \frac{\partial \mathscr{L}_R}{\partial g^{\nu\mu}} \equiv \frac{\partial \mathscr{L}_R}{\partial R_{\alpha\beta}} R_{\nu\beta}\, g_{\alpha\mu}. \tag{99}$$

References

1. W. Greiner, *Classical Mechanics*, 2nd edn. (Springer, Berlin, 2010)
2. J. Struckmeier, J. Muench, D. Vasak, J. Kirsch, M. Hanauske, H. Stoecker, Phys. Rev. D **95**, 124048 (2017). https://doi.org/10.1103/PhysRevD.95.124048
3. A. Einstein, *Private letter to Hermann Weyl* (ETH Zürich Library, Archives and Estates, 1918)
4. D. Kehm, J. Kirsch, J. Struckmeier, D. Vasak, M. Hanauske, Astron. Nachr./AN **338**(9–10), 1015 (2017). https://doi.org/10.1002/asna.201713421
5. F.W. Hehl, P. von der Heyde, G.D. Kerlick, J.M. Nester, Rev. Mod. Phys. **48**(3), 393 (1976). https://doi.org/10.1103/revmodphys.48.393
6. E. Noether, Nachrichten der Königlichen Gesesellschaft der Wissenschaften Göttingen. Mathematisch-Physikalische Klasse **57**, 235 (1918)
7. J. Struckmeier, A. Redelbach, Int. J. Mod. Phys. E **17**, 435 (2008). https://doi.org/10.1142/s0218301308009458
8. M. Gaul, C. Rovelli, Lect. Notes Phys. **541**, 277 (2000)
9. P. Matteucci, Rep. Math. Phys. **52**, 115 (2003). https://doi.org/10.1016/S0034-4877(03)90007-3, arXiv:gr-qc/0201079
10. M. Godina, P. Matteucci, Int. J. Geom. Methods Mod. Phys. **2**, 159 (2005). https://doi.org/10.1142/S0219887805000624, arXiv:math/0504366
11. J. Schouten, Ricci-Calculus (Springer. Berlin (1954). https://doi.org/10.1007/978-3-662-12927-2
12. F.W. Hehl, Rep. Math. Phys. **9**(3), 55 (1976)
13. F.W. Hehl, in *Proceedings, 15th Workshop on High Energy Spin Physics (DSPIN-13)*, Dubna, Russia, 8–12 October 2013 (2014)
14. C.W. Misner, K.S. Thorne, J.A. Wheeler, *Gravitation* (W. H. Freeman and Company, New York, 1973)
15. P. Jordan, Ann. der Phys. **428**(1), 64 (1939)
16. D. Sciama, Mon. Not. R. Astron. Soc. **113**, 34 (1953)
17. R. Feynman, W. Morinigo, W. Wagner, *Feynman Lectures On Gravitation, Frontiers in Physics* (Westview Press, Boulder, 2002)
18. S. Hawking, *The Theory of Everything* (New Millenium Press, 2003)
19. S. Carroll, *Spacetime and Geometry* (Prentice Hall, Englewood Cliffs, 2013)
20. T.W.B. Kibble, J. Math. Phys. **2**, 212 (1961)
21. D.W. Sciama, in *Recent Developments in General Relativity* (Pergamon Press, Oxford; PWN, Warsaw, 1962), pp. 415–439. Festschrift for Infeld
22. J. Plebanski, A. Krasinski, *An Introduction to General Relativity and Cosmology* (Cambridge University Press, Cambridge, 2006)
23. J. Struckmeier, P. Liebrich, J. Muench, M. Hanauske, J. Kirsch, D. Vasak, L. Satarov, H. Stoecker, Int. J. Mod. Phys. E **28**(1), 1950007 (2019). https://doi.org/10.1142/S0218301319500071, arXiv:1711.10333
24. K. Hayashi, T. Shirafuji, Prog. Theor. Phys. **64**(3), 866, 883, 1435, 2222 (1980). https://doi.org/10.1143/PTP.64.866

High-Resolution Experiments with Exotic Nuclei and Mesic Atoms

Hans Geissel, Gottfried Münzenberg and Christoph Scheidenberger

Abstract Pioneering heavy-ion research and the limits of stability of atoms and nuclei were central research topics of Walter Greiner. His ideas and directions have inspired and determined the experimental efforts to design novel separators and high-resolution spectrometers to produce and study the most exotic nuclei. This leads directly to the discovery of new nuclides, new decay modes and shapes, and unexspected shell effects near the driplines. Already in his early studies, Walter Greiner investigated mesic atoms which are still of basic interests to understand the strong interaction of matter and the mass modification of bound mesons. The existence of deeply-bound pionic states in heavy atoms was an important question in the Greiner School too. It was answered with a discovery experiment with the fragment separator FRS. In this article, we review characteristic experimental results inspired by the far-reaching ideas of Walter and his colleagues measured with the FRS at GSI. New experimental perspectives with the Super-FRS will also be adressed.

1 Introduction

This contribution consists of two articles dedicated to our friend and teacher Walter Greiner, who was one of the founders of the heavy-ion research centre GSI. His far-reaching ideas and experienced scientific leadership have led to many discoveries of new atomic and nuclear properties created in reactions with heavy ions.

The new ideas of Walter and his colleagues required from the experimentalists a novel type of heavy-ion accelerator facility [1, 2], fast and efficient separators and spectrometers equiped with electronics and detectors at the frontiers of technical

H. Geissel (✉) · G. Münzenberg · C. Scheidenberger
Gesellschaft für Schwerionenforschung GSI, Planckstraße 1,
64291 Darmstadt, Germany
e-mail: H.Geissel@GSI.de

G. Münzenberg
e-mail: G.Muenzenberg@GSI.de

C. Scheidenberger
e-mail: C.Scheidenberger@GSI.de

J. Kirsch et al. (eds.), *Discoveries at the Frontiers of Science*,
FIAS Interdisciplinary Science Series, https://doi.org/10.1007/978-3-030-34234-0_13

developments. These experimental tools were necessary to discover new short-lived nuclei towards the proton and neutron driplines and with maximum proton numbers beyond $Z > 100$ [3].

In this article, we review high-resolution spectrometer experiments performed with the fragment separator FRS [4]. Our previous article [5] is updated here for overlapping chapters. The category of spectrometer experiments has been the main scientific activity of the FRS group at GSI since the commissioning of the separator in 1990. In the final section of the present review, we describe new horizons of planned spectrometer experiments with the Super-FRS [6] which represent the core activity of the Super-FRS Experiment Collaboration, one of the NUSTAR sub-collaboration.

2 The In-Flight Separator FRS

The great scientific potential of relativistic exotic nuclei was pioneered at the Lawrence Berkeley National Laboratory (LBL) [7, 8]. In these experiments the formation of neutron haloes in the light dripline nuclei ^{11}Li and ^{11}Be were discovered and the modern concept of radioactive nuclear beam facilities was initiated [9]. The spectacular LBL results were a strong motivation for the experimental extension of the GSI facilities to relativistic energies. One of the conclusions was to build the combination of the synchrotron SIS-18 [2], the projectile fragment separator FRS [4] and the storage-cooler ring ESR [10, 11].

The FRS is a powerful and versatile in-flight facility for relativistic exotic nuclei with magnetic rigidities up to 18 Tm. The heavy-ion synchrotron SIS-18 provides beams of all stable isotopes from protons up to uranium which are converted via projectile fragmentation and fission into interesting exotic nuclei in the production targets placed at the entrance of the FRS. A selected fragment beam is spatially separated in flight from the primary beam and the contaminants within a few hundred nanoseconds, thus also yielding access to very short-lived nuclei. The FRS facility consists of three ion-optical branches dedicated to specific experimental categories. Besides its main function as a separator, the FRS also represents a multi-stage high-resolution magnetic spectrometer; the symmetric direct branch is especially well suited for this task. The layout of the GSI facilities is schematically presented in Fig. 1.

At low-velocities, both magnetic and electrostatic fields can be applied for the design of in-flight separators. However, electrostatic fields cannot be applied at relativistic energies, because the required electric rigidities exceeds the technical limits. Therefore, atomic slowing-down in shaped solid degraders, placed at dispersive focal planes, are employed instead to achieve a spatial mono-isotopic separation with the FRS [4]. The combination of slowing down in matter and two-fold magnetic rigidity analysis ($B\rho$-ΔE-$B\rho$) represents the spatial separation method of the FRS [4]. Typical degrader thicknesses are between 50 and 60% of the atomic range of the separated fragments. The lower limit of the operating domain of the $B\rho$-ΔE-$B\rho$ method is given by the minimum kinetic energies required to yield mainly bare fragments

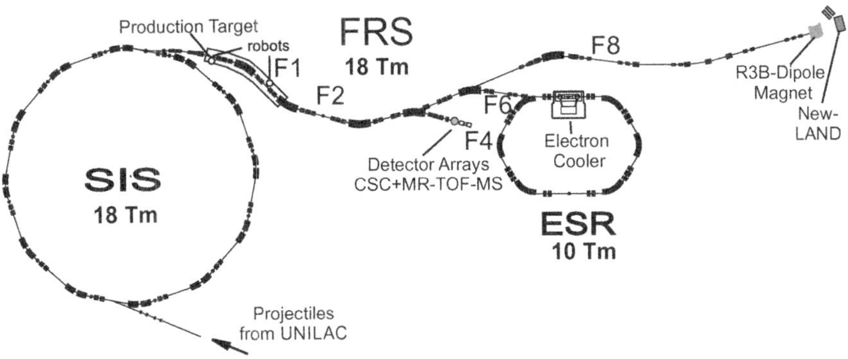

Fig. 1 The synchrotron SIS-18 accelerates projectiles from protons up to uranium with kinetic energies corresponding up to maximum 18 Tm. The exotic nuclei can be investigated either directly at the FRS or injected in the storage-cooler ring ESR, or matched with the NUSTAR R3B detector setup including a large superconducting dipole magnet. The versatile magnetic system of the FRS is ideal suited for high-resolution spectrometer experiments. The combination of the FRS with a Cryogenic Stopping Cell (CSC) and the Multiple-Reflection Tof-Flight Mass Spectrometer (MR-TOF-MS) is unique

emerging from the target and the degraders. The upper limit is determined by the loss due to secondary nuclear reactions.

The key role of atomic slowing of heavy ions in matter is taken into account by dedicated experiments with the FRS used as high-resolution magnetic spectrometer [12–14]. The basic knowledge of the slowing down of relativistic heavy ions, obtained from these FRS experiments, has been implemented in different computer programs [15–17]. The results have also been a base for heavy ion tumour therapy which was established at GSI a few years later [18–20]. The complete stopping of heavy ions in solid matter and the corresponding positron emission tomography (PET) for its application in heavy-ion tumour treatment has been also developed and refined in FRS experiments.

3 Discovery of New Isotopes and Decay Modes

A major design goal of the FRS was the realization of a pure isotopic spatial separation in flight for fragments of all elements up to uranium [21]. This requirement is very important for the access of rare isotopes, created with very low cross sections down to the pb range and for FRS-ESR decay and reaction experiments, where the option of only one stored nuclear species is essential.

A challenging performance criterion for each exotic nuclear beam facility is the discovery of new isotopes and the investigation of their properties. In this direction most of the new isotopes have been discovered with the FRS with uranium projectiles in the energy range between 700 and 1000 MeV/u. In total about 282 new isotopes

[22] have been discovered, in which the abrasion fission products represent a large fraction. Already in the pioneering experiments with uranium fission fragments [23, 24] about 120 new neutron-rich isotopes were identified and their production cross sections were measured. The isotopic separation and the unambiguous particle identification in mass and proton numbers were the keys for the success, more than 60 years after the discovery of the fission process. From this experience it became clear that for the next generation of in-flight facilities [6, 25, 26], the rich source of fission fragments with their inherent large phase-space population have to be included in the design goals.

In the recent years, the uranium beam intensity of the SIS-18 was improved to 10^9/s which enabled us to extend our knowledge to more neutron-rich isotopes created with ^{238}U projectile beams [27, 28]. About additional 100 new n-rich isotopes were discovered. The first properties measured for these isotopes were their production cross sections, half-lives and masses.

Among the nuclides, discovered with the FRS, were the two doubly-magic ones, ^{100}Sn [29] and ^{78}Ni [30]. In particular, the search for ^{100}Sn nuclei was a long-standing major goal in many laboratories. Many experimental methods were applied, but the rate of contaminates was simply too high. The doubly magic nucleus ^{100}Sn is close to the proton dripline and represents the heaviest known particle-stable $N = Z$ nucleus. It is a cornerstone for crucial tests of shell-model calculations, because it is expected to decay due to core polarisation via the purest Gamow-Teller transition without quenching.

The cross section for the production of ^{100}Sn ions via fragmentation of ^{124}Sn projectiles at 1000 MeV/u was a few pb. This tiny cross section and consequent experimental rate suggested the application of implantation-decay correlation, a technique established in heavy-element research [31].

The accelerator achievements of higher primary beam intensities, $1 \cdot 10^9$/s for ^{124}Sn projectiles, and the installation of an efficient gamma detector array at F4 [32] were the main motivation for the ^{100}Sn experiment in 2008 [33]. The results of the pioneer experiment were confirmed and the uncertainties strongly improved.

The high sensitivity and selectivity of the FRS experiments encouraged us to search for a new decay mode, the two-proton radioactivity. It was predicted many decades earlier [34] and should be an allowed decay from the ground state if the one-proton emission is energetically forbidden. After a few preparatory test experiments two-proton radioactivity was discovered for the ^{45}Fe nuclide [35, 36]. In follow-up experiments, the angular correlation of the simultaneously emitted two protons was elegantly measured with tracking [37]. An illustration of the powerful optical time projection chamber (OTPC) for discovering new decay modes is shown in Fig. 2. The beta decay of ^{31}Ar nuclei was measured with an OTPC at F4 of FRS. Thirteen events of the β-delayed 3 proton (β3p) emission were unambiguously recorded and yielded a branching ratio of 0.07(2)% for this decay mode. New structures and decay modes are expected near and beyond the driplines, which can be studied by the methods of invariant mass and in-flight decay [38, 39]. The objectives are experiments on: 2-proton radioactivity and other novel types of radioactivity (such as 4p, 2n, 4n),

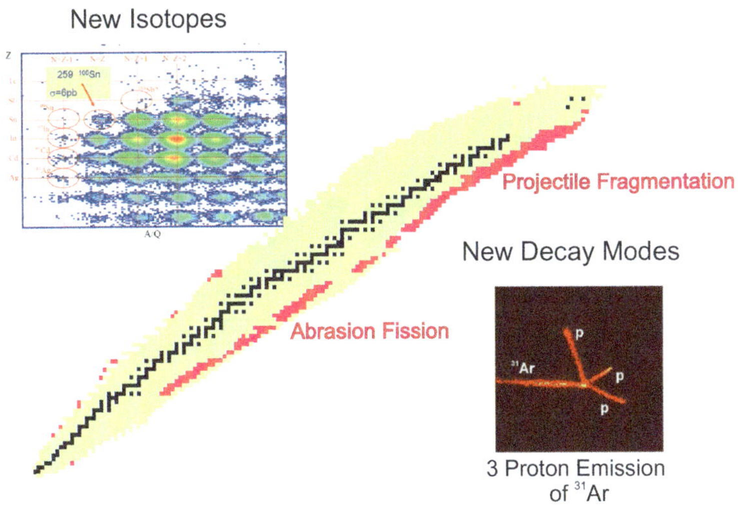

Fig. 2 Isotopes discovered with the FRS are indicated in the chart of nuclides. The new nuclei are created via projectile fragmentation or abrasion fission. An example of particle identification in flight for the ^{100}Sn experiment and the tracking of the 3 proton emission from the decay of ^{31}Ar ions in an optical time-projection chamber are inserted

resonance decays and studies of beta-delayed particle and multiple-particle emission from exotic nuclei.

A novel development is the combination of an in-flight separator with a gas-filled stopping cell which is connected with a high-resolution multiple-reflection time-of-flight mass spectrometer (MR-TOF-MS) [40–42]. A great challenge in such a system is for exotic nuclei produced at relativistic energies, because of the inevitable large range straggling compared to the small gaseous layer thickness. A technical improvement was realized via cooling down of the gas volume to increase the possible areal densities and freeze out the impurities. Such a Cryogenic Stopping Cell (CSC) [43, 44] characterized by an areal density of (3–5) mg/cm^2 pure He gas at 70–100 K has been operated in first pilot experiments at the FRS. Besides the initial high energy, also the creation process of the fragments in the target determines the final range straggling. An experimental solution is to apply the high optical resolving power of the FRS for energy compression with a monoenergetic degrader placed at a dispersive focal plane [45].

In the first experiments with the CSC [40, 41] by the use of uranium projectile fragments produced at 1000 MeV/u, the stopping efficiency after energy-bunching with a monoenergetic degrader system, placed at the central focal plane of the FRS, was about 15–27%. Including the survival and extraction efficiency, the total efficiency of the CSC in the fore-mentioned pioneering experiments was about 10%. The measured extraction time with ^{221}Ac ions was between 20 and 30 ms. Recent technical improvements of the MR-TOF-MS provided resolving powers of 600,000

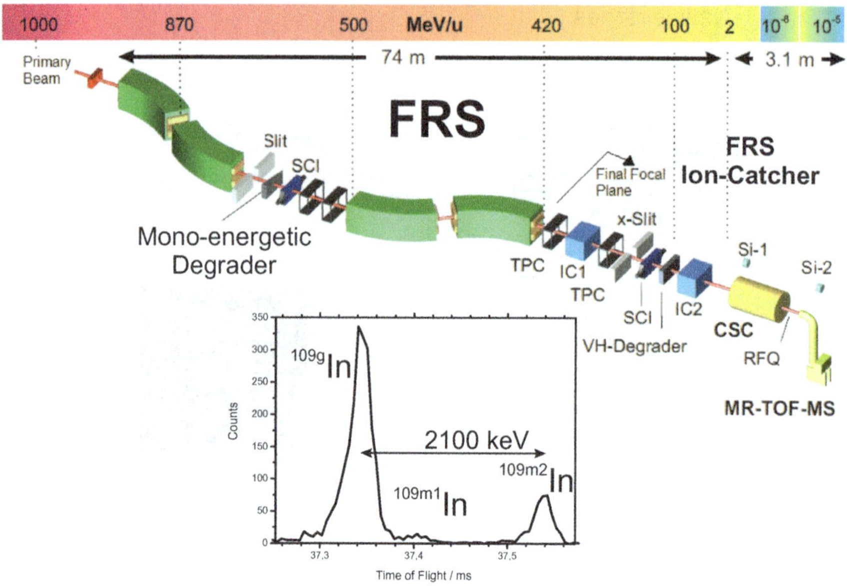

Fig. 3 Schematic representation of the combination of the FRS with the FRS-Ion-Catcher. The latter device consists of a cryogenic stopping cell (CSC) and a multiple-reflection time-of-flight mass spectrometer (MR-ToF-MS) [40]. The spatially, in-flight separated fragments created at 1000 MeV/u are energy-bunched with a 'mono-energetic' degrader and stopped in the with gas-filled CSC. The injected ions are extracted within about 20–30 ms for accurate high-resolution mass measurements with the MR-ToF-MS. A typical mass spectrum is shown with a mass resolving power of 440,000 for ^{109}In ions [47]

in a measurement time of less than 20 ms and with repetition rates exceeding 1000 Hz [46]. In the previous FRS experiments, the MR-TOF-MS has reached a maximum accuracy of $6 \cdot 10^{-8}$ in the measurement with ^{213}Fr ions (Fig. 3). The most short-lived measured nuclide was ^{213}Rn with a half-life of 19.5 ms. However, in a modified protype of the CSC, the mass and lifetime (1.7 ms) of ^{215}Po ions has been successfully measured [46]. The latter principle will be applied for the new CSC at the Super-FRS.

The ground-state masses of projectile fragments in the Pb region have been measured, for seven nuclides for which they were determined in the past only via decay spectroscopy. A comparison with previously well-known masses from experiments in the literature has demonstrated that the distribution of the observed MR-TOF-MS deviations are characterized by a Gaussian function with a mean value of (1.7 ± 4.4) keV and a standard deviation of (34.9 ± 8.9) keV [46]. These results are very promising and clearly show the great potential of the CSC combined with the MR-TOF-MS for the investigation of short-lived exotic nuclides. New research opportunities are opened up:

- Identification of new isotopes combined with mass, lifetime and decay measurements (in particular, for low-energy production reactions).
- New approach to synthesize and identify new neutron-rich isotopes up to heavy elements via deep-inelastic reactions in the stopping cell.
- Low-energy experiments with pure isomeric beams of all elements.

4 Discovery of New Nuclear Matter Distributions

From the pioneering experiments on nuclear matter distribution of radioactive isotopes at LBL Berkeley [9] it was concluded that the proton and neutron distributions may dramatically differ from nuclei close to the neutron dripline. This property was observed with radioactive beam facilities which had its origin at LBL. In particular, the ^{6,8}He isotopes were considered to have a thick neutron skin.

These conclusions were deduced from experimental interaction cross section measurements and model calculations based on Glauber multiple-scattering theory. Therefore, in the next generation of such experiments at the FRS, the goal was to measure the isotope chain of sodium nuclei for which the charge radii were experimentally known from hyperfine interaction measured with laser spectroscopy [48]. Indeed, with the FRS experiments at 950 MeV/u the neutron skin evolution as a function of the neutron number was experimentally discovered [49]. A monotonic increase was observed towards the neutron dripline of Na isotopes.

In the pioneer experiments at LBL it was observed that the extended neutron density of a halo nucleus manifests in a narrow momentum distribution of the core fragment after removal reaction of the halo nucleon [50]. The inclusive momentum measurements for ^{11}Li, ^{8}He, and ^{6}He nuclei demonstrated a narrow distribution for the removal of weakly-bound neutrons and a wider distribution for nucleon removal from the core. The momentum measurements were performed by angular measurements which yield the transverse components. After these experiments Hansen and Jonson proposed the first halo model [51] to explain the new nuclear matter distribution discovered in the Tanihata experiments.

Precise longitudinal momentum measurements are more difficult because of the dominant incident large momentum distribution of the primary radioactive beams, inherent from their creation process. Such longitudinal momentum measurements of the core fragments would require a high-resolution spectrometer operated as an energy-loss spectrometer which was not available at LBL.

A new era of spectroscopy of weakly bound nuclei was started with the operation of the ion-optical systems of the FRS and the S800 [52] as an energy-loss spectrometer. A measurement of the momentum distribution in coincidence with the γ-rays emitted from the de-excited core fragment provides state selectivity and yields the corresponding spectroscopic factors. In this way, the momentum distribution induced from secondary reactions can be precisely measured without suffering from the large incident phase space. Exclusive momentum measurements of secondary reaction products are ideally suited to study the population and correlations of nucleons in exotic nuclei [53]. A clear advantage of knock-out reactions at several hundred MeV/u is the rather simple reaction mechanism dominated by single nucleon collisions.

It was generally believed for many years that the formation of a halo distribution is only possible at and near the neutron dripline. However in 1994, a longitudinal momentum measurement with the FRS after proton removal led to the discovery of the first proton halo [54].

The proton halo in the ^{8}B nucleus was experimentally discovered via a narrow momentum distribution of ^{7}Be secondary fragments after one-proton removal reaction and by the strongly enhanced differential cross section [54]. These two signatures were consistently observed in different target materials. The experimental method and the ion-optical principle of measurement is illustrated in Fig. 4. The measured momentum distribution of ^{7}Be nuclei could be described with a theoretical prediction [55]. The follow-up experiment was an exclusive momentum measurement in coincidence with γ rays recorded with detectors at the central focal plane F2. It has provided new information on the removal from the pure ground and excited states. The coincidence measurement reveals that 13% core excitation is involved in the knock-out reaction with a carbon target [56]. The experimental results have guided

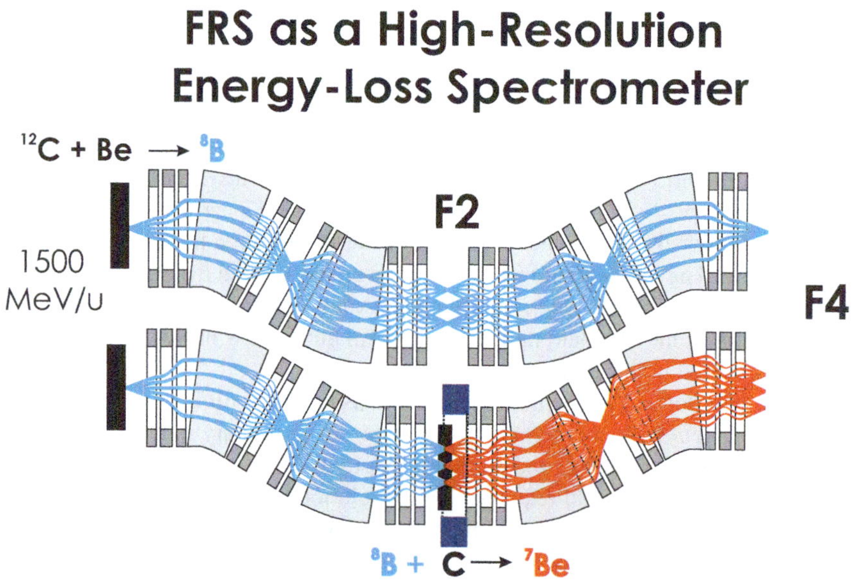

Fig. 4 Ion-optical operating mode applied in the discovery experiment of the proton halo in ^{8}B nuclei. Shown are the dispersive focal planes w/o the setup for secondary reaction measurements. The FRS was used as an energy-loss spectrometer for precise longitudinal momentum measurements of secondary fragments after one-nucleon removal reactions. The secondary break-up target at F2 is viewed with a gamma detectors (blue). The reaction example is for ^{8}B ions created with 1500 MeV/u ^{12}C projectiles. ^{7}Be secondary fragments, after one-proton removal reaction from the ^{8}B, are measured

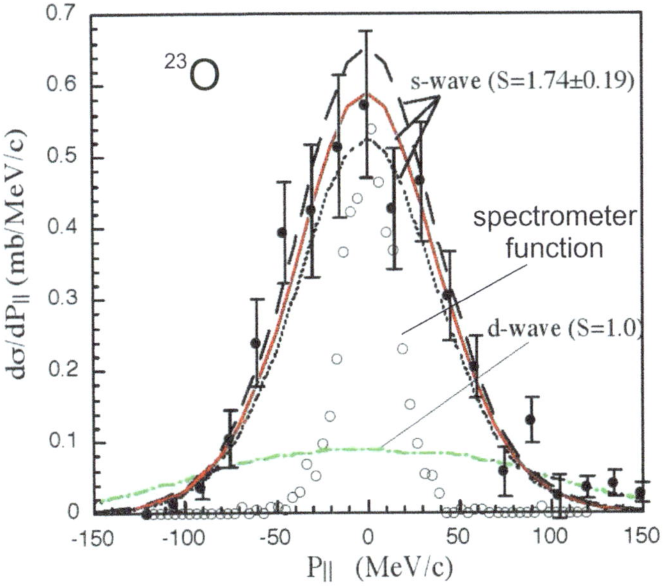

Fig. 5 Measured and calculated momentum distribution of ^{23}O secondary fragments after the one-neutron removal reaction from ^{24}O fragments at 920 MeV/u [57]. The measurement has been performed with the FRS operated as an energy-loss spectrometer, the principle with an ion-optical calculation is depicted in Fig. 4

the theoretical descriptions such that the ground state wave function of the halo proton in ^{8}B can now be well described with the three-body cluster model and the mean field QRPA theory [55]. Both theories can reproduce the measured momentum distribution. It is also interesting to note that the experimental and theoretical spectroscopic factors of 90% occupancy agree over a large energy range from 80 to 1440 MeV/u [53]. Momentum measurements analogous to the ^{8}B case have recently been performed with oxygen isotopes up to the neutron dripline [57]. The momentum distribution for one-neutron removal from ^{24}O at 920 MeV/u, see Fig. 5, revealed a large s-wave probability of the spherical shell closure, which confirms theoretical predictions that ^{24}O is a new doubly-magic nucleus.

Elastic scattering of protons at 800 MeV was successfully applied for the probing of nuclear-matter density distributions of stable nuclei in Gatchina [58]. Since the early FRS experiments we have investigated the nuclear matter distribution of exotic nuclei via elastic proton scattering in inverse kinematics at intermediate energies (700–800 MeV/u) [59, 60]. The hydrogen-filled ionization chamber IKAR served as an active target. The recoil-protons were detected in IKAR and the incident high energy exotic nuclear beam was identified and tracked in front and behind.

The absolute differential cross-sections for small-angle proton scattering were measured for several light neutron-rich nuclei (e.g. He, Li) up to the dripline. The high kinetic energies have the advantage that the application of the Glauber multiple-scattering theory is unique to deduce the nuclear-matter radius, and the halo structure of nuclei. The slope of the differential cross-section is directly correlated with the nuclear-matter radius and its curvature yields the radial shape of the halo structure. The secondary beams were produced by projectile fragmentation of ^{18}O ions at the entrance of the FRS. The first series of measurements studied the nuclear matter distribution of ^{4,6,8}He and ^{6,9,11}Li nuclei. Recent IKAR experiments with ^{8}B ions confirmed the discovery of the proton halo mentioned above.

In previous IKAR experiments we were restricted to radioactive beams of light elements because if secondary beams with higher ionic charge states would penetrate the hydrogen gas volume the exact proton tracking and identification inside the IKAR chamber would be difficult. The technical solution for experiments with heavier ions will be the partitioning of the IKAR anodes in many sections with a central small hole for the incident beams.

5 Exotic Atoms

An interesting and challenging experimental campaign of FRS spectrometer experiments was the investigation of deeply-bound states in heavy atoms. In exotic atoms, besides electrons also other heavier negative particles such as mesons or antiprotons could be bound in Bohr orbitals. Theoretically the existence of deeply bound pionic states was for a long period under controversal discussion. A description of the efforts and challenges is given in the review article by Yamazaki et al. [61].

In the present context, it is interesting to note that W. Greiner and his colleagues had very early calculated that there should exist 1s-bound pionic states with a narrow width for the heaviest stable elements [62]. However, for the superheavy elements ($Z \gg 100$) the quasi-stability of the 1s pionic state enabled by the interference of the Coulomb potential and the repulsive pion-nucleus potential should vanish.

Conventional experiments with exotic atoms have been carried out by slowing down mesons or antiprotons in matter accompanied by x-ray spectroscopy. However, this well-established experimental method failed to observe energy levels for deeply bound pionic states in heavy atoms ($Z > 12$). Therefore, at first, it was believed that the dominance of the strong interaction would totally hinder the formation of these states.

New experimental motivations were induced by refined calculations for the energy levels of deeply bound pionic states in heavy atoms [62, 66, 67]. They predicted narrow discrete pionic states even for the lowest orbits in the heaviest known atoms. The origin of these states is caused by the repulsive pion-nucleus potential which pushes the pionic wave-functions outwards from the nuclear matter radius and thus reduces the absorption due to the strong interaction. Based on these calculations pion-transfer in proton-pickup reactions was proposed, such as (n, d), (p, 2p) and (d, ^{3}He) reactions.

After several unsuccessful experimental trials in different laboratories, (d, ^{3}He) experiments have been carried out at the FRS with 300 MeV/u deuterons. The FRS

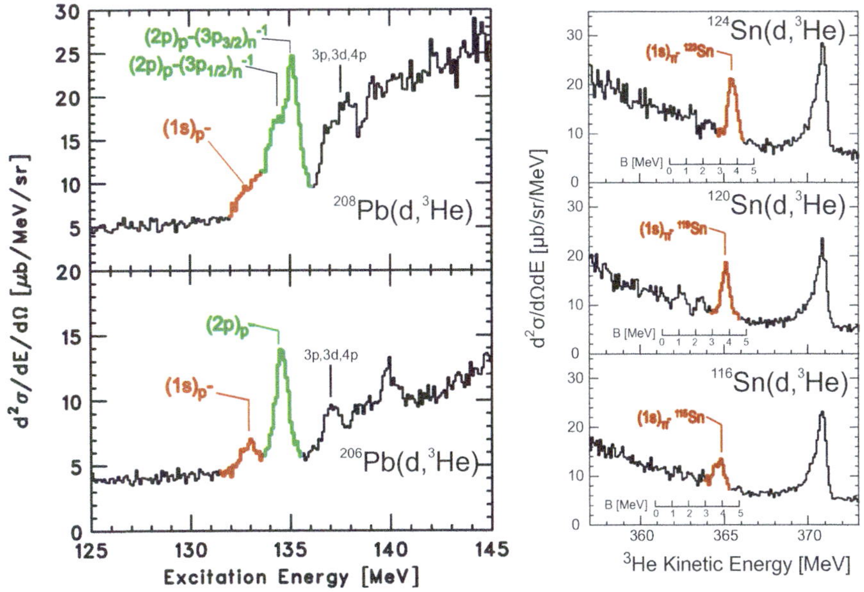

Fig. 6 Discovery of deeply bound pionic states in lead [63, 64] and tin [65] atoms. The pionic atoms have been created in the (d, ^{3}He) reaction at a kinetic energy matching the recoil-free condition

was used as a high-resolution 0-degree spectrometer for the (d, ^{3}He) reaction in different target materials. The momentum distribution of the ^{3}He ions was measured at the central focal plane F2 (Fig. 1). Already the pilot experiment with a ^{208}Pb target led to the discovery of deeply-bound pionic states in ^{207}Pb atoms [63, 64, 68], as shown in Fig. 6.

The FRS spectrometer optical conditions enabled a first-order momentum resolving power of 2860 for a 1 mm strip target. The experimental spectrum of ^{3}He ions clearly showed the narrow peak due to the formation of the bound pionic 2p state in ^{207}Pb coupled with the quasi-substitutional configurations from neutron particle hole states, as shown in Fig. 6. The 2p orbital peak of bound pions showed a strong skewness structure due the doublet structure of the two neutron-hole states ($3p_{1/2}$ and $3p_{3/2}$). The 1s state was observed only as a tail of the main 2p peak which made the determination of the 1s binding energy and width less accurate.

Therefore, in the follow-up experiment, a ^{206}Pb target was selected because the absence of the $3p_{1/2}$ neutron hole states in this reaction allowed the $1s_{\pi^-}$ peak to be well separated from the $2p_{\pi^-}$ components. In addition, several experimental improvements were carried out in the second campaign, such as a reduction of the target thickness and strip size, and electron cooling of the primary beam inside the synchrotron SIS-18, Schottky frequency measurements for the absolute energy determination of the primary beam and long-term drift corrections made possible by frequent π^0 measurements from p(d, π^0)^{3}He. Although the 1s and $2p_{\pi^-}$ groups were now well separated, the $1s_{\pi^-}$ group had still two main contributions from coupled neutron hole states. Nevertheless, with the theoretical line shapes the components were disentangled and the binding energies (B(1s), B(2p)) and the widths (Γ(1s), Γ(2p) were determined.

The new information on the optical potential of the pion-nucleus interaction and the evidence of partial chiral restoration in the nuclear medium was deduced from accurate measurements of B(1s) and Γ(1s). With the experience from the pionic lead atoms, this motivation led to a new series of experiments with different tin isotopes [65].

Indeed, from the experiments with tin isotopes the 1s binding energies and widths were measured with higher accuracy and resolution than obtained in the previous lead experiments. The tin results were used to deduce more accurately the isovector parameter (b_1) of the s-wave pion-nucleus potential [61, 69]. The b_1 value was $(-0.115 \pm 0.007)\,m_\pi^{-1}$, which indicates, compared with the corresponding free pion-nucleon interaction value of $-0.09\,m_\pi^{-1}$, a reduction of the chiral order parameter of 33% at the normal nuclear density [65]. The discovery of deeply-bound states in heavy atoms have opened an entirely new direction for the study the hadron properties in the nuclear medium [61]. The next generation of experiments aims to reduce further the uncertainties for the determination of the 1s binding energies and widths. This will reduce the uncertainties for the deduced parameters of the optical potential. Thinner targets and higher ion-optical resolving power are the main experimental tools to improve the results. The independence of the momentum spread of the primary beam can be compensated by cooling or by the use of dispersion-matched spectrometers. Experiments performed in inverse kinematics should also be more accurate and allow the investigation of radioactive atoms.

Following the first idea of experimental improvements recent pionic-atom experiment have been carried out with the RIBF RIKEN facilities [25]. The versatile high-resolution fragment separator BigRIPS takes in these experiments the role of the FRS. There are two main advantages compared to the pioneering experiments at GSI: 1. the angular and momentum acceptance of the BigRIPS spectrometer is substantially larger, 2. the superconducting ring cyclotron SCR provides by about a factor of 50 higher deuteron intensities. The obvious consequence is that measurements can be performed much faster and secondly the 2p and 1s states can be simultaneously measured with high statistics at different angles. The latter condition can substantially reduce the systematic errors for the determination of the binding energies and widths.

Indeed, the 1s and 2p states of bound pionic states in ^{121}Sn atoms have been measured over a larger angular range θ ($0 \leq \theta \leq 2°$) [70]. The angle-dependent population cross sections are measured for the first time with high statistics. The results show that the theoretical predictions [71] describe well the observed trend but the absolute values of the measured and calculated formation cross sections of the 1s pionic state differ significantly, as shown in Fig. 7.

The new RIBF facilities have, however, also some disadvantages compared to the experimental conditions at GSI. For example, the momentum spread of the incident deuteron beam is at least a factor of 5 worse, especially, when cooling has been applied for the synchrotron SIS-18. Again, an experimental solution can be the realization of

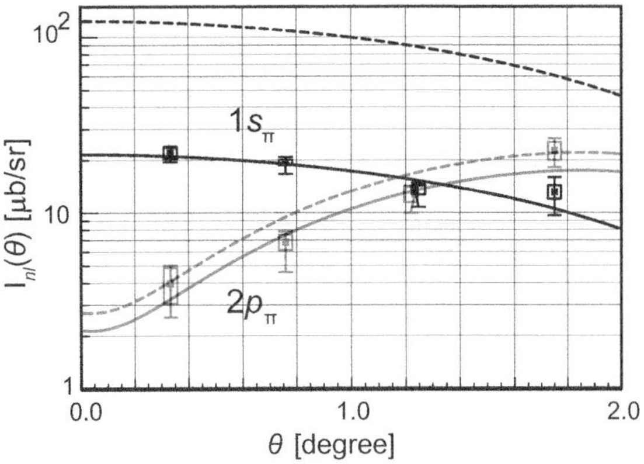

Fig. 7 Measured cross sections for the population of 1s and 2p pionic states [70] compared with the theoretical predictions (dashed lines) [71]. The solid lines represent the scaled theoretical values adjusted to the measured data

an ion-optical achromatic condition as it was used in the momentum measurements of the halo nuclei at the FRS. Indeed, much effort has been invested to reach these prime optical condition in front and behind the interaction target at the entrance of BigRIPS.

It is now a great challenge to search also for bound states of heavier mesons. In a recent FRS spectrometer experiment, the excitation spectrum of ^{11}C was measured in a ^{12}C(p, d) reaction near the η' emission threshold at an incident proton energy of 2500 MeV. This inclusive experiment could not confirm the present theoretical predictions for bound η' states [75–77]. Nevertheless, from the high-quality experimental data valuable constraints, see Fig. 8, on the meson-nucleus potential could be deduced. The next experimental goal is to increase the sensitivity with a detector system tagging the decay properties from possible η' formation.

6 Super-FRS, the Next-Generation In-Flight Facility at FAIR

In this section, we describe key features of the next-generation in-flight separator Super-FRS and scientific goals of the Super-FRS Experiment Collaboration (Super-EC) [38, 39], a sub-collaboration of NUSTAR. The Super-EC is a steady continuation of the FRS collaboration on the broader international base, NUSTAR. The Super-EC concentrates on experiments where the separator is used as a high-resolution spectrometer.

Presently, the intensity of the heavy-ion synchrotron SIS-18 is rather limited due to space-charge effects, in particular for heavy-ion beams. A direct solution is to perform the acceleration of the heaviest projectiles in lower charge states and increase simultaneously the maximum magnetic rigidity of the synchrotron in order to reach the same final velocity. This idea is applied in the future project FAIR by upgrading the UNILAC and adding a 100 Tm synchrotron, SIS-100, connected to SIS-18.

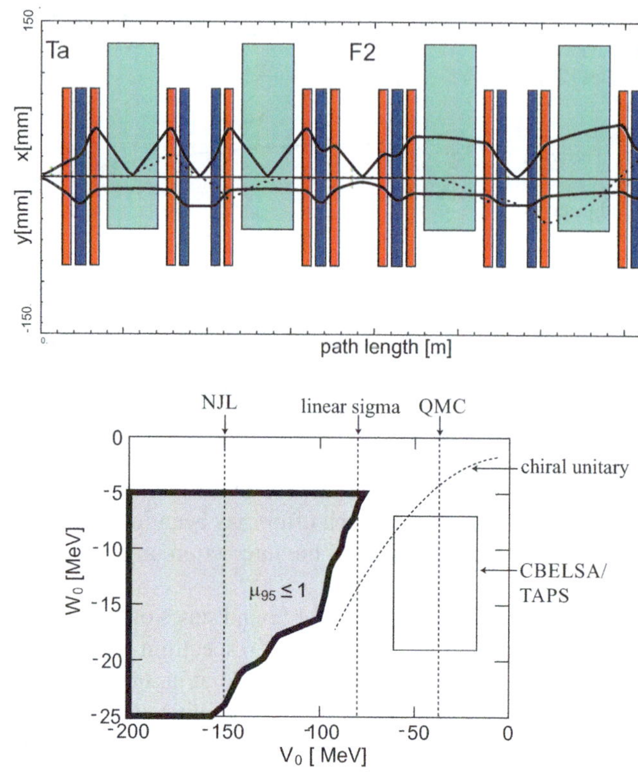

Fig. 8 Upper panel: The FRS spectrometer was operated in an ion-optical which removes the unreacted primary beam and is achromatic at F2. From F2 up to F4, the FRS is used as a dispersive spectrometer. The envelopes are shown in x- and y-direction (solid lines) and the dispersion coefficient for 1% momentum deviation. Lower panel: The experimental result of the non-observation of bound states can be used to constrain the possible eta-prime-nucleus potentials ($V_0 + iW_0$). The shaded region ($\mu_{95} \leq 1$) represents the region excluded within the present analysis. The area of allowed real and imaginary potentials from photoproduction experiments by the CBELSA/TAPS [72] is indicated. The range of different theoretical predictions are also depicted [73, 74]

The Super-FRS [6, 78] will be the most powerful in-flight separator for exotic nuclei up to relativistic energies corresponding to maximal 20 Tm. It is a large-acceptance superconducting fragment separator with three branches serving different experimental areas including a new storage ring complex. The layout of the Super-FRS is shown in Fig. 9.

The intensity gain of more than one order of magnitude for exotic nuclei compared to the present FRS is mainly for fragment beams with a large phase-space population, such as fission fragments or projectile fragments far off the mass from the primary beam. The gain factor of the fragment intensity due to the two-times larger apertures than the present FRS comes in addition to the expected gain from the new accelerator facilities [5].

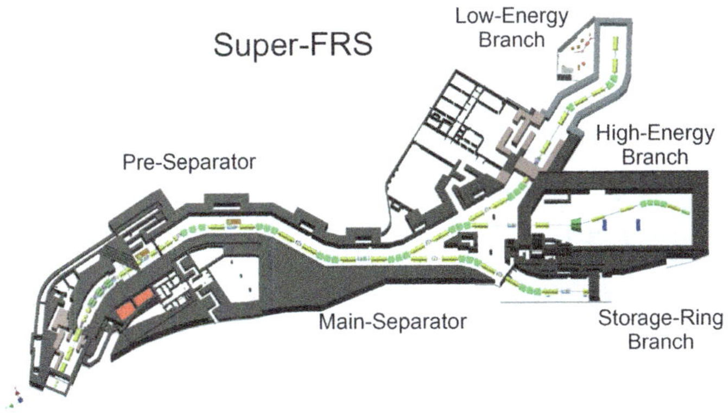

Fig. 9 The in-flight separator Super-FRS [6, 78] is the central facility for all NUSTAR experiments, which will be performed with the magnet systems directly as Spectrometer Experiments or at the focal planes in the different branches

The higher intensity and kinetic energy of the projectile fragments at FAIR are ideal conditions to produce and to study hypernuclei. As Walter Greiner used to say: 'With the new accelerators and experimental facilities at GSI-FAIR, a new dimension of the chart of nuclides should be explored.'

Hypernuclei can be produced by coalescence in peripheral collisions at energies around 1–2 A GeV [79]. This reaction mechanism can be employed with exotic nuclei, and thus a large expansion of the hypernuclear chart is expected, as it is shown in Fig. 10. The invariant mass of the final state, for example, a π^- and a nuclear fragment after weak decay of hypernuclei, gives an effective signature for identifying a hypernucleus. Such a method of production of hypernuclei has been demonstrated recently at GSI [79]. One of the unique features of hypernuclear spectroscopy with projectile fragmentation is that, due to a large Lorentz factor of the produced hypernuclei, the decay can be observed in flight behind the production target. Heavy-ion collisions also provide a possibility for forming multi-Λ hypernuclei. The Super-FRS, equipped with pion detectors just downstream of the secondary target at the central focal plane of the main separator, would provide efficient measurements with high-mass resolution.

In the Low-Energy Branch an Energy Buncher (EB) system [81, 82], consisting of dipole and multipole magnets, will be installed to reduce the energy spread of the fragment beam to enable an efficient stopping in a gas-filled CSC. The combination of the dispersive system with a monoenergetic degrader will be applied. For this task the ion-optical system of the EB must be per se a high-resolution dispersive device. Therefore, the Super-FRS can also be used as a high-resolution spectrometer, in particular when the Main-Separator is dispersion matched with the energy buncher in the Low-Energy Branch. Precise momentum measurements can then be performed after secondary reactions, such as nucleon-removal collisions, independent of the large initial momentum spread of the fragment caused in the production target [82].

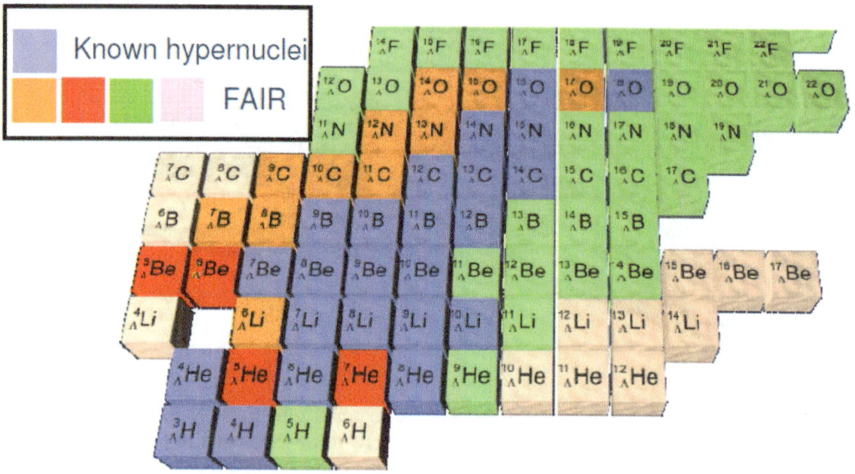

Fig. 10 Research with hypernuclei will substantially enrich the experimental program planned with the Super-FRS operated as a multi-stage separator-spectrometer. The combination of relativistic fragment beams and the hyperon production in secondary targets will substantially extend the studies of baryon-baryon interactions with nucleons and hyperons. A part of the 'new' chart of hypernuclei is shown, the blue colour indicates the presently known hypernuclei, whereas other colours indicate the new potential with exotic nuclei containing hyperons [80]

Although the Super-FRS is primarily a powerful in-flight separator, it will open also new research directions for high-resolution spectrometer experiments [38]. In particular, the combination of the Main Separator with the Energy Buncher in the Low-Energy Branch [6, 81, 82] or with the high-resolution spectrometer planned in the high-energy branch gives unique novel research opportunities:

- The search for new isotopes and the measurements of their production cross sections, kinematics and basic ground-state properties will be continued. In particular, the combination with the Ion-Catcher will present unique opportunities.
- The required accurate knowledge of the slowing down of heavy ions in matter, for most of the Super-FRS experiments, requires atomic-collision experiments in the new high-energy domain of FAIR and also in the kinetic energy range of (30–300) MeV/u in gases and solids.
- The energy-buncher spectrometer as a dispersion-matched facility enables precise longitudinal momentum measurements for studies of nucleon orbitals and spectroscopic factors, multi-nucleon transfer reactions with radioactive beams, secondary fragmentation combined with multiple-step Coulomb excitation, charge-exchange reactions a.o.
- Measurements of nucleon resonances in exotic nuclei. Pilot experiments with the FRS are promising and show new features. These peripheral reactions will also provide nuclear structure information.
- Experiments on the production and study of hyper-nuclei. Hypernuclei formation in fragmentation reactions proceeds via coalescence of Λ-particles, which allows

the population of exotic nuclei with hyperons. Spectroscopy experiments with mesic nuclei will provide basic information on the mass modification of mesons inside nuclear matter.
- Low-energy experiments based on the recombination of anti-protons with exotic nuclei. The interaction of exotic nuclei and antiprotons will be unique at FAIR.

Acknowledgements It is a great pleasure to thank T. Dickel, B. Franczak, K. Itahashi, R. Kanungo, W. R. Plaß, M. Pfützner, T. R. Saito, Y. K. Tanaka, I. Tanihata, H. Weick, J. S. Winfield, M. Winkler for fruitful collaboration and discussions of this article.

References

1. N. Angert, C. Schmelzer, Kerntechnik **19**(2), 57 (1977)
2. K. Blasche, B. Franczak, in *Proceedings of 3rd European Particle Accelerator Conference* (Berlin, 1992), p. 9
3. G. Münzenberg, H. Geissel, C. Scheidenberger contribution to this book
4. H. Geissel et al., Nucl. Instrum. Methods B **70**, 286 (1992)
5. H. Geissel, G. Münzenberg, C. Scheidenberger, FIAS Interdisciplinary Science Series ed. by W. Greiner (Springer, Switzerland, 2017)
6. H. Geissel et al., Nucl. Instrum. Methods B **204**, 71 (2003)
7. G.D. Westfall et al., Phys. Rev. Lett. **43**, 1859 (1979)
8. T.J.M. Symons et al., Phys. Rev. Lett. **42**, 40 (1979)
9. I. Tanihata et al., Phys. Rev. Lett. **55**, 2676 (1985); Phys. Lett. B **160**, 380 (1985)
10. B. Franzke, Nucl. Instrum. Methods B **24**(25), 18 (1987)
11. P. Kienle, Future accelerators and experimental facilities at GSI, report GSI-84-18 (1984). ISSN:0171-4546
12. C. Scheidenberger et al., Phys. Rev. Lett. **73**, 50 (1994)
13. C. Scheidenberger et al., Phys. Rev. Lett. **77**, 3987 (1996)
14. H. Weick et al., Phys. Rev. Lett. **85**, 2725 (2000)
15. T. Schwab, Ph.D. thesis, JLU Gießen, GSI report GSI-91-10 (1991); H. Geissel, C. Scheidenberger, H. Weick, https://web-docs.gsi.de/~weick/atima/
16. C. Scheidenberger et al., Nucl. Instrum. Methods B **142**, 441 (1998)
17. T. Schwab, Ph.D. thesis JLU Gießen, GSI report GSI-91-10 (1991); N. Iwasa et al., Nucl. Instrum. Methods B **126**, 284 (1997); Nucl. Instrum. Methods B **269**, 752 (2011)
18. G. Kraft, Progr. Part. Nucl. Phys. **45**, 475 (2000); G. Kraft et al., in *Proceeding of the EULIMA Workshop*, Nice (1988)
19. W. Enghardt et al., Phys. Med. Biol. **37**, 2127 (1992)
20. W. Enghardt et al., Onkologie **175**, (1999)
21. A. Magel et al., Nucl. Instrum. Methods B **94**, 548 (1994)
22. https://people.nscl.msu.edu/~thoennes/isotopes/
23. M. Bernas et al., Phys. Lett. B **331**, 19 (1994)
24. M. Bernas et al., Nucl. Phys. A **616**, 352 (1997)
25. T. Kubo, Nucl. Instrum. Methods Phys. Res. B **204**, 97 (2003)
26. M. Hausmann et al., Nucl. Instrum. Methods Phys. Res. B **317**, 349 (2013)
27. J. Kurcewicz et al., Phys. Lett. B **717**, 371 (2012)
28. T. Kurtukian-Nieto et al., Eur. Phys. J. A **50**, 135 (2014)
29. R. Schneider et al., Z. Phys. A **348**, 241 (1994)
30. Ch. Engelmann et al., Z. Phys. A **352**, 351 (1995)
31. S. Hofmann, G. Münzenberg, Rev. Mod. Phys. **72**, 733 (2000)
32. S. Pietri et al., Nucl. Instrum. Methods B **261**, 1079 (2007)

33. C.B. Hinke et al., Nature **486**, 341 (2012)
34. V.I. Goldansky, Nucl. Phys. **19**, 482 (1960)
35. M. Pfützner et al., Euro. Phys. J. A **14**, 279 (2002)
36. J. Giovinazzo et al., Phys. Rev. Lett. **89**, 102501 (2002)
37. K. Miernik et al., Phys. Rev. C **76**, 041304(R) (2007)
38. I. Tanihata, et al., Scientific program of the super-FRS collaboration, GSI-report 2014-4 (2014). https://doi.org/10.1520/GR-2014-4.
39. J. Äystö et al., Nucl. Instrum. Methods Phys. Res. B **376**, 111 (2016)
40. W.R. Plaß et al., Nucl. Instrum. Methods B **317**, 457 (2013)
41. T. Dickel et al., Nucl. Instrum. Methods A **777**, 247 (1989)
42. T. Dickel et al., Phys. Lett. B **744**, 137 (2015)
43. S. Purushothaman et al., Nucl. Instrum. Methods B **266**, 4488 (2008)
44. M. Ranjan et al., Europhys. Lett. **96**, 52001 (2011)
45. H. Geissel et al., Nucl. Instrum. Methods A **282**, 247 (1989)
46. S. Ayet et al., Phys. Rev. C **99**, 064313 (2019)
47. F. Greiner, Master thesis, JLU Gießen (2018)
48. G. Huber et al., Phys. Rev. C **18**, 2342 (1978)
49. T. Suzuki et al., Phys. Rev. Lett. **75**, 3241 (1995)
50. T. Kobayashi et al., Phys. Rev. Lett. **60**, 2599 (1988)
51. P.G. Hansen, B. Jonson, Europhys. Lett. **4**, 409 (1987)
52. P.G. Hansen, B.M. Sherrill, Nucl. Phys. A **693**, 133 (2001)
53. P.G. Hansen, J.A. Tostevin, Rev. Nucl. Part. Sci. **53**, 219 (2003)
54. W. Schwab et al., Z. Phys. A **350**, 283 (1995)
55. H. Lenske, F. Hofmann, C.M. Keil, Rep. Prog. Nucl. Part. Phys. **46**, 187 (2001)
56. D. Cortina-Gil et al., Phys. Lett. B **529**, 36 (2002)
57. R. Kanungo et al., Phys. Rev. Lett. **102**, 152501 (2009)
58. G.D. Alkhazov et al., Phys. Rep. C **42**, 89 (1978)
59. P. Egelhof et al., Eur. Phys. J. A **15**, 27 (2002)
60. A.V. Dobrovolsky et al., Nucl. Phys. A **766**, 1 (2006)
61. T. Yamazaki et al., Phys. Rep. **514**, 1 (2012)
62. E. Friedman, G. Soff, J. Phys. **G11**, L37 (1985)
63. T. Yamazaki et al., Z. Phys. A **355**, 219 (1996)
64. K. Itahashi et al., Phys. Rev. C **62**, 025202 (2000)
65. K. Suzuki et al., Phys. Rev. Lett. **92**, 072302 (2004)
66. H. Toki, T. Yamazaki, Phys. Lett. B **213**, 129 (1988)
67. H. Toki, S. Hirenzaki, T. Yamazaki, R.S. Hayano, Nucl. Phys. A **501**, 653 (1989)
68. H. Gilg et al., Phys. Rev. C **62**, 025201 (2000)
69. N. Kaiser, W. Weise, Phys. Lett. B **512**, 283 (2001)
70. T. Nishi et al., Phys. Rev. Lett. **120**, 152505 (2018)
71. N. Ikeno et al., Eur. Phys. J. A 47 (2011); PTEP **2015**, 033D01 (2015)
72. M. Nanova et al., Phys. Rev. C **94**, 025205 (2016)
73. Y.K. Tanaka et al., Phys. Rev. Lett. **117**, 202501 (2016)
74. Y.K. Tanaka et al., Phys. Rev. C **97**, 015202 (2018)
75. R.S. Hayano et al., Rev. Mod. Phys. **82**, 2949 (2010)
76. K. Itahashi et al., Prog. Theor. Phys. **128**, 601 (2012)
77. H. Nagahiro, Phys. Rev. C **87**, 045201 (2013)
78. M. Winkler et al., Nucl. Instrum. Methods B **266**, 4183 (2008)
79. C. Rappold et al., Nucl. Phys. A **913**, 170 (2013)
80. T.R. Saito, Priv. Commun. (2018)
81. J.S. Winfield et al., Nucl. Instrum. Methods A **704**, 76 (2013)
82. H. Geissel et al., Nucl. Instrum. Methods B **317**, 277 (2013)

Walter Greiner, a Pioneer in Super Heavy Element Research Historical Remarks and New Experimental Developments

Gottfried Münzenberg, Hans Geissel and Christoph Scheidenberger

Abstract With his theoretical work Walter Greiner pioneered super-heavy element research. He motivated the young scientists and actively shaped the profile of the *Gesellschaft für SchwerIonenforschung, GSI, Darmstadt*. We are happy that still during his lifetime we at GSI could prove some of his predictions: Fusion with magic nuclei and super heavy elements, the nuclear species existing only by shell stabilization. With the discovery of oganesson, $Z = 118$, the heaviest element known today, we have come to the end of super heavy-element production by the fusion of magic nuclei. In-flight separation and new experimental developments including Walter Greiner's new ideas for SHE synthesis will be discussed.

1 Introduction

Walter Greiner pioneered and supported strongly super-heavy element research. In his calculations, which are among the earliest ones, he predicted the existence of super-heavy elements: atomic nuclei existing only by shell stabilisation in the sea of liquid-drop instability [1]. He also worked out a model to explain the production of the heaviest elements in reactions with the doubly magic nuclei ^{208}Pb and ^{48}Ca [2, 3]. As member of the "Kernphysikalische Arbeitsgemeinschaft Hessen, KAH", Walter Greiner shaped the research program of *GSI, the Gesellschaft für Schwerionenforschung, mbH in Darmstadt*, at that time called "*SILAB*". Figure 1 displays the outline of the concept for the GSI accelerator and the research program. It is taken from the SILAB proposal. The main issues were the specifications for the accelerator

G. Münzenberg (✉) · H. Geissel · C. Scheidenberger
GSI Helmholtzzentrum für Schwerionenforschung mbH, Planckstr. 1, 64291 Darmstadt, Germany
e-mail: G.Muenzenberg@GSI.de

G. Münzenberg
Johannes Gutenberg-Universität Mainz, 55099 Mainz, Germany

Manipal Centre for Natural Sciences, Manipal University, Manipal 576104, Karnataka, India

H. Geissel · C. Scheidenberger
Justus-Liebig-Universität Gießen, Ludwigstr. 23, 35390 Gießen, Germany

J. Kirsch et al. (eds.), *Discoveries at the Frontiers of Science*,
FIAS Interdisciplinary Science Series, https://doi.org/10.1007/978-3-030-34234-0_14

2.2. Zweck des SILAB

Der Zweck des SILAB ist zunächst die Errichtung eines Linearbeschleunigers nach einem von den Entwicklungsgruppen Schmelzer (UNILAC) und Dänzer-Klein (HELAC) gemeinsam festzulegenden Konzept (vgl. Beschluß des Arbeitskreises Physik und des Niederenergie-Kernphysik-Ausschusses vom 29.10.68), zur Beschleunigung von positiven Ionen aller zugänglichen Nuklidsorten bis zu einer Endenergie von mehr als 6 MeV je Nukleon. Mit Hilfe derartiger hochbeschleunigter Ionen kann eine Vereinigung zweier Atomkerne (im Sinne einer Umkehrung der Kernspaltung) erreicht werden. Auf diese Weise sollte eine große Zahl neuer Nuklide herstellbar sein, und zwar sowohl neue überschwere Elemente als auch neue Isotope bereits bekannter Elemente.

Fig. 1 Original text from the SILAB proposal outlining the specifications of the GSI accelerator

to provide beams of all elements of the periodic table up to and including uranium and the production of new chemical elements and isotopes including Super Heavy Elements, SHE. They are a large part of success of GSI research: The uranium beam as a rich source for new isotopes discovered with the GSI fragment separator, and the new super-heavy elements at SHIP.

2 Super-Heavy Elements, the Presence

In March this year milestones in SHE research were celebrated. The official announcement of the names of the heaviest known chemical elements by Natalya Tarasova, the president of, the "International Union of Pure and Applied Chemistry, IUPAC" [4]. The people of Japan celebrated the new element 113, *nihonium* in the Japanese Academy of Science with a speech of Crown Prince Naruhito. Nihon is the original name of Japan. With nihonium the end of the cold fusion reaction to create super heavy elements with lead or bismuth targets has been reached. Earlier, in March, an International Colloquium dedicated to the naming of elements 115 Moscovium, 117 Tennesine, and 118 Oganesson was held in the Russian Academy of Sciences, Moscow. The elements carry the names of Moscow region, where Dubna, the place of discovery, is located, the US state of Tennessee where Oak Ridge is located, the place of manufacture of the actinide targets, and Yuri Oganessian, "for his pioneering contributions to transactinoid elements research". With element 118 the synthesis of super heavy elements using actinide targets and ^{48}Ca beams, the era of synthesis of super heavy elements with doubly magic nuclei as theoretically substantiated by Walter Greiner, has come to its end. At present, all chemical elements discovered so far are named officially. A complete overview on SHE research is given in a special issue of the journal Nuclear Physics A [5].

3 Historical Remarks, Our Discoveries Proving Greiner's Predictions

By irradiation of ^{208}Pb and ^{209}Bi targets with beams of ^{50}Ti and ^{54}Cr, Yuri Oganessian and collaborators claimed the observation of elements 104 to 107 formed by cold fusion in 2n and 3n channels [6, 7]. This assignment was based on the measurement of fission tracks. However, these results were debated. Spontaneous fission is not a safe identification method, especially for unknown species. Relying on the extra-push concept, the Berkeley group was convinced that the fission activities reported by Oganessian et al. cannot be attributed to heavy elements [8].

The first proof the concept of cold fusion was given by the observation of ^{257}Rf by irradiation of ^{208}Pb with ^{50}Ti, and formed by the evaporation of one neutron from the compound nucleus ^{258}Rf [9]. The identification was possible with the new concept of SHIP, the in-flight separation of the heavy elements with a velocity filter, and the identification by the $\alpha{-}\alpha$ correlation technique. Figure 2 displays an observed α decay sequence of an individual atom of ^{257}Rf compared to data published by Ghiorso [10].

With the synthesis of element 106 and the discovery of element 108 at SHIP, we could prove the concept of super heavy nuclei. We observed α-decay, which indicates stability against fission. In the nuclear liquid drop model, nuclei become unstable against fission beyond element 104. With the addition of shell corrections, an island of super heavy nuclei centred at element 114 is predicted. A "sea of instability" beyond element 104 separates the "super heavy island" of shell-stabilized nuclei from the trans-uranium elements. Our observation of α-decay is a clear indication of shell stabilization. The ultimate and nicely visible proof is given by the decay

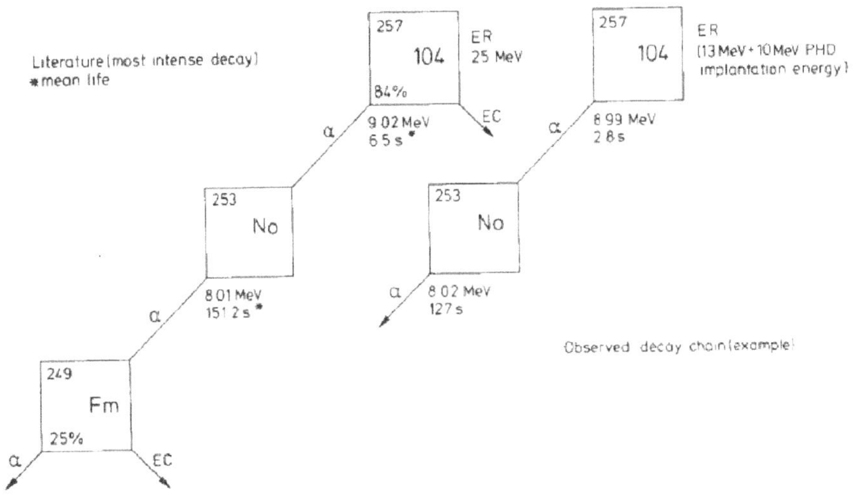

Fig. 2 Example of a single-atom decay chain (right) compared to literature data [9] (left)

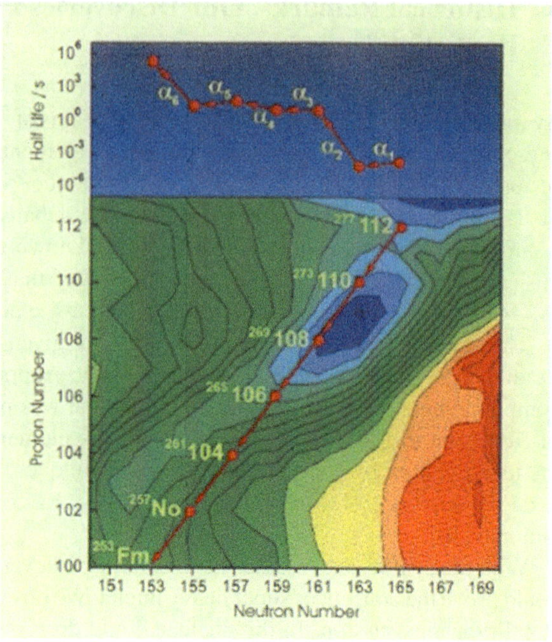

Fig. 3 Lower panel: The decay chain of element 112 plotted within the landscape of shell stability. The upper panel shows the measured decay-time intervals for this chain in a logarithmic scale

chain of element 112 [11], shown in Fig. 3 in the calculated shell landscape. This is the confirmation of the idea of super-heavy nuclei. Calculations of Möller and Nix [12] and of Adam Sobiczewski show that this region is centred at Z = 108 and N = 162 and that the origin of the enhanced stability is a hexadecapole deformation. This shell is the basis for the existence of the elements beyond rutherfordium, Z = 104. These elements are now commonly called "super heavy elements".

In the Nobel Symposium in 1974, Aage Bohr commented on a talk by Adam Sobiczewski: "What about the possibility of super-heavy nuclei in other shapes which are stabilized by shell structure?

4 In-Flight Separators for SHE Research

The next generation of SHE factories will open up new perspectives for SHE research including chemical, atomic, and nuclear studies as well as the discovery of new elements and isotopes. One central goal is the exploration of the region of super-heavies in the region of Z = 112 to Z = 120 and the approach of the magic neutron shell N = 184. The $\alpha - \alpha$ correlation technique will fail in this region for element identification, as already became clear with the new trans-copernicium elements. Regions of ß-decaying nuclei may be accessed. A next generation in-flight separator must include the capability of direct A and Z identification.

The Separators presently used in SHE research are listed in Table 1. We have two

Table 1 In-flight separators used for SHE research. The table displays the angular acceptance in msr, separation characteristics, and examples of use

	Acceptance/msr	Separation characteristics	Use e.g.
Vacuum separators			
SHIP, GSI	3	Velocity filer, reaction kinematics, sensitive to fusion	New elements
SHELS, Dubna	5.5	Same characteristics	Spectroscopy
VASILISSA, Dubna	15	Energy filter, reaction kinematics sensitive to fusion	Spectroscopy
Recoil mass separators			
FMA, ANL	8	Mass resolution, separates mass over ionic charge	In-beam spectroscopy
MARA, JYFL	9	Same characteristics	In-beam spectroscopy
S3, GANIL	9	Same characteristics	Spectroscopy
Gas filled separators			
BGS, Berkeley	45	Low mass resolution, collects all ionic charge states and velocities	
DGFRS, JINR	10	Same characteristics	New elements
GARIS, RIKEN	12	Same characteristics	New elements
GARIS II, RIKEN	18.5	Same characteristics	New elements
TASCA, GSI	13	Same characteristics	Chemistry
RITU, JYFL	8.5	Same characteristics	In-beam spectroscopy
SHANS, LANZHOU	13	Same characteristics	Spectroscopy

classes of in-flight separators: vacuum separators and gas filled separators. Vacuum separators include velocity and energy filters using the kinematic properties of the reaction products for separation and recoil mass separators for isotope separation. Gas filled separators have a mass separation: They separate light from heavy ions with poor resolution. The advantage of kinematic separators is their sensitivity to the specific reaction process. They separate fusion products from deep inelastic transfers and even α, xn from xn reaction channels, as proven at SHIP.

Discoveries of new elements have been made with SHIP, DGFRS, and GARIS. The criteria for SHE separators are good transmission and high background suppression. For light projectiles, such as carbon or neon, the transmissions are small, of the order of five to ten percent. For the fusion reactions with ^{48}Ca a great number of data exists from the SHE synthesis experiments at the gas filled separators DGFRS and TASCA. An evaluation by Hofmann [13] based on published data shows large fluctuations in the transmissions. The optimum transmission for all separators is approximately the same and around 40%.

The problem for designing gas filled systems is the inclusion of atomic processes such as electron stripping, capture, and scattering of the recoil ions passing the separator. An important parameter to be considered for all types of separators in the calculation of the transmission is target thickness. Scattering of the slow and heavy recoils in a thick target reduces the transmission. For SHIP the optimum target thickness is 0.5 mg/cm^2. Based on these data it is difficult to decide which separator type, gas filled or vacuum, is preferable. This is certainly a matter of experience. A technical challenge for vacuum separators is the high voltage.

5 The Future—SHE Factories

With the synthesis of element 113, nihonium, the cold—fusion concept using ^{208}Pb or ^{209}Bi targets has reached its limit because of the small production cross-section. The heavier elements were discovered by hot fusion using another magic nucleus, ^{48}Ca, a new concept developed by Yuri Oganessian at Dubna. Figure 4 compares the production cross sections for the trans fermium elements by cold and hot fusion [14]. Cold-fusion cross sections drop quickly towards the heaviest elements down to a level of 22 fb for nihonium, Z = 113. The reason is the fusion hindrance due to the fast-increasing entrance fissility or, in other words, the fast- increasing Coulomb repulsion between target and projectile. In hot fusion, the entrance fissility does not change so

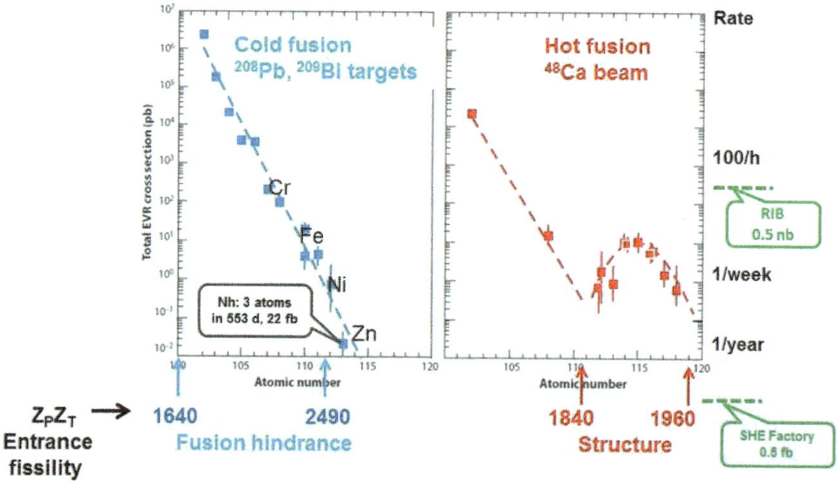

Fig. 4 Production cross sections for the trans fermium nuclei. Left panel: cold fusion, right panel: hot fusion. The sensitivity limits for SHE factory and RIB are given per year of beam time. Cross sections from Yu. Ts. Oganessian

fast. For heavy targets the relative change in the nuclear charge and consequently the entrance fissility is small. The cross section is dominated by structure effects. The bump in Fig. 4 is an indication for the $Z = 114$ shell. Compound nuclei are closer to the island of spherical SHE as compared to the evaporation residues, the nuclei finally produced in the reactions. The shell may show-up in the fusion cross sections but it is not observed in the decay data. The measured cross section are of the order of 10 pb.

With the production of oganesson, element 118, the heaviest available target has been reached. Going beyond needs a heavier projectile, ^{50}Ti or ^{54}Cr. These are not magic nuclei; the fusion cross sections will be much smaller. A reliable prediction is not available at present. First attempts to create element 120 have been made by Sigurd Hofmann. New prospects for SHE research will be opened up with the next generation SHE factories. Table 2 shows the beam intensities and annual doses available at present and with the future SHE-factories. Already the present accelerators such as the UNILAC at GSI, the RIKEN RILAC, and the Dubna U400 cyclotron deliver beams of 6×10^{12} ions/s on the average. With beam time of 100 d–300 d per year available for SHE research at JINR Dubna, the sensitivity level is 1 atom/10 fb. SHE factories including JINR Dubna with the new DC280 cyclotron as well as RIKEN and SPIRAL2 at GANIL with new powerful linear accelerators and ion sources will have a factor of 10 to 100 higher beam intensity. Taking into account the available beam time per year, the sensitivity is increased by a factor of about 50, reaching a sensitivity of 5 atoms/fb. Other accelerator labs such as Argonne National Lab, Canberra, FRIB at Michigan State University, IMP Lanzhou, LBNL Berkeley, Tokai, and Jyväskylä will work on special topics including reaction studies, nuclear structure, chemistry, and atomic physics.

New in the field are Rare-Isotope (RI) facilities SPIRAL2, FRIB, HIE ISOLDE, and NUSTAR at GSI. The RI beam intensities close to stability are up to 10^9/s on the average. They drop to 10^6/s five to ten isotopes away from stability. It turns out that the RIB intensities for all three schemes are of the same order of magnitude. FAIR intensities are higher for light beams such as neon and heavier. Because of the low beam intensity the use of RIB for SHE research is rather limited at present. Taking an optimistic value of 10^9 projectiles/s for isotopes close to stability the sensitivity is

Table 2 Intensities and sensitivities for fusion reactions with stable and radioactive beams for SHE production [15]

		Stable beams	RI beams close to stability for 30 d/a beam time
present	I =	$6\text{x}10^{12}$/s	
	annual dose	0.6 10^{20}	
	sensitivity	0.1 atom/fb a	
future	I=	10^{14}/s	10^{9}/s
	annual dose	$2\text{x}10^{21}$	$2.6\text{x}10^{15}$
	sensitivity	5 atoms/fb a	4 atoms/nb a

only 4 atoms/nb per year. This is about six orders of magnitude less than achievable with the SHE factories with stable beams! Here it must be taken into account that SHE factories are dedicated to SHE research whereas RI facilities have a broad and competitive research program. First generation of experiments will include reaction studies and isotope synthesis in the region up to rutherfordium.

6 Reaction Studies

Besides complete heavy ion fusion new reactions such as nuclear transfer will be investigated. Figure 5 shows a prediction for transfer cross sections for the reaction ^{338}U + ^{248}Cm by Zagrebaev and Greiner [16] compared to data from Schädel. He observed mendelevium as the heaviest element. Beyond, predicted cross sections drop fast, by about one order of magnitude per element. The power of the method for the creation of new trans- uranium isotopes is shown in Fig. 5 right panel. At SHIP in irradiations of ^{248}Cm with ^{48}Ca five new and trans uranium isotopes, marked by dots, were found [17]. The identified isotopes are shaded.

Large-scale investigations of nuclear transfer reactions are under way e.g. with VAMOS at GANIL. The problem of all in-beam methods is the sensitivity. Our main interest however are the most exotic transfers created with small cross sections. Even very rare processes can be observed with Multi-Reflection Time-Of Flight Mass Spectrometers MR-TOF-MS [18]. The high resolution of MR-TOF allows the identification of isotopes and elements by their isobaric mass and even by isomers, which indicate the angular momenta of the final nuclei—a crucial point for the survival of SHE. On the long range, we aim at the measurement of nuclear transfer around zero degrees to create transfer products with low angular momenta, thereby suppressing prompt disintegration by fission.

A test setup for first studies of nuclear transfer reactions is displayed in Fig. 6.

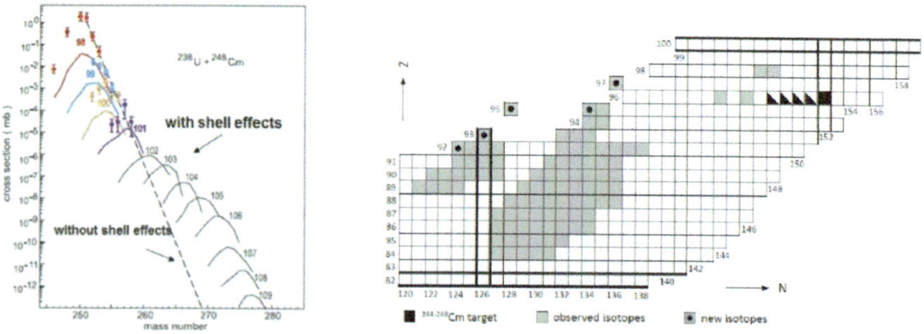

Fig. 5 Left panel experimental and predicted cross sections for heavy-element production by nuclear transfer [16], right panel observed transfer products, created by in transfer reactions irradiating ^{248}Cm with ^{48}Ca, new isotopes are marked by dots [17]

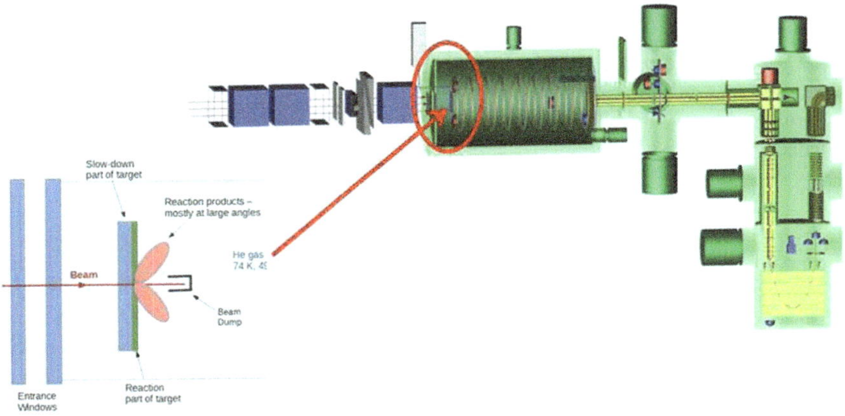

Fig. 6 Test setup for the investigation of transfer reactions at the FRS. The target is installed inside the ion catcher cell (see inset) can be irradiated with stable and radioactive beams at energies near Coulomb barrier [19]

It will be placed behind the FRS. The relativistic projectile fragments will be decelerated by energy degraders [18]. Stable and radioactive projectiles are available at near Coulomb-barrier energies. The target is placed inside the gas catcher where the reaction products are thermalized, extracted and guided to the MR-TOF-MS. A quadrupole mass filter will select the mass region of interest. With the MR-TOF-MS operated in the separator mode, the heavy nuclei can be directed to a specialized detector system for detailed spectroscopy.

7 Experiments and New Instrumentation

Next generation SHE factories will open up new possibilities for SHE research including atomic and more detailed nuclear studies as well as the discovery of new elements and isotopes. Certainly chemistry will play a major role in the future research programs, which will also rely on in-flight separators to achieve highest sensitivity and clean conditions. A main goal is the exploration of the region of super-heavies in the region of Z = 112 to Z = 120 and the approach to the magic neutron shell N = 184.

The $\alpha - \alpha$ correlation technique successfully applied to the discoveries of the heaviest elements will fail here as already became evident with the new trans-nihonium elements. Their α chains are not connected to the trans uranium region. We should in addition be able to identify ß-decaying nuclei and those decaying by spontaneous fission. The capability of direct A and Z identification is required. The next generation of in-flight separators will include gas filled separators and velocity filters

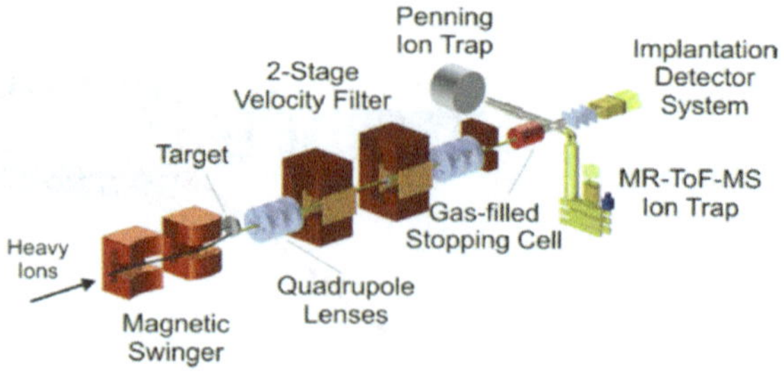

Fig. 7 A next generation compact velocity filter with stopper cell and MRTOF-MS which can also be used in combination with the commonly used implantation detector

with optimized transmission and separation quality; here one can mention GARIS-II and the gas filled separator at Dubna under construction. SHELS at Dubna is a new velocity filter. Within the Giessen-GSI-Manipal collaboration, calculations for a compact and optimized velocity filter, Fig. 7, with high transmission based on the experience with SHIP have been carried out. The transmission for ^{48}Ca + ^{208}Pb to produce ^{254}No has been calculated to 58%, a factor of 1.8 higher compare to SHIP [20].

The key to direct A,Z identification are gas-filled stopping cells combined with MR-TOF-MS or Penning Trap systems. A first step in this direction is the SHIP-SHIPTRAP combination. At RIKEN an MR-TOF-MS system has been coupled to GARIS II. With this system the A,Z identification of astatine-, polonium-, and bismuth isotopes has been achieved by isobaric mass analysis at a mass resolving power of more than 100 000 by measuring about 10 atomic nuclei [21]. At GSI such a system has been operated successfully at FRS with a resolution of 500 000. Isobaric A,Z identification and separation was achieved, in addition the α decay of ^{221}Ac was measured after the MRTOF. With this method in contrast to the presently used decay spectroscopy SHE are directly identified "still alive" [18]. These systems can be operated in a separator mode and coupled to detector system such as Si surface barrier detectors, germanium arrays, or beta spectrometers. They are ideally used to identify transfer products in large-scale survey experiments as discussed above.

8 Conclusion

With the discoveries of nihonium and oganesson, the era of fusion with magic nuclei ^{208}Pb and ^{48}Ca, theoretically substantiated by Walter Greiner has come to its end. With the discovery of the elements which exist only by shell stabilization, rutherfordium and beyond, the idea of super-heavy nuclei has been proven. To proceed further, high beam intensities are required and new reactions need to be explored. Prospects for SHE research are opened by the new SHE factories with beam intensities increased by one to two orders of magnitude, dedicated to SHE research. The search for trans-oganesson elements will be continued. More detailed structure investigations and chemistry will play a major role in the future. The SHE factories will be backed by laboratories with specialized research programs like in-beam spectroscopy and reaction studies including RIB. Studies of incomplete fusion, deep inelastic collisions are under way and need to be explored at small cross sections to see rare processes occurring with low probability. As a new development ion-catcher – ion-trap systems or MR-TOF-MS will play a major role in SHE research. They will allow for direct identification of new isotopes and elements and in addition open new perspectives for atomic physics including laser spectroscopy of super heavy atomic nuclei.

Acknowledgements We gratefully acknowledge support by the Giessen-GSI-Manipal collaboration M. Gupta, S. Heinz, S. Hofmann, H. M. Devaraja, T. Dickel, W. R. Plass, J. S. Winfield, and M. Winkler

References

1. W. Greiner, U. Mosel, Z. Phys. **222**, 261 (1969)
2. A. Sandulescu, W. Greiner, J. Phys. G **3**, 189 (1977)
3. R.K. Gupta, A. Sandulescu, W. Greiner, Z. Naturforsch. **32a**, 704 (1077)
4. https://IUPAC.org
5. Nucl. Phys. A **944**, 1 (2015). Special volume on SHE
6. Yu. Ts. Oganessian, Nucl. Phys. **A239**, 157 (1975)
7. Yu. Ts. Oganessian et al., Nucl. Phys. **A273**, 505 (1976)
8. V.E. Viola Jr., A.C. Mignerey, H. Breuer, K.L. Wolf, B.G. Glagola, W.W. Wilcke, W.U. Schröder, J.R. Huizenga, D. Hilscher, J.R. Birkelund, Phys. Rev. **C22**, 122 (1980)
9. G. Münzenberg, P Armbruster, W. Faust, K. Güttner, F.-P. Heßberger, S. Hofmann, W. Reisdorf, C.C. Sahm, K.-H. Schmidt, H.-J. Schött, B. Thuma, D. Vermeulen, Isotope identification in the transfermium region by α-α correlation after in-flight separation. in *Actinides in Perspective* ed. by N.E. Edelstein, vol. 223 (Pergamon, Oxford, 1981)
10. A. Ghiorso, M. Nurmia, J. Harris, K. Eskola, P. Eskola, Phys. Rev. Lett. **24**, 1317 (1969)
11. S. Hofmann, G. Münzenberg, Rev. Mod. Phys. **72**, 733 (2000)
12. P. Möller, J.R. Nix, At. Data Nucl. Data Tables **39**, 213 (1988)
13. S. Hofmann, General Program Advisory Committee (G-PAC) Meeting, GSI Darmstadt, Germany (2015)
14. Yu. Ts. Oganessian, Roadmap, JINR Dubna, Scientific Council (2010)
15. C. Scheidenberger, ECOS-EURISOL Town Meeting Orsay, France (2014)
16. V. Zagrebaev, W. Greiner, Phys. Rev. C **78**, 034610 (2008)
17. H.M. Devaraja et al., Phys. Lett. B **748**, 199 (2015)
18. T. Dickel et al. Nucl. Instrum. Methods Phys. Res. **A777**, 172 (2015)
19. T. Dickel et al., GPAC GSI (2017)
20. H.M. Devaraj, Thesis, Manipal University (2017)
21. P. Schury et al., Phys. Rev. C **95**, 011305 (2017)

Collective Motion and the Asymmetric-Matter Equation-of-State

Wolfgang Trautmann

Abstract The asymmetric-matter equation-of-state is a topic of highest current interest in nuclear structure and reactions as well as in astrophysics. The importance of the symmetry energy, i.e. of the difference between the energy densities of neutron matter and of symmetric nuclear matter, for many nuclear phenomena implies that a variety of observables can be used for investigating its strength. Most of them, however, probe densities below saturation. Higher densities can be accessed with heavy-ion reactions at sufficiently high energies. The observed collective flow of particles and fragments has appeared as a useful probe of the potential strengths during the high-density phase. Recent measurements of the elliptic-flow ratio of neutrons with respect to light charged particles at the GSI laboratory have led to a new constraint for the symmetry energy at suprasaturation density. It confirms, with higher precision, the moderately soft to linear density dependence of the symmetry energy previously deduced from the FOPI-LAND measurements. New calculations demonstrate the opportunities offered by future experiments at FAIR.

1 Introduction

Collective nuclear motion has always been a topic followed and advanced by Walter Greiner. This includes the nuclear compressibility which is of importance for many phenomena in nuclear structure and nuclear reactions as well as for astrophysics. Going beyond the small density intervals probed with giant resonances requires heavy-ion reactions at high energy. In their famous shock-wave paper of 1974, Scheid, Müller and Greiner have studied the new phenomena produced by the collective pressure in the shock-compressed overlap zone in the colliding system [1]. They have shown that heavy-ion reactions at sufficiently high energy offer a means to compress nuclear matter up to densities of several times the saturation value ρ_0 encountered in the nuclear interior. The properties of nuclear matter at suprasaturation densities may thus be studied in laboratory experiments [2–4].

W. Trautmann (✉)
GSI Helmholtzzentrum, Darmstadt, Germany
e-mail: w.trautmann@gsi.de

J. Kirsch et al. (eds.), *Discoveries at the Frontiers of Science*,
FIAS Interdisciplinary Science Series, https://doi.org/10.1007/978-3-030-34234-0_15

In a recent paper, Li and Han have documented that the many results obtained for the nuclear symmetry energy from terrestrial nuclear experiments and astrophysical observations are amazingly compatible, even though individual results scatter within considerable margins and are partly affected with large errors [5]. The average values deduced by the authors for the symmetry energy at saturation density and for the slope parameter describing its density dependence are $E_{sym}(\rho_0) = 31.6$ MeV and $L = 58.9$ MeV, respectively. The parameter L, defined as

$$L = 3\rho_0 \frac{\partial E_{sym}}{\partial \rho}|_{\rho=\rho_0}, \tag{1}$$

is proportional to the slope of the symmetry energy at the saturation density ρ_0 [6]. The authors quote also error margins representative for the variation of the individual results as $\Delta E_{sym}(\rho_0) = 2.7$ MeV and $\Delta L = 16$ MeV and conclude that L has a value about twice as large as $E_{sym}(\rho_0)$. A very similar conclusion can be drawn from the compilation of Lattimer and Steiner [7] adapted from the earlier work of Lattimer and Lim [8].

The nuclear symmetry energy governs many aspects of nuclear structure and reactions and determines very basic properties of neutron stars as, e.g., their radii [6, 9]. This implies that many, rather different, sources of information exist, in the laboratory and in the cosmos, that provide constraints for the equation of state (EoS) of asymmetric nuclear matter. The mentioned comparisons of obtained results are made possible by the fact that the forces identified as best describing a particular observation can be used in many-body calculations to determine the corresponding $E_{sym}(\rho_0)$ and L, the two quantities characterizing the asymmetric-matter EoS at saturation density. It does not require that the measurement or observation has actually tested the EoS at this density. The result may represent an extrapolation.

The predictions of microscopic models for the nuclear symmetry energy, obtained with realistic or phenomenological forces, appear to coincide at densities close to $\rho = 0.1$ fm^{-3}, i.e. at approximately two thirds of the saturation density [17]. It reflects the fact that the average density of atomic nuclei is below saturation and that the presence of the nuclear surface influences the properties that are chosen as constraints. The awareness that each observable carrying information on the nuclear EoS is connected to its proper range of density has to complement the interpretation of existing results.

Rather precise values for the symmetry energy and for the density to which it applies have recently been presented by Brown [12] and Zhang and Chen [13]. They are shown in Fig. 1 together with the low-density behavior of the symmetry energy obtained from heavy-ion collisions [14] and from the analysis of isobaric analog states [15] as reported in Ref. [16]. In the study of Brown, a set of selected Skyrme forces is used whose parameters are fitted to properties of doubly magic nuclei. By using particular values for the neutron skin of ^{208}Pb nuclei within a given range as additional constraints, new sets of these forces with slightly adjusted parameters are obtained. It is found that all predictions coincide at a density $\rho = 0.1$ fm^{-3}, independent of the choice made for the thickness of the neutron skin, but that the

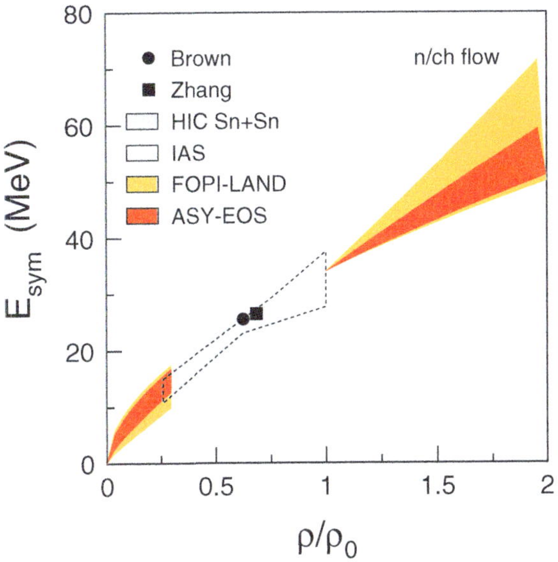

Fig. 1 Constraints deduced for the density dependence of the symmetry energy from the ASY-EOS data [10] in comparison with the FOPI-LAND result of Ref. [11] as a function of the reduced density ρ/ρ_0. The low-density results of Refs. [12–15] as reported in Ref. [16] are given by the symbols, the grey area (HIC), and the dashed contour (IAS). For clarity, the FOPI-LAND and ASY-EOS results are not displayed in the interval $0.3 < \rho/\rho_0 < 1.0$ (from Ref. [10]; Copyright (2016) by the American Physical Society)

slopes at this density depend on this choice. Only a precise knowledge of the neutron skin of ^{208}Pb will permit the extrapolation to the saturation point. It underlines the importance of this quantity determined by the balance of pressures felt by neutrons in the neutron-enriched interior of the ^{208}Pb nucleus at approximately saturation density and in the low-density neutron-rich surface [18, 19].

Sensitivities to higher densities can be expected from observables related to the early phases of heavy-ion collisions at sufficiently high energies. Calculations predict that densities up to three times the saturation value are reached for short times ($\approx$20 fm/c) in the central zone of heavy-ion collisions with energies of up to $\approx$1 GeV/nucleon [20]. The resulting pressure produces a collective outward motion of the compressed material whose strength will be influenced by the symmetry energy in asymmetric systems [4]. A measurement differentiating between the collective flows of neutrons and protons can thus be expected to provide information on the high-density symmetry energy [21].

The ratio of elliptic flow strengths observed for neutrons and light charged particles has been proposed as an observable sensitive to the EoS of asymmetric matter [11]. It is approximately linearly correlated with the slope parameter L (Eq. 1) as shown by Wang et al. [22]. From a reanalysis of the earlier FOPI-LAND data [23, 24] and the comparison with calculations performed with the UrQMD transport model [25] a moderately soft to linear symmetry term, characterized by a coefficient $\gamma = 0.9 \pm 0.4$ for the power-law parametrization of the density dependence of the potential part of the symmetry energy, was obtained [11, 26–28]. It is represented by the yellow band in Fig. 1. Motivated by this finding, an attempt has been made to improve the accuracy with a new experiment that was conducted at the GSI laboratory in 2011 (ASY-EOS experiment S394). The new result, reported in Ref. [10], is represented by the narrower red band in the figure. Both results are displayed over the range of densities up to twice saturation. It will be shown in this talk that the range of sensitivity of this observable reaches even beyond that point.

2 The ASY-EOS Experiment

The experimental setup of the ASY-EOS experiment at the GSI laboratory followed the scheme developed for FOPI-LAND by using the Large Area Neutron Detector (LAND [29]) as the main instrument for neutron and charged particle detection (Fig. 2). Several detection systems with azimuthal symmetry with respect to the beam axis served in determining the orientation of the reaction plane. Upstream of the target, a thin plastic scintillator foil viewed by two photomultipliers was used to record the projectile arrival times and served as a start detector for the time-of-flight measurement. LAND was positioned at a laboratory angle close to 45° with respect to the beam direction. A veto wall of thin plastic scintillators in front of LAND was

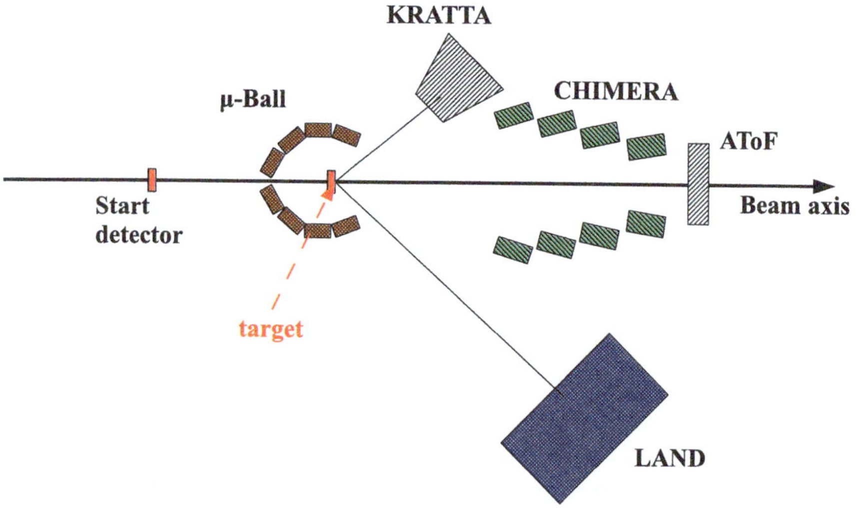

Fig. 2 Schematic view of the experimental setup used in the ASY-EOS experiment S394 at the GSI laboratory showing the six main detector systems and their positions relative to the beam direction. The dimensions of the symbols and the distances are not to scale (from Ref. [31])

used for discriminating between neutrons and charged particles. This configuration permitted the measurement of directed and elliptic flows of neutrons and charged particles near mid-rapidity within the same angular acceptance. Opposite of LAND, covering a comparable range of polar angles, the Kraków Triple Telescope Array (KRATTA [30]) was installed to permit flow measurements of identified charged particles under the same experimental conditions.

For the event characterization and for measuring the orientation of the reaction plane, three detection systems had been installed. The ALADIN Time-of-Flight (AToF) Wall [32] was used to detect charged particles and fragments in forward direction at polar angles up to $\theta_{\rm lab} \leq 7°$. Its capability of identifying large fragments and of characterizing events with a measurement of $Z_{\rm bound}$ [32] permitted the sorting of events according to impact parameter. Four double rings of the CHIMERA multidetector [33, 34] carrying together 352 CsI(Tl) scintillators in forward direction and four rings with 50 thin CsI(Tl) elements of the Washington University Microball array [35] surrounding the target provided sufficient coverage and granularity for determining the orientation of the reaction plane from the measured azimuthal particle distributions. A detailed description of the experiment is available in Ref. [10].

3 Experimental Results

As in the FOPI-LAND experiment, the reaction ^{197}Au+^{197}Au at 400 MeV/nucleon was studied. The elliptic flows of neutrons and light charged particles were determined from the azimuthal distributions of these particles with respect to the reaction plane reconstructed from the distributions of particles and fragments recorded with the three arrays AToF, CHIMERA, and Microball. Methods developed and described in Refs. [36, 37] for correcting the finite dispersion of the reaction plane were applied. The coefficients v_1 and v_2 representing the strengths of directed and elliptic flows, respectively, were deduced from fits with the Fourier expansion

$$f(\Delta\phi) \propto 1 + 2v_1\cos(\Delta\phi) + 2v_2\cos(2\Delta\phi). \tag{2}$$

Here $\Delta\phi$ represents the azimuthal angle of the momentum vector of an emitted particle with respect to the angle representing the azimuthal orientation of the reaction plane. Constraints for the symmetry energy were determined by comparing the ratios of the elliptic flows of neutrons and charged particles (ch), v_2^n/v_2^{ch}, with the corresponding UrQMD predictions for soft and stiff assumptions.

For the analysis, the UrQMD model was employed in the version adapted to the study of intermediate energy heavy-ion collisions [38]. The chosen isoscalar EoS is soft and different options for the dependence on isospin asymmetry were implemented. Two of them are used here, expressed as a power-law dependence of the potential part of the symmetry energy on the nuclear density ρ according to

$$E_{\rm sym} = E_{\rm sym}^{\rm pot} + E_{\rm sym}^{\rm kin} = 22\ {\rm MeV}(\rho/\rho_0)^{\gamma} + 12\ {\rm MeV}(\rho/\rho_0)^{2/3} \tag{3}$$

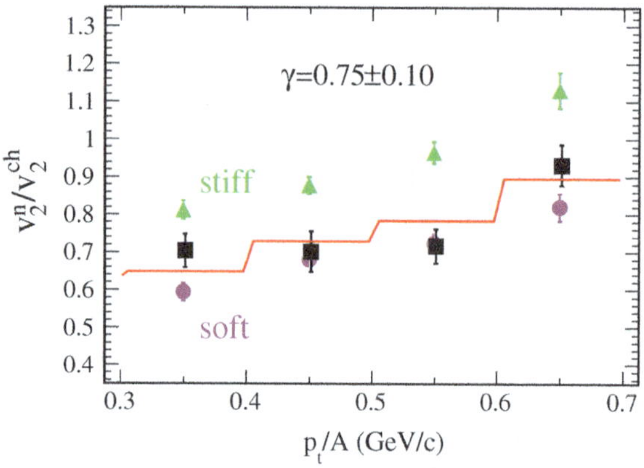

Fig. 3 Elliptic flow ratio of neutrons with respect to all charged particles for central ($b < 7.5$ fm) collisions of ^{197}Au+^{197}Au at 400 MeV/nucleon as a function of the transverse momentum/nucleon p_t/A. The black squares represent the experimental data, the green triangles and purple circles represent the UrQMD predictions for stiff ($\gamma = 1.5$) and soft ($\gamma = 0.5$) power-law exponents of the potential term, respectively. The solid line is the result of a linear interpolation between the predictions, weighted according to the experimental errors of the included four bins in p_t/A, and leading to the indicated $\gamma = 0.75 \pm 0.10$ (from Ref. [10]; Copyright (2016) by the American Physical Society)

with $\gamma = 0.5$ and $\gamma = 1.5$ corresponding to a soft and a stiff density dependence, respectively.

The predictions obtained with these assumptions for the measured flow ratio are shown in Fig. 3 together with the experimental result. The histogram represents the linear interpolation between the predictions giving a best fit of the flow ratios presented here as a function of the transverse momentum p_t. The corresponding power-law coefficient is $\gamma = 0.75 \pm 0.10$ with a purely statistical error $\Delta\gamma = 0.10$. In comparison with FOPI-LAND, this represents an improvement of the accuracy by more than a factor of two.

The necessary corrections and the methods used for estimating systematic uncertainties are described in detail in Ref. [10]. With all the resulting errors included, the acceptance-integrated elliptic-flow ratio leads to a power-law coefficient $\gamma = 0.72 \pm 0.19$. This is the result displayed in Fig. 1 as a function of the reduced density ρ/ρ_0. The new result confirms the former and has a considerably smaller uncertainty. It is also worth noting that the present parametrization is compatible with the low-density behavior of the symmetry energy from Refs. [12–15] that are included in the figure. The corresponding slope parameter describing the variation of the symmetry energy with density at saturation is $L = 72 \pm 13$ MeV. The sharp value $E_{sym}(\rho_0) = 34$ MeV is a consequence of the chosen parametrization (Eq. 3). Using values lower than the default $E_{sym}^{pot}(\rho_0) = 22$ MeV, as occasionally done in other UrQMD studies [39], will lower the result for L. Performing the present UrQMD analysis with $E_{sym}^{pot}(\rho_0) = 19$ MeV, corresponding to $E_{sym}(\rho_0) = 31$ MeV, yields a power-law coefficient $\gamma = 0.68 \pm 0.19$ and a slope parameter $L = 63 \pm 11$ MeV. The observed changes remain both within the error margins of these quantities. However, the precise results of Brown [12] and Zhang and Chen [13] are not equally met with this alternative parametrization of the symmetry energy.

4 Sensitivity to Density

The range of densities probed with the elliptic-flow ratio was explored in a study using the Tübingen version [40] of the QMD model (TüQMD). The applied method consisted of performing transport calculations for the present reaction with two parametrizations of the symmetry energy that were chosen to be different for a selected range of density and identical elsewhere. The magnitude of the obtained difference between the two predictions for the elliptic flow ratio is interpreted as a quantitative measure of the sensitivity to the selected density region.

In the momentum-dependent one-body potential introduced by Das et al. [41], the stiffness of the symmetry energy is controlled with a parameter x and choices ranging from soft ($x = +1$) up to rather stiff ($x = -2$) density dependences are commonly selected in model studies. For the present case, a mildly stiff ($x = -1$) and a soft ($x = +1$) parametrization were chosen for the density range with different symmetry energies while the nearly linear case with $x = 0$ was chosen for the common part.

To quantify the results, a quantity DEFR (Difference of Elliptic-Flow Ratio) was defined as

$$\mathrm{DEFR}^{(n,Y)}(\rho) = \frac{v_2^n}{v_2^Y}(x = -1, \rho) - \frac{v_2^n}{v_2^Y}(x = 1, \rho). \tag{4}$$

It represents the differences of the elliptic flow ratios calculated with the TüQMD transport model for two different density dependences of the symmetry energy. Here Y indicates a charged particle or a group of charged-particle species and x the stiffness parameter that is used in the calculations at densities smaller than the argument ρ. The difference of the $x = \pm 1$ potentials is thus effectively only tested at densities up to the particular ρ, the argument of DEFR. This choice leads to $\mathrm{DEFR}^{(n,Y)}(0) = 0$ and to the full stiff-soft splitting for large values of the argument ρ. The slope of DEFR at intermediate densities is a measure of the impact on elliptic flow observables of that particular region of density.

In the upper panel of Fig. 4, the density dependence of $\mathrm{DEFR}^{(n,Y)}$ for the choice Y = all charged particles is presented. It is seen that DEFR increases monotonically up to density values in the neighborhood of $2.5\,\rho_0$, close to the maximum density probed by nucleons in heavy-ion collisions at 400 MeV/nucleon incident energy [20]. The distribution of the sensitivity as a function of density is obtained by forming the derivative of DEFR with respect to density. It is presented in the lower panel of Fig. 4 for three choices of Y, only protons (n/p), sum of all hydrogen isotopes (n/H), and all charged particles (n/ch).

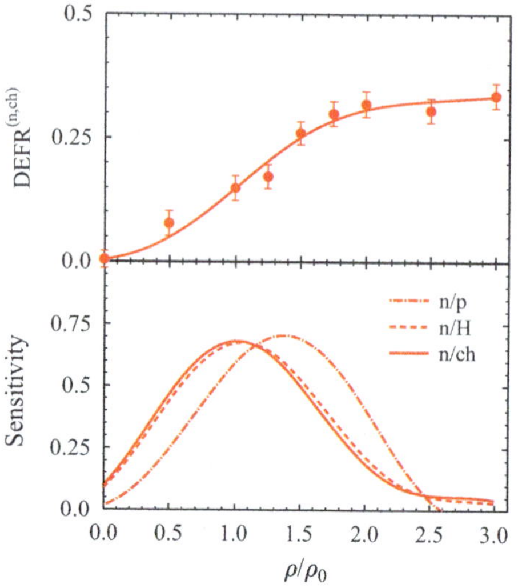

Fig. 4 Density dependence of the difference of the elliptic flow ratio (DEFR) of neutrons over charged particles, defined by Eq. 4, for ^{197}Au+^{197}Au collisions at 400 MeV/nucleon obtained with the TüQMD transport model [40] and the FOPI-LAND acceptance filter (top) and the corresponding sensitivity density (bottom panel, solid line) together with sensitivity densities obtained from elliptic-flow ratios of neutrons over all hydrogen isotopes (dashed) and neutrons over protons (dash-dotted) (from Ref. [10]; Copyright (2016) by the American Physical Society)

The figure shows that the sensitivity achieved with the elliptic-flow ratio of neutrons over charged particles, the case studied here, reaches its maximum close to saturation density and extends beyond twice that value. It is compatible with the conclusions reached by Le Fèvre et al. in their study of the symmetric matter EoS, based on FOPI elliptic-flow data and calculations with the Isospin Quantum Molecular Dynamics transport model [42]. For ^{197}Au+^{197}Au collisions at 400 MeV/nucleon, the broad maximum of the force-weighted density defined by the authors is spread over the density range $0.8 < \rho/\rho_0 < 1.6$.

The sensitivity of the neutron-vs-proton flow ratio has its maximum in the 1.4–1.5 ρ_0 region, i.e. at significantly higher densities than with light complex particles being included (Fig. 4). This observation carries the import potential for determining the curvature of the symmetry energy at saturation, in addition to the slope [10]. With very recent calculations Cozma has, in fact, demonstrated this potential by using the FOPI-LAND neutron-vs-proton flow ratio [11] in comparison with the ASY-EOS neutron-vs-charged particles flow ratio to determine a curvature parameter $K_{sym} = 96 \pm 315(\text{exp}) \pm 170(\text{th}) \pm 166(\text{sys})$ MeV [43]. The dominating experimental error is caused by the lower statistical accuracy of the FOPI-LAND data if isotopically pure protons are selected. The theoretical errors are the result of exploring the full parameter space, including the isoscalar sector, of the transport model while the systematic error reflects shortcomings of the model in correctly reproducing the multiplicities of intermediate-mass fragments. The corresponding result for the slope parameter is $L = 85 \pm 22(\text{exp}) \pm 20(\text{th}) \pm 12(\text{sys})$ MeV.

5 Conclusion and Outlook

Several analyses of the FOPI-LAND and the more recent ASY-EOS elliptic-flow ratios agree on the moderately soft to linear density dependence of the symmetry energy. The original result $L = 83 \pm 26$ MeV obtained with the UrQMD transport model and the FOPI-LAND data [11] has been confirmed by the analysis of Wang et al. [22] using selected Skyrme forces within the UrQMD ($L = 89 \pm 45$ MeV, 2-σ uncertainty) and, with higher precision, by the ASY-EOS result $L = 72 \pm 13$ MeV obtained with the UrQMD transport model [10]. The recent result reported by Cozma [43], $L = 85 \pm 22(\text{exp}) \pm 20(\text{th}) \pm 12(\text{sys})$ MeV, has a larger error because the full parameter space, including the isoscalar sector, has been explored in the calculations. It is unique in the sense that the slope and curvature parameters are uncorrelated in the MDI2 force used in this particular version of the TüQMD transport model, a feature emphasized by the author [43]. The obtained slope parameters are all compatible with each other. They are slightly larger than the value $L = 59$ MeV calculated by Li and Han [5] by averaging over 28 terrestrial and astrophysical studies but touch this value with the lower end of their 1-σ error margins.

The ASY-EOS value $\gamma = 0.72 \pm 0.19$ obtained for the power-law coefficient of the potential part in the UrQMD parametrization of the symmetry energy and the slope parameter $L = 72 \pm 13$ MeV are equivalent to a symmetry pressure $p_0 = \rho_0 L/3 = 3.8 \pm 0.7\,\text{MeVfm}^{-3}$. The latter may be used to estimate the pressure in neutron-star matter at saturation density [10]. For an assumed asymmetry $\delta = (\rho_n - \rho_p)/\rho = 0.9$ in that part of a neutron star, it amounts to $3.4\,\text{MeVfm}^{-3}$, a value that compares well with the pressure obtained by Steiner et al. [44] from neutron-star observations.

The sensitivity study shows that suprasaturation densities are effectively probed with the elliptic flow ratio of neutrons with respect to charged particles. Because the interpretation of the FOPI pion ratios [45] is not yet conclusive (see, e.g., Refs. [46, 47] and references given therein), this observable is presently unique as a terrestrial source of information for the EoS of asymmetric matter at high densities. The different sensitivities probed by flow ratios of neutrons with respect to protons and with respect to light charged particles permit a determination of the curvature parameter K_{sym}. Its uncertainty is large at present but, as shown by Cozma, errors close to ± 150 MeV may be ultimately within reach, provided that higher precision and isotopic resolution for charged particles can be achieved [43]. The presented experimental results and theoretical studies thus provide a strong encouragement for the continuation of flow measurements of the present kind with improved detection systems at FAIR, the Facility for Antiproton and Ion Research.

Acknowledgements The author acknowledges fruitful discussions and the very constructive collaboration with his colleagues within the ASY-EOS Collaboration. See Ref. [10] for the complete list of authors.

References

1. W. Scheid, H. Müller, W. Greiner, Phys. Rev. Lett. **32**, 741 (1974)
2. H. Stöcker, W. Greiner, Phys. Rep. **137**, 277 (1986)
3. J. Aichelin, A. Rosenhauer, G. Peilert, H. Stoecker, W. Greiner, Phys. Rev. Lett. **58**, 1926 (1987)
4. P. Danielewicz, R. Lacey, W.G. Lynch, Science **298**, 1592 (2002)
5. B.-A. Li, X. Han, Phys. Lett. B **727**, 276 (2013)
6. For a review, see B.-A. Li, L.-W. Chen, C.M. Ko, Phys. Rep. **464**, 113 (2008)
7. J.M. Lattimer, A.W. Steiner, Eur. Phys. J. A **50**, 40 (2014)
8. J.M. Lattimer, Y. Lim, Astrophys. J. **771**, 51 (2013)
9. B.-A. Li, À. Ramos, G. Verde, I. Vidaña, Topical issue on nuclear symmetry energy. Eur. Phys. J. A **50**, 9 (2014)
10. P. Russotto et al., Phys. Rev. C **94**, 034608 (2016)
11. P. Russotto et al., Phys. Lett. B **697**, 471 (2011)
12. B.A. Brown, Phys. Rev. Lett. **111**, 232502 (2013)
13. Z. Zhang, L.-W. Chen, Phys. Lett. B **726**, 234 (2013)
14. M.B. Tsang, Y. Zhang, P. Danielewicz, M. Famiano, Z. Li, W.G. Lynch, A.W. Steiner, Phys. Rev. Lett. **102**, 122701 (2009)
15. P. Danielewicz, J. Lee, Nucl. Phys. A **922**, 1 (2014)
16. C.J. Horowitz, E.F. Brown, Y. Kim, W.G. Lynch, R. Michaels, A. Ono, J. Piekarewicz, M.B. Tsang, H.H. Wolter, J. Phys. G **41**, 093001 (2014)
17. C. Fuchs, H.H. Wolter, Eur. Phys. J. A **30**, 5 (2006)
18. X. Roca-Maza, M. Centelles, X. Viñas, M. Warda, Phys. Rev. Lett. **106**, 252501 (2011)
19. S. Abrahamyan et al., Phys. Rev. Lett. **108**, 112502 (2012)
20. X. Jun, L.-W. Chen, C.M. Ko, B.-A. Li, Y.-G. Ma, Phys. Rev. C **87**, 067601 (2013)
21. B.-A. Li, Phys. Rev. Lett. **88**, 192701 (2002)
22. Y. Wang, C. Guo, Q. Li, H. Zhang, Y. Leifels, W. Trautmann, Phys. Rev. C **89**, 044603 (2014)
23. Y. Leifels et al., Phys. Rev. Lett. **71**, 963 (1993)
24. D. Lambrecht et al., Z. Phys. A **350**, 115 (1994)
25. Q. Li, Z. Li, S. Soff, R.K. Gupta, M. Bleicher, H. Stöcker, J. Phys. G **31**, 1359 (2005)
26. M.D. Cozma, Phys. Lett. B **700**, 139 (2011)
27. W. Trautmann, H.H. Wolter, Int. J. Mod. Phys. E **21**, 1230003 (2012)
28. P. Russotto et al., Eur. Phys. J. A **50**, 38 (2014)
29. T. Blaich et al., Nucl. Instrum. Methods Phys. Res. A **314**, 136 (1992)
30. J. Łukasik et al., Nucl. Instrum. Methods Phys. Res. A **709**, 120 (2013)
31. P. Russotto et al., EPJ WebConf. **88**, 00022 (2015)
32. A. Schüttauf et al., Nucl. Phys. A **607**, 457 (1996)
33. A. Pagano et al., Nucl. Phys. A **734**, 504 (2004)
34. E. De Filippo, A. Pagano, Eur. Phys. J. A **50**, 32 (2014)
35. D.G. Sarantites et al., Nucl. Instrum. Methods A **381**, 418 (1996)
36. J.-Y. Ollitrault (1997), arXiv:nucl-ex/9711003
37. A. Andronic, J. Łukasik, W. Reisdorf, W. Trautmann, Eur. Phys. J. A **30**, 31 (2006)
38. Q. Li, C. Shen, C. Guo, Y. Wang, Z. Li, J. Łukasik, W. Trautmann, Phys. Rev. C **83**, 044617 (2011)
39. Y. Wang, C. Guo, Q. Li, H. Zhang, Z. Li, W. Trautmann, Phys. Rev. C **89**, 034606 (2014)
40. M.D. Cozma, Y. Leifels, W. Trautmann, Q. Li, P. Russotto, Phys. Rev. C **88**, 044912 (2013)

41. C.B. Das, S.D. Gupta, C. Gale, B.-A. Li, Phys. Rev. C **67**, 034611 (2003)
42. A. Le Fèvre, Y. Leifels, W. Reisdorf, J. Aichelin, C. Hartnack, Nucl. Phys. A **945**, 112 (2016)
43. M.D. Cozma, Eur. Phys. J. A **54**, 40 (2018)
44. A.W. Steiner, J.M. Lattimer, E.F. Brown, Astrophys. J. Lett. **765**, L5 (2013)
45. W. Reisdorf et al., Nucl. Phys. A **781**, 459 (2007)
46. M.D. Cozma, Phys. Lett. B **753**, 166 (2016)
47. W. Trautmann, H.H. Wolter, in *Walter Greiner Memorial Volume*, ed. by P.O. Hess, H. Stöcker (World Scientific, 2018), arXiv:1712.03093 [nucl-ex] (2017)

Photon Absorption and Electron Scattering by Endohedrals

Miron Ya. Amusia

Abstract We concentrate here on photon absorption as well as electron and positron scattering upon endohedrals that consist of a fullerenes shell and an inner atom A. The aim is to understand the effect of fullerene electron shell in formation of corresponding cross-section. We consider the problem substituting the action of a complex multiatomic fullerenes shell by a combination of static pseudopotential and dynamic polarization potential. The electron correlations in the atom A are taken into account in the frame of the random phase approximation with exchange (RPAE). We demonstrate that the fullerenes shell strongly affects the cross-sections, bringing in a number of peculiarities, such as confinement resonances and giant-endohedral resonances and partial wave Ramsauer-type minima. Numerical data are obtained for endohedrals $A@C_{60}$ and $A@C_{60}@C_{240}$, where A are noble gas atoms He, Ar and Xe.

1 Introduction

The year 1985 has been marked by discovery of a rather exotic in shape multiatomic molecule C_{60} [1]. It presented an almost spherically symmetric construction of 60 carbon atoms with an empty interior. This discovery opened the door for detection of other "empty" molecules, constructed from both carbon and non-carbon atoms. All these objects received the name fullerenes. As to the carbon constructions, in includes now even giants, such as C_{540}!

Among other unusual features of fullerenes, one is of particular interest. Namely, it appeared that fullerenes can be "stuffed" by almost any atom A of the Periodic table. One can put inside a fullerene also a small molecule. Inside a big fullerene a small one could be placed also. All such construction received a general name

I dedicate this article to the blessed memory of my friend and colleague professor Walter Greiner.

M. Ya. Amusia (✉)
Racah Institute of Physics, The Hebrew University, 91904 Jerusalem, Israel
e-mail: miron.amusia@mail.huji.ac.il

A. F. Ioffe Physical-Technical Institute, St. Petersburg 194021, Russia

J. Kirsch et al. (eds.), *Discoveries at the Frontiers of Science*,
FIAS Interdisciplinary Science Series, https://doi.org/10.1007/978-3-030-34234-0_16

endohedrals and are denoted for atoms as A@C_N, presenting an atom A trapped inside a fullerene C_N. First endohedral La@C_{60} has been observed one week after the discovery of fullerenes, and presented in [2].

Endohedral is a very complex multi-atomic and many-electron object. Its ab initio calculation is very complicated if possible currently at all. So, in this paper we will use a simplifying approach and simulate the fullerene shell by a spherical potential, adding to it a dynamic polarization potential. Of importance are the electron correlations in atoms that we take into account in the frame of so-called random phase approximation with exchange (RPAE). So, we will treat A@C_N as a "big atom" [3]. In this small review we will present also some results on two-layer endohedrals A@C_{N1}@C_{N2}, where fullerene C_{N1} is placed inside C_{N2}.

The studies of structure and properties of endohedrals are of interest, since they are scientifically exciting objects, they could exist in Nature and have a whole variety of technological applications. The inner atom A in an endohedral serves as a lamp that illuminates C_N from the inside. As a concrete example, we consider almost spherical C_{60} with a noble gas atom, in most cases, He, Ar and Xe, placed inside. It is essential and simplifying the consideration that noble gas atomic nuclei are located at the center of the fullerene sphere. It is also essential that the fullerene radius R_F is considerably bigger than the atomic radius R_A.

The fullerene shell affects the inner atom, modifying its radius and energy levels. The Atom A and fullerene C_N can also exchange electrons, transferring them in both direction and even collectivizing them, totally, or only to some extent. There are good evidences, however, at least for noble gas endohedrals, that these effects are inessential and the inner structure of both objects, C_N and A, are not altered, when one puts A inside C_N. However, as we will demonstrate below, C_N strongly affects the processes that took place with participation of A.

We will consider here photoionization, low-energy electron (positron) scattering and decay of vacancies in A, concentrating on the role of C_N upon all these processes. Among the most important effects in this area is distortion of the atomic Giant resonance, formation of Giant endohedral and Interference endohedral resonances [3] and demonstration and analyses of so-called quantum phase additivity in the $e^{\mp} + A@C_N$ scattering.

It deserves to be mentioned also that an "empty" multi-particle construction could in principle be formed from a very big number of nucleons, since in them the Coulomb repulsion is much weaker than the nuclear attraction. It is in place to mention here that the option of a long linear nucleus I have discussed with W. Greiner already in the early ninetieth. One could imagine that an "empty" nucleonic construction could be "staffed" by an ordinary nucleus, at least by a small one.

2 Fullerene Shell Action

The action of C_N includes static action of the fullerene upon atomic A photoelectron or incoming electrons (positrons) in the scattering process $e^{\mp} + A@C_N$. This action is accounted for by introducing pseudopotential $U_F(r)$

$$U_F(r) = \begin{cases} -U_0 & \text{at } R_{in} \le r \le R_{out} = R_{in} + b \\ 0 & \text{at } r < R_{in} \text{ and } R_{out} < r \end{cases} \tag{1}$$

Here b is the thickness of the fullerene shell that is close to a single-atomic carbon diameter and R_{in} is the inner radius of the fullerene. The concrete values of U_0 and b for C_{60} were chosen to reproduce the experimental value of the binding energy of the extra electron in the negative ion C_{60}^- and the low- and medium-energy photoionization cross-section of C_{60} [4].

Fullerenes are polarizable objects. Therefore, an electron that collides with a fullerene shell has to be affected by so-called polarization potential $W(r)$, the simplest form of which is

$$W_F(r) = -\alpha_F / 2(r^2 + d^2)^2 \tag{2}$$

Here α_F is the fullerene dipole static polarizability and d is the length parameter. In our calculations we put $d = (R_{in} + R_{out})/2 \equiv R_F$.

In photoionization of endohedral atoms, potentials (1) and (2) affect the shape of the cross-section, by adding resonance structure that corresponds to reflection of the photoelectron wave by the fullerenes potentials. Maxima in the cross-sections that appear due to the action of potentials (1) and (2) are called confinement resonances.

3 Polarization Factor

Due to big size and relatively big distance between fullerene nuclei and its electron shell, C_N are highly polarizable objects. This is reflected in its big polarizability. The incoming beam of electromagnetic radiation, in order to ionize the atom A, has to go via the fullerene shell. In dipole approximation that is valid for all photon frequencies in interesting for us energy range, an expression can be derived that connects the electric field $\mathbf{E}_{in}$ inside the fullerene with that of the outside $\mathbf{E}$. To simplify this expression, one has to assume that the radius of fullerene is not simply bigger than the atomic radius, but is much bigger i.e. $R_F/R_A \gg 1$. Applying this inequality, one obtains the following relation (see e.g. in [5]):

$$\mathbf{E}_{in} \equiv \mathbf{E}G_F(\omega) = \mathbf{E}\left[1 - \frac{\alpha_F(\omega)}{R_F^3}\right] \tag{3}$$

Here $\alpha_F(\omega)$ is the fullerenes dipole dynamic polarizability that at $\omega = 0$ is equal to α_F from (2). The function $G_F(\omega)$ is called polarization factor.[1]

[1] Here, as in all the rest of this paper the atomic system of units is used, with $e = m_e = \hbar = 1$.

It is obvious that if one neglects the potentials (1) and (2), the account of fullerene leads to a very simple relation between photoionization cross section of an endohedral $A@C_N$—$\sigma^{A@C_N}(\omega)$ and that of an isolated atom A—$\sigma^{A}(\omega)$:

$$\sigma^{A@C_N}(\omega) = |G_F(\omega)|^2\sigma^{A}(\omega) \tag{4}$$

The factor $|G_F(\omega)|^2$ can have maxima that corresponds to those regions of ω, where $\left|1 - \alpha_F(\omega)/R_F^3\right|$ is big. According to (4), this feature leads to a maximum in the photoionization cross-section.

4 Destruction and Formation of Resonances

In this section using concrete examples we demonstrate that fullerene shell can both destroy atomic resonances and form new ones. Let us start with the case of resonance destruction. As a concrete example, let us compare the photoionization cross-section of atom Xe, endohedral $Xe@C_{60}$ and endohedral $Xe@C_{60}@C_{240}$ in the area above the ionization threshold of the $4d^{10}$subshell. It is known since relatively long ago that there the cross-section has a high and broad maximum called atomic Giant resonance [6]. Its existence is a manifestation of very strong collective effects in the photoionization of this subshell. In fact, photons in the maximum's frequency region are absorbed by the whole 10-electron 4d-subshell.

The theoretical description of the Giant resonance for atom was achieved by RPAE (see [5] and references therein). Considering photoionization of $Xe@C_{60}$ and two-shell endohedral, e.g. $Xe@C_{60}@C_{240}$ [7], one has to take into account that all Xe one-electron excited states are affected by potentials (1) and (2). At ω bigger than the ionization energy of $4d^{10}$ sub-shell, $G_F(\omega)$ is close to 1. In calculations, we put the radiuses of C_{60} and C_{240} equal to $R_{60} = 6.72$ and $R_{240} = 13.5$; the depth of the potential wells for C_{60} is $U_0^{60} = 0.44$ and for C_{240} is $U_0^{240} = 0.53$.

The results of calculations are presented in Fig. 1. We see that the atomic Giant resonance under the action of fullerene shell or shells is destroyed and substituted by a number of narrower and higher resonances.

Note that the total oscillator strength, i.e. the area under the photoionization curve, of the group of $Xe@C_{60}$ resonances is almost the same as that for $4d^{10}$ Xe itself.

Let us consider photoionization of endohedrals in the region of lower ω. There one has to take into account also the polarization factor $G_F(\omega)$, introduced in (4). The dynamic polarizability $\alpha_F(\omega)$ one can find using the well-known dispersion relation that connects the real and imaginary part of polarizability and takes into account that the imaginary part is simply proportional to the photoionization cross-section of the system under consideration [8]:

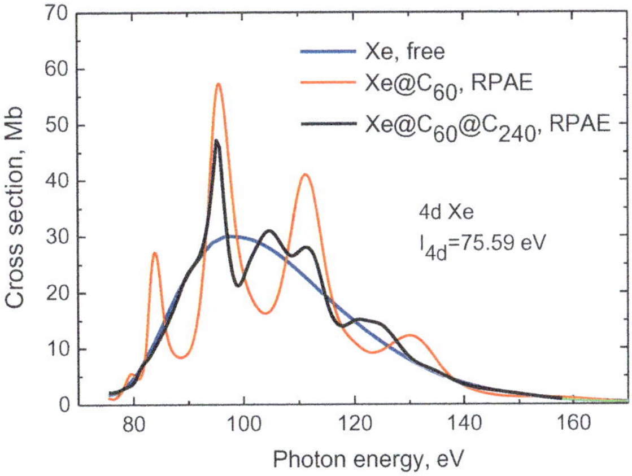

Fig. 1 Destruction of $4d^{10}$Xe resonance

$$\operatorname{Im}\alpha_F(\omega) = c\sigma_F(\omega)/4\pi\omega,\ \operatorname{Re}\alpha_F(\omega) = \frac{c}{2\pi^2}\int_{I_F}^{\infty}\frac{\sigma_F(\omega')d\omega'}{\omega'^2-\omega^2} \quad (5)$$

Here c is the speed of light. We take $\sigma_F(\omega)$ from experiment. A number of quite accurate measurements exist, starting with published quite long ago in [9]. The dependence of $\alpha_F(\omega)$ upon ω one can easily understand, having in mind that $\sigma_F(\omega)$ for C_{60} is a powerful broad maximum located at $\omega \approx 22eV$. The so-called total oscillator strength of this maximum is close to 240—the number of collectivized electrons in C_{60}. This has to be compared to the total Xe $4d^{10}$subshell oscillator strength, the biggest for atoms that is equal to about 8. This means that $G_F(\omega)$ factor will be of great importance in this ω region.

Figure 2 presents the results for photoionization cross-section calculations for

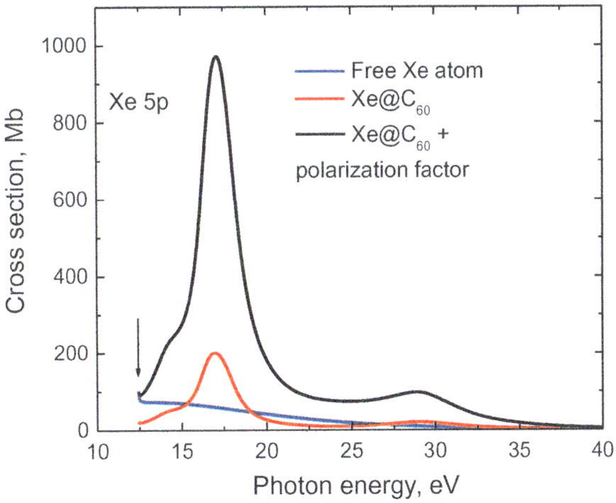

Fig. 2 Endohedral resonance in $5p^6$ of Xe

outer $5p^6$ subshell of Xe. We include the effects of fullerene shell potentials (1) and (2) as well as polarization factor (3).

The potentials (1) and (2) transform the monotonically decreasing cross section considerably, creating a maximum at 17 eV. But it is the inclusion of the $G_F(\omega)$ factors that leads to dramatic changes. We see that a powerful maximum is formed at about 16 eV and a prominent second maximum appear in the Xe@C_{60} cross-section. The cross-section is more than 900 Mb that is about 45 times bigger than for the isolated atom. We call the respective structure *endohedral resonance* [10]. It is important to note that the total oscillator strength of this resonance is close to 30, that is, by a factor of six more than the corresponding value for the $5p^6$ subshell in the isolated Xe atom. This extra oscillator strength can be "borrowed" only from the oscillator strength of the C_{60}. This is an additional evidence of the fact that in the photoionization process in the frequency range of the Xe $5p^6$ subshell, the fullerene and atomic electrons interact very strong, in fact, they became common, to large extent.

5 Decay of Vacancies

Here we consider the effect of fullerene shell upon the probability of atom's A vacancy decay. Let us start with radiative decay. It can proceed directly by the atom A, emitting a photon. It is also possible that due to interaction between atomic and fullerene electrons the fullerene shell becomes virtually or even really excited and then emits a photon. The amplitudes of two these processes have to be summed determining together the decay probability. We perform calculations assuming as before that $R_F \gg R_A$ [11]:

$$\Gamma_{\gamma,if}^{A@C_N} = \Gamma_{\gamma,if}^{A}\left|1 - \frac{\alpha_F(\omega_{if})}{R^3}\right|^2. \tag{6}$$

Here $\Gamma_{\gamma,if}^{A}$ and $\Gamma_{\gamma,if}^{A@C_N}$ are the radiative width of the vacancy i due to its transition to the vacancy f in an isolated atom and endohedral; ω_{if} is the energy of the emitted photon. We see that the effect of fullerene shell upon radiative decay width is determined by the enhancement factor $G_F(\omega)$ that was introduced in (3) in connection to Giant endohedral resonances.

The presence of fullerene shell can open a new, non-radiative or Auger, decay channel. As an example of such a situation, consider the decay of a subvalent vacancy ns^2 in a noble gas atom. The transition of an electron from an outer subshell np^6 into a vacancy in ns^2 leads to emission of a photon with the energy $\omega_{ns,np}$. It cannot decay via emitting another np electron, since the transition has not enough energy to ionize atom A. However, this energy is enough to ionize the fullerenes shell, thus opening an Auger-decay channel and therefore increasing by many orders of magnitude the width of a vacancy ns^2 in an endohedral, as compared to that in an free atom. In the frame of the same assumptions that leads to (6), one can obtain [11]:

$$\Gamma_{A,if}^{A@C_N} = \Gamma_{\gamma,if}^{A} \frac{3}{8\pi}\left(\frac{c}{\omega_{if}}\right)^4 \frac{\sigma_F(\omega_{if})}{R_F^6}. \tag{7}$$

Here $\Gamma_{A,if}^{A@C_N}$ is the Auger-width of the subvalent vacancy in an endohedral. The ratio $\eta^{A\gamma} \equiv \Gamma_{A,ns,np}^{A@C_N}/\Gamma_{\gamma,ns,np}^{A@C_N}$ varies from 0.5×10^5 to 0.5×10^6 for noble gas endohedrals from Ne to Xe.

6 Electron and Positron Scattering

At first glance, low-energy elastic scattering cross section of a slow electron should be determined by the size of C_N only, being independent upon the presence or absence of the atom A inside the fullerene. It means that the elastic scattering cross-section $\sigma_{el}(E)$ as a function of incoming electron energy E, at $E \to 0$ should be determined only by R_F^2, $\sigma_{el}(0) \sim R_F^2$. This should be correct if the low-energy scattering is a classical process.

Direct calculations did not support this assumption [12]. It appeared that the cross-section even at low energies is essentially different from a constant value, depending upon inner structure of the target, namely, upon whether it is an empty fullerene or an endohedral. This difference signals that the low-energy electron (positron) $e^{\mp}$ scattering process is entirely quantum-mechanical. To find the respective scattering phases, one has to solve the following equation for the l partial scattering wave [13]:

$$\left(-\frac{1}{2}\frac{d^2}{dr^2} - \frac{Z}{r} + V_{H_{\mp}F_{-}}(r) + U_{F\mp}(r) + W_{A,\mp}(r) + \right.$$
$$\left. W_{F,\mp}(r) + \Delta W_{FA,\mp}(r) + \frac{l(l+1)}{2r^2} - E\right) P_{El,\mp}^{A@C_N}(r) = 0 \tag{8}$$

Here Z is the atom A nuclear charge, $V_{H_{\mp}F_{-}}$ is the Hartree-Fock potential for e^-(Hartree—for e^+), $U_{F\mp}(r)$ is the fullerene potential (1) for $e^{\mp}$, $W_{A\mp}(r)$ is the polarization potential of the atom A, and $W_{F\mp}(r)$ is the fullerene polarization potential, determined by (2) for an electron. It proved to be essential to take into account the mutual influence of atomic and fullerenes polarizabilities that we named interference of polarizabilities and contribution of which denoted in (8) as $\Delta W_{FA,\mp}(r)$.

As is known, the scattering cross-section is expressed via the scattering phases $\delta_{l\mp}(E)$. To simplify the problem, let us at first neglect the interference of polarizabilities, i.e. put $\Delta W_{FA,\mp}(r) = 0$. As an example, Fig. 3 presents the elastic scattering p—phase of electrons upon Ar, C_{60}, and endohedral Ar@C_{60} [11]. We see that the rule of additivity of phases takes place in this case, namely

$$\delta_l^{A@C_N}(E) = \delta_l^{C_N}(E) + \delta_l^{A}(E) \tag{9}$$

Fig. 3 Rule of additivity of phases in $e^{-} + Ar@C_{60}$ process

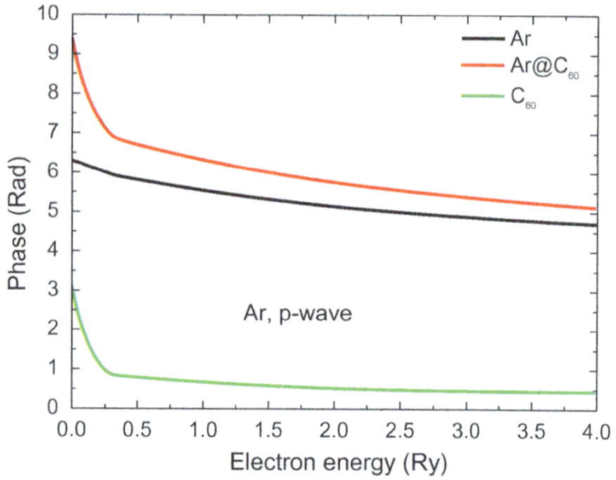

Note that the polarizability of the atom A is taken into account in the frame of simplified version of RPAE, using the many-body diagram technique (see e.g. [5]). Details on how to calculate $\Delta W_{FA,\mp}(r)$ are presented in [13], but its contribution is important leading to some violation of the phase additivity. We see in Fig. 4 that the electron elastic scattering cross-section by an endohedral that consist of sixty atoms is strongly modified due to presence of a single additional atom inside the fullerenes shell. The presence of an inner Ar leads to Ramsauer-type minima in the s-wave partial cross-section [14].

At first glance, the scattering of positrons is much simpler to treat than the electron scattering.

The only thing what is needed is to neglect the exchange between incoming positron and target electrons. However, the situation is much more complex. Indeed, the incoming positron strongly interacts with virtually excited in the scattering process atomic electrons. They can even form a sort of a temporary bound state called *virtual positronium* $\tilde{P}s$. Its role was recognized in atomic physics long ago [15]. This same effect has to be taken into account in positron—endohedral scattering.

Fig. 4 S-wave contribution to the $e^{-} + Ar@C_{60}$ cross-section

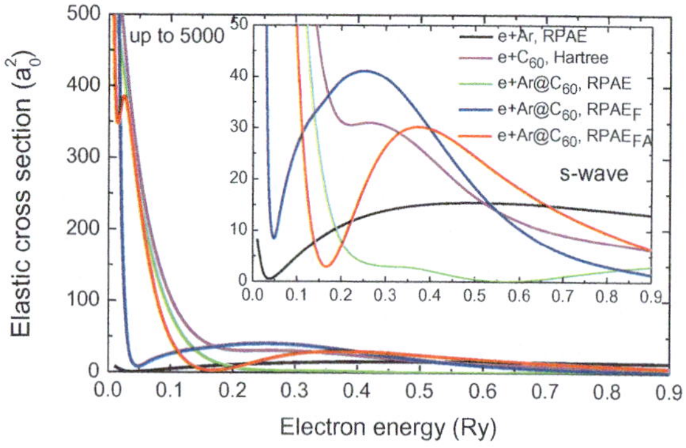

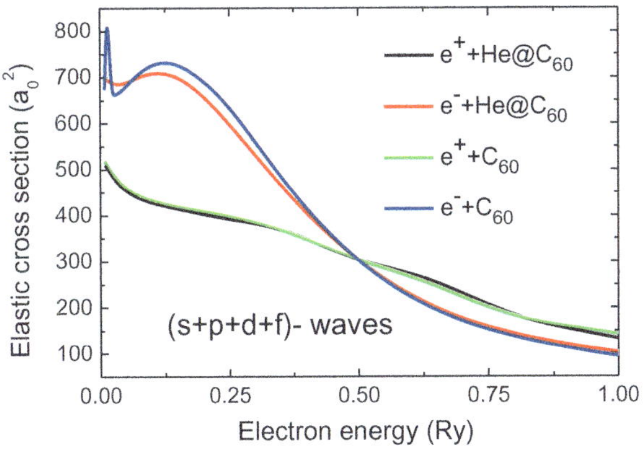

Fig. 5 $e^{\mp}$ scattering upon He and He@C_{60}

Formation of $\tilde{P}s$ modifies the polarization potential. The simplest way to include it is to shift the energy in $\alpha_F(\omega)$ from 0, as it is in (2), to $\omega = I_{Ps}$, where I_{Ps} is the real positronium binding energy. We assume also that. $W_{F,+}(r) = -W_{F,-}(r)$. The results of calculations are presented in Fig. 5 [16]. We took into account first four scattering phases, s, p, d, f and present results for $e^{\mp} + C_{60}$ and $e^{\mp} + He@C_{60}$ cross-section, considering them as the sum of all these four contributions.

7 Conclusions and Perspectives

In this short review we present a number of results in investigation of photon absorption, electron (positron) scattering, and vacancy decay with participation of single- and double-shell endohedrals.

We demonstrate destruction of the atomic Giant resonance and formation, in another photon frequency region, of endohedral Giant resonances due to effect of fullerene shell upon the photoionization cross-section.

We demonstrate that the fullerene shell opens new channels in atomic vacancies decay.

In elastic scattering of slow electrons or positrons upon endohedrals the inner single atom plays, unexpectedly, a prominent role. Due to polarization of the fullerene, the elastic electron scattering cross-sections acquire very big resonances at low energy, and a deep "Ramsauer-type" minimum. The difference between electron and positron cross-sections is very big, the latter being to large extent determined by virtual positronium formation.

We plan to improve the quality of calculations. Experimental investigations to check the validity of predictions are certainly desirable.

Walter Greiner—A Colleague and Friend

(Personal recollections based on half a century long acquaintance)
With the presented article I pay tribute to my old friend professor Walter Greiner, for me Walter, since we were "av du" for more than half a century. Best tribute for a big scientist is discussion of new scientific results. Indeed, each of us is mortal, but Science is immortal.

I vividly remember our first meeting at the Ioffe Institute in former Leningrad, I guess, in the very middle of sixties.

Just at that time, I was pushed off nuclear physics that was my topic before, and have concentrated on applying many-body theory approach to atomic instead of nuclear physics. Walter was interested mainly in nuclear physics, but many-body diagram technique seemed to be good in both domains, namely in studies of multi-nucleon nuclei and multi-electron atoms.

We discussed concrete scientific problems of common interest, such as Giant resonances in both nuclei and atoms. However, particularly great attention we paid to the theory of groups, planning to concentrate in the nearest future on starting experimental activity in this direction of research. It was not mathematics. We discussed the way of organization of big scientific groups that would work not as a simple sum of researchers but as a coherent well-organized team. We both agree that scientific productivity without loss of quality of research could be greatly increased in this case. However, it was clear that one has to introduce many organization modifications. For example, Walter suggested regular communication via, perhaps, even written directives from the group leader to his subordinates. Do not know, whether he managed to materialize this idea.

To organize a big and stable group cemented not administratively, but by interest to work under a respected group leader, with members being motivated by love to science, was a challenge. The members of such a group have to be able to sacrifice, at least to some extent, their individuality to the love of collective work. This was a challenge both in the USSR with its kolkhoz traditions and habits, and in the Western individualistic community.

We understood that collective work would increase the number of co-authors in any publication. Only by including many co-authors, one could organize a brainstorm that would increase the effectiveness of the research considerably. In such an approach, almost each participant of the brainstorm would become an author. Of course, in case of failure, such an approach could lead to friction among participants and eventually to decay of the group. The organization of collective research with coherent effect belongs to a weakly developed domain of organization of creativity, addressing at first the fundamental issues: whether this is possible at all, and could it be useful. If we come to two "yes" than one has to find the way "how" to implement all this in the real life.

The great experiment of Walter permitted to give very impressive positive answers to all these questions. Walter managed to organize such a group that many people considered as impossible and even counter-productive. I remember numerous discussions between prominent physicist in that time USSR on the subject that was ignited by Walter's visit. Not only I, but also my considerably older and experienced colleagues were very much impressed by his deep intuition as a physicist and the broad range of his ideas. Let me remind you that at that time he was only about thirty years old. My discussions about him with older colleagues, including A. Migdal and Ya. Smorodinsky demonstrated that he was accepted as a promising scientific leader, however all but Migdal were sure that his experiment on new way of organizing scientific research would fail. Fortunately, the big majority proved to be wrong.

With time, Walter managed to build a great pyramid. A number of his former students and group members became prominent scientists. All this is an outstanding achievement.

During our memorable conversations, Walter described interesting formal relations that he planned to have with his students. Twenty years later, I had a privilege almost a year to be a witness of such relations and found that they were useful and fruitful.

During these discussions, Walter has made his prediction of Great Reunification or Great Merge (do not mix it with Great purge!). In spite of its importance, this prediction was never published and even not widely publicized, so, perhaps, I am the only one who can confirm now that indeed such a prediction took place. It was experimentally confirmed almost a quarter of a century later, in 1989. I have in mind the Reunification of Germany. Already during our first discussions, we touch political issues. When I mentioned the now late GDR as an independent German state, Walter answered sharply, with a knock by his palm upon the table: "It can be and will be only one German state that unites all Germans". This was a prophetic prediction of historic process that seemed to me simply impossible giving the power of the USSR and the East block at that time. Let me mention that during this conversation I was looking with fear at the telephone on the table in my room that could inform "interested" what we dare to discuss!

More or less by chance, I have visited Frankfurt in the spring of 1989. My hosts were Greiner and G. Soff. Walter suggested me to spend in Frankfurt a longer period and nominated for Humboldt research prize. I have received it in 1990 and spent in Germany five months in 1991 and five – in 1992. This prize modified my life powerfully and positively in many aspects. I was very much impressed by the enthusiastic scientific atmosphere that I felt during all my stay. I acquired a number of new friends, and not only among scientists, both Germans and foreigners. This stay was very fruitful for me and I am grateful to Walter for this.

When Walter invited me to come, some of my friends predicted that I would have a hard time since my host will force me to co-authorship. How they were wrong! Walter and I had many discussions, common interests in a number of directions. It appeared that we have almost the same views on a number of scientific (as well as non-scientific) problems, but did not presented even a single common talk or submitted a publication. I did not feel a smallest push toward co-authorship. I attended very many seminars at the Greiner Institute. What I saw on the weekly bases was that in all presented at seminars talks, later published as articles with the speaker as the first author and Walter as the last, in all seen by me cases, he was the main source of the principal idea of the work.

Walter indirectly, by himself as a positive example, taught me how to conduct a seminar to make it a source of inspiration to the speaker. He managed always to concentrate on something interesting and useful in the presented work, finding a reason what for to thank the speaker. His politeness during pallavers, careful praising each speaker, even if the presented work or the talk itself from my point of view did not deserve anything but strong criticism, amazed me. Literally, his motto was "Do not forget to say "thank you" not only to those who are in power and above you, but also to those who depend upon you". For me it was quite different from what I saw in the USSR, particularly among members of the famous Landau school, and what I followed conducting my own seminar during more then twenty years.

Several years after we met for the first time, he wrote me that moves to atomic physics, having in mind the process of generating of positrons from vacuum under the action of strong, so-called critical, electric field. As far as I know, he was one of the motors, if not the strongest one, behind the idea to create such a field in heavy ion collisions. This was one of the most important ideas implemented in GSI. Although difficulties with separating "overcritical" positrons from that created because of collision process itself proved to be impossible to overcome, the creation of GSI became a great and long lasted success.

During my stay, Walter several times turned to the problem of Jewish Catastrophe under the Nazi regime. He literally felt personal responsibility for the tragedy that happened, and wanted at least somehow to repair the damage not only to Jewish people but also to German science and culture. It was not only words, but also some concrete actions, e.g. organization of long stays and collaboration with Judah Eisenberg, and convincing Walter Meyerhof, son of Otto Meyerhof, Nobel prize winner, to come back to Germany at least for a shot stay. Note that Meyerhof promised never visit Germany after he and his father fled this country after Nazis came to power. Greiner was several times in Israel and became honorary professor of the Tel Aviv University. In 1999 he has a plan to divide the International Symposium *Nuclear Matter-hot and dense*, dedicated to the memory of J. Eisenberg, into two parts, in Tel Aviv and Bethlehem. However, information about planned terrorist attack in Bethlehem prevented materialization of this plan, and all the conference was in Tel Aviv.

Walter was deeply interested in understanding what is going on in Soviet Union during the period that was called perestroika. He took close to his heart the unexpected hardships that almost overnight made the life of Soviet scientist so difficult. He felt not only abstract co-passion but also actively helped in establishing new relations between German and USSR scientists. At first, I want to mention that he managed to bring to Germany for long stays a number of that time young Soviet scientists, to mention only a few of them, M. Gorenshtein, I. Mishustin and my former students A. Soloviev and A. Korol. They were well accepted by the Greiner Institute at the Frankfurt University and after at FIAS that was organized to large extent by Walter.

His help included not only invitations to a number of Soviet scientists to spend in Germany a considerable period. Together, we wrote letters to Riesenhuber, to some other people in the Ministry of science and Education with the aim that could be formulated as "Save Soviet Science". Having in mind much broader cooperation than science, he connected me to the Deutsche Bank Vice-President.

Walter's kindness and attentiveness had no limits and not only I, but a number of people of the former USSR are grateful to him forever.

References

1. H.W. Kroto, J.R. Heath, S.C. O'Brien, R.F. Curl, R.E. Smalley, Nature **318**(6042), 162–163 (1985)
2. J.R. Heath, S.C. O'Brien, Q. Zhang, Y. Liu, R.F. Curl, H.W. Kroto, F.K. Tillel, R.E. Smalley, J. Am. Chem. Soc. **107**, 7779 (1985)
3. M.Ya. Amusia, Chem. Phys. **414**, 168–175 (2013)
4. V.K. Dolmatov, J.L. King, J.C. Oglesby, J. Phys. B **45**, 105102 (2012)
5. M.Ya. Amusia, L.V. Chernysheva, V.G. Yarzhemsky, *Handbook of Theoretical Atomic Physics, Data for Photon Absorption, Electron Scattering, Vacancies Decay* (Springer, Berlin, 2012), pp. 1–806
6. M.Ya. Amusia, *Atomic Photoeffect* (Plenum Press, New York, London, 1990), pp. 1–303
7. M.Ya. Amusia, L.V. Chernysheva, E.Z. Liverts, Phys. Rev. A **80**, 032503-1-12 (2009)
8. M.Ya. Amusia, A.S. Baltenkov, Phys. Let. A **360**, 294–298 (2006)
9. R.K. Yoo, B. Ruscic, J. Berkowitz, J. Chem. Phys. **96**, 911 (1992)
10. M.Ya. Amusia, A.S. Baltenkov, L.V. Chernysheva, JETP **134**, 2(8), 221–230 (2008)
11. M.Ya. Amusia, A.S. Baltenkov, Phys. Rev. A **73**, 063206 (2006)
12. M.Ya. Amusia, L.V. Chernysheva, JETP Lett. **101**(7), 503–506 (2015)
13. M.Ya. Amusia, L.V. Chernysheva, JETP Lett. **103**(4), 260–264 (2016)
14. V.K. Dolmatov, M.Ya. Amusia, L.V. Chernysheva, Phys. Rev. A **95**, 012709 (2017)
15. M.Ya. Amusia, N.A. Cherepkov, L.V. Chernysheva, S.G. Shapiro, J. Phys. B: At. Mol. Phys. **9**(17), L531–L534 (1976)
16. M.Ya. Amusia, L.V. Chernysheva, JETP Lett. **106**(1), 1–6 (2017)

Power Flows in Complex Renewable Energy Networks

Mirko Schäfer, Bo Tranberg and Martin Greiner

Abstract The transition towards a sustainable, clean energy infrastructure is strongly dependent on the efficient integration of the fluctuating renewable power generation from wind and solar. With a focus on power flows, in this contribution we review complex renewable energy networks as a weather-data driven modelling approach to a highly renewable future electricity system. The benefit of cross-border transmission between the European countries in such a scenario is discussed, taking into account the role of spatial coarse-graining for the modelling results. Flow allocation methods are presented as a tool set to analyse the spatio-temporal flow patterns and to allocate both transmission and generation capacity costs.

1 Introduction

In line with the Paris agreement, in their 2030 Climate and Energy Framework the European Union has committed to reduce Greenhouse gas emissions by 40% compared to 1990 levels [1]. In order to reach this goal and also more ambitious future targets, our energy system has to undergo a fundamental transformation. All over Europe we witness the first steps towards such a sustainable future system: Centralised power generation from fossil and nuclear fuels by now already has to some degree been replaced by decentralised renewable generation, primarily from onshore wind and solar PV power generation. This growing integration of heterogeneously distributed, fluctuating renewable power generation, and the foreseeable stronger

M. Schäfer (✉)
INATECH, University of Freiburg, Emmy-Noether-Strasse 2,
79110 Freiburg, Germany
e-mail: mirko.schaefer@inatech.uni-freiburg.de

B. Tranberg
Ento Labs ApS, Inge Lehmanns Gade 10,
8000 Aarhus C, Denmark

M. Schäfer · B. Tranberg · M. Greiner
Department of Engineering, Aarhus University, Inge Lehmanns Gade 10,
8000 Aarhus C, Denmark

J. Kirsch et al. (eds.), *Discoveries at the Frontiers of Science*,
FIAS Interdisciplinary Science Series, https://doi.org/10.1007/978-3-030-34234-0_17

coupling of the electricity, heating and transport sector represent a major technological, economic and societal challenge.

Large-scale models of the energy system support the transition towards a sustainable energy system by providing policy guidelines for efficient infrastructure layouts, optimal operational concepts, and the corresponding economic framework. Ever more detailed models allow to resolve the necessary spatio-temporal scales, for instance with respect to load patterns or the weather-dependent renewable generation from wind and solar, represent different technologies, or incorporate market and investment dynamics driving the energy system [2–6].

Applied Theoretical Physics can contribute another perspective to this challenge. By focussing on simplified, more abstract models, this approach aims to identify the fundamental relationships and key dynamics of complex systems, and to develop and test new concepts and methods. The research scheme which we term 'complex renewable energy networks' addresses the analysis and design of energy systems with a high share of renewable generation. The renewable generation as well as the electricity consumption in these models is based on realistic data, whereas the system representation (constituents and network structure), the backup power dispatch and the power flow equations in general are simplified. In this context, different studies for instance have addressed the optimal mix between solar and wind power generation [7], the influence of storage and an associated phase transition [8, 9], the benefit of power transmission [10, 11], the role of backup flexibility classes [12], or principal mismatch patterns between generation and load [13].

A cost efficient layout for a future sustainable electricity system will make strong use of locations with good resource qualities for renewable generation, that is for instance onshore wind in countries like Denmark bordering the North Sea, and solar PV in South European countries [14, 15]. Long-range power transmission will connect this capacity to the load centres, and will help to smooth out the spatio-temporal fluctuations inherent in weather-dependent renewable generation from wind and solar [15]. From an economic perspective, the increasing integration of European energy markets will intensify the cross-border trade of electricity, with non-trivial connections between the commercial flow of transactions and the corresponding physical power flows in the intermeshed European transmission grid. Due to the geographical heterogeneity of future cost-efficient infrastructure layouts, the location of power generation will often be at a large distance to the location of its consumption, with the corresponding power flow patterns mirroring the variability in the underlying fluctuating renewable generation patterns. This increasing interconnectedness underlines the fundamental importance of the network perspective on large-scale energy systems, not only with respect to the usage and possible expansion of transmission capacities, but also concerning the distribution of costs in particular during the build-up of a sustainable future system.

In this contribution we focus on some aspects of power flows in complex renewable energy networks. In Sect. 2 we briefly review the general modelling framework, with the subsequent Sect. 3 focussing on the representation of power flows. Section 4 discusses the influence of the spatial scale for transmission capacity measures in coarse-grained electricity system models. Flow allocation methods are reviewed in the subsequent Sect. 5. The article is concluded by an outlook in Sect. 6.

2 Complex Renewable Energy Networks

We consider a networked electricity system model in which the node n represents a given region, for instance a whole country or a cell in a segmentation of a country. We demand that for every node n at any time t the following balancing condition is fulfilled:

$$G_n^R(t) - L_n(t) = P_n(t) + B_n(t) \,. \tag{1}$$

Here $G_n^R(t)$ denotes the renewable generation and $L_n(t)$ the load, while $P_n(t)$ refers to the nodal import or exports, and $B_n(t)$ to the nodal balancing. We emphasise that the left side of this equation is data-driven: The aggregated load per node $L_n(t)$ at hour t is given by historical electricity consumption data, which is for instance provided by the European Transmission System Operators [7]. The renewable generation $G_n^R(t)$ is based on data from a Renewable Energy Atlas, which converts historical weather data into a corresponding potential wind and solar generation time series [7, 16]. These time series are scaled in such a way that the renewable generation covers *on average* a share γ_n of the load in each node:

$$\langle G_n^R \rangle = \langle G_n^W \rangle + \langle G_n^S \rangle = \gamma_n \langle L_n \rangle \,. \tag{2}$$

Here G_n^W denotes the wind power generation, whereas G_n^S gives the solar power generation from photovoltaics. The parameter γ_n represents the renewable penetration—the choice $\gamma_n = 1$ corresponds to the case in which node n on average covers 100% of its load from wind and solar power generation. A second set of parameters α_n constitutes the mix between wind and solar:

$$\langle G_n^W \rangle = \alpha_n \langle G_n^R \rangle \,, \tag{3}$$

$$\langle G_n^S \rangle = (1 - \alpha_n) \langle G_n^R \rangle \,. \tag{4}$$

The mismatch $\Delta_n(t)$ denotes the instantaneous difference between the renewable generation $G_n^R(t)$ and the load $L_n(t)$ at time t:

$$\Delta_n(t) = G_n^R(t) - L_n(t) \,. \tag{5}$$

According to Eq. (1), at every time step this mismatch has to be balanced by a combination of nodal power imports or exports $P_n(t)$, and nodal balancing $B_n(t)$. The latter represents flexible backup power generation $G_n^B(t) = -\min(B_n(t), 0)$ when $B_n(t) < 0$, or curtailment of excess renewable power generation $C_n(t) = \max(B_n(t), 0)$ when $B_n(t) > 0$. The nodal power injection $P_n(t)$ is associated with exports $P_n^+(t)$ for $P_n(t) > 0$, or with imports $P_n^-(t) = -P_n(t)$ for $P_n(t) < 0$.

For a given time series of renewable generation and load, different balancing schemes (i.e. combinations of nodal balancing power and imports and exports) are consistent with the balancing equation (1). With the abbreviation $\Delta(t) = \sum_n \Delta_n(t)$ for the global mismatch, the result of the balancing scheme can be written as $B_n(t) = s_n(t)\Delta(t)$:

$$P_n(t) = \Delta_n(t) - B_n(t) = \Delta_n(t) - s_n(t)\Delta(t) \, , \tag{6}$$

For the coefficients $s_n(t)$ it has to hold $\sum_n s_n(t) = 1$, such that the set of nodal imports and exports is balanced:

$$\sum_n P_n(t) = \sum_n [\Delta_n(t) - s_n(t)\Delta(t)] = \Delta(t) - \left[\sum_n s_n(t)\right]\Delta(t) = 0 \, . \tag{7}$$

In various studies the simplified balancing scheme denoted as 'synchronised balancing' has been used [17]. This approach distributes the backup proportionately to the load of the various nodes, that is

$$B_n(t) = s_n(t)\Delta(t) = \frac{\langle L_n \rangle}{\sum_k \langle L_k \rangle}\Delta(t) \, . \tag{8}$$

Note that here the coefficient s_n does not depend on time. In contrast, for a balancing scheme which for instance minimises the total power flows in the system, the coefficients $s_n(t)$ will be time-dependent and have to be determined from an optimisation algorithm [17].

For a given network with load and renewable generation time series, and a choice of parameters γ_n, α_n and balancing scheme, the infrastructure needs of the system can be assessed. The nodal backup energy is defined as the time average of the backup power generation:

$$E_n^B = \langle G_n^B \rangle \, . \tag{9}$$

The backup capacity is given as the 99% quantile of the backup generation:

$$0.99 = \int_0^{\mathscr{K}_n^B} dG_n^B \, p_n(G_n^B) \, . \tag{10}$$

Here $p_n(G_n^B)$ denotes the distribution of backup generation events at node n. The definition of the backup capacity excludes extreme events that are assumed to be covered by emergency equipment or flexible demand outside the model. Analogously, the transmission capacity $\mathscr{K}_l^T$ is defined as the 99% quantile of the power flow distribution on link l:

$$0.99 = \int_{-\mathscr{K}_l^T}^{\mathscr{K}_l^T} dF_l \, p_l(F_l) \, , \tag{11}$$

Here we assume identical transmission capacity in both directions of the links. The total backup/transmission capacity is calculated as the sum over the nodal/link capacities:

$$\mathcal{K}^B = \sum_n \mathcal{K}_n^B \tag{12}$$

$$\mathcal{K}^T = \sum_l \mathcal{K}_l^T . \tag{13}$$

A measure $\mathcal{T}$ of the transmission system costs is defined as the weighted sum

$$\mathcal{T} = \sum_l d_l \mathcal{K}_l^T , \tag{14}$$

which takes into account the length d_l of the link l.

Several studies have applied the concept reviewed in this section to a simplified network representation of the European electricity system with countries as nodes, and links between the nodes representing the corresponding cross-border transmis-

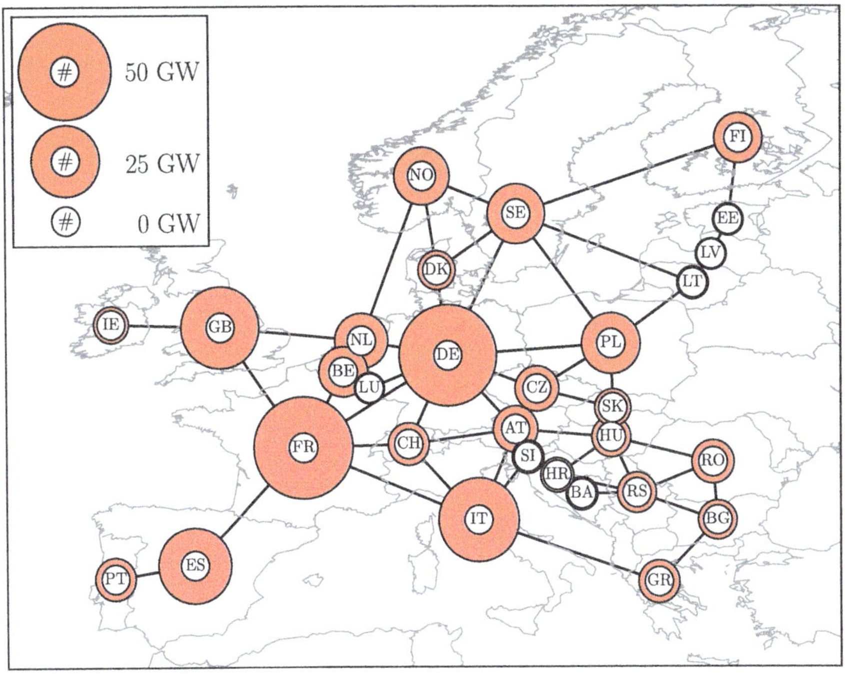

Fig. 1 A simplified network representation of the European electricity system with countries as nodes. The circles are drawn with an area proportional to the country's average load (for this figure the average load refers to the years 2010–2014)

sion capacity (see Fig. 1). For the original data set the renewable generation time series is based on eight years of hourly weather data from 2000 to 2007 with a spatial resolution of $50 \times 50\,\text{km}^2$, with the resulting potential generation from wind and solar spatially aggregated to country level [7]. The load time series per country is obtained with the same temporal resolution from data provided by the transmission system operators [7]. Other studies have applied this approach to the US electricity system [18, 19], or have studied a global renewable electricity grid [20].

3 Power Flows

On transmission level, the electricity grid in general is built as an AC network. Although the power flows accordingly follow the full AC power flow equations, for stable network operation these equations can be linearised using the so-called DC approximation [21, 22]. For an injection pattern P_n represented as a vector $\mathbf{P}$, the linearised DC power flow equations read

$$P_n = \sum_m B_{nm} \Theta_m \tag{15}$$

or $\mathbf{P} = \mathbf{B}\Theta$, where Θ is the vector of voltage phase angles, and $\mathbf{B}$ is the nodal admittance matrix with entries

$$B_{nm} = \begin{cases} -b_{nm} & n \neq m \\ \sum_m b_{nm} & n = m \end{cases} . \tag{16}$$

In this definition b_{nm} are the line susceptance of the link between node n and m. Assuming for simplicity unit line susceptances $b_{nm} = 1$, the nodal admittance matrix $\mathbf{B}$ corresponds to the network Laplacian $\mathbf{L}$, and the voltage phase angles can be calculated as $\Theta = \mathbf{L}^{\dagger}\mathbf{P}$ with $\mathbf{L}^{\dagger}$ the Moore-Penrose pseudo inverse of $\mathbf{L}$. The vector $\mathbf{F}$ comprising the power flows F_l on the links l of the network is then calculated as $\mathbf{F} = \mathbf{K}^T\Theta$, where $\mathbf{K}^T$ is the transposed incidence matrix with

$$K^T_{ln} = \begin{cases} 1 & \text{if link } l \text{ starts at node } n \, , \\ -1 & \text{if link } l \text{ ends at node } n \, , \\ 0 & \text{otherwise} \, . \end{cases} \tag{17}$$

The resulting linear relation between the injection pattern P_n and the power flow pattern F_l can be expressed as

$$F_l = \sum_n H_{ln} P_n \, , \tag{18}$$

where H_{ln} are the entries of the matrix $\mathbf{H} = \mathbf{K}^T \mathbf{L}^\dagger$ of power transfer distribution factors (PTDF matrix). In general form the PTDF matrix reads $\mathbf{H} = \Omega \mathbf{K}^T \mathbf{B}^\dagger$, where the diagonal matrix Ω contains the line susceptances on the links l.

From the injection pattern $P_n(t)$ at time step t of the time series thus follows via the power flow equations the corresponding flow pattern $F_l(t)$. The time series of flow patterns then determines via the flow distribution $p_l(F_l)$ the transmission capacity needs of the system (see Eq. (11)). Note that this definition assumes unconstrained transmission, with the corresponding infrastructure measures determined retroactively. In case of restricted transmission (expressed as a set of limits $|F| \leq F^{max}$ to the line flows), the balancing mechanism introduced in the previous section has to be extended to take into account the corresponding constraints.

In Ref. [10] the transmission needs for a fully renewable European power system with $\gamma_n = 1$ for all countries have been assessed. It has been shown that unconstrained transmission between the European countries reduces the necessary amount of balancing energy $E^B = \sum_n E_n^B$ from 24 to 15% of the total load $\sum_n \langle L_n \rangle$. The corresponding power flows require cross-border transmission capacities which are approximately 11.5 times larger than today's values. Nevertheless, it has been found that 98% of this benefit is already provided by interconnector capacities 5.7 times larger than the current values. This benefit of transmission has been confirmed using a more detailed model [15], and also has been studied in the context of the transition period towards a highly renewable system [11]. It should be emphasised that these studies use coarse-grained network models with countries as nodes, which tend to overestimate the transmission capacity needs of the system due to two factors: Firstly, on this aggregation level, the interconnector transmission infrastructure is represented by the net transfer capacities (NTC) provided by the transmission system operators, which are the capacities available to the market. Due to inefficient market coupling, the NTCs are in general considerably lower than the real physical thermal capacities [23]. Secondly, the transmission infrastructure between different countries is weaker than in the countries themselves, which leads to an overestimation of the transmission needs projected to the total infrastructure in coarse-grained models [24]. These issues indicate the need to understand the role of spatial coarse-graining in electricity system modelling, which will be further discussed in the next section.

4 Scaling of Transmission Capacities in Coarse-Grained System Models

A highly renewable electricity system is influenced by several spatial scales. Both wind and solar power generation depend on weather patterns, which are subject to both temporal and spatial correlations, and which depend on the geographical conditions of the system. This generation side faces a heterogeneous spatial distribution of consumers and thus load patterns for the demand side. The power grid as the network connecting generation and load also operates on different spatial scales, from the local distribution level to the long-range power flows on the high-voltage transmission level. Due to lack of data or computational limits, models of the electricity system in general are not able to cover all these factors in detail by using a fine spatial resolution, but rather have to coarse-grain the information below a certain spatial scale. In particular, renewable generation patterns are smoothed by such a spatial aggregation, whereas the non-consideration of the lower-level network topology might hide bottlenecks affecting the efficient working of the system. Only recently the influence of the level of spatial coarse-graining on the modelling results for a cost-optimised European electricity system has been addressed. Whereas Ref. [24] uses a numerical approach, Ref. [25] addresses this problem from a complex networks perspective. Using different levels of approximations it has been shown that for complex renewable energy networks the total transmission capacity and cost (see Eqs. (12) and (14)) under spatial coarse-graining scale as

$$\mathcal{K}_M^T \approx \sqrt{\frac{M \ln M}{N \ln N}} \mathcal{K}_N^T \approx \left(\frac{\langle d \rangle_M}{\langle d \rangle_N} \right)^{-1} \sqrt{1 - \frac{2 \ln \left[\frac{\langle d \rangle_M}{\langle d \rangle_N} \right]}{\ln N}} \mathcal{K}_N^T , \tag{19}$$

$$\mathcal{T}_M \approx \sqrt{\frac{\ln M}{\ln N}} \mathcal{T}_N \approx \sqrt{1 - \frac{2 \ln \left[\frac{\langle d \rangle_M}{\langle d \rangle_N} \right]}{\ln N}} \mathcal{T}_N . \tag{20}$$

In these equations N (M) stands for the number of nodes in a detailed (coarse-grained) representation of the system, and $\langle d \rangle$ denotes the average link length in the corresponding network. In particular the scaling of the transmission infrastructure cost measure is interesting, indicating only a weak dependence on the spatial scale of the system. Although the derivation in Ref. [25] ignores most of the intricacies inherent in highly renewable electricity system (for instance correlations in the generation patterns, or the influence of different dispatch schemes), such results obtained using methods from Applied Theoretical Physics can provide important analytical insights complementing the purely numerical approach applied in more detailed models.

5 Flow Allocation Methods

The complex fluctuating power flow patterns in renewable energy networks emerge from the collective set of nodal imports and exports. Although these flow patterns directly follow from the underlying injection patterns through the power flow equations, for a given flow pattern in general it is difficult to trace back the specific nodal influence on a particular line flow. Recently the structure of power flow patterns and the corresponding network usage in renewable energy networks have been addressed using methods of flow allocation [26–29]. Classically, such allocation methods attribute shares of the power flow on a transmission line to the corresponding nodal injections, thus allowing to dissect the global pattern into contributions associated with the different network nodes. The application of respective techniques has for instance been discussed in the context of loss allocation schemes [30] or the European inter transmission system operator compensation mechanism [31]. Flow tracing [32, 33], an algorithm which follows the nodal injection downstream through the network, has been applied to a simplified model of a highly European electricity system in [27]. Here also a novel statistical usage measure has been introduced to account for the fluctuating injection and flow patterns. The method has been reformulated to account for instance for different generation types in [28, 34]. If we decompose the nodal export P_n^+ into contributions associated with different labels α, that is

$$P_n^+ = \sum_\alpha P_{n,\alpha}^+ , \tag{21}$$

the conversation of α-flow in a node n reads

$$P_{n,\alpha}^+ + \sum_m q_{m,\alpha} F_{m\to n} = q_{n,\alpha} P_n^- + \sum_k q_{k,n} F_{n\to k} . \tag{22}$$

Here $F_{m\to n}$ denotes the power flow from node m to node n, and the principle of proportional sharing assumes that all flow components are perfectly mixed in a node and thus the mix $q_{n,\alpha}$ applies to the out-flow from node n and to the imports P_n^-. Equation (22) can be solved both iteratively or as a matrix equation, yielding the composition of nodal imports and power flows on the links of the network. This tracing method in particular connects the location of consumption of electricity with the location of its generation, which allows a flow-based allocation of operational and capital costs [35]. In [29] the tracing approach has been compared to an analysis based on a decomposition method, which feeds elementary injections patterns (for instance associated with commercial trades) into the power flow equations to obtain a dissection of the global power flow patterns. It has been shown that while for individual time steps different methods yield significantly different results, aggregating measures over long time series show only slightly varying results. Figure 2 visualises power flows in a simplified model of a highly renewable electricity system at a certain time step, and their decomposition according to the two allocation method described above.

6 Conclusion and Outlook

Inspired by Applied Theoretical Physics, complex renewable energy networks represent a weather-data driven approach to modelling a highly renewable future electricity system. This research scheme puts the focus on the key relations between the spatio-temporal scales and correlations in the weather and generation data, and their relation to the underlying network structure representing the power grid. Using this frame-

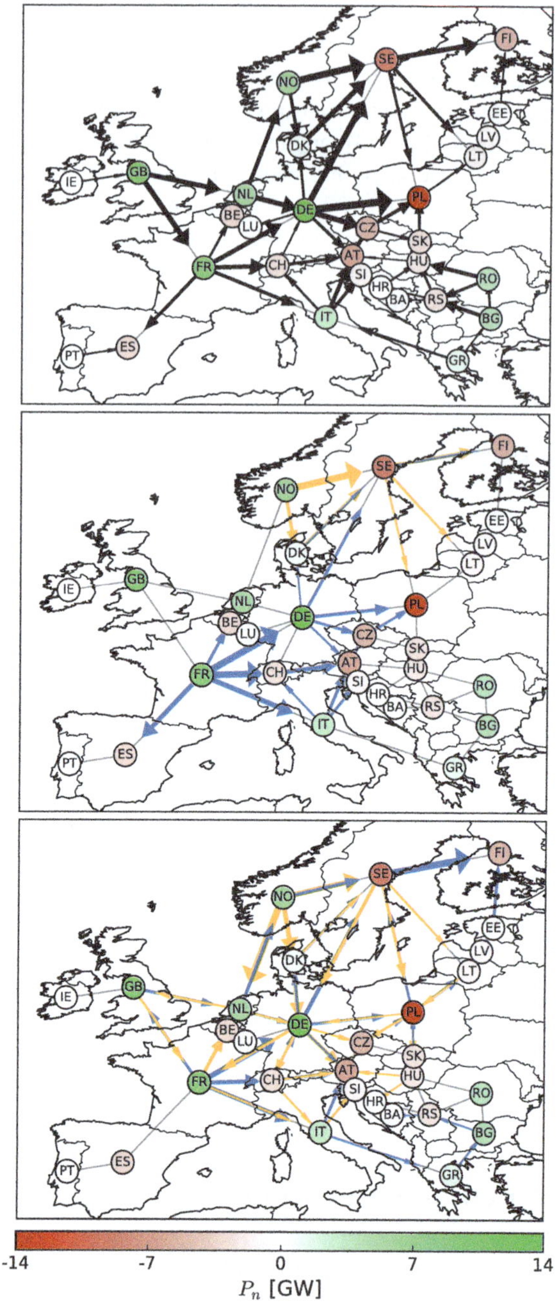

Fig. 2 Three figures showing power flows in a simplified model of the European electricity system for the same exemplary hour. The top panel shows the total power flows in the system. The two bottom panels show the partial flows assigned to Norway (yellow) and France (blue) based on the flow tracing method (middle panel) and the partial flows assigned based on the random contract decomposition (bottom panel). Nodes are coloured according to their export (green) and import (red) during this hour. The size of the arrows for partial flows are scaled to the maximal partial flow of the respective country during the given hour. For clarity only partial flows larger than a small threshold of 5% of the largest partial flow caused by Norway or France are shown

work amongst other results it has been shown that in scenarios with a high share of renewable power generation, transmission between the European countries provides a significant system benefit in terms of reduced backup energy and capacity. A realistic interpretation of the corresponding necessary transmission expansion requires to take into account the spatial coarse-graining of the system model. Flow allocation methods can provide a deeper understanding of the flow patterns in renewable energy networks, serving as a tool set for a flow-based allocation of nodal system costs.

A future sustainable energy infrastructure will not be built on a 'green field', but has to emerge from today's system. One major current research challenge is to investigate this transition. Given today's liberalised regulatory framework, the system evolution results from the complex interactions between technological progress, political, societal and economic boundary conditions, and the individual consumer and investment decisions. Accordingly, there is no central omnipotent planning authority realising a design which has been identified to be optimal with respect to certain criteria—a concept which is often employed in energy system modelling. Integrating this evolutionary system dynamics into complex renewable energy networks will not only advance the field of energy system modelling and stimulate fundamental research on the dynamics in complex systems, but will also contribute new insights regarding the design of and in particular the transition towards a future renewable sustainable system.

References

1. European Commission. A policy framework for climate and energy in the period from 2020 to 2030 (2014)
2. H.-M. Henning, A. Palzer, A comprehensive model for the German electricity and heat sector in a future energy system with a dominant contribution from renewable energy technologies. Part I: methodology. Renew. Sustain. Energy Rev. **30**, 1003–1018 (2014)
3. Y. Scholz, Renewable energy based electricity supply at low costs: development of the REMix model and application for Europe. Ph.D. thesis, University of Stuttgart (2012)
4. F.U. Leuthold, H. Weigt, C. von Hirschhausen, A large-scale spatial optimization model of the european electricity market. Netw. Spat. Econ. **12**(1), 75–107 (2012)
5. S. Pfenninger, A. Hawkes, J. Keirstead, Energy systems modeling for twenty-first century energy challenges. Renew. Sustain. Energy Rev. **33**, 74–86 (2014)
6. T. Brown, J. Hösch, D. Schlachtberger, PyPSA: python for power system analysis (2017), arXiv:1707.09913 [physics]
7. D. Heide, L. von Bremen, M. Greiner, C. Hoffmann, M. Speckmann, S. Bofinger, Seasonal optimal mix of wind and solar power in a future, highly renewable Europe. Renew. Energy **35**(11), 2483–2489 (2010)
8. D. Heide, M. Greiner, L. von Bremen, C. Hoffmann, Reduced storage and balancing needs in a fully renewable European power system with excess wind and solar power generation. Renew. Energy **36**(9), 2515–2523 (2011)
9. T.V. Jensen, M. Greiner, Emergence of a phase transition for the required amount of storage in highly renewable electricity systems. Eur. Phys. J. Spec. Top. **223**(12), 2475–2481 (2014)
10. R.A. Rodriguez, S. Becker, G.B. Andresen, D. Heide, M. Greiner, Transmission needs across a fully renewable European power system. Renew. Energy **63**, 467–476 (2014)

11. S. Becker, R.A. Rodriguez, G.B. Andresen, S. Schramm, M. Greiner, Transmission grid extensions during the build-up of a fully renewable pan-European electricity supply. Energy **64**, 404–418 (2014)
12. D. Schlachtberger, S. Becker, S. Schramm, M. Greiner, Backup flexibility classes in emerging large-scale renewable electricity systems. Energy Convers. Manag. **125**, 336–346 (2016)
13. M. Raunbak, T. Zeyer, K. Zhu, M. Greiner, Principal mismatch patterns across a simplified highly renewable european electricity network. Energies **10**(12) (2017)
14. E.H. Eriksen, L.J. Schwenk-Nebbe, B. Tranberg, T. Brown, M. Greiner, Optimal heterogeneity in a simplified highly renewable European electricity system. Energy **133**, 913–928 (2017)
15. D. Schlachtberger, T. Brown, S. Schramm, M. Greiner, The benefits of cooperation in a highly renewable European electricity network. Energy **134**, 469–481 (2017)
16. G.B. Andresen, A.A. Søndergaard, M. Greiner, Validation of Danish wind time series from a new global renewable energy atlas for energy system analysis. Energy **93**, 1074–1088 (2015)
17. R.A. Rodriguez, M. Dahl, S. Becker, M. Greiner, Localized vs. synchronized exports across a highly renewable pan-European transmission network. Energy, Sustain. Soc. **5**, 21 (2015)
18. S. Becker, B.A. Frew, G.B. Andresen, M.Z. Jacobson, S. Schramm, M. Greiner, Renewable build-up pathways for the US: generation costs are not system costs. Energy **81**, 437–445 (2015)
19. S. Becker, B.A. Frew, G.B. Andresen, T. Zeyer, S. Schramm, M. Greiner, M.Z. Jacobson, Features of a fully renewable US electricity system: optimized mixes of wind and solar PV and transmission grid extensions. Energy **72**, 443–458 (2014)
20. M. Dahl, R.A. Rodriguez, A.A. Søndergaard, T. Zeyer, G.B. Andresen, M. Greiner, Infrastructure estimates for a highly renewable global electricity grid, *New Horizons in Fundamental Physics*, FIAS Interdisciplinary Science Series (Springer, Cham, 2017), pp. 333–356
21. A.J. Wood, B.F. Wollenberg, P. Generation, *Operation, and Control* (Wiley, New York, 2012)
22. K. Purchala, L. Meeus, D. Van Dommelen, R. Belmans. Usefulness of DC power flow for active power flow analysis, in *IEEE Power Engineering Society General Meeting*, vol. 1 (2005), pp. 454–459
23. ACER and CEER. Annual report on the results of monitoring the internal electricity and gas markets in 2016. Technical report (2017)
24. J. Hörsch, T. Brown, The role of spatial scale in joint optimisations of generation and transmission for European highly renewable scenarios, in *14th International Conference on the European Energy Market (EEM)*, June 2017
25. M. Schäfer, S. Bugge Siggaard, Kun Zhu, C. Risager Poulsen, M. Greiner, Scaling of transmission capacities in coarse-grained renewable electricity networks. EPL (Europhys. Lett.) **119**(3), 38004 (2017)
26. T. Brown, Transmission network loading in Europe with high shares of renewables. IET Renew. Power Gener. **9**(1), 57–65 (2015)
27. B. Tranberg, A.B. Thomsen, R.A. Rodriguez, G.B. Andresen, M. Schäfer, M. Greiner, Power flow tracing in a simplified highly renewable European electricity network. New J. Phys. **17**(10), 105002 (2015)
28. M. Schäfer, S. Hempel, J. Hörsch, B. Tranberg, S. Schramm, M. Greiner, Power flow tracing in complex networks, *New Horizons in Fundamental Physics*, FIAS Interdisciplinary Science Series (Springer, Cham, 2017), pp. 357–373
29. M. Schäfer, B. Tranberg, S. Hempel, S. Schramm, M. Greiner, Decompositions of injection patterns for nodal flow allocation in renewable electricity networks. Eur. Phys. J. B **90**(8), 144 (2017)
30. A.J. Conejo, J.M. Arroyo, N. Alguacil, A.L. Guijarro, Transmission loss allocation: a comparison of different practical algorithms. IEEE Trans. Power Syst. **17**(3), 571–576 (2002)
31. L.O. Camacho, I.J. Pérez-Arriaga, An assessment of inter-TSO compensation algorithms in the internal electricity market of the european union. Int. J. Electr. Power Energy Syst. **29**(10), 699–712 (2007)
32. J. Bialek, Tracing the flow of electricity. Transm. Distrib. IEE Proc. - Gener. **143**(4), 313–320 (1996)

33. D. Kirschen, R. Allan, G. Strbac, Contributions of individual generators to loads and flows. IEEE Trans. Power Syst. **12**(1), 52–60 (1997)
34. J. Hörsch, M. Schäfer, S. Becker, S. Schramm, M. Greiner, Flow tracing as a tool set for the analysis of networked large-scale renewable electricity systems. Int. J. Electr. Power Energy Syst. **96**, 390–397 (2018)
35. M. Schäfer, L.J. Schwenk-Nebbe, J. Hörsch, B. Tranberg, M. Greiner, Allocation of nodal costs in heterogeneous highly renewable European electricity networks, in *14th International Conference on the European Energy Market (EEM)* (2017)

On Configuration Space, Born's Rule and Ontological States

Hans-Thomas Elze

Abstract It is shown how configuration space, possibly encompassing ordinary spatial structures, Born's rule, and ontological states aiming to address an underlying reality beyond Quantum Mechanics relate to each other in models of Hamiltonian cellular automata.

This paper is dedicated to the memory of Walter Greiner. – Always Walter shared freely his wise suggestions concerning topics worth exploring in physics and beyond, all the way to 'career moves' to be done by his students. Yet, rarely did I follow his fatherly advice. – It happened now and then during the weekly "Palaver" seminars that Walter quickly dismissed foundational or interpretational issues as irrelevant for physics, no matter who dared to mention them. – Yet, quite recently, following my talk at Walter's *F*ranfurt *I*nstitute of *A*dvanced *S*tudy, about new approaches to understand Quantum Mechanics as emerging from something 'intelligible' beneath, he encouraged me strongly to pursue these ideas ...

1 Introduction

It shall *not* be our concern to derive Quantum Mechanics (QM) from somehow physically motivated and more or less parsimonious sets of axioms, such as recent information theoretical *reconstructions* of QM [2, 8–10], nor to propose yet another *interpretation* of QM, of which there are already too many (Copenhagen, Many Worlds, Qbism, ...) to sort out seemingly incompatible aspects [15]. We will explore *deformations of quantum mechanical models*, due to the presence of a finite discrete-

H.-T. Elze (✉)
Dipartimento di Fisica "Enrico Fermi", Universitá di Pisa,
Largo Pontecorvo 3, 56127 Pisa, Italy
e-mail: elze@df.unipi.it

J. Kirsch et al. (eds.), *Discoveries at the Frontiers of Science*,
FIAS Interdisciplinary Science Series, https://doi.org/10.1007/978-3-030-34234-0_18

ness scale l (of length or time, choosing $c = \hbar = 1$, henceforth). And what can be learnt from them regarding basic concepts of QM.

Presently, we shall reconsider *configuration space* and the *Born rule* in relation to the foundational hypotheses introduced recently by Gerard't Hooft, in particular, the existence and relevance of *ontological states* of cellular automata that may give rise to the reality described by physics [11].

Our discussion will refer to the class of *Hamiltonian Cellular Automata* (CA) [4]. See Ref. [6] for a wider perspective on striving for the understanding of QM as an emergent structure, the *raison d'être* for Hamiltonian CA, and numerous references to earlier work.

2 Hamiltonian Cellular Automata—Some Essentials

The *Hamiltonian CA* describe discrete linear dynamical systems that show quantum features—especially, but not only in the continuum limit [4, 5].

These are deterministic classical CA with denumerable degrees of freedom ("bit processors"). The state of such a CA is described by *integer valued* coordinates x_n^α and momenta p_n^α, where $\alpha \in \mathbf{N}_0$ labels different degrees of freedom and $n \in \mathbf{Z}$ successive states. – We mention that a generalization studied earlier consists in introducing also a dynamical discrete time coordinate together with its associated momentum, which must be distinguished from the 'CA clock' time n [4].

Only *finite differences* of variables can play a role in the dynamics here, where *no infinitesimals* or ordinary derivatives are available!

Dynamics and symmetries of such systems are all contained in a suitable *action principle* [4], the consistency of which places severe constraints on its detailed form. Ultimately, this is responsible for the essential *linearity* of the QM models, which result in the continuum limit within this class of CA, as we have shown earlier.

This action principle yields *finite difference* equations of motion:

$$\dot{x}_n^\alpha = S_{\alpha\beta} p_n^\beta + A_{\alpha\beta} x_n^\beta \quad , \tag{1}$$

$$\dot{p}_n^\alpha = -S_{\alpha\beta} x_n^\beta + A_{\alpha\beta} p_n^\beta \quad , \tag{2}$$

with given integer-valued symmetric and antisymmetric matrices, $\hat{S} \equiv \{S_{\alpha\beta}\}$ and $\hat{A} \equiv \{A_{\alpha\beta}\}$, respectively, defining the model under consideration (summation over repeated Greek indices applies, except where stated otherwise). Note that we introduced the notation $\dot{O}_n := O_{n+1} - O_{n-1}$. – This may guide the eye and indicate an analogy with Hamilton's equations in the continuum, suggesting the name *Hamiltonian CA*; however, in distinction to usually first-order derivatives, we encounter second-order finite difference operators here.

This leads to discrete time reversal invariance of the Eqs. (1)–(2); i.e. the equations are invariant under reversal of the updating direction. The state of such a CA can be updated in both directions, $(n \mp 1, n) \rightarrow (n \pm 1)$, given appropriate initial values of its variables.

The pair of equations of motion can be combined into one equation:

$$\dot{\psi}_n^\alpha = -iH_{\alpha\beta}\psi_n^\beta \; , \qquad (3)$$

together with its adjoint, by introducing the self-adjoint "Hamiltonian" matrix, $\hat{H} := \hat{S} + i\hat{A}$, and complex integer-valued (*Gaussian integer*) state variables, $\psi_n^\alpha := x_n^\alpha + ip_n^\alpha$. This obviously looks like the Schrödinger equation, despite involving only Gaussian integer quantities.

The resemblance between Eq. (3) and the Schrödinger equation in the continuum, for a whole class of models, can hardly be accidental. Indeed, we have constructed an invertible map between the discrete equation describing Hamiltonian CA in terms of the variables ψ_n^α and a continuous time equation describing the *same* CA in terms of a complex "wave function" $\psi^\alpha(t)$ [4]. This has been achieved with the help of Sampling Theory [12, 16], introducing the finite discreteness scale l in terms of a bandwidth limit or high-frequency cut-off for all such wave functions.

Thus, we obtain the Schrödinger equation, yet modified to incorporate a power series in higher-order time derivatives with l-dependent coefficients:

$$2\sinh(l\partial_t)\psi^\alpha(t) = -i\hat{H}_{\alpha\beta}\psi^\beta(t) \; . \qquad (4)$$

This implies corresponding modifications of other standard results of QM and may have interesting phaenomenological consequences.

Naturally, known features of QM are reproduced in the continuum limit, $l \rightarrow 0$, of the *deformation* of QM implied here [5, 6].

3 Conservation Laws, Configuration Space and a Glimpse of the Born Rule

There are l-dependent conservation laws in one-to-one correspondence with those of the corresponding quantum mechanical model obtained for $l \rightarrow 0$. With the help of Eq. (3) and its adjoint one easily verifies the theorem:

- For any matrix $\hat{G}$ that commutes with $\hat{H}$, $[\hat{G}, \hat{H}] = 0$, there is a *discrete conservation law*:

$$\psi_n^{*\alpha}G_{\alpha\beta}\dot{\psi}_n^\beta + \dot{\psi}_n^{*\alpha}G_{\alpha\beta}\psi_n^\beta = 0 \; . \qquad (5)$$

  For self-adjoint $\hat{G}$, defined in terms of Gaussian integers, this statement concerns real integer quantities.
- Rearrangement of Eq. (5) gives the related *conserved quantity* $q_{\hat{G}}$:

$$q_{\hat{G}} := \psi_n^{*\alpha}\hat{G}_{\alpha\beta}\psi_{n-1}^\beta + \psi_{n-1}^{*\alpha}\hat{G}_{\alpha\beta}\psi_n^\beta = \psi_{n+1}^{*\alpha}\hat{G}_{\alpha\beta}\psi_n^\beta + \psi_n^{*\alpha}\hat{G}_{\alpha\beta}\psi_{n+1}^\beta \; , \qquad (6)$$

i.e. a complex (real for $\hat{G} = \hat{G}^\dagger$) integer-valued correlation function which is invariant under a shift $n \to n + m$, $m \in \mathbf{Z}$.

- For $\hat{G} := \hat{1}$, the conservation law states a constraint on the state variables:

$$q_{\hat{1}} = 2\mathrm{Re}\ \psi_n^{*\alpha}\psi_{n-1}^{\alpha} = 2\mathrm{Re}\ \psi_{n+1}^{*\alpha}\psi_n^{\alpha} = \mathrm{const}\ . \tag{7}$$

The latter replaces for discrete CA the familiar *normalization* of state vectors in QM, to which it reduces in the limit $l \to 0$. This follows from mapping the discrete to the continuum version by applying Shannon's reconstruction theorem, to which we alluded above [4]. Here, we obtain:

$$\mathrm{const} = q_{\hat{1}} = \mathrm{Re}\ \psi^*(t)\cosh\left[l\frac{\mathrm{d}}{\mathrm{d}t}\right]\psi(t) \tag{8}$$

$$= \psi^{*\alpha}(t)\psi^{\alpha}(t) + \frac{l^2}{2}\mathrm{Re}\ \psi^{*\alpha}(t)\frac{\mathrm{d}^2}{\mathrm{d}t^2}\psi^{\alpha}(t) + \mathrm{O}(l^4)\ , \tag{9}$$

displaying the l-dependent corrections to the continuum limit, namely the normalization $\psi^{*\alpha}\psi^{\alpha} = \mathrm{const}$, which is usually conserved but not in the present case of CA.

In order to interpret the *conserved 'two-time' correlation* function $q_{\hat{1}}$ of Eq. (7), we recall that the greek indices abstractly label different CA *degrees of freedom*. These could relate to an internal space or to localized sites of a spatial structure. However, the entire formalism describing CA and demonstrating their quantum features, so far, is independent of attaching such a traditional meaning to the degrees of freedom. This observation is valid for multipartite CA as well [5, 6]. In this sense, the *primacy of configuration space* appears naturally here.[1]

This will not lead us to immediate practical consequences or new theoretical insights, however, we speculate that the physics of matter *in* space or *in* spacetime and the physics *of* space or *of* spacetime itself emerge together from underlying discrete structures and CA like processes (possibly of a kind totally unexpected). Loosely speaking, some evolving complex aggregates of CA degrees of freedom will show up as matter and some as space(time) and must be linked in intricate ways.

Experiencing effects of these phenomena only at scales some nineteen or more orders of magnitude away from the Planck scale, say, QM plays the role of an effective description that has been found most successful when applied to atomic or subatomic particles and the forces that influence their motion. This should include, in principle, the description of macroscopic amounts of solid, liquid, or gaseous matter.

However, since the early days of QM doubts shroud the beauty of this picture where it is supposed to reflect, within QM, the whole process of a measurement undertaken in a laboratory or elsewhere (let alone the alive physicist involved). There has been a puzzling gap in understanding this *measurement problem*, unless one is

[1] Mathematically speaking, the Gaussian integer wave functions ψ, with components ψ^{α}, $\alpha \in \mathbf{N}_0$, can be seen as elements of a linear space endowed with an integer-valued scalar product, $\psi^*\psi := \psi^{*\alpha}\psi^{\alpha}$, i.e. a unitary space. Taking its incompleteness into account, the space of states can be classified as a pre-Hilbert module over the commutative ring of Gaussian integers [5].

willing to undersign one of the available "interpretations" of QM which eliminate this puzzle, at the expense of introducing others [15] (for an overview and proposal of a solution, see [1]). We shall come back to this issue when discussing ontological CA states.

Besides that Eq. (7) replaces the normalization of state vectors, a necessary ingredient of the *Born rule* in QM, we can rephrase its content by relating it to a *counting procedure* as follows.

Recall that $\psi_n^\alpha := x_n^\alpha + ip_n^\alpha$. Thus, for a pair of successive states labelled by n and $n+1$ and for each degree of freedom α, there is a pair of integer components x_n^α and p_n^α that may be called the (numbers of) x^α- and p^α-*initials*, respectively. Correspondingly, the integer components x_{n+1}^α and p_{n+1}^α are called the (numbers of) x^α- and p^α-*finals*, respectively. This includes the possibility of zero or missing (i.e. negative) numbers of initials or finals. Summing the product of x^α-initials and -finals and the product of p^α-initials and -finals defines the number of α-*links*, L_α:

$$L_\alpha := x_{n+1}^{*\alpha} x_n^\alpha + p_{n+1}^{*\alpha} p_n^\alpha \quad \in \mathbf{Z} \ , \tag{10}$$

dropping the summation convention henceforth. Then, the total number of links is given by $L := \sum_\alpha L_\alpha$ and Eq. (7) states:

$$L = q_{\hat{1}}/2 = \text{const} \ , \tag{11}$$

i.e., the *link number is conserved*. – The effect of the CA evolution according to the equations of motion (1) and (2) is to change or redistribute the numbers of initials and finals over the available degrees of freedom, while keeping the total number of links constant.[2]

The relative *weights*, $w_\alpha := L_\alpha/L$ ($L \neq 0$), with $\sum_\alpha w_\alpha = 1$, generally, are not confined to $[0, 1]$. However, they correspond uniquely to the QM probabilities $p_\alpha := \psi^{*\alpha}(t)\psi^\alpha(t)/\psi^*\psi$, since $w_\alpha \rightarrow p_\alpha$ for $l \rightarrow 0$, in accordance with Eqs. (7)–(9).[3] – In distinction, the case $L = 0$ does not allow a meaningful continuum limit. We will encounter an example and its interpretation in Sect. 4, introducing *ontological states*.

Thus, we recover attributes of the *Born rule*. Of course, our description of deterministic Hamiltonian CA, so far, says as little as quantum theory about the origin of the *randomness of experimental outcomes*. However, the hypothesis of ontological states and their generally statistical relation to the (pre-)quantum states of CA will add a new element changing this situation.

---

[2]One may ponder the possibility that the counting of links and their conservation law could be generalized such that α-initials are linked to a *finite set* of α'-finals, including the α-finals. This could be pictured as a *forward lightcone*-like structure, considering spatial sites labelled by index α etc. Could there correspondingly exist Hamiltonians that would conform with *locality* in the usual special relativistic sense?

[3]The possibility of *negative* link numbers L_α or weights w_α reminds of "probabilities" falling outside of $[0, 1]$ appearing in QM as discussed by Wigner, Feynman, and others; see, e.g., the reviews by Khrennikov [13] and by Mückenheim *et al.* [14].

4 The Hypothesis of Ontological States

Ontological states that underlie quantum and, *a fortiori*, classical states of physical objects are central to the *Cellular Automaton Interpretation* (CAI) of QM [11].

There is ample motivation to reexamine the foundations of quantum theory in perspective of essentially classical concepts—above all, determinism and existence of ontological states of reality—which stems from observations of quantum features in a large variety of deterministic and, in some sense, "classical" models [4–6]. – It is worth emphasizing that quantum states here are considered to form part of the mathematical language used, they are "templates" for the description of the "reality beneath", including ontological states and their deterministic dynamics.

Finite and discrete CA may provide the necessary versatility to accommodate ontological states and their evolution, besides (proper) quantum "template" states (especially in the continuum limit). The following general remarks serve to obtain an operational understanding of what we are looking for in such models:

ONTOLOGICAL STATES ($\mathcal{OS}$) are states a deterministic physical system can be in. They are denoted by $|A\rangle,\ |B\rangle,\ |C\rangle,\ \dots$. The set of all states may be very large, but is assumed to be denumerable, for simplicity.

There exist *no superpositions* of $\mathcal{OS}$ "out there" as part of physical reality.

The $\mathcal{OS}$ evolve by *permutations* among themselves, $\dots \to |A\rangle \to |D\rangle \to |B\rangle \to \dots$, for example. Apparently this is the only possibility, besides producing a growing set of states or superpositions, which do not belong to the initial set of $\mathcal{OS}$.

By declaring the $\mathcal{OS}$ to form an orthonormal set, fixed once for all, a *Hilbert space* can be defined. – Operators which are diagonal on this set of $\mathcal{OS}$ are *beables*. Their eigenvalues describe physical properties of the $\mathcal{OS}$, corresponding to the abstract labels $A,\ B,\ C,\ \dots$ above.

QUANTUM STATES ($\mathcal{QS}$) are superpositions of $\mathcal{OS}$. These are templates for doing physics with the help of mathematics. – The amplitudes specifying superpositions need to be interpreted, when applying the formalism to describe experiments. Here the *Born rule is built in*, i.e., by definition! By experience, interpreting amplitudes in terms of probabilities has been an amazingly useful invention.[4]

From here the machinery of QM can be seen to depart, incorporating especially the powerful techniques related to unitary transformations. The latter exist, in general, only in a rudimentary discrete form on the level of $\mathcal{OS}$, due to the absence of superpositions.

While quantum theory has been very effective in describing experiments, its *linearity* is the characteristic feature of the unitary dynamics embodied in the Schrödinger equation.[5] For a prospective ontological theory, it is of interest that

[4]It is indeed possible to change the proportionality between *absolute values squared* of complex amplitudes and probabilities into a more complicated relation, however, only at the expense of mathematical simplicity [11].

[5]This linearity is reflected by the Superposition Principle and entails interference effects and the possibility of nonclassical correlations among parts of composite objects, i.e. entanglement in multipartite systems.

QM remains notoriously indifferent to any reduction or collapse process one might be tempted to add on, in order to modify the collapse-free linear evolution and, in this way, solve the measurement problem.

Concluding this brief recapitulation of some essential points of CAI, in particular how QM fits into this wider realistic picture, one more remark is in order, concerning the *absence of the measurement problem*:

CLASSICAL STATES of a macroscopic deterministic system, including billiard balls, pointers of apparatus, planets, are *probabilistic distributions* of $\mathcal{OS}$, since any kind of repeated experiments performed by physicists, with only limited control of the circumstances, pick up different initial conditions regarding the $\mathcal{OS}$. Hence, different outcomes of apparatus readings must generally be expected. Yet any reduction or collapse to a δ-peaked distribution, say, of pointer positions is only an apparent effect, induced by the intermediary use of quantum mechanical templates in describing the evolution of $\mathcal{OS}$ during an experiment. Ontologically speaking, there were/are no superpositions, to begin with, which could possibly collapse [11]!

This provides a strong motivation for pursuing an approach to understand reality as based on ontological states.

4.1 Avoiding Superpositions

We recall that $\mathcal{OS}$ evolve by permutations among themselves. This is quite different from the behaviour commonly found in QM, namely the dynamical formation of *superposition states* (except for stationary states).

In order to assess the formation, or not, of superpositions by an evolving Hamiltonian CA, we first remind ourselves that the Schrödinger equation is formally solved by:

$$\partial_t \psi(t) = -i\hat{H}\psi(t) \quad \Rightarrow \quad \psi(t) = \mathrm{e}^{-i\hat{H}t}\psi(0) \;,$$

given the initial state $\psi(0)$, and consider the analogous formal solution of the CA equation of motion (3), $\dot{\psi}_n = \psi_{n+1} - \psi_{n-1} = -i\hat{H}\psi_n$. In terms of an auxiliary operator $\hat{\phi}$, defined by $2\sin\hat{\phi} := \hat{H}$, one finds indeed:

$$\psi_n = \frac{1}{2\cos\hat{\phi}}\big(\mathrm{e}^{-in\hat{\phi}}[\mathrm{e}^{i\hat{\phi}}\psi_0 + \psi_1] + (-1)^n\mathrm{e}^{in\hat{\phi}}[\mathrm{e}^{-i\hat{\phi}}\psi_0 - \psi_1]\big) \;. \tag{12}$$

where *two initial states* are required, ψ_0 and ψ_1, corresponding to the fact that Hamiltonian CA are described by a second-order finite difference equation.

With the help of the general solution (12), one obtains:

$$\psi_n = \hat{T}(n-m+1)\psi_{m+1} + \hat{T}(n-m)\psi_m \;, \tag{13}$$

where $\hat{T}$ is a transfer operator that can be read off by comparing with the explicit form of this relation. This generalizes the composition law for the unitary time evolution operator in QM. – Furthermore, the simple exponential expression for the solutions in QM can be recovered from Eq. (12) by taking the appropriate limits $n \to \infty$ and $l \to 0$, keeping $n \cdot l$ fixed, *and* choosing initial conditions such that $\psi_1 \equiv \psi_0$. In this case, we have:

$$\psi_n = \left[\hat{T}(n+1) + \hat{T}(n)\right]\psi_0 \ . \tag{14}$$

The Eq. (13) and especially Eq. (14) tell us to generally expect the formation of superposition states and, therefore, *not* ontological states which evolve by permutations among themselves.

While this seems to severely obstruct the search for $\mathcal{OS}$ from the outset, we now present a first simple example illustrating that evolving $\mathcal{OS}$ are possible in a *two-state* CA. – Consider the CA described by ψ_n^α, $\alpha = 1, 2$, with equation of motion given by:

$$\psi_n = \psi_{n-2} - i\hat{H}_2\psi_{n-1} \ , \quad \psi_n \equiv \begin{pmatrix} \psi_n^1 \\ \psi_n^2 \end{pmatrix} \ , \quad \hat{H}_2 := \begin{pmatrix} 0 & 1 \\ 1 & 0 \end{pmatrix} \equiv \hat{\sigma}_1 \ . \tag{15}$$

Furthermore, we choose two orthogonal initial states, $\psi_0 = (1, 0)^t$ and $\psi_1 = (0, 1)^t$. By solving the equation of motion most simply by iteration, we obtain the following sequence of states:

$$\begin{aligned} &\psi_0 \ , \ \psi_1 \ , \ \psi_2 = (1-i)\psi_0 \ , \ \psi_3 = -i\psi_1 \ , \\ &\psi_4 = -i\psi_0 \ , \ \psi_5 = -(1+i)\psi_1 \ , \ \psi_6 = -\psi_0 \ , \ \psi_7 = -\psi_1 \ , \ \dots \ , \end{aligned} \tag{16}$$

which after four more steps begins to reproduce the initial pair of states.

Here the normalization of the states, considered as if of vectors embedded in a Hilbert space for a moment, changes dynamically. This would be a disaster in QM! However, for Hamiltonian CA this norm is not conserved. Instead, it is replaced by a *conserved 'two-time' correlation function*, *cf.* Eqs. (6)–(7), reproducing the norm conservation only in the continuum limit [4].

Thus, apart from the change of normalization, the evolution here essentially swaps two orthogonal input states, once per updating step. This provides a very simple example of a CA evolving $\mathcal{OS}$, in agreement with CAI.

4.2 More Interesting CA and Their $\mathcal{OS}$

Following the primitive example just given, the question arises, whether there exists any generalization describing something more interesting.

Besides systems with block diagonal $\hat{H}$ for multiple two-state components, one may try a higher-dimensional state space for generalizations of the model of Eqs. (15)–(16). Indeed, we find easily that the Hamiltonians:

$$\hat{H}_3 := \begin{pmatrix} 0 & -i & 1 \\ i & 0 & -i \\ 1 & i & 0 \end{pmatrix}, \quad \hat{H}_4 := \begin{pmatrix} 0 & -i & 0 & 1 \\ i & 0 & -i & 0 \\ 0 & i & 0 & -i \\ 1 & 0 & i & 0 \end{pmatrix}, \tag{17}$$

for three- and four-state CA, respectively, lead to analogous evolution-by-permutation of $\mathcal{OS}$ as the previous example, $\hat{H}_2$ of Eq. (15). We may generally consider the m-dimensional state space with Hamiltonian:

$$\hat{H}_m := \begin{pmatrix} 0 & -i & 0 & \cdots & 0 & 1 \\ i & 0 & -i & 0 & \cdots & 0 \\ & & & & & \\ & & & \ddots & & \\ 0 & \cdots & 0 & i & 0 & -i \\ 1 & 0 & \cdots & 0 & i & 0 \end{pmatrix}, \tag{18}$$

which works like the previous examples for 'neighbouring' pairs of orthogonal initial states, $\psi_0 = \psi^{(k)} := (0, \ldots, 0, 1, 0, \ldots, 0)^t$, with nonzero k-th entry ($1 \leq k \leq m - 1$), and correspondingly $\psi_1 = \psi^{(k+1)}$.

The CA evolution with Hamiltonian $\hat{H}_m$, Eq. (18), does not change the normalization of these states, but can introduce phases ($\pm 1, \pm i$) when permuting them. To give an explicit example, choosing $\psi_0 = \psi^{(m-1)}$ and $\psi_1 = \psi^{(m)}$, the result of one updating step is $\psi_2 = \psi_0 - i\hat{H}_m\psi_1 = -i\psi^{(1)}$. Such phases are carried on by further updating steps until they are eventually cancelled and the initial configuration reappears, only after $4m$ updates.

Therefore, all the $4m$ $\mathcal{OS}$, which eventually *differ by phases*, must be considered as *different states* here. Note that in $x' + ip' \equiv \psi' = i\psi \equiv i(x + ip)$, for example, the roles of coordinates (real parts) and momenta (imaginary parts) are exchanged, $x' = -p$, $p' = x$. Such states cannot be seen as embedded in a projective space, which would be the case of normalized states in QM. Loosely speaking, they are 'more classical'.

All pairs of (initial) states of this kind have the conserved *link number* $L = 0$, cf. Eqs. (10)–(11) Whereas initial configurations with $\psi_0 \equiv \psi_1$ have $L > 0$—they could serve as quantum mechanical templates (in the continuum limit), *cf.* introduction of this Sect. 4 and discussion of Eqs. (13)–(14). Thus, the conserved link number illustrates the *ontology conservation law* [11].

The dynamics described by $\hat{H}_m$ resembles the *cogwheel model* discussed in Refs. [3, 7, 11]—first introduced by 't Hooft as a 'particle' making uniform jumps over fixed positions on a circle, one per fixed time interval—which has been shown to have surprising quantum features (providing a discrete representation of the quantum harmonic oscillator). However, while evolution was given by a unitary first-order updating rule in those models, it is a second-order process determined by a self-adjoint operator, $\hat{H}_m$, in the present case.

Before closing, a remark is in order here. Namely, *interacting multipartite Hamiltonian CA* [6], in particular those consisting of two-state "Ising spin" subsystems, offer an alternative to look for more complex behaviour concerning ontological states than in the one-component examples chosen here for simplicity. Such systems have been considered recently [3], with further results to be presented elsewhere. Generalization in this direction seems necessary, in order to develop ontological models that possibly can serve as a realistic base from which the theory of interacting relativistic quantum fields can emerge in analogous ways as the QM models we described.

5 Conclusion

In retrospect, one could subsume our results, *cf.* especially the summary given in Sect. 2, as pertaining to a particular *discretization* of QM, which introduces the finite scale l, conceivably the Planck scale.

However, we have reported in Sect. 3 the resulting conservation laws in such models of discrete *Hamiltonian CA*, which are entirely described in terms of integer valued quantities, and illustrated the one-to-one correspondence with those of continuum models of quantum theory, which are recovered for $l \to 0$. In particular, the conserved quantity, a two-time correlation, replacing here the conserved norm of a QM state vector, has led us to a simple interpretation in terms of a *counting procedure*—unnoticeable in quantum theory, since there only the coincidence limit of the correlation matters![6] This seems to shed a different light on the *Born rule*.

Up to this point, though, we still did not encounter the randomness eventually seen in experimental outcomes.

As we have argued, following 't Hooft, the *probabilistic features of QM* can be understood to result from the available mathematical description of the underlying deterministic reality. The unavoidable mismatch between the two can be precisely traced to the *nonexistence of superposition states* of "stuff" that is ontologically there and the powerful use that is made of such formal superpositions in quantum theory.

By the *hypothesis of ontological states* and by illustrating their existence within the present class of discrete models, of Hamiltonian CA kind, one has left standard QM, as suggested by the *Cellular Automaton Interpretation* [11], *cf.* the introduction to Sect. 4. This may provide some indication that reality can be understood to exist "out there", sometimes misnamed "Einstein's dream", that ontological states describe states in which a deterministic physical system can be and how it evolves. Yet QM is neither abandoned nor has quantum theory been changed, but one begins to understand it as a most effective mathematical construct/language to describe the reality of what we perceive. Much more is left to be done.

[6]Similar observations hold for all conservation laws and could be of phenomenological interest.

Acknowledgements It is a pleasure to thank Dirk Rischke and Horst Stöcker for the invitation to the *International Symposion on Discoveries at the Frontiers of Science* in honour of Walter Greiner (FIAS, Frankfurt, June 26–30, 2017), for support, and especially for kind hospitality.

References

1. A.E. Allahverdyan, R. Balian, T.M. Nieuwenhuizen, Understanding quantum measurement from the solution of dynamical models. Phys. Rep. **525**(1), 1 (2013)
2. G. Chiribella, G.M. D'Ariano, P. Perinotti, Informational derivation of quantum theory. Phys. Rev. A **84**, 012311 (2011)
3. H.-T. Elze, Emergent discrete time and quantization: relativistic particle with extradimensions. Phys. Lett. A **310**, 110 (2003)
4. H.-T. Elze, Action principle for cellular automata and the linearity of quantum mechanics. Phys. Rev. A **89**, 012111 (2014)
5. H.-T. Elze, Multipartite cellular automata and the superposition principle. Int. J. Quantum Inf. (IJQI) **14**, 1640001 (2016)
6. H.-T. Elze, Ontological states and dynamics of discrete (pre-)quantum systems. Int. J. Quantum. Inf. (IJQI) **15**(8), 1740013 (2017)
7. H.-T. Elze, O. Schipper, Time without time: a stochastic clock model. Phys. Rev. D **66**, 044020 (2002)
8. P.A. Hoehn, Reflections on the information paradigm in quantum and gravitational physics. J. Phys. Conf. Ser. **880**, 012014 (2017)
9. P.A. Hoehn, Quantum theory from rules on information acquisition. Entropy **19**(3), 98 (2017)
10. P.A. Hoehn, C. Wever, Quantum theory from questions. Phys. Rev. A **95**, 012102 (2017)
11. G. 't Hooft, *The Cellular Automaton Interpretation of Quantum Mechanics*. Fundamental Theories of Physics, vol. 185 (Springer, Heidelberg, New York, 2016)
12. A.J. Jerri, The Shannon sampling theorem—its various extensions and applications: a tutorial review. Proc. IEEE **65**, 1565 (1977)
13. A. Khrennikov, *Interpretations of Probability* (VSP, Utrecht, Boston, 2003)
14. W. Mückenheim et al., A review of extended probabilities. Phys. Rep. **133**(6), 337 (1986)
15. M. Schlosshauer (ed.), *Elegance and Enigma—The Quantum Interviews*. The Frontiers Collection (Springer, Berlin, Heidelberg, 2011)
16. C.E. Shannon, Communications in the presence of noise. Proc. IRE **37**, 10 (1949)

Photon Scattering Off Nuclei

Hartmuth Arenhövel

Abstract The study of nuclear and subnuclear structure by means of photon scattering is outlined. Besides a brief exposition of the formalism a few illustrative examples are discussed.

1 Introduction

I would like to begin with a brief personal remark: It was during the fall of 1964—I just had completed my diploma thesis in experimental physics at the University of Freiburg—, when Walter Greiner offered me to work on a Ph.D. thesis in his new established theory group at the University of Frankfurt. As subject of the thesis he had proposed to investigate the structure of heavy deformed nuclei in the region of the giant resonances by means of photon scattering. By the end of 1965 Hans-Jürgen Weber and myself were Walter's first Ph.D.-students to complete their Ph.D. Since then, my interest in the study of electromagnetic reactions on nuclei in general, e.g. photo absorption and scattering, electron scattering and meson production, has continued up to present times.

Electromagnetic reactions on nuclei provide an excellent tool to investigate nuclear structure. In addition, they also lead to valuable insights into the electromagnetic properties of the nuclear constituents, proton and neutron, like for example, electric and magnetic polarizabilities and electromagnetic form factors. In this context photon scattering experiments are a particularly interesting source of information on off-shell properties of the nuclear constituents. On the other hand genuine microscopic calculations of photon scattering cross sections are rather complicated since in principle the complete excitation spectrum has to be taken into account.

Here I will give a brief account of photon scattering reactions on nuclei with emphasis on my own work and that of my collaborators. It is not intended as a

Dedicated to the memory of Walter Greiner.

H. Arenhövel (✉)
Institut für Kernphysik Johannes Gutenberg-Universität, 55099 Mainz, Germany
e-mail: arenhoevel@kph.uni-mainz.de

J. Kirsch et al. (eds.), *Discoveries at the Frontiers of Science*,
FIAS Interdisciplinary Science Series, https://doi.org/10.1007/978-3-030-34234-0_19

general review, rather a personal view on this interesting reaction which over many years has fascinated myself and which was and still is subject of my own research. Early reviews on nuclear photon scattering may be found in Refs. [1–3] and a more recent one in Ref. [4].

In the next section I will present a short summary of the basic scattering formalism, in particular the expansion of the scattering matrix in terms of generalized polarizabilities as basic quantities. It will be followed by a few illustrative examples, partly on earlier work on medium and heavy weight nuclei within the dynamic collective model of the giant dipole resonances and partly on the lightest nucleus, the deuteron, with emphasis on subnucleon degrees of freedom like meson exchange and isobar currents.

2 Formalism of Photon Scattering

I will briefly describe the formal features of the photon scattering process

$$\gamma_\lambda(k) + N_i(P_i) \longrightarrow \gamma'_{\lambda'}(k') + N_f(P_f)\,, \tag{1}$$

where an incoming photon with four momentum $k = (k_0, \vec{k})$ and circular polarization $\vec{e}_\lambda$ ($\lambda = \pm 1$) is scattered off a nucleus in the initial intrinsic state $|i\rangle$ with total four momentum $P_i = (E_i, \vec{P}_i)$ making a transition to a final intrinsic state $|f\rangle$ with total four momentum $P_f = (E_f, \vec{P}_f)$ while emitting a final photon with four momentum $k' = (k'_0, \vec{k}\,')$ and polarization $\vec{e}\,'_{\lambda'}$.

2.1 The Photon Scattering Amplitude

Since the electromagnetic interaction is weak one makes a Taylor expansion with respect to the electromagnetic field A_μ up to second order, because at least two photons are involved in the scattering process. Thus in this lowest, i.e. second order in the e.m. coupling, the scattering amplitude is given by two terms, the contact or two photon amplitude (TPA or seagull) $B_{\lambda'\lambda}(\vec{k}\,', \vec{k})$, arising from the second order term of the Taylor expansion, and the resonance amplitude (RA) $R_{\lambda'\lambda}(\vec{k}\,', \vec{k})$ from the iterated linear interaction term. A diagrammatic illustration is shown in Fig. 1.

Accordingly, the total scattering amplitude is the sum of these two contributions

$$T^{fi}_{\lambda'\lambda}(\vec{k}\,', \vec{k}) = B^{fi}_{\lambda'\lambda}(\vec{k}\,', \vec{k}) + R^{fi}_{\lambda'\lambda}(\vec{k}\,', \vec{k})\,, \tag{2}$$

where the two-photon amplitude (diagram (a) of Fig. 1) has the form

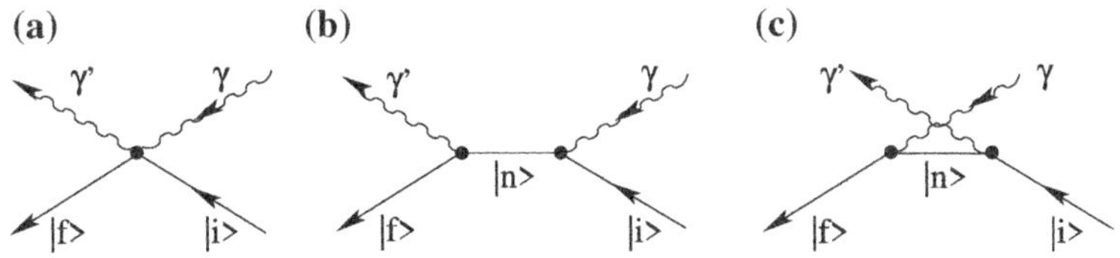

Fig. 1 Diagrammatic representation of photon scattering: The two-photon amplitude (**a**) and the resonance contribution (direct (**b**) and crossed (**c**)). To be read from right to left

$$B^{fi}_{\lambda'\lambda}(\vec{k}', \vec{k}) = \sum_{l,m=1}^{3} e'^{*}_{\lambda',l} \langle f | \widehat{B}_{lm}(\vec{k}', \vec{k}) | i \rangle e_{\lambda,m} , \tag{3}$$

with $\widehat{B}_{lm}(\vec{k}', \vec{k})$ as Fourier transform of the second order coefficient of the Taylor expansion $\widetilde{B}_{lm}(\vec{x}, \vec{y})$, i.e.

$$\widehat{B}_{lm}(\vec{k}', \vec{k}) = -\int d^3x\, d^3y\, e^{i\vec{k}'\cdot\vec{x}} \widetilde{B}_{lm}(\vec{x}, \vec{y}) e^{-i\vec{k}\cdot\vec{y}} . \tag{4}$$

Assuming a nonrelativistic description, the overall center-of-mass (c.m.) motion can be separated and the Hamiltonian splits accordingly into an intrinsic part and the c.m. kinetic energy

$$H = H_{int} + \frac{\vec{P}^2}{2M_A} , \tag{5}$$

where $H_{int} = T_{int} + V$ denotes the Hamiltonian of the internal motion, $\vec{P}$ the total c.m. momentum and M_A the mass of the nucleus. An intrinsic state is denoted by $|n\rangle$ with intrinsic energy e_n. Then the resonance amplitude (RA) (diagrams (b) and (c) of Fig. 1) is given by the following matrix element between intrinsic states

$$\begin{aligned} R^{fi}_{\lambda'\lambda}(\vec{k}', \vec{k}) = \langle f | \Big[& \vec{e}'^{*}_{\lambda'} \cdot \vec{J}(-\vec{k}', 2\vec{P}_f + \vec{k}') \\ & \times G_{int}\Big(k_0 - \frac{\vec{k}\cdot(2\vec{P}_i + \vec{k})}{2M_A} + i\epsilon\Big) \vec{e}_\lambda \cdot \vec{J}(\vec{k}, 2\vec{P}_i + \vec{k}) \\ & + \big(\vec{e}_\lambda \leftrightarrow \vec{e}'^{*}_{\lambda'} , k_\mu \leftrightarrow -k'_\mu\big) \Big] | i \rangle , \end{aligned} \tag{6}$$

with the resolvent or propagator

$$G_{int}(z) = (H_{int} - e_i - z)^{-1} . \tag{7}$$

The current operator in Eq. (6)

$$\vec{J}(\vec{k}, \vec{P}) = \vec{j}(\vec{k}) + \frac{1}{2M_A} \vec{P} \rho(\vec{k}) \tag{8}$$

consists of the intrinsic current density operator $\vec{j}(\vec{k}\,)$ plus a term taking into account the convection current of the separated c.m. motion, where $\rho(\vec{k}\,)$ denotes the Fourier transform of the intrinsic charge density operator of the nucleus. The intrinsic charge and current density operators consist of a kinetic or one-body ($\rho_{[1]}, \vec{j}_{[1]}$) and a two-body meson exchange part ($\rho_{[2]}, \vec{j}_{[2]}$)

$$\rho(\vec{k}\,) = \rho_{[1]}(\vec{k}\,) + \rho_{[2]}(\vec{k}\,)\,, \tag{9}$$

$$\vec{j}(\vec{k}\,) = \vec{j}_{[1]}(\vec{k}\,) + \vec{j}_{[2]}(\vec{k}\,)\,, \tag{10}$$

with

$$\rho_{[1]}(\vec{k}) = \sum_l e_l\, e^{-i\vec{k}\cdot\vec{r}_l}\,, \tag{11}$$

$$\vec{j}_{[1]}(\vec{k}\,) = \frac{1}{2M}\sum_l \Big(e_l\{\vec{p}_l, e^{-i\vec{k}\cdot\vec{r}_l}\} + \mu_l \vec{\sigma}_l \times \vec{k}\, e^{-i\vec{k}\cdot\vec{r}_l}\Big)\,. \tag{12}$$

Here, e_l and μ_l denote charge and magnetic moment of the l-th particle and $\vec{r}_l$, $\vec{p}_l$ and $\vec{\sigma}_l$ its internal coordinate, momentum and spin operators, respectively. The expressions for the corresponding exchange operators depend on the interaction model. At least in the nonrelativistic limit, the exchange contribution to the charge density vanishes (Siegert's hypothesis). Furthermore, also the TPA consists of a kinetic one-body contribution and a two-body exchange amplitude

$$\widehat{B}_{lm}(\vec{k}\,', \vec{k}\,) = \widehat{B}_{[1],lm}(\vec{k}\,', \vec{k}\,) + \widehat{B}_{[2],lm}(\vec{k}\,', \vec{k}\,)\,, \tag{13}$$

where the kinetic one-body operator is given by the sum of the individual proton Thomson scattering amplitudes

$$\widehat{B}_{[1],lm}(\vec{k}\,', \vec{k}\,) = -\frac{1}{M}\sum_j e_j^2\, e^{-i(\vec{k}-\vec{k}\,')\cdot\vec{r}_j}\,\delta_{lm}\,. \tag{14}$$

It is important to note that the splitting of the scattering amplitude into a resonance and a two-photon amplitude is gauge dependent. This gauge dependence is reflected in gauge conditions for the current and the two-photon amplitude which follow from the Gauge invariance of the electromagnetic interaction. In detail one finds the following gauge conditions for the e.m. operators

$$\vec{k}\cdot\vec{j}(\vec{k}\,) = [H, \rho(\vec{k}\,)]\,, \tag{15}$$

$$\sum_l k_l' B_{lm}(\vec{k}\,', \vec{k}\,) = [\rho(-\vec{k}\,'), j_m(\vec{k}\,)]\,. \tag{16}$$

The first condition in (15), connecting the charge density with the current, describes current conservation, while the second relates the TPA to the commutator of charge and current densities. Separating the one-body and two-body (exchange) contributions, one finds the following conditions

$$\vec{k} \cdot \vec{j}_{[1]}(\vec{k}) = [T, \rho_{[1]}(\vec{k})] , \tag{17}$$

$$\vec{k} \cdot \vec{j}_{[2]}(\vec{k}) = [V, \rho_{[1]}(\vec{k})] + [T, \rho_{[2]}(\vec{k})] , \tag{18}$$

$$\sum_l k_l' \widehat{B}_{[1],lm}(\vec{k}', \vec{k}) = [\rho_{[1]}(-\vec{k}'), j_{[1],m}(\vec{k})] , \tag{19}$$

$$\sum_l k_l' \widehat{B}_{[2],lm}(\vec{k}', \vec{k}) = [\rho_{[1]}(-\vec{k}'), j_{[2],m}(\vec{k})] + [\rho_{[2]}(-\vec{k}'), j_{[1],m}(\vec{k})] . \tag{20}$$

Important consequences are the low energy limits [5, 6]

$$\vec{j}(0) = [H, \vec{D}] , \tag{21}$$

the Siegert theorem, where $\vec{D}$ denotes the unretarded dipole operator, and

$$\widehat{B}^{ii}_{[1],lm}(0,0) = -\frac{Ze^2}{M}\delta_{lm} , \tag{22}$$

$$\widehat{B}^{ii}_{[2],lm}(0,0) = -\langle i|[D_l, [V, D_m]]|i\rangle , \tag{23}$$

$$\widehat{R}^{ii}_{lm}(0,0) = \frac{NZe^2}{AM}\delta_{lm} - \widehat{B}^{ii}_{[2],lm}(0,0) , \tag{24}$$

resulting in the classical Thomson limit for the total nuclear scattering amplitude

$$T^{ii}_{\lambda'\lambda}(0,0) = -\vec{e}^{\,\prime *}_{\lambda'} \cdot \vec{e}_{\lambda} \frac{(Ze)^2}{AM} , \tag{25}$$

with the approximation $M_A = AM$.

To close this section, I would like to mention the optical theorem, which relates the forward elastic scattering amplitude to the total photo absorption cross section

$$\sigma_{tot}(k, \rho) = \frac{4\pi}{k} Im\Big[Tr\Big(\rho\, T^{ii}(\vec{k}, \vec{k})\Big)\Big] , \tag{26}$$

where ρ denotes the photon-nucleus polarization density matrix of the initial state.

2.2 Generalized Nuclear Polarizabilities

In the scattering process the incoming photon and the scattered one transfer various angular momenta via the e.m. multipole operators L and L' to the nucleus which can be coupled to a total angular momentum transfer J. It is very useful to expand the scattering amplitude with respect to this total angular momentum transfer. It leads to the concept of generalized polarizabilities $P^{(L'L)J}_{fi,\lambda'\lambda}(k',k)$, first introduced by Fano [7] for the case of pure $E1$ transitions and later generalized in Refs. [8–11]. A review may be found in Ref. [1]. A further generalization to the $(e,e'\gamma)$ reaction of electron scattering (virtual compton scattering) is given in Ref. [12].

The polarizabilities allow one to separate geometrical aspects related to the angular momentum properties from dynamical effects as contained in the strength of the various polarizabilities. The expansion of the total scattering amplitude in terms of these polarizabilities reads

$$T^{fi}_{M_f M_i,\lambda'\lambda}(\vec{k}',\vec{k}) = (-)^{I_f-M_i} \sum_{L',M',L,M,J,m} (-)^{L+L'}\widehat{J}^2 \begin{pmatrix} I_f & J & I_i \\ -M_f & m & M_i \end{pmatrix} \times \begin{pmatrix} L & L' & J \\ M & M' & -m \end{pmatrix} P^{(L'L)J}_{fi,\lambda'\lambda}(k',k) D^L_{M,\lambda}(R) D^{L'}_{M',-\lambda'}(R'), \quad (27)$$

where the abbreviation $\widehat{J} = \sqrt{2J+1}$ is used, and (I_i, M_i) and (I_f, M_f) refer to the angular momenta of the initial and final states and their projections on the quantization axis, respectively. Furthermore, R and R' describe the rotations which carry the quantization axis into the directions of the photon momenta $\vec{k}$ and $\vec{k}'$, respectively, and $D^L_{M,\lambda}(R)$ and $D^{L'}_{M',-\lambda'}(R')$ denote the corresponding rotation matrices in the convention of Rose [13]. The polarizabilities contain the dynamic properties of the system and depend on the absolute values of the photon momenta only, whereas the geometrical aspects, i.e. the angular dependencies, are contained in the rotation matrices.

The general definition of the Polarizabilities is then obtained by the inversion of Eq. (27), i.e.

$$P^{(L'L)J}_{fi,\lambda'\lambda}(k',k) = \frac{(-)^{L'-L-I_f}}{8\pi^2}\widehat{L'}^2\widehat{L}^2 \sum_{M_f M_i M' M m} \begin{pmatrix} I_f & J & I_i \\ -M_f & m & M_i \end{pmatrix} \begin{pmatrix} L & L' & J \\ M & M' & -m \end{pmatrix} \times \int dR' \int dR\, D^{L*}_{M,\lambda}(R) D^{L'*}_{M',-\lambda'}(R')\, T^{fi}_{M_f M_i,\lambda'\lambda}(\vec{k}',\vec{k}). \quad (28)$$

One should note that J is bound by the multipole orders L and L' and the spins I_i and I_f of the initial and final nuclear states, respectively, i.e.

$$|L-L'| \le J \le L+L' \quad \text{and} \quad |I_i - I_f| \le J \le I_i + I_f. \quad (29)$$

For example, to elastic scattering off a spin-zero nucleus only the scalar polarizabilities ($J=0$) contribute.

It is furthermore useful to classify the polarizabilities according to the total parity transfer in case that parity is conserved. This leads to the introduction of

$$P^J_{fi}(M^{\nu'}L', M^{\nu}L, k', k) = \frac{1}{4}\sum_{\lambda',\lambda=\pm 1} \lambda'^{\nu'}\lambda^{\nu}\, P^{(L'L)J}_{fi,\lambda'\lambda}(k',k). \quad (30)$$

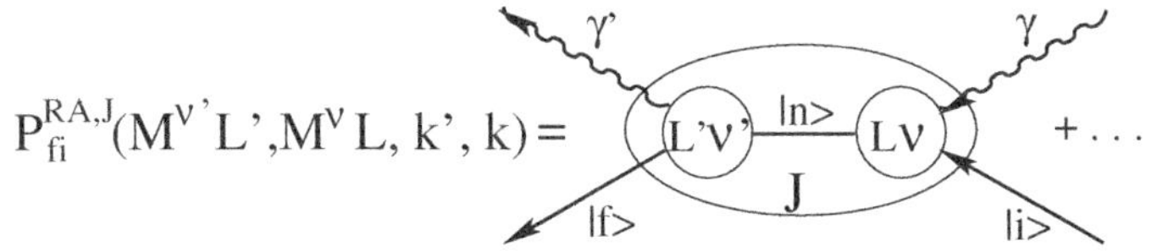

Fig. 2 Graphical representation of the contribution of the direct term of the resonance amplitude to the generalized polarizability $P^J_{fi}(M^{\nu'}L', M^{\nu}L, k', k)$, where the angular momentum transfer L of the incoming photon is coupled with the angular momentum transfer L' of the outgoing photon to a total angular momentum transfer J

In terms of these one has

$$P^{(L'L)J}_{fi,\lambda'\lambda}(k', k) = \sum_{\nu',\nu=0,1} \lambda'^{\nu'} \lambda^{\nu} P^J_{fi}(M^{\nu'}L', M^{\nu}L, k', k)\,, \tag{31}$$

where ν and ν' classify the type of multipole transition, i.e. $\nu = 0$ means electric ($M^0L = EL$) and $\nu = 1$ magnetic ($M^1L = ML$).

For parity conservation a simple selection rule follows

$$P^J_{fi}(M^{\nu'}L', M^{\nu}L, k', k) = 0\,, \quad \text{if} \quad (-)^{L'+\nu'+L+\nu} \neq \pi_i \pi_f\,, \tag{32}$$

with π_i and π_f denoting the parities of initial and final states, respectively. A graphical visualization of the generalized polarizability is shown in Fig. 2 for the direct resonance term.

The polarizabilities $P^J_{fi}(M^{\nu'}L', M^{\nu}L, k', k)$ can be separated into a TPA and a resonance contribution

$$\begin{aligned} P^J_{fi}(M^{\nu'}L', M^{\nu}L, k', k) = {} & P^{TPA,J}_{fi}(M^{\nu'}L', M^{\nu}L, k', k) \\ & + P^{RA,J}_{fi}(M^{\nu'}L', M^{\nu}L, k', k)\,, \end{aligned} \tag{33}$$

where for the resonance amplitude one has

$$\begin{aligned} P^{RA,J}_{fi}(M^{\nu'}L', M^{\nu}L, k', k) = {} & 2\pi(-)^{L+J}\frac{\hat{L}\hat{L}'}{\hat{J}} \\ & \times \langle I_f e_f || \Bigg(\Big[M^{\nu'[L']}(k') G_{int}(k_0 - \frac{\vec{k}\cdot(2\vec{P}_i + \vec{k})}{2M_A} + i\varepsilon) M^{\nu[L]}(k) \Big]^{[J]} \\ & + \Big\{ M^{\nu[L]}(k) \leftrightarrow M^{\nu'[L']}(k')\,, k_\mu \leftrightarrow -\vec{k}'_\mu \Big\} \Bigg) || I_i e_i \rangle\,. \end{aligned} \tag{34}$$

The superscript $[J]$ indicates a spherical tensor of rank J, and "$[\ldots]^{[J]}$" means that two spherical tensors are coupled to a spherical tensor of rank J. Furthermore, $M^{\nu'[L']}(k')$ denotes a standard electromagnetic current multipole operator [13]

$$M_m^{\nu'[L']}(k') = \begin{cases} \nu = 0 \ \text{electric}: \int d^3x\, \vec{A}_m^{[L]}(E; k, \vec{x}) \cdot \vec{J}(\vec{x}), \\ \nu = 1 \ \text{magnetic}: \int d^3x\, \vec{A}_m^{[L]}(M; k, \vec{x}) \cdot \vec{J}(\vec{x}), \end{cases} \tag{35}$$

with the multipole fields

$$\vec{A}_m^{[L]}(M; k, \vec{x}) = i^L j_L(kx) \vec{Y}_m^{(L1)[L]}(\hat{x}), \tag{36}$$

$$\vec{A}_m^{[L]}(E; k, \vec{x}) = \frac{1}{k} \vec{\nabla} \times \vec{A}_m^{[L]}(M; k, \vec{x}). \tag{37}$$

The two-photon contribution to the polarizability is given by

$$P_{fi}^{TPA,J}(M^{\nu'}L', M^{\nu}L, k', k) = 2\pi(-)^{L+J+1} \frac{\hat{L}\hat{L}'}{\hat{J}}$$
$$\langle I_f e_f || \int d^3x\, d^3y \sum_{lm} \Big[A_l^{[L']}(M^{\nu'}; k, \vec{x}) B_{lm}(\vec{x}, \vec{y}) A_m^{[L]}(M^{\nu}; k, \vec{y}) \Big]^{[J]} || I_i e_i \rangle . \tag{38}$$

The evaluation of the TPA contribution to the polarizabilities is straightforward once the TPA operator $B_{lm}(\vec{x}, \vec{y})$ is given.

For the resonance contribution, one finds by evaluating the reduced matrix element in standard fashion (see e.g. [14])

$$P_{fi}^{RA,J}(M^{\nu'}L', M^{\nu}L, k', k) = 2\pi(-)^{L+I_f+I_i} \hat{L}\hat{L}'$$
$$\times \sum\!\!\!\!\!\!\int_{e_n, I_n} \Bigg[\begin{Bmatrix} L & L' & J \\ I_f & I_i & I_n \end{Bmatrix} \frac{\langle I_f e_f || M^{\nu'[L']}(k') || I_n e_n \rangle \langle I_n e_n || M^{\nu[L]}(k) || I_i e_i \rangle}{e_n - e_i - k_0 + \frac{\vec{k} \cdot (2\vec{P}_i + \vec{k})}{2M_A} - i\varepsilon}$$
$$+(-)^{L+L'+J} \begin{Bmatrix} L' & L & J \\ I_f & I_i & I_n \end{Bmatrix} \frac{\langle I_f e_f || M^{\nu[L]}(k) || I_n e_n \rangle \langle I_n e_n || M^{\nu'[L']}(k') || I_i e_i \rangle}{e_n - e_i + k_0' - \frac{\vec{k}' \cdot (2\vec{P}_i - \vec{k}')}{2M_A} - i\varepsilon} \Bigg]. \tag{39}$$

Obviously, the calculation of the resonance part is more involved because of the summation over all possible intermediate states $|I_n e_n\rangle$ with angular momentum I_n and intrinsic energy e_n.

The optical theorem in (26) allows one to relate the imaginary part of the scalar polarizabilities of elastic scattering $P_{ii}^0(M^{\nu}L, M^{\nu}L, k, k)$ to the partial contribution $\sigma(M^{\nu}L)(k)$ of the multipole $M^{\nu}L$ to the total unpolarized photo absorption cross section $\bar{\sigma}_{tot}$

$$\bar{\sigma}_{tot}(k) = \sum_L \Big(\sigma(EL)(k) + \sigma(ML)(k) \Big)$$
$$= \frac{4\pi}{2k(2I_i + 1)} \sum_{\lambda M} Im\, T_{\lambda\lambda MM}^{ii}(\vec{k}, \vec{k})$$

$$= \frac{4\pi}{k\widehat{I}_i} \sum_L \frac{(-)^{L+1}}{\widehat{L}} Im\Big[P_{ii}^{RA,0}(EL, EL, k, k) + P_{ii}^{RA,0}(ML, ML, k, k)\Big]. \tag{40}$$

From this relation follows

$$Im[P_{ii}^0(M^\nu L, M^\nu L, k, k)] = \frac{k}{4\pi}(-)^{L+1}\widehat{I}_i \widehat{L}\sigma(M^\nu L)(k)\,. \tag{41}$$

An important property of the scattering amplitude is the low energy theorem [5] according to which up to terms linear in the photon momentum k the scattering amplitude is completely determined by global properties like charge, mass and magnetic moment. This means for the polarizabilities that in the limit $k = 0$ only the scalar E1-polarizability is nonvanishing, i.e.

$$P^J(E1, E1)|_{k=0} = -\delta_{J0}\,\widehat{I}_i\,\sqrt{3}\,\frac{e^2 Z^2}{M_A}\,, \tag{42}$$

with I_i as ground state spin. It corresponds to the Thomson scattering amplitude. Internal properties like static electric and magnetic polarizabilities contribute in the next nonvanishing order (k^2) only. An extension of this theorem by a more general low energy expansion of the polarizabilities has been discussed in Ref. [6] with the result that for electric multipole transitions EL with even L the contributions to $P^J(EL, EL)|_{k=0}$ up to order $2L - 2$ vanish.

2.3 The Elastic Scattering Cross Section

The elastic scattering cross section for unpolarized photons and targets is given by

$$\frac{d\sigma_{elastic}}{d\Omega} = \frac{c(P_i, k, k')}{2(2I_i + 1)} \sum_{\lambda,\lambda',M_i,M_f} |T^{ii}_{M_f M_i,\lambda'\lambda}(\vec{k}', \vec{k})|^2\,, \tag{43}$$

with a kinematic factor for collinear intial momenta

$$c(P_i, k, k') = \frac{E_i(k_0 + E_i - k_0')}{(\widehat{k}' \cdot (k + P_i))^2}\,, \tag{44}$$

where $\widehat{k}' = k'/k_0'$. For elastic scattering in the c.m. frame ($\vec{k} + \vec{P}_i = 0$ and $k_0 = k_0'$) one has $c_{c.m.}(P_i, k, k') = (M_A^2 + k^2)/W^2$ with $W = k_0 + E_i$ as invariant mass.

In terms of the polarizabilities one finds for the unpolarized cross section [1]

$$\frac{d\sigma_{elastic}}{d\Omega} = \frac{c(P_i,k,k')}{2I_i+1} \sum_{L',L,K',K,J} \sum_{\nu',\nu,\bar{\nu}',\bar{\nu}} P^J_{ii}(M^{\nu'}L', M^{\nu}L, k', k) \times P^J_{ii}(M^{\bar{\nu}'}K', M^{\bar{\nu}}K, k', k)^* g_J^{\nu'L'\nu L;\bar{\nu}'K'\bar{\nu}K}(\theta), \tag{45}$$

where the angular functions depend on the scattering angle θ only and are given by

$$\begin{aligned} g_J^{\nu'L'\nu L;\bar{\nu}'K'\bar{\nu}K}(\theta) = & \frac{(-)^J}{2}\widehat{J}^2(-)^{L+K+\nu'+\bar{\nu}'} \\ & \times \sum_j \widehat{j}^2(1+(-)^{L+K+j+\nu+\bar{\nu}})(1+(-)^{L'+K'+j+\nu'+\bar{\nu}'}) \\ & \times \begin{pmatrix} L' & K' & j \\ 1 & -1 & 0 \end{pmatrix} \begin{pmatrix} L & K & j \\ 1 & -1 & 0 \end{pmatrix} \begin{Bmatrix} L & K & j \\ K' & L' & J \end{Bmatrix} P_j(\cos\theta), \end{aligned} \tag{46}$$

with $P_j(\cos\theta)$ denoting a Legendre polynomial. For pure $E1$ transitions this expression simplifys considerably and one obtains with scalar, vector and tensor polarizabilities

$$\frac{d\sigma_{elastic}(E1)}{d\Omega} = \frac{c(P_i,k,k')}{(2I_i+1)} \sum_{J=0}^{2} |P^J_{ii}(E1,E1)|^2 g_J^{E1}(\theta), \tag{47}$$

where in an abbreviated notation the angular functions are

$$g_0^{E1}(\theta) = \frac{1}{6}(1+\cos^2\theta), \tag{48}$$

$$g_1^{E1}(\theta) = \frac{1}{4}(2+\sin^2\theta), \tag{49}$$

$$g_2^{E1}(\theta) = \frac{1}{12}(13+\cos^2\theta). \tag{50}$$

In this case the angular distribution is symmetrical around 90° in the c.m. system.

The generalization to polarized photons and oriented nuclei is presented in Refs. [1, 15].

3 Applications

Now I will discuss several applications: (i) Heavy deformed nuclei within the dynamic collective model of Danos and Greiner, (ii) Photon scattering off ^{12}C for energies in the region of the first nucleon resonance, the $\Delta(1232)$, and finally (iii) photon scattering off the deuteron as a means to study subnuclear degrees of freedom, for example meson exchange currents or the static polarizabilities of the neutron.

3.1 *Photon Scattering Off Complex Nuclei in the Giant Dipole Resonance Region*

The giant dipole resonance (GDR) in medium and heavy weight nuclei can well be explained as a collective phenomenon in the framework of the hydrodynamical model of Steinwedel and Jensen as an oscillation of a proton fluid against a neutron fluid. According to this model, as pointed out independently by Danos and Okamoto [16], the GDR will be split into two peaks for an axially symmetric deformed nucleus corresponding to different frequencies for oscillations along and perpendicular to the symmetry axis.

In view of the additional collective surface degrees of freedom, Danos and Greiner [17] proposed in 1964 a unified dynamic collective model of the giant resonances (DCM) which includes the coupling between the rotation-vibration surface degrees of freedom and the giant resonance d.o.f. leading to additional dynamic effects. A weak point of this approach is that it provides energies and strengths of the GDR states at discrete energies only, but not the shape nor the widths of the states. Usually a Lorentzian shape is assumed with some simple adhoc model for the width, the parameters of which are used for a fit procedure to the absorption cross section.

As a result of this dynamic coupling considerably strong dipole transition strengths from the GDR states to the low lying rotational and vibrational states appear leading to sizeable Raman scalar and tensor scattering into these low lying collective states (see Fig. 3, left panel). Indeed, such Raman scattering has been measured for ^{238}U, ^{232}Th and ^{209}Bi by Jackson and Wetzel [18] although these authors found a significantly

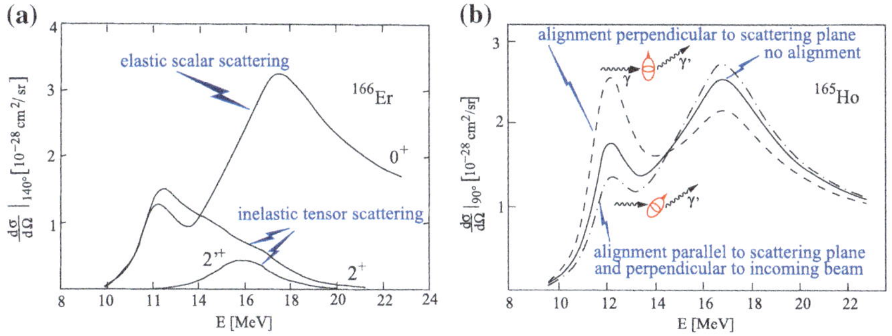

Fig. 3 Left panel: Calculated elastic and inelastic photon scattering cross sections at 140° for ^{166}Er (from Ref. [9]). Right panel: Elastic photon scattering cross sections for ^{165}Ho (from Ref. [15]): unoriented target: solid curve; aligned target: **a** perpendicular to scattering plane: dashed curve, **b** parallel to scattering plane and perpendicular to incoming photon beam: dash-dot curve

weaker Raman cross section (about 40 %) than predicted by the DCM. A similar reduction of the inelastic DCM strengths was found later for a series of vibrational medium weight nuclei by Bowles et al. [19]. A possible explanation of this reduction could be a 10–15% nonresonant contribution of direct transitions into the continuum which would appear in the scalar but not in the tensor polarizability as has been discussed in Ref. [20].

Another interesting feature of the DCM is the fact that a deformed nucleus with a nonvanishing ground state spin becomes dynamically triaxial and thus optically anisotropic (nonvanishing elastic tensor polarization) and thus its absorption and scattering cross sections depend on the nuclear orientation. The reason for this feature is the fact that for a nonvanishing ground state spin $I \geq 1$ there is a tensor contribution to elastic scattering which, however, does not show up for an unoriented nucleus. This is shown in the right panel of Fig. 3 for ^{165}Ho having a ground state spin $I = 7/2$. Experimentally such a dependence of the photon absorption cross section of ^{165}Ho on the nuclear orientation was found by Ambler et al. [21].

Subsequently, the DCM was further developed in order to describe a much more general class of potential energy surfaces for the low energy collective d.o.f. allowing a unified description of the GDR for nuclei with quite different collective character-

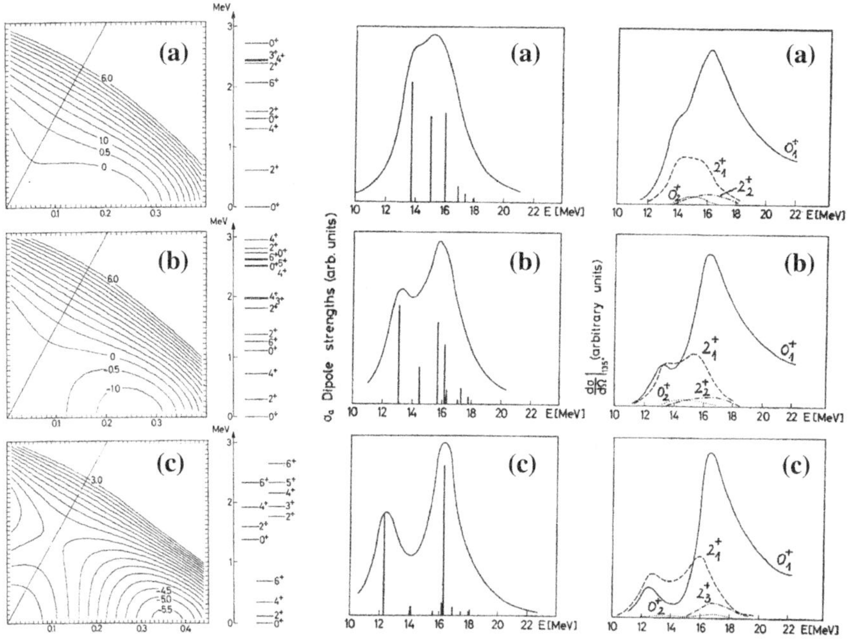

Fig. 4 Transition study from a vibrational (**a**) to a strongly deformed nucleus (**c**) (from Ref. [22]): Left panels: Collective potential energy surfaces and low-energy spectra. Middle panels: Dipole strengths and γ-absorption cross sections. Right panels: Elastic and Raman γ-scattering cross sections to low lying collective states

istics by Rezwani et al. [22, 23]. An example, taken from Ref. [22], is shown in Fig. 4 displaying the collective potential energy surface, the absorption strength and cross section and the elastic and inelastic scattering cross sections for three different types of nuclei, resembling a transition from a nucleus with an anharmonic vibrational character (a) to a strongly deformed rotational nucleus (c).

Case A represents a vibrational nucleus with strong anharmonicities, resulting in a shift and splitting of the surface two-phonon triplet. Furthermore, a slight axially symmetric deformation leads to a nonvanishing intrinsic quadrupole moment. Since the nucleus is easily deformable in β- and γ-directions, it appears dynamically triaxial resulting in a slight splitting of the dipole strength into three equally spaced states. Moreover, considerable inelastic tensor γ-scattering into the low-lying 2^+-states is observed.

The opposite situation of a strongly deformed nucleus is displayed by case C. It corresponds to a good axially symmetric rotator with clearly separated ground, β- and γ-rotational states. Accordingly the GDR is split into two distinct peaks, and strong Raman-scattering via the tensor polarizability into the 2^+-state of the ground state rotational band appears. The transition between these two extreme cases is represented by case B.

3.2 *Photon Scattering Off Complex Nuclei in The* $\boldsymbol{\Delta}$**(1232)** *Region*

The influence of internal nucleon degrees of freedom constitutes an important field of research in nuclear and medium energy physics. In particular, the role of the lowest excited state of the nucleon, the $\Delta(1232)$ resonance, has been studied in the low energy domain [24] as well as in the energy region of real Δ excitation in pion production and photo absorption.

For photon scattering in the energy region of about 300 MeV the simplest model is a static approach assuming pure $M1$-scattering off the individual nucleons, whereas the nuclear structure is manifest only via the elastic form factor with respect to the momentum transfer. Then the scattering matrix has the simple form [25]

$$T^{fi}_{\lambda'\lambda} = \langle f| \sum_{l=1}^{A} t_{\lambda'\lambda}(l) e^{-i(\vec{k}-\vec{k}')\cdot\vec{r}_l} |i\rangle \,, \tag{51}$$

where the elementary scattering operator is given by

$$\begin{aligned} t_{\lambda'\lambda}(l) = c^2 \Bigg(& \frac{\vec{e}^{\,\prime *}_{\lambda'} \cdot [\vec{\sigma}^{\,l}_{N\Delta} \times \vec{k}\,'\,] \, \vec{e}_{\lambda} \cdot [\vec{\sigma}^{\,l}_{\Delta N} \times \vec{k}\,]}{M_\Delta - M_N - k_0 - i\Gamma_\Delta/2} \\ & + \frac{\vec{e}_{\lambda} \cdot [\vec{\sigma}^{\,l}_{N\Delta} \times \vec{k}\,] \, \vec{e}^{\,\prime *}_{\lambda'} \cdot [\vec{\sigma}^{\,l}_{\Delta N} \times \vec{k}\,'\,]}{M_\Delta - M_N + k'_0 - i\Gamma_\Delta/2} \Bigg) \,. \end{aligned} \tag{52}$$

The $N \to \Delta$ spin transition matrix $\vec{\sigma}^{\,l}_{N\Delta}$ is defined as in Ref. [24]. Furthermore,

$$c = G^{\Delta N}_{M1}(M_\Delta + M_N)/(4M_\Delta M_N)\,, \tag{53}$$

where M_N, M_Δ and Γ_Δ denote respectively nucleon and Δ mass and width. The latter and the magnetic transition strength $G^{\Delta N}_{M1}$ are fit to the experimental photo absorption cross section of the nucleon in the Δ resonance region.

In this simple approach nuclear structure enters only via the nuclear form factors of mass and spin (for more details see Ref. [25]). A comparison of this approach with experiment for ^{12}C and ^{208}Pb is shown in Fig. 5.

It turns out, that the resulting calculated cross section largely underestimates the experiment by orders of magnitude for both nuclei. Collective effects in a more refined Δ-hole model for ^{12}C by Koch et al. [27] lead to a slight enhancement but the discrepancy remained essentially. However, Hayward and Ziegler [26] already had pointed out that the finite energy resolution with respect to the measured scattered photons of about 10% would lead to the inclusion of corresponding inelastic contributions in the experimental data.

Subsequently, such inelastic contributions have been studied in Ref. [25] within two simple models: (i) Inelastic scattering into 1-particle-1-hole excitations with excitation energies within an energy interval Δk, and (ii) a sum rule approach to the giant resonances as final states (see Ref. [25] for further details). The results of these two models are also shown in Fig. 5. Both approaches show an enhancement due to

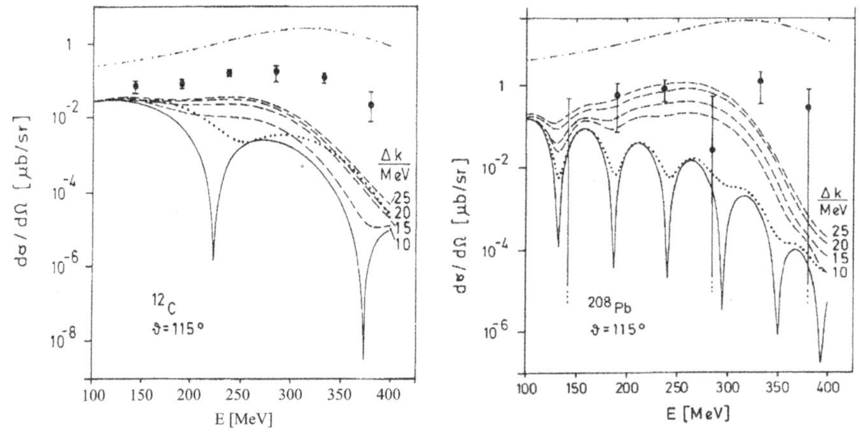

(a) ^{12}C (b) ^{208}Pb

Fig. 5 Elastic and quasielastic photon scattering cross sections at $\theta = 115°$ for ^{12}C and ^{208}Pb (from Ref. [25]). Experimental data from Ref. [26]. Solid curve: elastic scattering in static approach; dashed curves include inelastic contributions up to an excitation energy Δk; dotted curve: sum rule approach (see Ref. [25] for further details); dash-dotted curve: incoherent sum of elastic elementary $\gamma - N$ scattering cross section inelastic contributions, particularly strong for the 1p–1h approach. However, for ^{12}C still sizeable strength is missing and a more thorough theoretical treatment is needed.

3.3 Photon Scattering Off the Deuteron

Now I would like to turn my attention to more fundamental studies with respect to the role of subnuclear degrees of freedom like meson exchange currents (MEC) and internal nucleon degrees of freedom in terms of nucleon resonances (isobar currents IC) in electromagnetic reactions. Lightest nuclei present ideal laboratories for such studies and I will concentrate on the deuteron for which extensive studies on electromagnetic reactions exist [2, 3, 28, 29].

In photon scattering meson exchange effects do not only appear via MEC in the resonance amplitude but also as additional contributions in the TPA as derived in [30] for a NN one-pion exchange potential (see Fig. 6 for the corresponding diagrams) as well as for isobar contributions.

It is intriguing, and indeed had been suggested, that diagram (d) of Fig. 6 allows the extraction of a density form factor of charged mesons in the nucleus (see references in Ref. [31]). However, a careful analysis has shown that such an interpretation is not possible [31]. It turns out that the TPA of an exchanged pion (diagram (d) of Fig. 6) reflects the form factor of a pion transition density between the two nucleons and not the virtual pion density inside the deuteron. This feature is illustrated in Fig. 7, where the monopole part of this transition density is displayed. Moreover, as already mentioned above, the diagrams in Fig. 6 are not by itself gauge invariant and thus not separately measurable.

A first realistic calculation of elastic deuteron photon scattering with inclusion of subnuclear effects has been carried out in Ref. [32] taking as reference frame the photon deuteron Breit frame, then $k' = k$. Furthermore, small contributions from the c.m. motion have been neglected. For unpolarized photons and deuterons only the scalar polarizability $P^0_{ii}(M^\nu L, M^\nu L, k, k)$ contributes to the elastic scattering cross

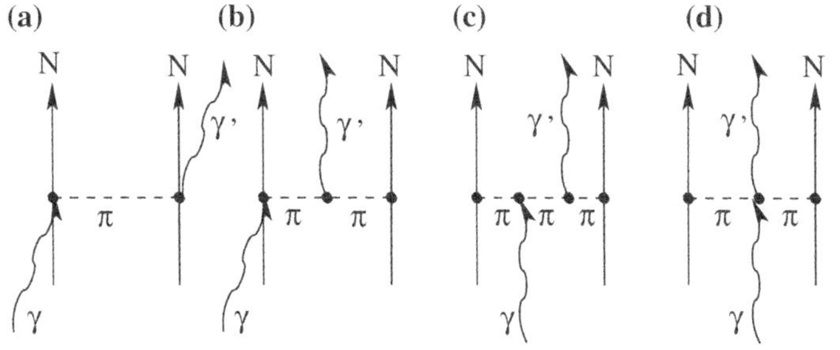

Fig. 6 Diagrammatic representation of pion exchange TPA contributions to the two-photon amplitude

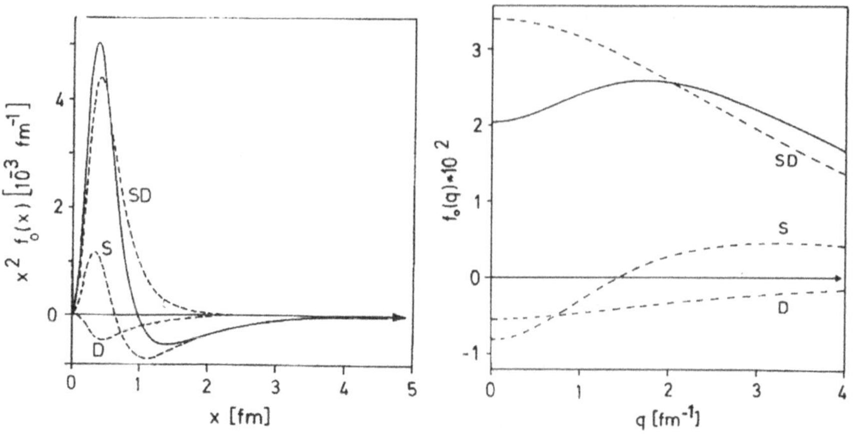

Fig. 7 Left panel: Monopole part of π-exchange transition density for the deuteron (solid curve) from the π-TPA diagram (d) of Fig. 6. Dashed curves show separate contributions from S- and D-states and $S-D$ interference. Right panel: Form factor of transition density displayed in the left panel, again with separate contributions (from Ref. [31])

section. The calculation of the resonance amplitude requires the summation over all excited energies and all possible intermediate states.

As detailed in Ref. [32], a subtracted dispersion relation for the individual polarizabilities of given multipolarity has been assumed using Eq. (41) for the imaginary part of the resonance amplitude (the TPA is real). Thus the real part of the scalar polarizability is determined from

$$Re(P_{ii}^{RA,0}(M^{\nu}L, M^{\nu}L, k, k)) - \delta_{\nu 0}\delta_{L1}\, P_{ii}^{RA,0}(E1, E1, 0, 0) = (-)^{L+1}\widehat{L}\widehat{I}_i \frac{k^2}{2\pi^2} P \int_{k_{\text{th}}}^{\infty} dk' \frac{\sigma(M^{\nu}L)(k')}{k'^2 - k^2}\,. \quad (54)$$

Here, k_{th} denotes the threshold energy for photo absorption. Since the integral extends in principle up to infinity, one has to include also contributions from particle production above the corresponding threshold energy, e.g. from the total photo pion production cross section for energies above the pion production threshold. These were neglected limiting this approach to the low energy region. For the evaluation of the partial contributions $\sigma(M^{\nu}L)(k)$ to photo disintegration the Reid soft core potential had been used. Besides the one body current and TPA π-meson exchange currents also $N\Delta$ and $\Delta\Delta$ isobar configurations have been included.

A more recent realistic calculation from my group by Wilbois et al. [33] has been based on the solution of the off-shell NN-scattering matrix instead of using dispersion relations. Again inelasticities, which appear above pion production threshold, are excluded, limiting this approach also to low photon energies. For the deuteron bound state and the final state interaction (FSI) the realistic Bonn OBEPQ-B potential has been used. MEC contributions from pion- and rho-meson exchange to the current

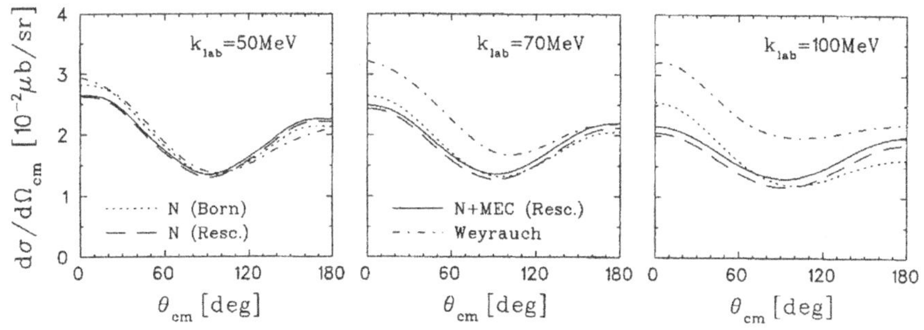

Fig. 8 Differential scattering cross sections for photon energies 50, 70 and 100 MeV from Ref. [33]. Notation of curves: dotted (labeled "N (Born)") without FSI and MEC beyond Siegert, dashed ("N (Resc.)") including FSI but no MEC, and full ("N+MEC (Resc.)") with FSI and MEC. Dash-dot curves ("Weyrauch") represent a calculation with a separable interaction from Ref. [35]

and the TPA amplitude have been included. For the electric transitions so-called Siegert operators have been used, incorporating thus implicitly the major part of MEC contributions [34]. Figure 8 displays the resulting differential scattering cross section at three photon energies, 50, 70, and 100 MeV. For comparison the results of another approach by Weyrauch [35] using a separable NN-interaction are also shown. Contributions from internal nucleon structure as manifest, e.g. in nucleon polarizabilities, were neglected in both calculations.

Comparing the dotted with the dashed curves one notes a sizeable increase from FSI with increasing energy, at forward angles a reduction and in the backward region an increase, thus reducing the strong asymmetry of the case without FSI considerably. The additional MEC effects beyond the Siegert operators are quite small for the lowest energy, but become more pronounced at 100 MeV, leading to an overall increase of the cross section. The increasing difference to Weyrauch's result [35] indicate that the separable interaction used in that work is not appropriate at higher energies.

A comparison with experimental data is shown in Fig. 9. One readily notes a systematic overestimation of the experimental results (dashed curves). An improved description is achieved if the influence of the internal nucleon structure in terms of nucleon polarizabilities is considered (full curves). The additional introduction of the free neutron and proton polarizabilities reduces the cross sections sizeably and gives a better description of the experiment. In view of the fact, that the neutron polarizabilities are not directly measurable, one turns the argument around in order to determine them from photon scattering off the deuteron. However, this procedure is not completely free from model dependencies.

Since then quite a few more theoretical investigations of this reaction using various approaches have been published [36], where the major emphasis had been laid on the extraction of the neutron polarizabilities (for recent results see Ref. [37]).

In fact, the deuteron is often used to determine internal neutron properties like polarizabilities and electromagnetic form factors, which otherwise are not available, since free neutron targets are absent. The question of off-shell modification of such

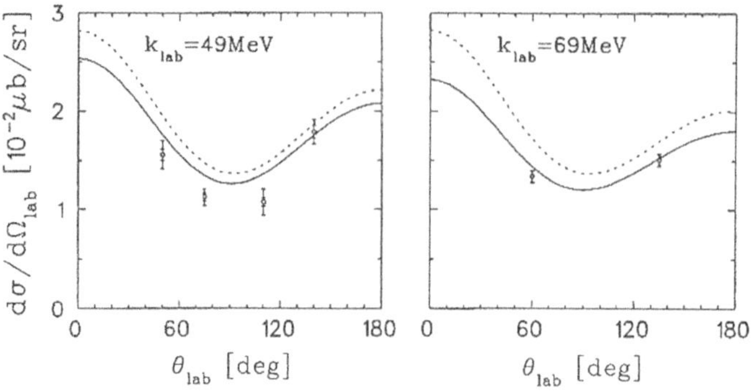

Fig. 9 Differential scattering cross sections for photon energies of 49 and 69 MeV (from Ref. [33]). Experimental data from Ref [38]. Solid curves: complete calculation with nucleon polarizabilities included. Dashed curves: without nucleon polarizabilities

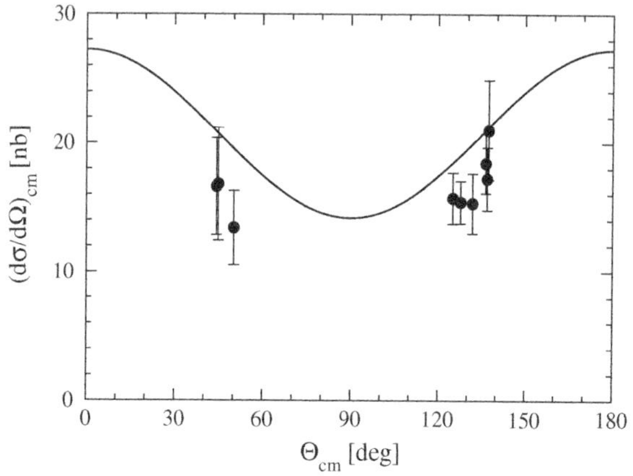

Fig. 10 Differential scattering cross section in the unretarded dipole approximation calculated with the Lorentz integral transform method for a photon energy of 55 MeV (from Ref. [40]). Experimental data from Ref. [41]

internal nucleon properties is thereby left open with the hope that such effects are small since the deuteron is quite a loosly bound system.

Another interesting aspect which reactions on the deuteron offer is that the deuteron can also serve as a test ground for new theoretical methods. As an illustrative example, I will consider recent work based on the Lorentz integral transform (LIT) method (for a review of this method see Ref. [39]). The LIT method is particularly suited for theoretical studies of reactions on complex nuclei. It allows the calculation of reactions without the need of determining the complex final scattering states, reducing this problem to the solution of a bound state equation. Therefore, as a test case for the application of the LIT method to photon scattering reactions, the LIT has been applied to the deuteron for a pure unretarded $E1$ radiation by Bampa et al. [40].

The resulting cross section at one energy is shown in Fig. 10 together with recent experimental data. Nucleon polarizabilities were not included. The slight difference to the more complete calculation at 50 MeV in Fig. 8 is caused by the restriction to $E1$ radiation which is also the reason for the completely symmetric angular distribution around 90° (see Eq. (48)) in contrast to the slight asymmetry at 50 MeV in Fig. 8.

4 Conclusions

With these few examples, I hope to have convinced the reader, that nuclear photon scattering reactions has been and still is a wide and interesting field of research in nuclear structure studies from low energy collective properties to the influence of subnuclear degrees of freedom as manifest e.g. in meson exchange and isobar currents. Hopefully, in the future more experimental results will be available, although the cross sections are quite small compared to hadronic reactions. Of particular interest are scattering experiments on complex nuclei at higher energies in the region of the first nucleon resonance, in order to study the behavior of a nucleon resonance in a nuclear environment.

References

1. H. Arenhövel, W. Greiner, *Progress in Nuclear Physics* (Pergamon Press, Oxford, 1969), p. 167
2. H. Arenhövel, *Workshop on Perspectives in Nuclear Physics at Intermediate Energies*, ed. by S. Boffi, C. Ciofi degli Atti, M.M. Giannini (World Scientific, Singapore, 1984), p. 97
3. H. Arenhövel, *New Vistas in Electro-Nuclear Physics*, ed. by E.L. Tomusiak, H.S. Caplan, E.T. Dressler (Plenum Press, New York, 1986), p. 251
4. M.-Th. Hütt, A.I. L'vov, A.I. Milstein, M. Schumacher, Phys. Rep. **323**, 457 (2000)
5. J.L. Friar, Ann. Phys. (N.Y.) **95**, 1428 (1975). References therein
6. H. Arenhövel, M. Weyrauch, Nucl. Phys. A **457**, 573 (1986)
7. U. Fano, NBS technical note 83 (1960). Reprinted in *Photonuclear Reactions*, ed. by E.G. Fuller, E. Hayward (Dowden, Hutchinson & Ross, Stroudsburg, 1976), p. 338
8. H. Arenhövel, Ph.D. thesis, University Frankfurt (1965)
9. H. Arenhövel, M. Danos, W. Greiner, Phys. Rev. **157**, 1109 (1967)
10. R. Silbar, H. Überall, Nucl. Phys. A **109**, 146 (1968)
11. R. Silbar, Nucl. Phys. A **118**, 389 (1968)
12. H. Arenhövel, D. Drechsel, Nucl. Phys. A **233**, 153 (1974)
13. M.E. Rose, *Elementary Theory of Angular Momentum* (Wiley, New York, 1957)
14. A.R. Edmonds, *Angular Momentum in Quantum Mechanics* (Princeton University Press, Princeton, 1957)
15. H. Arenhövel, W. Greiner, Nucl. Phys. **86**, 193 (1966)
16. M. Danos, *Nucl. Phys.* **5**, 23 (1958); K. Okamoto, *Phys. Rev.* **110**, 143 (1958)
17. M. Danos, W. Greiner, Phys. Rev. **134**, B284 (1964)
18. H.E. Jackson, K.J. Wetzel, Phys. Rev. Lett. **28**, 513 (1972)
19. T.J. Bowles et al., Phys. Rev. C **24**, 1940 (1981)
20. H. Arenhövel, Phys. Rev. C **6**, 1449 (1972)
21. E. Ambler, E.G. Fuller, H. Marshak, Phys. Rev. **138**, B117 (1965)
22. V. Rezwani, G. Gneuss, H. Arenhövel, Phys. Rev. Lett. **25**, 1667 (1970)
23. V. Rezwani, G. Gneuss, H. Arenhövel, Nucl. Phys. A **180**, 254 (1972)

24. H.J. Weber, H. Arenhövel, Phys. Rep. C **36**, 277 (1978)
25. H. Arenhövel, M. Weyrauch, P.-G. Reinhard, Phys. Lett. B **155**, 22 (1985)
26. E. Hayward, B. Ziegler, Nucl. Phys. A **414**, 333 (1984)
27. J.H. Koch, E.J. Moniz, N. Ohtsuka, *Ann. Phys.* (N.Y.) **154**, 99 (1984)
28. H. Arenhövel, M. Sanzone, Few-Body Syst. Suppl. **3**, 1 (1991)
29. H. Arenhövel, W. Leidemann, E.L. Tomusiak, Eur. Phys. J. A **23**, 147 (2005)
30. H. Arenhövel, Z. Phys. A **297**, 129 (1980)
31. M. Weyrauch, H. Arenhövel, Phys. Lett. B **134**, 21 (1984)
32. M. Weyrauch, H. Arenhövel, Nucl. Phys. A **408**, 425 (1983)
33. T. Wilbois, P. Wilhelm, H. Arenhövel, Few-Body Syst. Suppl. **9**, 263 (1995)
34. H. Arenhövel, Z. Phys. A **302**, 25 (1981)
35. M. Weyrauch, Phys. Rev. C **41**, 880 (1990)
36. M.I. Levchuck, Few-Body Syst. Suppl. **9**, 439 (1995); J.J. Karakowski, G.A. Miller, Phys. Rev. C **60**, 014001 (1999); M.I. Levchuck, A.I. L'vov, Nucl. Phys. A **674**, 449 (2000); Nucl. Phys. A **684**, 490 (2001); R.P. Hildebrandt, H.W. Grießhammer, T.R. Hemmert, D.R. Phillips, Nucl. Phys. A **748**, 573 (2005); R.P. Hildebrandt, H.W. Grießhammer, T.R. Hemmert, Eur. Phys. J. A **46**, 111 (2010)
37. L.S. Myers et al., Phys. Rev. C **92**, 025203 (2015)
38. M. A. Lucas, Ph.D. thesis, University of Illinois at Urbana-Champaign (1994). Unpublished
39. V.D. Efros, W. Leidemann, G. Orlandini, N. Barnea, J. Phys. G: Nucl. Phys. **34**, R459 (2007)
40. G. Bampa, W. Leidemann, H. Arenhövel, Phys. Rev. C **84**, 034005 (2011)
41. M. Lundin et al., Phys. Rev. Lett. **90**, 192501 (2003)

A Confluence of Ideas and Experiments—A Tribute to Professor Walter Greiner

John W. Harris

Abstract This article is a tribute to Professor Walter Greiner that describes, from my personal perspective and experience, the profound influence of his ideas in two specific subfields of experimental nuclear physics: "nuclear molecules" and relativistic heavy-ion physics. In relating the former, I use the close relationship between Walter as the theorist and D. Allan Bromley the experimentalist who together pioneered that field. I then describe the richness of exciting physics and the tremendous growth of the field of relativistic heavy-ion physics as a result of Walter and the "Greiner School".

1 Professor Walter Greiner

This is a tribute to Professor Walter Greiner. It will in no way do justice to his tremendous accomplishments and impact in physics, but will hopefully provide a somewhat unique perspective on his life and work. Professor Greiner was Professor of Physics at the JW Goethe-Universität. He influenced nuclear physics in so many ways, that in this tribute I will only mention a few of those that have had direct impact on my work and the direction of this field. His work extended from the theoretical concepts for the formation of nuclear molecules, superheavy elements, shock compression in collisions of nuclei and beyond, to the education of physicists at all stages resulting in the production of more than 13 textbooks covering all aspects of graduate physics.

J. W. Harris (✉)
D. Allan Bromley Professor of Physics, Department of Physics, Yale University,
PO Box 208120, New Haven, CT 06520-8120, USA
e-mail: john.harris@yale.edu

J. Kirsch et al. (eds.), *Discoveries at the Frontiers of Science*,
FIAS Interdisciplinary Science Series, https://doi.org/10.1007/978-3-030-34234-0_20

Professor Greiner served in many capacities over his career, to name a few, including Chair of the Institute of Theoretical Physics at the JW Goethe-Universität. He served as Dean of the Physics Faculty of JW Goethe-Universität, and Director of the Frankfurt Institute for Advanced Studies (FIAS). He was a Fellow or Honorary Member of many physics societies including Fellow of the Royal Society of Arts (FRSA) in London, Honorary Member of the Roland Eotvos Society of Hungary, Honorary Member of the Romanian Academy, and Officier dans l'Ordre Palmes Academiques in France. Professor Greiner was awarded the Max Born Prize and Medal, the Otto Hahn Prize and the Alexander von Humboldt Medal.

Walter believed that collaboration between theorists and experimentalists was essential. He played an important role in establishing the GSI. All heavy ion facilities world-wide have benefited from his innovative ideas. He supervised more than 187 "Diplomarbeit" and 141 Ph.D. theses in Frankfurt. He held guest professorships at 22 universities and national laboratories in the US, Canada, Russia, and Egypt, and earned many honorary doctoral degrees from universities throughout the world.

2 Professor Walter Greiner and Professor David Allan Bromley—Lifelong Parallels

I am compelled, in my honor of being the D. Allan Bromley Professor of Physics at Yale University, to point out the strong parallels in the lives and research of Professors Walter Greiner and D. Allan Bromley, shown in Fig. 1, as well as elucidate details of their close personal relationship.

Fig. 1 Professors Walter Greiner (left) and D. Allan Bromley (right)

D. Allan Bromley, as described by Walter [1]:

> He was a leading figure in nuclear physics, who made seminal contributions to heavy ion physics by discovering nuclear molecules, studying deuteron stripping reactions and the production of negative ions. Allan was elected to the National Academy of Sciences in 1990. Above all he created a school for experimental nuclear physics at Yale, guiding others worldwide. And in the area of public policy he will be remembered as one of the nation's most effective science advisers to the President of the United States.

Professor Bromley was the 1st Sterling Professor of the Sciences at Yale University, and served as Chair of Physics Department and the Dean of Engineering at Yale. He was elected to the U.S. National Academy of Sciences. Among other roles, he was elected President of the American Association for Advancement of Science (multiple times), President of the Union of Pure and Applied Physics, and President of the American Physical Society. He was the 1st Cabinet-level Assistant to the President of U.S. for Science and Technology, the Senate-confirmed Director of Science and Technology Policy in the White House, and was awarded the U.S. National Medal of Science.

3 In Walter's Own Words

In recalling his relationship with Allan Bromley, Walter said [2]:

> I had become a professor quite young and from various possibilities I had chosen the Johann Wolfgang Goethe University in Frankfurt am Main. On January 1, 1965 I took on a chair in theoretical physics and my interests shifted from photonuclear physics to heavy ion physics. The possibility of forming molecules between nuclei, eventually even fusing them to become larger and larger nuclear complexes, i.e. superheavy nuclei, fascinated me. This naturally led me straight to Allan Bromley. He studied experimentally nuclear molecules for quite some time, had become a great promoter of nuclear physics with heavy ions world-wide, and thus I met him first on an international conference at Heidelberg in 1966. A year later, again, our world lines crossed at the University of Virginia in Charlottesville, Va., where he came for a colloquium and I was there looking around and checking out the place, because they had offered me a professorship.

Fig. 2 The exuberant young Walter (left) and the dapper Allan dressed in his usual attire (right)

Another view of the two can be seen in Fig. 2, as Walter continued:

> I told Allan about our - at that time crude - ideas on nuclear molecules, how nuclear matter might get compressed when heavy nuclei encounter: while they approach occurs first some binding, then a strong repulsion, thus a potential pocket will appear and nuclei are bound for a short time in quasi-molecular form. He invited me up to Yale. It was my first visit to this distinguished place. Allan showed me this famous laboratory – The Wright Nuclear Structure Laboratory – and his experimental results. This was the starting point of rich scientific collaboration and the beginning of a wonderful friendship. Together with my former student and friend Werner Scheid.... I invented the so-called "double-resonance-mechanism" and could explain many precision measurements from the Wright Nuclear Structure Laboratory at Yale quite well. Not everything was quantitatively in full agreement but all the many observations were well understood. Most, if not all, of this physics is summarized in the monography on "Nuclear Molecules", which I published with Werner Scheid and Jae Park in 1995 at World Scientific. We dedicated this book to Allan Bromley.

A drawing from an article by Walter [3] describing the idea of nuclear molecules is shown in Fig. 3. The cover of the Greiner, Park and Scheid monograph "Nuclear Molecules" [4] that was dedicated to Bromley is shown in Fig. 4 (left) with a picture of Walter and Allan (right) on the occasion of the Symposium held at Yale in 1987 honoring Allan Bromley's 60th birthday.

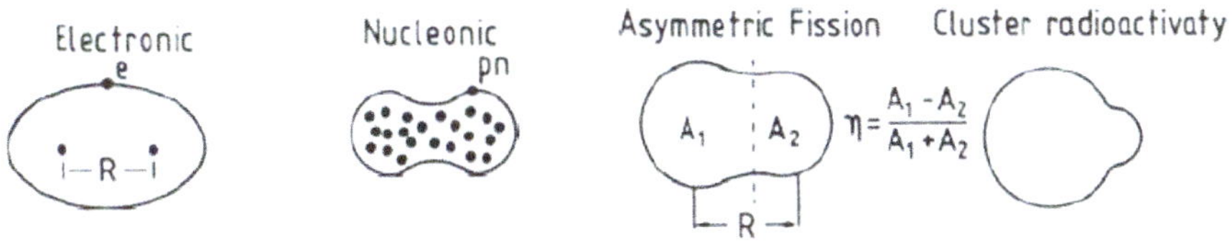

Fig. 3 Drawings of electronic and nucleonic (nuclear) molecules from [3], where Walter noted that shape isomers are also nuclear molecules

Fig. 4 Left: the cover of the Greiner, Park and Scheid book "Nuclear Molecules" [4] that was dedicated to Allan Bromley. Right: a photo [5] of Walter and Allan at Yale in 1987 on the occasion of Allan Bromley's 60th birthday

Walter continued [1]:

> It was in the late sixties / early seventies when I had the idea that in the collision of two heavy atoms intermediate electronic molecules, so-called quasi-molecules or quasi-atoms were formed, and thus quasi-molecular x-rays should be observed. Allan was very excited; he grasped the idea immediately and he . . . and his colleagues performed first-class experiments here at the Yale-tandem and proved the theory of quasi-molecular x-rays was correct. I came for a semester to Yale, brought Bernd Müller along and a very productive and exciting research time began.

W. Greiner and Neil Lane subsequently wrote [1]:

> No doubt, the experimental discovery and theoretical description of nuclear molecules have been key stimuli for the development of the research field of heavy ion physics.

4 Crossroads—Nuclear Molecules

My first contact with the work of Walter Greiner and the "Greiner School" was during my first year of graduate school. I was analyzing resonance structure in excitation functions of various inelastic scattering channels in $^{12}C + {}^{16}O$ scattering. The structures were more than obvious. I was instructed to read a paper about nuclear molecules [6] by Walter Greiner and his colleagues. The concept of a "di-molecular" nuclear system in that paper peaked my interest since it described the resonance structures in the data that I was analyzing, it was understandable, and perhaps most of all it involved some of the quantum mechanics that I was studying in class at the time. The relative motion of two nuclei in a quasi-molecular potential can be described by a quasi-molecular potential for different angular momentum states as seen in Fig. 5 (left). From today's perspective, and of course unbeknownst to me at the time, that paper firmly depicts the close collaboration between Greiner on the theoretical side and Bromley in his experiments. We later published the paper on the quasi-molecular resonances in the inelastic scattering in $^{12}C + {}^{16}O$ [7].

Clearly, the similarity between theory and data is striking as can be seen in Fig. 5 (left and right) [8]. Many additional studies, some of which I was involved in, both in the $^{12}C +^{12} C$ system [9] and in other systems [10, 11], confirmed the theory describing the quasi-molecular nuclear resonance structure seen in heavy-ion collisions in this lower energy range.

Walter continued his story [2]:

> In Germany we had since 1965 striven for establishing an "Inter-University Nuclear Research Centre". We decided Heavy Ion Physics to be the central subject of research. Later, we had great political fights with our colleagues from Karlsruhe und Heidelberg ..... In 1969, we finally succeeded in founding the "Gesellschaft für Schwerionenforschunng (GSI)" and in realizing the Laboratory according to our ideas at a location between Frankfurt and Darmstadt. Allan Bromley helped tremendously with encouraging remarks, letters to important people and colloquia about the scientific value of heavy ion physics. He came often to Germany, in particular to Frankfurt.

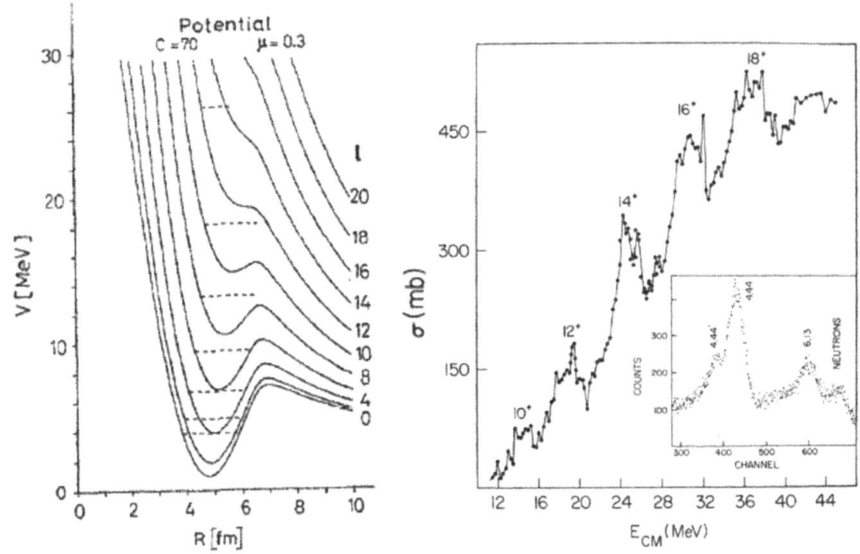

Fig. 5 Left: quasi-molecular potential of quasi-bound (solid lines) and virtual (dashed lines) states in ${}^{12}C + {}^{12}C$ inelastic scattering [6]. Right: energy dependence of the total cross section for the $2^+ \rightarrow 0^+$ γ transition in ${}^{12}C$ in ${}^{12}C + {}^{12}C$ scattering [8]. Inset is a γ singles spectrum

Figure 6 is a picture of Rudolf Bock and Allan Bromley in a typical discussion during the late 1960s. These discussions often involved how to obtain a new heavy-ion facility in Germany, as Walter explains [2]:

> The figure shows him in action at that time at some of the weekly Inter-University Heavy Ion Colloquia which I organized. . ....Rudolf Bock, a man of the early days of the "Arbeitsgemeinschaft Hessischer Kernphysiker" and the only experimenter at that time who had experience with heavy ions, was most important in building GSI. He was the principal organizer knowing what and how things should be experimentally done in the early days. He was also a very good friend of Allan Bromley and had, as you can see from this picture, intensive discussions with him.
> My university awarded him on May 12, 1978 the highest academic degree: the doctor of science honoris causa - . . .... Since that time his portrait shines from the wall of the seminar room of our Institut für Theoretische Physik in Frankfurt.

Fig. 6 Rudolf Bock and Allan Bromley discussing at a German Inter-University Heavy-Ion Colloquium

And Allan said about Walter [1]:

> ... my good friend Walter Greiner, a brilliant and distinguished theorist with whom I have worked throughout my career, at various times, I am sure had much to do with this honor.

In 1988 D. Allan Bromley was awarded the U.S. National Medal of Science by President Ronald Reagan—the highest honor available to a scientist in the United States. The citation reads: "For seminal work on nuclear molecules, for development of tandem accelerators and semi-conductor detectors for charged particles, for his contributions to particle-gamma correlation studies, and for his role in founding the field of precision heavy-ion physics."

Allan and Walter were both awarded an honorary degree from the University of the Witwatersrand at Johannesburg in South Africa. Walter spoke at Bromley's 60th birthday celebration at Yale in 1987 (the two are shown on that occasion in Fig. 4 (right)). After many more complimentary remarks about Allan, Walter co-wrote Bromley's memoir for the National Academy of Sciences saying [1]:

> But most important to me is: Allan was to me a wonderful friend. I miss him so much! As long as I live I will never forget him and always think about him and thank the Lord that I had him as a scientific companion, as reliable and wise fellow and a dear friend.

You can see from this quote that Walter was a very compassionate person. Walter went on to produce 13 textbooks in the Springer Classical and Theoretical Physics Series, of which Allan was co-author of 5.

5 Relativistic Heavy Ion Physics

I was very fortunate, fresh out of graduate school, to attend "The First Workshop on Ultra-Relativistic Nuclear Collisions" [12] and was particularly interested in a talk on nuclear compression in relativistic heavy ion collisions that was listed on the program as a presentation by Professor Walter Greiner. Instead, a young postdoc and former student of Professor Greiner, Dr. Horst Stöcker gave the presentation. Their work fascinated me, and I quote from their abstract "*we develop the essential tools for the treatment of the dynamics of High Energy Heavy Ion Collisions....We study the influence of the nuclear equation of state....we investigate the possibility of a transition from nuclear to quark matter in High Energy Heavy Ion Collisions.*" [13]. What more of a challenge could a young experimentalist (me) have than to enter this new field and to investigate such exciting effects?[1]

5.1 Compression Effects and the Nuclear Equation of State

Looking back, it's interesting to me that as Walter had just published the seminal paper on nuclear molecular structure [6], that he was working on a paper on the possibility of observing nuclear compression and producing shock waves in heavy-ion collisions [18]. The latter had a tremendous impact on the field, as it discusses the implications of high-energy head-on heavy-ion collisions reaching 3–5 times normal nuclear matter densities. It also suggests stronger compression for higher collision energies above 100 MeV per nucleon in the laboratory and the potential to investigate nuclear compressibility by studying the energy dependence of the elastic and inelastic modes. It states the need to treat such collisions with relativistic hydrodynamics and proposes the possibility of forming shock waves.

[1]As an aside—I wish to mention a few other theory influences that excited me. First was a 1973 paper by Chapline et al. [14] that proposed that hot and dense nuclear matter may be formed in head-on collisions of very energetic heavy ions. Then there was a series of lectures by Feshbach [15] that I attended, which intrigued me about a new field called relativistic heavy ion physics. There was the tome by Bethe [16] that I first learned about in taking a summer course from him on "Intermediate Quantum Mechanics," which I would later use as a reference to questions I had in nuclear physics. Later correspondences with him on a paper [17] we were writing was flattering and helpful. Likewise, much later I invited E. Teller to Berkeley to discuss our heavy ion results some 20 years after his paper [14]. Teller, then in his mid-80's, surprised and humbled (naive) me with his intellect and knowledge.

As noted above, Greiner and Stöcker subsequently presented the case for investigating the nuclear equation of state (EOS) in these collisions, the influence of the compressibility in such collisions on the EOS, and the possibility of a change in the EOS from nuclear to quark matter in high-energy heavy-ion collisions [13]. They elaborated further on this in their paper on collective sidewards flow [19] in which they perform three-dimensional nonrelativistic fluid-dynamical calculations for collisions of asymmetric systems, e.g. a light ion (Ne) on a heavy ion (U). In performing these calculations, the dependence of the internal energy and pressure on the local density and temperature, i.e. an EOS, was required as input. For the EOS, they separate the internal energy of the system into compressional and thermal components. The results of these calculations in terms of collective flow will be discussed in the next section.

This led to a series of investigations [17, 20, 21] that proposed to extract the EOS from the measured pion multiplicity through comparison with various model calculations. By simply partitioning the center-of-mass energy into thermal and potential (compressional) energy fractions, and using the pion multiplicity to determine the thermal energy fraction, the compressional energy was derived from energy conservation. A first comparison was made with a Cascade model [22] that reproduced cross sections in the proton-nucleus case, where there is no need to include potential energy effects. However, if compression were present in nucleus-nucleus collisions, any energy going into compression would reduce the energy available for pion production and produce a deficit of pions in the data compared to such model calculations, as seen in Fig. 7 (left). Thus, a comparison of these model calculations with the measured pion multiplicity in nucleus-nucleus collisions allowed extraction of the compressional energy in the collision.

Assuming the total multiplicity of produced pions reflects the maximum density reached in central nucleus-nucleus collisions at laboratory energies of a few GeV/nucleon, densities for different interacting systems at different energies were first derived from the Cascade calculation. A systematic discrepancy between the pion multiplicities predicted by the Cascade model calculations and those in the experiment was found to depend on the density and attributed to a bulk compressional effect not present in the calculations. The first results of this experimental investigation [21] found an extremely stiff EOS as seen in Fig. 7 (right).

In refining this approach for head-on collisions of medium-to-heavy nuclei [17, 20], the cascade model was replaced by a thermal-chemical equilibrium model [23] with shock compression [14, 24], and thus corresponds to using a schematic hydrodynamic model [25]. In this approach it was found that the pion-like degree of freedom reflects the high density stage and freezes out at the onset of expansion [20]. This was a feature of chemical equilibrium being maintained among the most abundant particles—pions, nucleons and delta resonances—through the high density and the chemical freezeout stages with little effect on the pion multiplicity from the expansion after that. It allowed extraction of the equation of state by equating the over-production of the pion multiplicity in the models with their lack of accounting for any compressional energy. This comparison resulted in a stiff equation of state verifying more generally that measured initially seen in Fig. 7 (right).

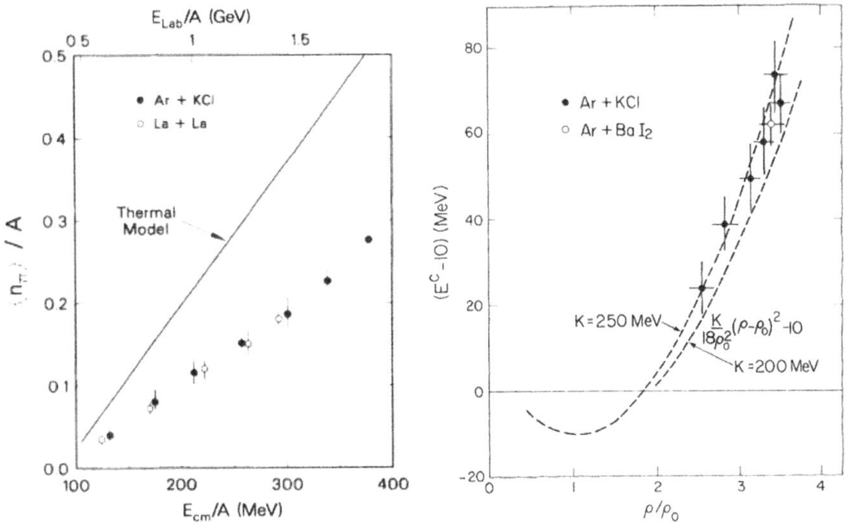

Fig. 7 Left: data on pion multiplicity per participant nucleon as a function of center-of-mass energy per participant nucleon for Ar+KCl and La+La. The line represents a thermal model calculation described in the text [20]. Right: compressional energy, less 10 MeV for nuclear binding, as a function of density in units of normal nuclear density derived from a comparison of the pion multiplicity measured in Ar+KCl and Ar+BaI$_2$ [21] and a Cascade model calculation [22]. See text for details

It was also noted at the time [17] that the approximations inherent in the above approach ignored the flow energy in the high density stage, any dilute surface effects, medium corrections to the specific heat, and the role of the Fermi degeneracy energy. Therefore, the resulting compressional energies could be considered only as a first estimate of the nuclear matter equation of state, which appeared to be fairly stiff and in agreement with relativistic field theoretical predictions [26].

5.2 *Compression Effects—Collective Flow in Relativistic Heavy Ion Collisions*

As already pointed out above, Walter Greiner, and his colleagues Stöcker and Maruhn, predicted a collective sidewards flow [19] from their three-dimensional nonrelativistic fluid-dynamical calculations for collisions of asymmetric systems, for different impact parameter collisions, and at various c.m. energies. Two schematic results of these calculations can be seen in Fig. 8 for asymmetric beam collisions of Ne+U with the Ne beam entering from the right. The top figure is for head-on collisions and the bottom figure for impact parameter b = 6 fm collisions. They also predicted angu-

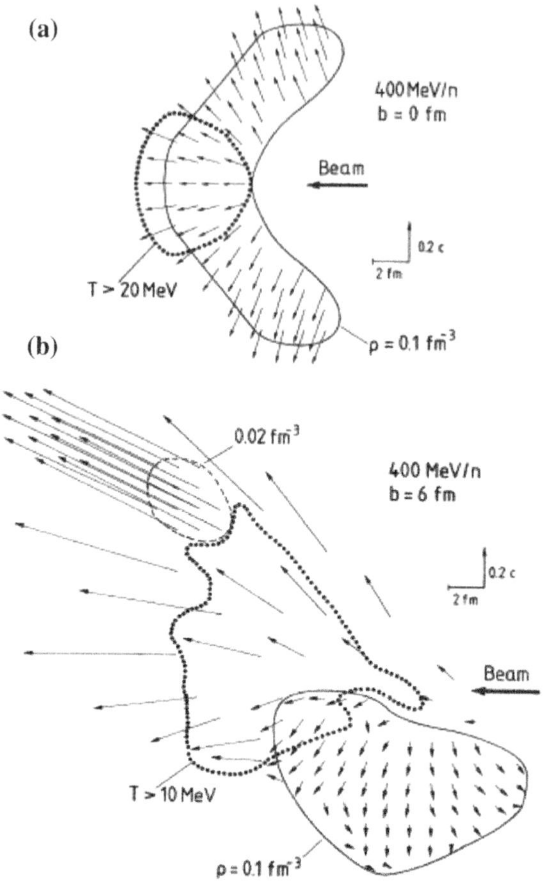

Fig. 8 Projections onto the reaction plane with the beam incident from the right. Arrows are velocity field vectors and contours are of constant density and temperature as labeled. A central collision (b = 0) (top) and an intermediate impact parameter collision (b = 6 fm.) (bottom) for $E_{lab} = 400$ MeV/nucleon Ne + U collisions from [19]

lar distributions of particles for various colliding configurations and nuclei, which exhibited a clear sidewards collective flow of the velocity vectors as seen in Fig. 8.

Soon after, data from the Plastic Ball at the BEVALAC for symmetric colliding systems (Ca+Ca and Nb+Nb), shown in the top two diagrams of Fig. 9 (left), exhibited the predicted sidewards flow [27]. The sidewards flow can be seen on the right side of the two diagrams ($y_{cm} > 0$) as an off-axis vertical asymmetry (away from the beam direction indicated by $p_x/m = 0$), where the beam enters from the left. The left side of the diagrams ($y_{cm} < 0$) shows a depletion due to losses of low energy particles in the target region. The bottom left diagram is the result of a Cascade model calculation [22], where no off-axis vertical asymmetry is observed, indicating the expected lack of sidewards flow in this model. The experimental results in [27] also show more details in terms of the angular distributions as a function of collision centrality and the size (mass) of the colliding system. These exhibit clear sidewards flow that increases with centrality and system size as predicted in [19].

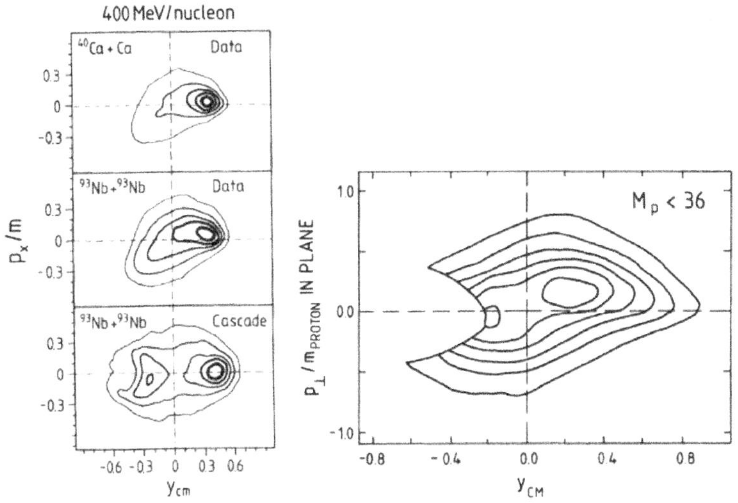

Fig. 9 Projections onto the reaction plane of contours of constant yield with the beam incident from the left. Left: Plastic Ball data [27] for two symmetric collision systems (top Ca+Ca, center Nb+Nb) and Cascade calculations (bottom Nb+Nb). Depletion at $y_{cm} < 0$ is due to losses of low energy particles in the detector. Right: contours of constant invariant proton cross section from the Streamer Chamber [28] for intermediate impact parameter events (b $\sim 0.5 b_{max}$) in the asymmetric collision system Ar+Pb at $E_{lab} = 800$ MeV/nucleon. Depletion at $y_{cm} < 0$ is due to target absorption losses. See text for discussion

Data in an asymmetric collision system (Ar+Pb) from the Streamer Chamber at the BEVALAC also exhibited sidewards flow [28]. These data are presented in Fig. 9 (right) for intermediate multiplicities (impact parameters) as a function of the same variables as in the left-side of the figure. The off-axis vertical asymmetry is again apparent on the right side of the diagram representing the presence of collective sidewards flow as predicted in [19] originally for asymmetric systems.

Subsequently, the flow of identified light and intermediate-mass nuclei ($1 \leq Z \leq 9$) was measured over a large solid angle in the Plastic Ball [29] for peripheral and central collisions of $E_{lab} = 200$ MeV/nucleon Au+Au. The fragments were found to exhibit stronger flow than lighter particles. Both the transverse momentum per nucleon and the alignment of the fragments in position and momentum space relative to the reaction plane were observed to increase with the mass of the fragment as predicted by theory [30–32].

This first-time observation of enhanced flow of fragments, along with the earlier measurements of collective sidewards flow at the BEVALAC and its predictions paved the way for future experiments to identify and study collective nuclear effects in relativistic heavy ion collisions.

Experiment E877 at the Brookhaven AGS used calorimetry for the first time to determine the reaction plane orientation. They observed directed flow and higher harmonics of the event shape for Au+Au collisions at $E_{lab} = 11.4$ GeV/c per nucleon

[33] by incorporating a Fourier expansion of the azimuthal distributions of the transverse energy in various pseudo-rapidity intervals. Azimuthal event anisotropies indicating directed sidewards flow were obtained. E877 later reported measurements of both transverse energy and charged particle azimuthal distributions for different centralities relative to the reaction plane [34], where they observed the strongest directed flow for semi-central collisions and a stronger anisotropy for slower particles. They reported a small but nonzero elliptic anisotropy focused in the reaction plane, reflecting an influence of the initial collision geometry, in addition to a directed flow oppositely directed at forward compared to backward pseudo-rapidities. An azimuthal anisotropy in event-by-event particle distributions had been mentioned previously in [35] as a signature of transverse collective flow, and later called elliptic flow.

A study of directed and elliptic flow as a function of rapidity and transverse momentum was conducted by NA49 [36] for Pb+Pb collisions at $E_{lab} = 158$ GeV/nucleon at the CERN SPS. Protons and pions were found to exhibit directed flow on opposite sides of the beam to each other at large rapidities. Elliptic flow was found to peak at medium-to-large rapidities for pions and closer to mid-rapidity for the protons. The flow axes of the elliptic flow for both pions and protons was found to be in the plane of the directed flow, i.e. in the reaction plane. Shadowing by spectator matter was therefore excluded by these data as the origin of elliptic flow. It was concluded that the elliptic flow must therefore be an effect resulting from the high density stage of the collision.

E895 measured proton elliptic flow in Au+Au over the laboratory energy range of 2–8 GeV per nucleon and found a transition from a negative to positive elliptic flow at a 4 GeV per nucleon [37]. They proposed that this suggested a softening of the nuclear equation of state that was stiff ($K = 380$ MeV) at low laboratory energies and soft ($K = 210$ MeV) at laboratory energies above 4 GeV per nucleon.

With the advent of heavy-ion colliders, STAR measured the elliptic flow of charged particles at mid-rapidity in Au+Au collisions at a c.m. energy of 130 GeV per nucleon pair [38] at the Relativistic Heavy Ion Collider at Brookhaven National Laboratory. The elliptic flow was observed to reach a maximum for near-peripheral collisions and to decrease for the most central collisions. The elliptic flow was larger than had been measured at lower energies and increased as a function of the transverse momentum. This was interpreted as representing a higher degree of thermalization than at lower collision energies, since the most central collisions were found to reach the hydrodynamical limit for the elliptic flow.

STAR later measured the directed flow in Au+Au collisions at 62.4 GeV per nucleon pair in the mid-pseudorapidity region and in the forward pseudo-rapidity region [39]. Using the three-particle cumulant method, the event plane method with mixed harmonics, and the standard method with the event plane reconstructed from spectator neutrons, the three methods were shown to exhibit good agreement. The charged particle directed flow was found to be in the opposite direction from that of the fragmentation neutrons as shown in Fig. 10 (left), allowing determination of the sign of the directed flow (v_1).

First measurements of charged particle elliptic flow at the CERN Large Hadron Collider were reported by ALICE [40] for Pb–Pb collisions at the c.m. energy of

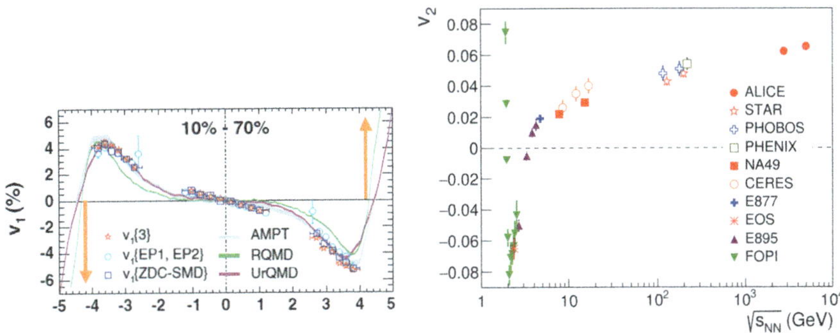

Fig. 10 Left: directed flow (v_1) of charged particles measured in STAR as a function of pseudo-rapidity for 62.4 GeV per nucleon pair Au+Au collisions and centrality selection 10–70% [39]. The direction of flow for spectator neutrons is represented by the arrows, whose location along the pseudo-rapidity axis corresponds to the rapidity of the beam in each direction. Right: recent compilation of data on the integrated elliptic flow (v_2) as a function of c.m. energy per nucleon pair for near-central AA Collisions (~20–30% centrality, A ~ 200). See [40, 43] for references to all data presented

2.76 TeV per nucleon pair. The elliptic flow increased by ~30% compared to RHIC collisions of Au+Au at 200 GeV c.m. energy per nucleon pair as can be seen in Fig. 10 (right). The LHC and RHIC elliptic flow results have similar behavior as a function of transverse momentum and are consistent with hydrodynamic model predictions that include viscous corrections [41, 42]. The larger integrated elliptic flow at the LHC appears to be due to an increase in the mean transverse momentum.

Subsequently, ALICE reported results of elliptic (v_2), triangular (v_3), and quadrangular (v_4) flow of charged particles at mid-rapidity in Pb+Pb collisions at 5.02 TeV c.m. energy per nucleon pair [43]. The anisotropic flow was measured using two-particle correlations with a pseudo-rapidity gap greater than one unit using the multi-particle cumulant method. Increases in the elliptic, triangular, and quadrangular flows were observed when going from 2.76 to 5.02 TeV, and was attributed primarily to the increase in the average transverse momentum of particles between the two energies. The increase in the integrated elliptic flow can be seen in Fig. 10 (right). The measurements were found to be compatible with hydrodynamic model calculations. Also, the transition from a negative to positive elliptic flow at 4 GeV per nucleon as seen by E895 [37] and mentioned previously is apparent.

6 Closing Remarks

Professor Walter Greiner has had a profound impact on physics. His seminal papers have started new fields of research in nuclear physics. His teaching and educational physics books have informed generations. His former students continue to carry out his rich legacy. I thank Walter for all of this, as well as befriending me at a time in my young career when I thought that he did not even recognize me.

Acknowledgements I would like to thank Walter for the bold and broad physics perspective that he imparted to the world and that impacted my research. I thank Lee Schroeder and Reinhard Stock for giving me my first opportunity to perform research in relativistic heavy-ion physics. I sincerely appreciate Reinhard Stock for his lifelong friendship and mentorship. I thank Paula Farnsworth for her assistance in the preparation of this document. This work was supported in part by the Office of Science and Office of Nuclear Physics of the U.S. Department of Energy.

References

1. W. Greiner, N. Lane, *Biographical Memoir of David Allan Bromley* (National Academy of Sciences, Washington, D.C., 2009)
2. P.A. Fleury, F. Iachello (eds.), *D. Allan Bromley, Nuclear Scientist and Policy Innovator* (World Scientific, 2006)
3. W. Greiner, in *Clustering Phenomena in Atoms and Nuclei*, ed. by M. Brenner, T. Lönnroth, F.B. Malik (Springer, Berlin, Heidelberg, 1992)
4. W. Greiner, J. Park, W. Scheid, *Nuclear Molecules* (World Scientific, Hackensack, NJ, 1995)
5. Permission for use granted by the American Institute of Physics' Emilio Segre Visual Archives
6. H.-J. Fink, W. Scheid, W. Greiner, Nucl. Phys. A **188**, 259 (1972)
7. R.E. Malmin, J.W. Harris, P. Paul, Phys. Rev. C **18**, 163 (1978)
8. T.M. Cormier, J. Applegate, G.M. Berkowitz, P. Braun-Munzinger, P.M. Cormier, J.W. Harris, C.M, Jachcinski, L.L. Lee, Jr., J. Barrette, H.E. Wegner, Phys. Rev. Lett. **38**, 940 (1977)
9. T.M. Cormier, C.M. Jachcinski, G.M. Berkowitz, P. Braun-Munzinger, P.M. Cormier, M. Gai, J.W. Harris, J. Barrette, H.E. Wegner, Phys. Rev. Lett. **40**, 924 (1978)
10. P. Braun-Munzinger, G.M. Berkowitz, T.M. Cormier, C.M. Jachcinski, J.W. Harris, J. Barrette, M.J. Levine, Phys. Rev. Lett. **38**, 944 (1977)
11. J. Barrette, M.J. Levine, P. Braun-Munzinger, G.M. Berkowitz, M. Gai, J.W. Harris, C.M. Jachcinski, Phys. Rev. Lett. **40**, 445 (1978)
12. *The First Workshop on Ultra-Relativistic Nuclear Collisions*, Lawrence Berkeley Laboratory, Berkeley, CA, 21–24 May 1979
13. H. Stöcker, J. Hofmann, J.A. Maruhn, W. Greiner, *The First Workshop on Ultra-Relativistic Nuclear Collisions*. Lawrence Berkeley Laboratory Report (unpublished, 1979), pp. 355–419
14. G.F. Chapline, M.H. Johnson, E. Teller, M.S. Weiss, Phys. Rev. D **8**, 4302 (1973)
15. R. Balian et al. (eds.), *Nuclear Physics with Heavy Ions and Mesons—Proceedings of the Les Houches Summer School Session XXX*, 4 July–20 August 1977. Lectures on Relativistic Heavy Ions (Amsterdam, North–Holland, 1978)
16. H.A. Bethe, Annu. Rev. Nucl. Sci. **91**, 93 (1971)
17. J.W. Harris, R. Bock, R. Brockmann, A. Sandoval, R. Stock, H. Ströbele, G. Odyniec, H.G. Pugh, L.S. Schroeder, R.E. Renfordt, D. Schall, D. Bangert, W. Rauch, K.L. Wolf, Phys. Lett. **153B**, 377 (1985)
18. Werner Scheid, Hans Müller, Walter Greiner, Phys. Rev. Lett. **32**, 741 (1974)
19. Horst Stöcker, Jouchim A. Maruhn, Walter Greiner, Phys. Rev. Lett. **44**, 725 (1980)
20. J.W. Harris, G. Odyniec, H.G. Pugh, L.S. Schroeder, M.L. Tincknell, W. Rauch, R. Stock, R. Bock, R. Brockmann, A. Sandoval, H. Ströbele, R.E. Renfordt, D. Schall, D. Bangert, J.P. Sullivan, K.L. Wolf, A. Dacal, C. Guerra, M.E. Ortiz, Phys. Rev. Lett. **58**, 463 (1987)
21. R. Stock, R. Bock, R. Brockman, J.W. Harris, A. Sandoval, H. Ströebele, K.L. Wolf, H.G. Pugh, L.S. Schroeder, M. Maier, R.E. Renfordt, A. Dacal, M.E. Ortiz, Phys. Rev. Lett. **49**, 1236 (1982)

22. J. Cugnon et al., Nucl. Phys. A **352**, 505 (1981) and **379**, 553 (1982)
23. R. Hagedorn, J. Ranft, Suppl. Nuovo Cimento **6**, 169 (1968); J.I. Kapusta, Phys. Rev. C **16**, 1493 (1977)
24. G.F. Chapline et al., Phys. Rev. D **8**, 4302 (1973); M.I. Sobel et al., Nucl. Phys. A **251**, 502 (1975)
25. H. Stöcker, W. Greiner, W. Scheid, Z. Phys. A **286**, 121 (1978)
26. J. Boguta, H. Stöcker, Phys. Lett. **120B**, 289 (1983)
27. H.A. Gustafsson, H.H. Gutbrod, B. Kolb, H. Löhner, B. Ludewigt, A.M. Poskanzer, T. Renner, H. Riedesel, H.G. Ritter, A. Warwick, F. Weik, H. Wieman, Phys. Rev. Lett. **52**, 1590 (1984)
28. R.E. Renfordt, D. Schall, R. Bock, R. Brockmann, J.W. Harris, A. Sandoval, R. Stock, H. Ströbele, D. Bangert, W. Rauch, G. Odyniec, H.G. Pugh, L.S. Schroeder, Phys. Rev. Lett. **53**, 763 (1984)
29. K.G.R. Doss, H.-A. Gustafsson, H. Gutbrod, J.W. Harris, B.V. Jacak, K.-H. Kampert, B. Kolb, A.M. Poskanzer, H.-G. Ritter, H.R. Schmidt, L. Teitelbaum, M. Tincknell, S. Weiss, H. Wieman, Phys. Rev. Lett. **59**, 2720 (1987)
30. H. Stöcker, A.A. Ogloblin, W. Greiner, Z. Phys. A **303**, 259 (1981)
31. L.P. Csernai, H. Stöcker, P.R. Subramanian, G. Graebner, A. Rosenhauer, G. Buchwald, J.A. Maruhn, W. Greiner, Phys. Rev. C **28**, 2001 (1983)
32. L.P. Csernai, G. Fai, J. Randrup, Phys. Lett. **140B**, 149 (1984)
33. J. Barrette et al. (E877 Collaboration), Phys. Rev. Lett. **73**, 2532 (1994)
34. J. Barrette et al. (E877 Collaboration), Phys. Rev. C **55**, 1420 (1997); Erratum Phys. Rev. C **56**, 2336 (1997)
35. J.-Y. Ollitrault, Phys. Rev. D **46**, 229 (1992)
36. H. Appelshäuser et al. (NA49 Collaboration), Phys. Rev. Lett. **80**, 4136 (1998)
37. C. Pinkenburg et al. (E895 Collaboration), Phys. Rev. Lett. **83**, 1295 (1999)
38. K.H. Ackermann et al. (Star Collaboration), Phys. Rev. Lett. **86**, 402 (2001)
39. J. Adams et al. (STAR Collaboration), Phys. Rev. C **73**, 034903 (2006)
40. K. Aamodt et al. (ALICE Collaboration), Phys. Rev. Lett. **105**, 252302 (2010)
41. G. Kestin, U.W. Heinz, Eur. Phys. J. C **61**, 545 (2009)
42. H. Niemi, K.J. Eskola, P.V. Ruuskanen, Phys. Rev. C **79**, 024903 (2009)
43. J. Adam et al. (The ALICE Collaboration), Phys. Rev. Lett. **116**, 132302 (2016)

Pions in Matter Matter

Christoph Hartnack

Abstract The Isospin Quantum Molecular Dynamics Model (IQMD) is used to analyse the properties of pions in heavy ion collisions in the range beyond 2 AGeV. We find a strong rescattering causing the pions to show a signature of rather low freeze out densities even if they are initiated at high densities. The analysis of the pion sideward flow may yield insights into the delta-nucleon interaction. The mass distribution of the final delta does not need a medium mass shift of the pole in order to be understood. The analysis of pion isospin ratios may reveal information on the neutron skin of the nucleus.

1 Introduction

One of the main interests of the study of relativistic heavy ion collisions is the investigation of the properties of nuclear matter at extreme densities and excitation energies. These investigations include the production of secondary particles, the properties of particles in a (dense) nuclear medium, the compression and repulsion of dense nuclear matter, its equilibration during the reaction and its decay into fragments and single particles. The most prominent secondary particle is the pion, a pseudoskalar meson which due to its very small mass can already be detected in heavy ion collisions of a few hundred Mev/nucleon of incident energies in the laboratory frame. This article will focus on observables related to that particle in an energy range of several hundred AMeV to a few AGeV.

C. Hartnack (✉)
SUBATECH, UMR 6457 IMT Atlantique, IN2P3/CNRS,
Université de Nantes, 4 rue A. Kastler, 44070 Nantes, France
e-mail: hartnack@subatech.in2p3.fr

J. Kirsch et al. (eds.), *Discoveries at the Frontiers of Science*,
FIAS Interdisciplinary Science Series, https://doi.org/10.1007/978-3-030-34234-0_21

1.1 A Very Personal Introduction

In this—very personally written—subsection the author will try to explain how he came into those activities and how he got into the group of Walter Greiner, who was his lab director, master and Ph.D. thesis advisor, who became his mentor in the first steps of physical research and a friend—and to whom the author wants to dedicate this article. A reader only interested in the pions themselves may skip this and the following subsection and continue at Sect. 1.3. If he needs a general introduction to the field he is recommended to read the Physics Report of Stöcker and Greiner [1], which in my starting time was regarded as some sort of "bible" for heavy ion physics. However, since in the institute of Greiner the term "bible" was already reserved to the three volumes on Nuclear Physics by Eisenberg and Greiner—and regarding the blue colour of the coverage of a Physics Report reprint—we used to call it the "blue catechism".

First of all I have to confess that my entry into the field of heavy ion collisions was not at all a straight line. I decided to study physics only at the last minute before closure of the applications of the University (finally the study consultants in mathematics and chemistry did not invoke my enthusiasm) and I could not really decide in between theoretical nuclear physics or experimental atomic physics. It was only on a Friday at noon time, when I accompanied another student to the "Mensa" (the student's restaurant) when he stopped by at the office of Prof. Greiner, who came out and greeted us: "Well, Mr. Rischke, you already told me that you wanted to join my institute. Nice! And Mr. Hartnack, we already know each other from your interviews for the Studienstiftung, you also decided to join us!" I was quite puzzled and stammered that up to now I had not decided between Schmidt-Boecking and him. However, Greiner cut my words: "You are decided and from now on—being a member of the institute—you will have to attent our palaver[1] regularly!" I did not imagine at that moment all the consequences which that decision could have for my life, but a first direct result was already applying next day since I inherited immediately the duty of transporting the tea carrier for the Friday palaver (each Friday seminar was preceded by a cup of tea), a task for which it took me about one year to find another victim of heritage.[2]

When arriving in the institute I was proposed to work on the production of positron peaks at GSI and I had already started to try to digest a book on "Quantum Electro Dynamics of Strong fields" when the swing-by of a young dynamical professor freshly arriving from Michigan State University kicked me out of my orbit. He was working on a model called "Vlasov-Uehling-Uhlenbeck" [2, 3] and it seemed

[1]A palaver was an internal seminar organised only by the Greiner institute itself, which was programmed on every Monday afternoon and on each Friday afternoon. It was obligatory for all members of the institute and missing it could result in a personal convocation by Prof. Greiner. Of course these duties were supplemented by seminars each Tuesday (GSI seminar), Wednesday (Physikalisches Kolloquium) and Thursday (Kernphysikalisches Kolloquium). The reader may thus imagine the high level of seminar culture in the institute.

[2]This had to be typically the most recent entrant into our institute, as Greiner liked to indicate "it is an old tradition that the youngest apprentice has to go for the beer...".

to be fascinating, even if I did not really understand, what it was doing. However, I decided to change my topic of study and went to Greiner's office: "Professor, no more positrons, I want to work with Stöcker on that Flasche-Uehling-sonstwie theory![3]". Greiner stared at me and said: "Mister Hartnack, you want to do a theory which you even do not know how to pronounce?" When confirmed it, he only said "go on!!!" and this it how it started (and it is for that reason that from now on I will abbreviate Vlasov-Uehling-Uhlenbeck by VUU). My task was in principle to marry the VUU model with another recent ansatz called "Quantum Molecular Dynamics" (QMD) [4]. While the first model was a mesoscopic model including isospin and pions the latter model was microscopic but lacking exactly these ingredients. This was the birth of a model, which later on was called "Isospin Quantum Molecular Dynamics" model or IQMD [5].

1.2 *Some Nostalgic Remarks Containing a Genealogy of QMD*

It would be a hypocrisis to neglect the importance of the Institute of Theoretical Physics of Frankfurt for the dynamics of the development of QMD models at that time. Thanks to Walter Greiner the institute disposed of excellent conditions for working on this field and a high culture of collaboration and scientific exchange,[4] for instance by contact to people working on relativistic hydrodynamics.[5]

The re-arrival of Horst Stöcker back from MSU brought us the first version of the VUU model, made by Hans Kruse, extended by Barbara Jacak, implemented on the IBM main frame (the computer at GSI used by us) by Joe Molitoris,[6] which lateron was handed to me via Ludwig Neise.[7]

[3] A direct translation would be "Bottle-Uehling-something else" but this only reflects that I did not manage the correct pronounciation at that time.

[4] Not only scientific. I remember one day when my colleague in office offered me a chocolate. Just at this moment Horst Stöcker came in, saw us chewing, and threatened us to pillage the room, if we did not show him the source of our chocolate. When we decided to offer him some chocolates as well, Walter Greiner showed up, in search for Horst but discovering three peoples eating chocolates. Evidently he had to be invited as well but the box of my room mate got completely empty.

[5] However, I have to admit that in the beginning it was a little bit confusing for me to enter a world of $\hbar = c = k_B = 1 = \text{me} = \text{you} = \text{everything}$ and to use the word "Bazooka" for describing heavy ion collisions and not in a militarian context.

[6] Joe should also have been in the institute around the time of my arrival but I honestly do not remember—perhaps I was still too much taken by my attempt to digest the book on QED of strong fields. Overall for me this part is sourced by "oral tradition". I never met Hans Kruse and Barbara Jacak only a couple of years later. However, from some remnants in the source code I assume, that it was originally developed on Vax machines, thus it might be him who transferred it.

[7] However, I have to admit that—due to the incompatibility of working times of my tutor and me—the initiation happened one day at 5 p.m. on the staircase of the institute, when I was just leaving and he was just coming: "The source code is on UF36.NFM.FORT(APVUU). Have fun." However,

In parallel to this, Jörg Aichelin, who had moved from MSU to Heidelberg, brought a first version of QMD, that he had developed with Horst Stöcker at MSU. Since that code was based in the collision term on a BUU, that he had developed with George Bertsch, the code did not include isospin yet, while VUU did. Besides the isospin BUU and VUU differed also in another point, even if they solved the same VUU equation using pointlike particles overlayed in parallel events: BUU was determining the particle densities as points in a grid (which runs faster for dense systems), VUU surrounded each particle with a sphere (which avoids to run out of the grid in the final state).

The freshly arrived QMD was handed to a young student named Georg Peilert.[8] He used the code for analysing the "sideward flow" of fragments measured by the Plastic Ball collaboration. Here the first signatures of collectivity were found ("the fragments go with the flow") as well as an influence of p_X/p_T on the nuclear equation of state. Lateron he worked on the topic of a Pauli potential.

In parallel the same QMD was rewritten in Heidelberg by Achim Bohnet, a student of Jörg Aichelin. That code was lateron used especially for the analysis of fragmentation data measured by ALaDiN, Indra or ALaDiN-Indra. Most of the publications mentioning "QMD" without any prefix refer to that code. Recently sometimes the term "BQMD" has been also used for that code to make the situation clearer.

IQMD, as already indicated, was built on VUU, using the isospin and the collisions of VUU but adding potentials and Pauli blocking inspired by QMD and taking into account the isospin explicitly.[9] Besides the pion production, a point we will focus lateron, one of the first applications of IQMD was the analysis of the so-called squeeze-out. This effect was first seen by the Diogène collaboration[10] and lateron by the Plasticball collaboration. It was found that this effect was a collective effect and sensitive to the nuclear equation of state. Furtheron it was assumed that the normalised squeeze $\langle (p_X^2 + p_Y^2)/p_T^2 \rangle$ may show a plateau between 600 and 800 MeV when a soft eos is used.[11] This region is seen today as the minimum of v_2.

A big advantage of IQMD/VUU with respect to QMD/BUU was the propagation of pions in the first one, while the latter one only propagate deltas. Therefore IQMD was naturally in the first line for analysing pion data. It first application was the

the lineprinter-listing of the code helped me to fight boredom in two weeks of hospitalisation and assured me some—strange—admiration from the medical staff, head physician included.

[8]A franconian boy from the Main, who—by neglection of that subtlety—was regarded by us as a Bavarian. Today I understand better, why he was unhappy about this.

[9]In principle it is still possible to run old VUU out of the IQMD code. But it is a little bit tricky to do it. To cite the Diploma thesis of Steffen Bass: "Quod licet Jovi, non licet bovi.".

[10]Legend tells, that in a conference the speaker of that collaboration showed an azimuthal distribution that they did not understand and tried to claim it as fluctuation when Horst Stöcker raised up, shouting: "This is the squeeze-out, that I have predicted!".

[11]This variable would today be called $2v_2$, but that name was not invented at this time. Lacking a name for such self-suggesting quantity we called it P_{XYT}, neglecting that it is impossible to pronounce such a variable in a conference talk. From this point of view it was an advantage to create lateron v_1 and v_2, even if v_1 corresponds in a similar way to a flow variable already used before: $v_1 = \langle p_X/p_T \rangle$.

analysis of pion data from the Diogène collaboration [6]. The collaboration had measured a positive mean transverse momentum of pions towards the reaction plane, which could also be confirmed by calculations of IQMD [7], while the "standard model" of the collaboration, the intranuclear cascade model of Cugnon (INC) [8] got zero mean momentum. Since the cascade lacked nucleon potentials, especially the nuclear equation of state (eos), the first assumption was, that signal could be useful for determining the nuclear equation of state (eos). However, in IQMD the influence of the eos was quite small and even calculations without potentials yielded positive momenta. It took a couple of years to understand, what really had been the reason [9].

In the mean time a young diploma student, Steffen Bass, had joined the team and Horst asked me to tutor his master thesis, while I was just finishing my Ph.D. Steffen was very enthusiastic on the subject[12] and did a lot of calculations on pion dynamics and the pion-delta-cycle. He extended IQMD in several parts of the pion sector, like detailed balance and was the driving force for converting the code from IBM mainframe to Unix type machines, which opened us to the upcoming computer clusters. He lateron became one of the central persons of the UrQMD project,[13] but to keep the chronological order, there should be first some words on RQMD.

In the end of the 80s Heinz Sorge joined the institute with the idea of creating a relativistic QMD. It was decided that it should be based on IQMD and so we started to collaborate.[14] Lateron he built up an own group with young diploma students like Raffaele Mattiello, Andreas von Keitz, Andre Jahns and Luke Winckelmann. The results were very promising and helped to understand the experiments performed at AGS and CERN-SPS.[15] Lateron Heinz Sorge went to Stony Brook, developing furtheron on this code there, while the Frankfurt part continued under the guidance of Thomas Schönfeld. However, a real collaboration on that distance was becoming more and more difficult on the long term.

In the meantime—tired of the differences between different QMDs—Steffen Bass, Jens Konopka (who had developed another QMD with time equations on the particle width) and me decided together with Horst Stöcker to unify all QMDs into one unified version: the UQMD. This new code should be remade completely from scratch. First

[12] I suffered somehow from this, since I was in pressure of finishing my thesis while he wanted to learn more and more details on IQMD. This led to the famous hide-and-seek at GSI: one day I hided myself in the darkest corner of the NIM-Pool of the Unilac, hoping to advance my thesis silently. 10 minutes later the door opened and a voice whispered "I think to have seen him here...". 30 seconds later Steffen stood in front of me grinning: "You forgot an important thing. I have done an internship at EPOS and I have many friends nearby...".

[13] I was told that his car in the US was immatriculated with the characters URQMD, but that this caused strange speculations in the non-specialised part of the population. I had similar experience in France when naming a folder containing QMD results as "My Q-Files" (mes fichiers Q).

[14] Only few people may know, but the first RQMD group consisted of Heinz and me. I remember many interesting discussions with him around a conference in Peniscola (Spain).

[15] I remember, that one day Horst Stöcker had just left for a conference, when newest RQMD calculations were finished. I made a photocopy of the print onto a slide and rushed to the tram station where I was able to catch Horst for giving him a slide (still warm from the copymachine) with literally brandnew results.

analysis of the differences had been done [10] and first concepts delivered, when one day Horst Stöcker convoked an urgent meeting of the “UQMD-steering board”, consisting of the 4 persons named above. Since Steffen and Jens were actually on a conference and me being in Nantes, the remaining representative[16] decided with unanimity to enhance the project by merging the group with the Frankfurt part of the RQMD group and to change the focus on ultrarelativistic energies. This fusion of UQMD and the Frankfurt RQMD part was the birth of UrQMD [11]. However, a long pathway was before us, before the first version was running. This new code was rewritten independently of all the preceding models—but of course inspired by them. Major changes included the enhancement of mesons and resonances from the π, Δ pair to many other species, the inclusion of strings and many other things necessary for going up to highest energies.[17]

In the “low energy” sector around 1–2 GeV, where higher resonances play no important role, IQMD and UrQMD, both independent models, yield compatible results. This is the domain of energy we want to study in the following sections.[18]

1.3 The Production of Pions

The philosophy of IQMD follows the idea used by many other microscopic transport models [1, 12, 13] to decompose the interaction of nucleons into a longrange part described by local central force two-body potentials and into a shortrange part described by stochastic collisions. Let us first focus on the collisions, while the potential part will be discussed lateron. However, a detailed description of both parts and their application in IQMD can be found in [10]. Two particles collide if their minimum distance d, i.e. the minimum relative distance of the centroids of the Gaussians during their motion, in their CM frame fulfills the requirement:

$$d \leq d_0 = \sqrt{\frac{\sigma_{\text{tot}}}{\pi}}, \qquad \sigma_{\text{tot}} = \sigma(\sqrt{s}, \text{ type}). \tag{1}$$

[16]The singular is intended! However, we had no quota reglementation for our board. But in retrospective I have to admit that it was a visionary—and wise—decision, even if in a first reaction I had been rather disappointed since I felt being absorbed by another group following another topic. Indeed, the workload would have been to heavy to build up two codes from scratch in parallel and the new project was clearly of top priority.

[17]In this short description I could not name all companions of our adventure. I remember close work together with Maria Berenguer-Vidal, Sven Soff, Christian Spieles, Henning Weber, Markus Bleicher and many many others. I apologize to those not mentioned as well to those who might feel to have been displayed in a wrong light.

[18]In relecture of this part the authors feels urged to highlight—in deep gratitude—the person behind all the students: Horst Stöcker has always been the “heart” of the project, guiding with longrange visions, having the feeling for important points and being extremely talented in motivating his students by sharing his enthusiasm.

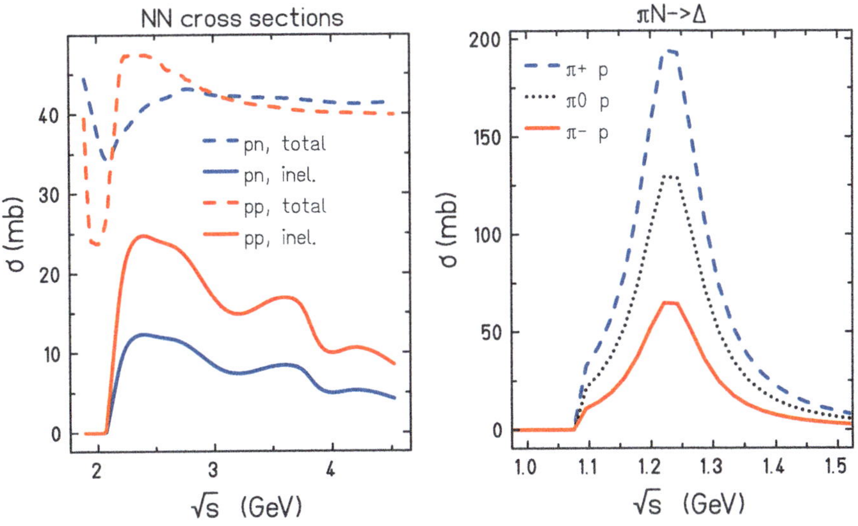

Fig. 1 Left: Inelastic and elastic cross sections for nucleon-nucleon collisions. Right: Cross sections for pion-nucleon reactions

where the cross section is assumed to be the free cross section of the regarded collision type ($N-N$, $N-\Delta$, ...).

The cross sections for elastic and inelastic collisions are obtained by a table lookup using experimentally measured cross sections (when available) or derived from available cross sections using symmetry assumptions and detailed balance. The energy dependence of the cross sections in nucleon-nucleon and pion-nucleon collisions can be depicted from Fig. 1. In the nuclear medium the free cross sections are supplemented by a test of the Pauli blocking in the final state.

The pion production in IQMD is done via the Δ-channel, where deltas can be produced in nucleon-nucleon (NN) collisions and be reabsorbed in $N\Delta$ collisions. The Δ decays and produces a free pion, which can be reabsorbed in collisions with a nucleon and form a Δ again:

$$NN \leftrightarrow N\Delta \qquad \Delta \leftrightarrow N\pi \tag{2}$$

These reactions have to comply with detailed balance and isospin effects have to be taken into account by the use of Clebsch-Gordon coefficients. For more details see [14].

While the production and absorption of deltas are assumed to happen at the point of nearest distance of the collision, the production of pions is following the exponential decay law of the delta, where the decay constant is determined by the width of the resonance. Different assumptions on the dependence of the decay width can be done, several of them are shown on the left hand side of Fig. 2. However, the influence of these parametrizations on the pion yield is assumed to be less than 10%.

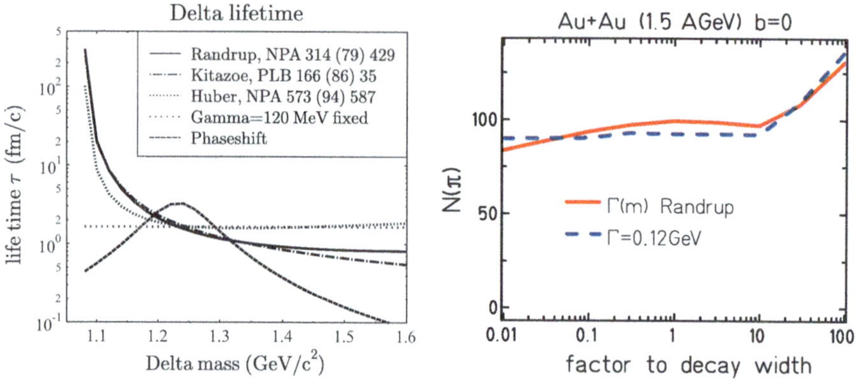

Fig. 2 Left: Different parametrisations for the mass dependence of the delta decay width. Right: Influence of an additional factor to the decay width on the pion yield

In IQMD it is possible to apply an additional factor to the decay width, which influences the interplay between the presence of pions and deltas during the high density phase. A larger factor to the decay width reduces the decay time: the delta lives for shorter time and the pion enters the reaction zone quite earlier. On the contrary, a smaller factor enhances the period of the presence of the deltas and delays the arrival of the pion. Since the pion number is directly related to the number of deltas and since a larger lifetime of the deltas enhances also the availability for being absorbed, there is some influence of that factor on the total pion yield. However, when remaining in a reasonable range, the effect on the absolute pion yield is quite moderate as it can be seen on the right hand side of Fig. 2. It should also be noted that a pion, being rapidly produced in the high density phase, has a very high chance of being reabsorbed in a collision with a nucleon and thus to create a delta again. This may induce a high number of rescattering of pions in dense nuclear matter. We will come back to this point lateron.

The production of pions in IQMD has successfully been tested by various comparisons with experimental measurements performed by the FOPI collaboration at GSI [15]. As an example Fig. 3 presents on the left hand side the excitation function of the pion yield in Au+Au collisions measured by FOPI in comparison to IQMD calculations. For this purpose, the events calculated by IQMD have undergone the same analysis procedures as the experimental data. We see that the multiplicities obtained by IQMD are slightly higher than those measured by FOPI and that the excitation functions show nicely the same behaviour. The right hand side of Fig. 3 compares the rapidity distributions of negative pions in central collisions. As already discussed before, IQMD shows slightly larger absolute pion yields, thus the absolute numbers of the rapidity distribution are also higher than the experimental points. However, the structure of the distribution is quite similar. The distribution is peaked at midrapidity underlining that most of the pions are produced by the first collisions or in collisions of the stopped participant matter. We will also discuss this later.

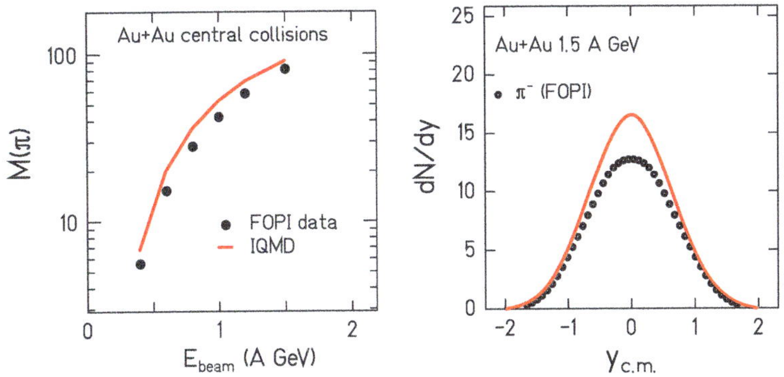

Fig. 3 Left: Comparison of the excitation function of pions measured be FOPI and IQMD calculations. Right: Comparison of the rapidity distribution of pions between experimental data of the FOPI collaboration and IQMD calculations

2 Pions and the Nuclear Equation of State

One important question in nuclear physics is, how many energy is needed to compress nuclear matter. This subject, which is directly related to the density dependence of the volume term in a Bethe-Weizsäcker mass formula, is dubbed as "search for the nuclear equation of state" (eos). Since detailed derivations can be found widely in literature (e.g. [1, 12, 13]) we will directly focus on its application in IQMD: here particles can interact via two-body potentials, which are besides Coulomb interactions and Yukawa type interactions (the latter corresponding to the surface term in Bethe-Weizsäcker) there are also interactions of Skyrme-type, momentum dependent interactions and asymmetry potentials, the first two corresponding to the nuclear equation of state, the latter to that of asymmetric matter. Details can be found in [10].

The Skyrme-type interactions correspond to the most simple ansatz that the density dependence of the potential for the volume term in Bethe-Weizsäcker can be described using 3 parameters α, β, γ.

$$U = \alpha \left(\frac{\rho}{\rho_0}\right) + \beta \left(\frac{\rho}{\rho_0}\right)^{\gamma} \tag{3}$$

Since 2 of the 3 constraints for fixing these parameters are taken by the condition that the energy per nucleon should have a minimum at ground state density ρ_0 with a value corresponding to the Bethe-Weizsäcker mass formula, the remaining third condition is linked to the incompressibility modulus and thus to the repulsion of nuclear matter against compression. This will be called the "stiffness" of the equation of state. A "hard eos" requires more energy for a given compression than a "soft eos".

Since the concept of the nuclear equation of state is in principal related to infinite equilibrated nuclear matter, while in heavy ion collisions one may encounter streaming finite matter, we may add additional momentum dependent forces which take

Table 1 Parameter sets for the nuclear equation of state used in the QMD model. S and H refer to the soft and hard equations of state, M refers to the inclusion of momentum dependent interaction

	α (MeV)	β (MeV)	γ	δ (MeV)	$\varepsilon\left(\frac{c^2}{\mathrm{GeV}^2}\right)$
S	−356	303	1.17	–	–
SM	−390	320	1.14	1.57	500
H	−124	71	2.00	–	–
HM	−130	59	2.09	1.57	500

into account that inhomogenity in momentum space. This leads to local potentials of the following type [13].

$$U_i(\mathbf{r_i}, t) = \alpha\left(\frac{\rho_{int}}{\rho_0}\right) + \beta\left(\frac{\rho_{int}}{\rho_0}\right)^{\gamma} + \delta \ln^2\left(\varepsilon\,(\Delta\mathbf{p})^2 + 1\right)\left(\frac{\rho_{int}}{\rho_0}\right), \quad (4)$$

where ρ_{int} is the interaction density obtained by convoluting the distribution function of a particle with the distribution functions of all other particles of the surrounding medium. $\Delta\mathbf{p}$ is the relative momentum of a particle with respect to the surrounding medium. The parameters δ, and ϵ in Eq. (4) can be fitted to the optical potential in nuclei. In IQMD these momentum dependent interactions can be supplemented optionally. In order to obtain the same density dependence in infinite equilibrated matter (where the momentum distribution will be a Fermi-distribution), the parameters α, β, γ will have to be readjusted to meet again the same constraints as indicated previously. Values of these parameters for the different model choices can be found in Table 1.[19]

2.1 *Pions, a Signature for the High Density Region?*

Pions have been regarded quite early as possible candidates for exploring the properties of dense matter, since according to transport models the number of pions coming out finally seemed to be determined quite early.

Figure 4 presents on the left hand side the time evolution of the central density (full lines) in a central Au+Au collision at 400 MeV/nucleon for a soft (red line) and a hard (blue line) equation of state. With both equations of state we can reach densities higher than twice times normal density, nevertheless the soft eos allows for higher densities than the hard one. This is related to the fact that a soft eos requires less compressional energy and thus allow for higher compression. However, if we regard for the freeze-out densities of the pions (dotted lines), the values become much smaller and the differences between a hard eos and a soft eos vanish. This related

[19]We have been already informed that the abbreviation “SM” for soft + momentum dependence could reveal other associations, but we keep it for some kind of tradition.

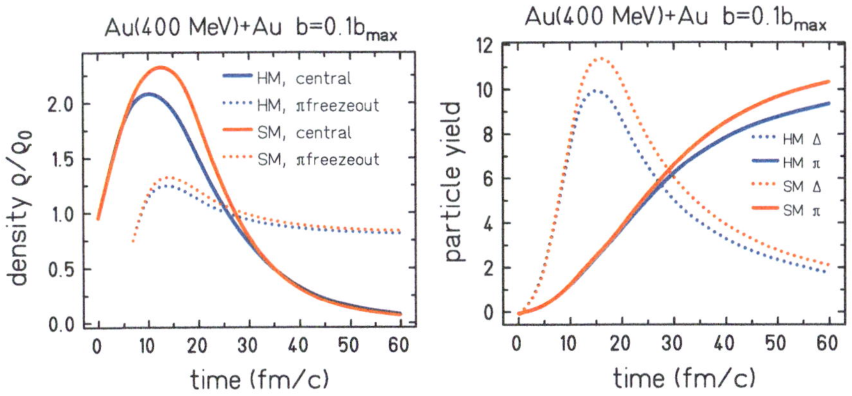

Fig. 4 Left: Time evolution of the central density (full line) and the freeze-out density (dotted) line for a hard and a soft eos. Right: Time evolution of the yields of deltas (dotted lines) and pions (full lines) for a hard and a soft eos

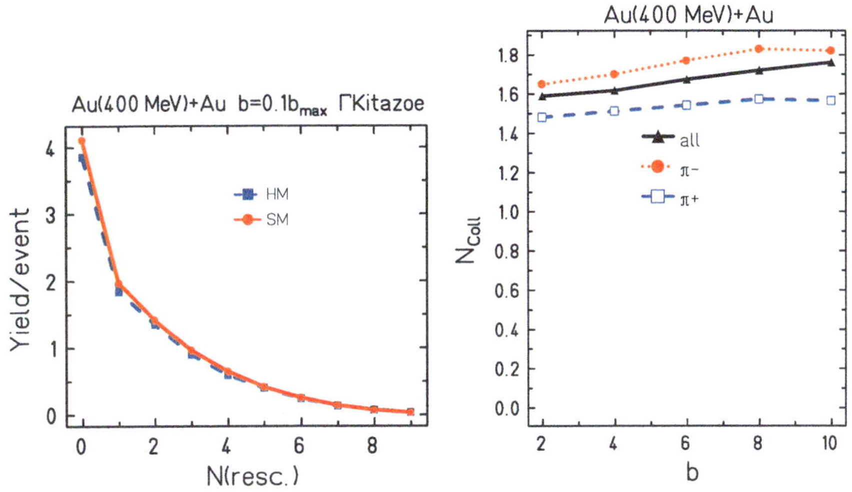

Fig. 5 Left:The yield of pions as a function of the number of rescattering for a hard and a soft eos. Right: Mean rescattering number of π^- and π^+ as function of the impact parameter

to the high absorption cross section of pions which prevents most of the pions from escaping directly from the high density zone. Thus, in dense matter the pions will be mostly remain inside the deltas. This can be seen on the right hand side of Fig. 4 where the time evolution of the yields of deltas and pions are plotted for a hard and a soft eos. We see a steep rise of the delta yield during the compression phase reaching a maximum around maximum compression. The pions come out quite late and the effect of the eos to the final pion yield is quite moderate.

The previously described effect should result in a significant number of rescatterings of the pions. This can be seen in Fig. 5 where on the left hand side the pion yield is plotted as a function of the rescattering number for a central collision of Au+Au. Even if a large number of pions (about 40%) did not undergo any collisions, most of the pions underwent at least one collision, some of them even more than 5! The difference of the nuclear equation of state is quite small: A soft eos (red line) and a hard eos (blue line) yield very similar results. The right hand side shows the average value of collisions as a function of the impact parameter for negative (red line) and positive pions (blue line). The values are in the range of about 1.6–1.8 and support thus the assumption of the importance of rescattering. It should also be noted that π^- show a slightly stronger rescattering than π^+. This can be explained by the effect that the Au nucleus is quite neutron rich and that π^- have a larger cross section with neutrons than π^+, as it has been shown on the right hand side of Fig. 1. This point will become important lateron.

The rescattering is also responsible for the fact that pions freeze out at quite low densities. Figure 6 compares therefore on the left hand side the density pions have seen at the very first collisions (dotted lines) with those seen at the very final production (full lines). While the first reactions producing the first Δ happen at quite high densities, the final freeze out happens at densities below ground state density. Therefore the signature of the equation of state is washed out: while for a soft eos (red dotted line) the first production testifies higher densities than for a hard eos (blue dotted lines) the differences for the freeze out densities become quite small. When looking on the mean freeze out densities as function of the impact parameter (right hand side) we find that π^- (red line) show slightly lower densities than π^+ (blue line). Analogously this is related to the higher rescattering of neutrons in neutron rich matter.

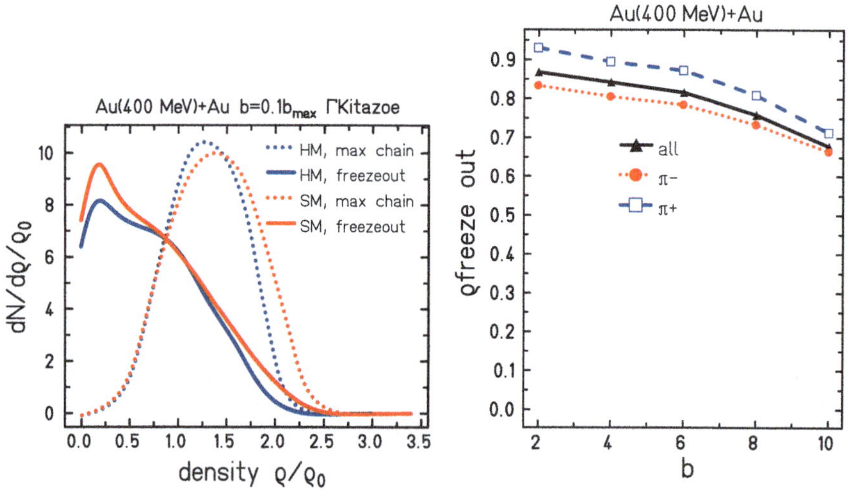

Fig. 6 Left: Distribution of maximum density (dotted line) and freezeout-density (full line) of pions for a hard and a soft eos Right: The freeze out density of π^- and π^+ as function of the impact parameter

2.2 *Pion Flow: We See an Eos, but Which One?*

However, despite the effect of rescattering, there is one observable of pions that is sensitive to the nuclear equation of state. Bass reported that the sideward flow of pions in very central collisions is higher for a hard eos than for a soft one [16]. In order to describe the effect of the sideward flow within one variable, we will use the "directed flow value" p_X^{dir} defined as the average value of the in-plane momentum p_X taken with the sign of the rapidity relative to the centre of mass:

$$p_X^{\text{dir}} = \frac{1}{N} \sum_{i=1}^{N} p_X^i \quad \text{sign}(y_i - y_{\text{CM}}) \tag{5}$$

If we now study the impact parameter dependence of p_X^{dir} of pions for a hard (blue line) and a soft eos (red line), as presented on the left hand side of Fig. 7, we find indeed higher flow values for the hard eos in central reactions than for a soft one. This behaviour is already known for nucleons where it had been extensively studied [1]. Concerning the pions, we have already stated that during the high density phase they were mostly "hidden" inside the Δ. However, in IQMD the Δ is treated like a heavy nucleon and it is exposed to the same forces. If we now cut off the forces on the Δ (black line) the strong sideward flow is significantly reduced. The effect of the eos thus only enters into the pion flow via the Δ. This is supported on the right hand side of Fig. 7 showing the influence of the Δ decay width: changing the width

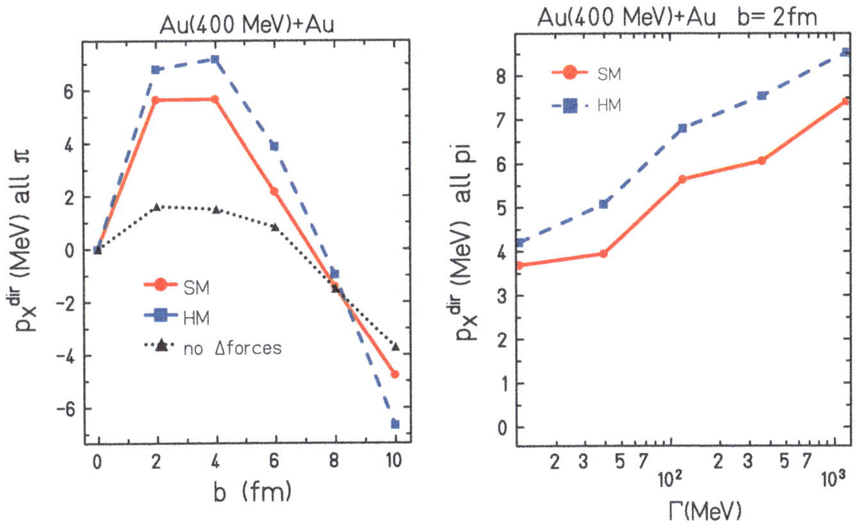

Fig. 7 Left: Directed flow p_X^{dir} of pions as a function of impact parameter for a hard and a soft eos and for a calculation without forces on the Δ. Right: The influence of the Delta decay width on p_X^{dir} for a hard and a soft eos

changes the lifetime of the Δ and changes thus directly the directed flow of the pion in a central collision. We can thus conclude that the sideward flow may give us a hint on the nuclear equation of state—but it is rather the interaction of the Δ with nuclear matter which is analysed.

2.3 Coming Back to Diogène Data

As mentioned in Sect. 1.2 the analysis of pion data from the Diogène collaboration [6] raised the question, whether the positive mean transverse momentum of pions towards the reaction plane was a signature of the eos, since this behaviour could not be confirmed by the intranuclear cascade model of Cugnon (INC) [8], a model lacking potential interactions of the nucleons, while IQMD calculations could reproduce the effect qualitatively [7]. However, the problem of IQMD was that the effect remained, even when switching off the potential part.

Figure 8 illustrates on the left hand side that problematic: in IQMD the mean flow $\langle p_X \rangle$ (averaged over all rapidities without any regard to the centre of mass rapidity) stays positive for all impact parameters. Calculations with a soft eos (red line) and a hard eos (blue line) show quite similar values. A calculation in cascade mode (no potentials, black line) only differs slightly at very high impact parameters. It is again the effect of the Δ that can explain this: it is the rescattering of the pion in matter, as shown on the right hand side of Fig. 8. If we increase the lifetime of the delta to infinity (blue line) and thus suppress any possibility of rescattering, the mean flow

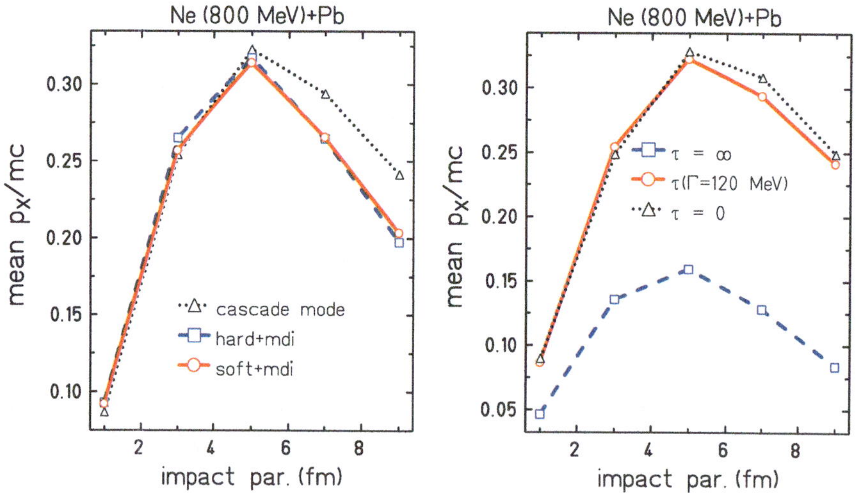

Fig. 8 Left: Mean flow $\langle p_X \rangle$ of pions as a function of impact parameter for a hard and a soft eos and for a calculation without forces. Right: The influence of the Delta decay width on the mean flow

reduced to quite small values. It can thus be assumed that the differences between INC and IQMD were not related to the inclusion of nuclear potentials but to a different treatment of the delta.

3 What the Pion Tells Us About the Nucleus

As we have previously seen, the pion is a particle strongly interacting with nucleons and strongly coupled to nuclear matter via the delta. The rescattering of pions is important for understanding several observables that have already been discussed. Let us now study some questions related to the properties of nuclear matter.

3.1 Can Pions Measure the In-Medium Mass of Deltas?

Several experiments already tried to extract the medium mass of the delta by reconstruction from pion-nucleon pairs in low multiplicity events. The Diogène collaboration [6] analysed π^-p pairs in few prong events of the reaction Ne(600 AMeV)+NaF and found a maximum lower than the delta pole mass of 1232 MeV, which triggered the discussion of an in-medium change of the delta mass. However, it should be stated that the Breit-Wigner distribution of the delta should only be expected in $NN \rightarrow N\Delta \rightarrow NN\pi$ reactions with sufficient available energy. In heavy ion collisions we have a mixture of very different available collisions and the delta mass spectrum should correspond to the convolution of the available energy with a Breit-Wigner function. Additionally, we should also consider the contribution of rescattering $\pi N \rightarrow \Delta \rightarrow \pi N$ to the mass spectra. In the latter case the delta mass is completely determined by the reaction cinematics. It should be noted that IQMD assumes a free Breit-Wigner distribution for the delta mass without any medium modifications.

Figure 9 shows on the left hand side the mass distribution of the delta producing the final pion, i.e. of the last delta before freeze-out. We see that for both central reactions (red line) and peripheral reactions (blue line) in Ne(600 AMeV)+Ne the distributions are quite similar: none of them shows a peak at the pole but at a maximum a quite smaller values. As already indicated below, due to manyfold rescattering the nucleons show a broad momentum spectrum, thus the centre of mass energy of the collision system also shows a broad spectrum. There are lots of collisions with insufficient energy to reach the pole which only can create deltas at lower masses. Therefore the maximum is shifted to lower values. The effect of rescattering can be seen on the right-hand-side: pions without rescattering (red line) show some kind of plateau for m_Δ below the pole mass. This is due to the contribution of collisions with insufficient energy. Due to the Breit-Wigner distribution they will tend to give the delta the maximum available mass. Collisions with sufficient energy will tend to

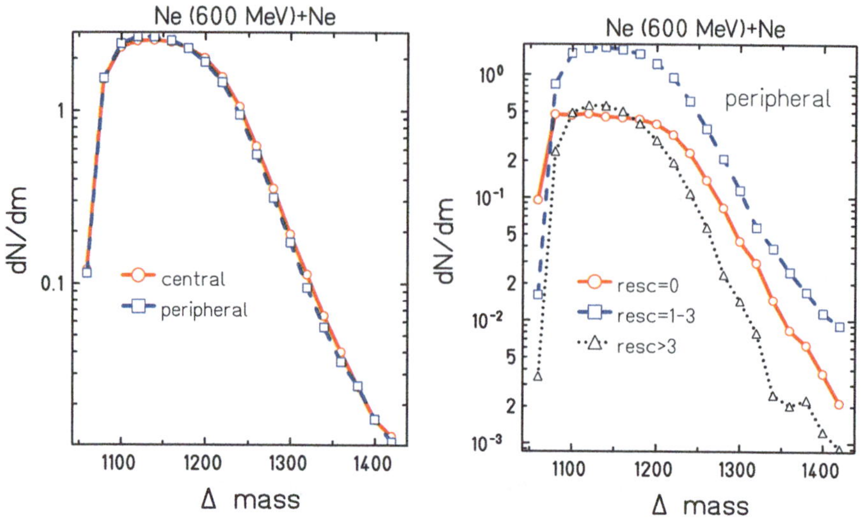

Fig. 9 Left: The delta mass of the final pion production for central and peripheral collisions. Right: The delta mass of the final pion production for different rescattering

give the delta a mass nearby the pole. Therefore the plateau stops at the delta pole mass.

For pions, which have performed rescattering, the mass of the delta is completely determined by the reaction cinematics. Therefore, the mass distribution corresponds to the convolution of the spectrum of available $\sqrt{s}$ with the absorption cross section. We also see that pions, which have performed many collisions (black line), show a peak at slightly lower values than those having collided moderately (blue line). This corresponds probably to the loss of energy in the subsequent collisions. The total mass distribution is the sum of all distributions and is dominated by the pions with moderate rescattering (blue line). The "mass shift" can thus be explained without assuming any medium effect.

3.2 Why Do We See Different Temperatures in Pion Spectra?

A method frequently used to describe a pion spectrum is the fit with a thermal spectrum using one (for small systems) or two temperatures. Of course one may pose the question of the physical meaning of that temperature. Is it the mean temperature, some kind of "chemical freeze-out"? We should remember that the system is not thermalised—we will show this point in the next subsection—and that a freeze out would require that all components—even with different number of rescattering—should show the same freeze out temperature.

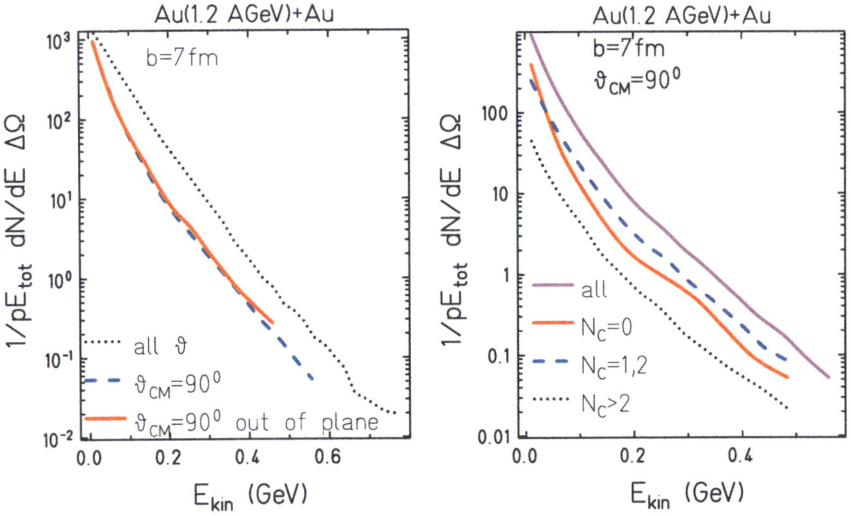

Fig. 10 Left: Spectra of pions with and without a cut to cm angles perpendicular to the beam and with an additional cut out of plane. Right: Contribution of the rescattering to the spectra perpendicular to the beam

Figure 10 shows on the left hand side the pion spectrum in a semi-central collision of Au(1.2 AGeV)+Au. While the spectrum taken over 4π (black line) could be nearly fit with one temperature, the spectrum perpendicular to the beam ($60^\circ \leq \vartheta \leq 120^\circ$, blue line) is comparable with two temperatures which is even slightly more pronounced when looking only for its out of plane component (red line). The gap between the 4π spectrum (black line) and the perpendicular spectrum (red line) can already be interpreted as an indicator for an angular distribution significantly peaked forward-backward along the beam axis. The reason for this is rescattering of the pions in the spectator matter [17]. If we now decompose the spectrum perpendicular to the beam according to the number of rescattering (right hand side), we see that the pions without rescattering (red line) dominate at the low energy region while those with one or two collisions (blue line) at higher energies. Pions that did rescatter frequently (black line) seem to show similar behaviour than those with one or two collisions but with smaller yields. However, this decomposition also changes with incoming energy and centrality of the system. The "temperature" of the spectrum may thus more correspond to some "available energy after rescattering" than to a real temperature in a pure thermodynamical sense.

3.3 Pions and Thermodynamics: A Happy Coincidence?

Thermodynamical models have been applied quite successfully for describing particle ratios down to very low energies. However, this seems to be quite astonishing since thermal equilibrium seems not be established in heavy ion collisions at low energies even at most central collisions. In the following we will study central collisions at 1.5 AGeV incident energy and quantify the isotropy as the ratio of the total transverse energy divided by two time longitudinal energy. IQMD normally uses free cross sections for collisions. We allow to apply a scaling factor to the total cross section (i.e. all cross sections) or only to the pion absorption cross section $\pi N \rightarrow \Delta$ or only to the inelastic delta production cross section $NN \rightarrow N\Delta$. In the last case we apply the same factor to the reverse reaction in order to keep detailed balance.

Figure 11 presents on the left hand side the effect of the scaling of the total cross section (red line), the pion absorption cross section (black line) and the delta production cross section (blue line) as well as to both pion absorption and delta production cross sections (magenta line) on the isotropy in a central collision of Au (1.5 AGeV)+Au. The thick green line marks the isotropy value of 1 which would be a requirement for full thermalisation. We see that without scaling (i.e. factor equals 1) we are still far from that purpose. In order to achieve full isotropy we would nearly need a factor of 2 for the total cross section (red line) or even 3 if we would only scale the inelastic delta production cross section (blue line). The pion absorption cross section (black line) does not show any influence, therefore the combination of pion absorption and delta production scaling (magenta line) follows exactly the behaviour of the delta production only. We thus see, that even for a heavy system like Au+Au we are far away from complete thermalisation.

Of course, the system size is important for thermalisation, therefore the right hand side of Fig. 11 presents the isotropy as function of the radius of the nucleus if we

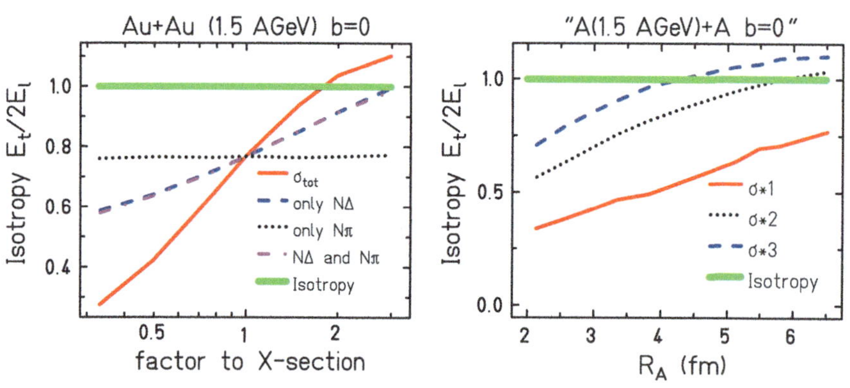

Fig. 11 Left: Momentum isotropy in a central collision of Au(1.5 AGeV)+Au as a function of a scaling of the cross sections. Right: Momentum isotropy as a function of the nuclear radius for different scalings of the total cross sections

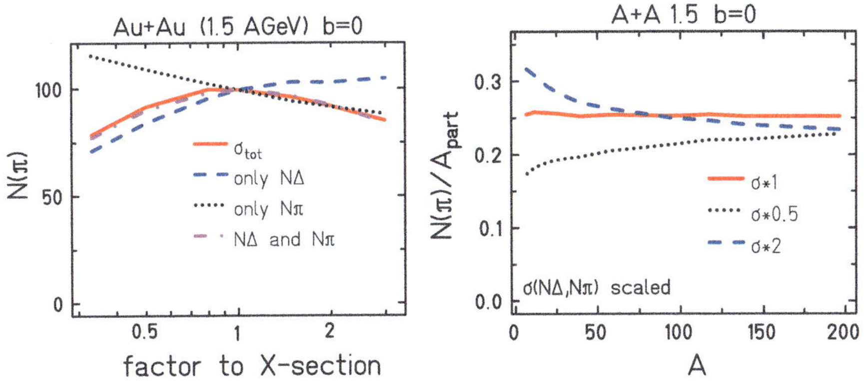

Fig. 12 Left: Pion number as a function of the scaling factor to the cross sections in a central collision of Au(1.5 AGeV)+Au. Right: Pion number as a function of A for a common scaling factor to pion absorption and delta production

use normal cross sections (red line) and if we scale all cross sections by a factor of 2 (black line) or 3 (blue line). We see that we would need at least a factor of 2 for approaching isotropy for heavy systems or even 3 for intermediate systems. Therefore it seems to be astonishing, why thermal models should work.

Figure 12 shows on the left hand side the influence of a scaling factor on the pion yield: the pion absorption (black line) and the delta production (blue line) show opposite behaviour, thus the combination of both (magenta line) shows a weaker dependence, with a maximum of the pion number nearby the normal cross sections (factor 1). A scaling of the total cross section (red line) yields the same result as the combination of pion absorption and delta production. The actual cross sections seem thus to allow for a maximum pion number.

The right hand side shows the effect of a scaling factor on the combination of pion absorption and delta production on the A-dependence of pion production. The constant π/A ratio can only be assured when we use unscaled cross sections (red line). An enhancement (blue line) or reduction (black line) of the cross sections would break this property. Therefore, we can conclude that the actual situation of pion production and absorption cross section is very favorable for the success of thermodynamics [18].

The production of K^+ at energies around 1–2 AGeV is coupled to the pions, since pions and deltas are needed to store the required energy temporarily. Since in the violent early stage pions are mostly "hidden" in deltas, it is the $N\Delta$ channel which dominates the production of K^+ [19]. Due to the need of two steps in this processus the dense phase plays an important role for the production: only partners in the near neighborhood can contribute to the two-step process. Thermal models take care of that by applying some volume reduction factor γ_s.

Figure 13 shows on the left hand side the influence of a scaling factor on the number of K^+: if we only scale pion absorption (black line) or only delta production

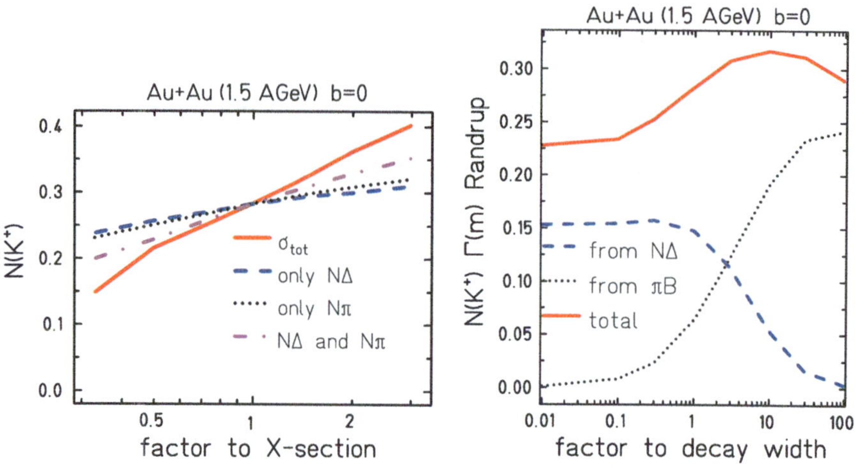

Fig. 13 Left: Number of K^+ as a function of the scaling factor to the cross sections in a central collision of Au(1.5 AGeV)+Au. Right: Number of K^+ as a function of a scaling factor to the delta decay width

(blue line) the effect is quite moderate. However, a combination of both scaling factors (magenta line) enhances the effect, which becomes maximum for a scaling of all cross sections (red line). It should be kept in line, that a scaling of all cross sections influences the density reached in the reaction and that the production of K^+ is very sensitive on that.

It should also be noted that in contrast to the delta production, the pion absorption has no inverse reaction channel governed by cross sections, since the delta decay is governed by the decay width. It has already been shown in Fig. 2 that the decay width does not change the pion yield if we do not scale it too tremendously. For the kaons we might assume a stronger dependence, since the decay width will effect the interplay between deltas and pions in the dense region. However, the right hand side of Fig. 13 indicates a rather moderate effect: even if the decay width influences strongly the individual channels for kaon production via the Δ (blue line) or via the pion (black line) the total yield (red line) profits from some kind of cancellation of both effects.

The production of K^- However, is strongly dominated by the charge exchange of a hyperon $\pi\Lambda \leftrightarrow K^-N$. Of course the inverse channel is also strong. When scaling the charge exchange channel, both directions will be scaled by the same factor. Since the hyperon production is linked to the production of K^+ (the energetical most favorable channel for K^+ is via the production of a $K^+\Lambda$ pair), the production of K^- is strongly linked to that of the K^+. Every penalty applying to the production of K^+ also penalises the production of K^-.

From this point of view, it is quite understandable that on the left hand side of Fig. 14 the ratio K^-/K^+ is quite independent from scaling factors applied to the pion absorption and delta production (blue line). Additionally this ratio is on a wide range

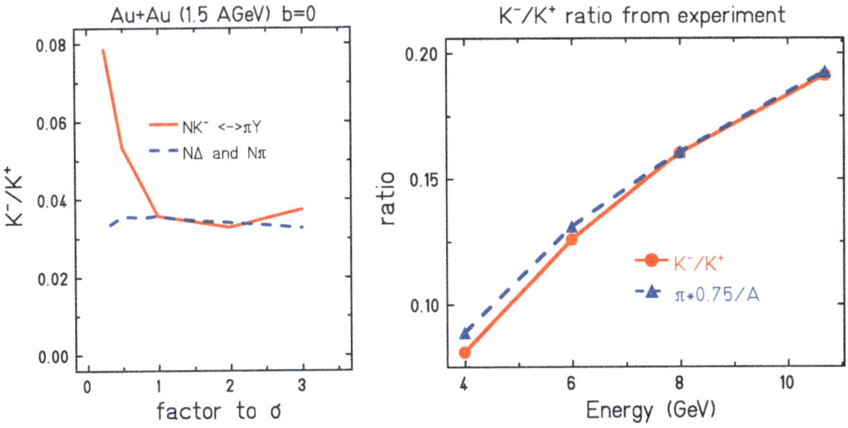

Fig. 14 Left: Influence of the scaling of $\pi\Lambda$ and pion absorption + delta production on the K^-/K^+ ratio in a central collision of Au(1.5 AGeV)+Au. Right: Scaling of K^-/K^+ as function of energy in comparison to π/A

also quite insensitive to a scaling of the strangeness charge exchange cross section (red line) in both directions. This is due to the strong coupling in the strangeness exchange channel [20].

Applying these properties we can thus define a law of mass action on the ratio of K^-/K^+ with $[K^-]\cdot[N] = const \cdot [\pi]\cdot[\Lambda]$ where [] stands for a concentration, resp. yield. Taken into account that [] = [K$^+$] we arrive finally to

$$\frac{[K^-]}{[K^+]} = const \cdot \frac{[\pi]}{[N]} \tag{6}$$

which denotes that the ratio of K^-/K^+ should scale with π/A, where the latter one is (by happy coincidence) independant of the system size A. The right hand side of Fig. 14 proves the validity of this relation by the compilation of AGS data [18]: the curves for K^-/K^+ (red line) as a function of incident energy correspond to the values of π/A scaled by a factor of 0.75 (blue line). Here again the combination of facts plays very confortable for the application of thermal models, independent of the question whether the requirements for thermalisation are fulfilled.

3.4 Pions Can Measure the Neutron Skin!

The isospin ratio π^-/π^+ of pions has revealed recently great interest. The starting point was the finding that the energy dependence of this ratio in central collisions of Au+Au measured by FOPI [15] could be well reproduced by IQMD for energies above 600 AMeV, but was completely underestimated by IQMD at 400 AMeV incident energies. Lateron it was claimed that this discrepancy could be explained by the asymmetry potential between neutrons and protons and that this ratio would be very sensitive to the equation of state of asymmetric matter [21]. However, other ingredients may also play a role, one very interesting question is the neutron skin of nuclei.

In standard IQMD calculations, but also in many other models, the initialisation of a nucleus is done that way that the centroids of the Gaussians are distributed inside a sphere in the rest frame of the nucleus according to

$$(\mathbf{r}_i - \mathbf{r}_{CM})^2 \leq R_A^2 \qquad R_A = R_0 \cdot A^{1/3} \tag{7}$$

where $\mathbf{r}_i$ and $\mathbf{r}_{CM}$ are the the position vectors of particle i and of the centre-of-mass of the nucleus respectively. Furthermore $A = Z + N$ is the number of nucleons of the nucleus and $R_0 = 1.12$ fm is the radius parameter. We will call this initialisation "R_P=R_N". It assures the same rms radius for protons and neutrons even for heavy isospin-asymmetric systems. As a consequence, the neutrons have systematically a higher density than the protons in the whole nucleus. If we want to assure the same density of protons and neutrons at least at the centre of the nucleus we have to allow protons and neutrons to have different rms radii. This can be obtained by the distribution of the centroids of the Gaussians according to

$$R_P = R_0 \cdot (2Z)^{1/3} \qquad R_N = R_0 \cdot (2N)^{1/3} \tag{8}$$

where R_P and R_N denote the radii for protons and neutrons and which we will call "R_P<R_N".

Figure 15 shows the density profiles of neutrons (full lines) and protons (dotted lines) for both initialisations: R_P=R_N on the left hand side and R_P<R_N on the right hand side. For the first initialisation the charge ratio $Z/A(R)$ (dashed line) is always constant, while it decreases with increasing radius for the second one. As it has been previously stated, π^- have a stronger rescattering in neutron rich matter. However, this rescattering will not change the isospin of the pion since $\pi^- n \rightarrow \Delta^- \rightarrow \pi^- n$. However, when rescattering π^+ in neutron rich matter, only one third of the pions will be reconverted into π^+ again, while in the other case a π^0 will be produced. Therefore a neutron skin may influence the π^-/π^+ due to rescattering in the skin.

Another effect may be the Pauli blocking in the delta decay. This Pauli blocking may penalize channels producing a neutron again. In case of a $\pi^- n \rightarrow \Delta^- \rightarrow \pi^- n$ this penalty will only delay the reaction without penalty since no concurring channel exists. In case of the rescattering of a π^+, however, this will add an additional penalty on the production of π^+ and therefore enhance the π^-/π^+ ratio.

Figure 16 shows on the left hand side this effect for a collision of Au(400 AMeV)+Au. For a calculation with R_P<R_N (red line) the ratios are systematically higher than for a calculation with R_P=R_N (blue line). A calculation neglecting the Pauli blocking of the delta decay (black line) yields even lower ratios. It should be noted that especially for the R_P<R_N-option the ratios increase strongly for peripheral collisions. This effect is even stronger when going to a higher energy of 1.2 AGeV as

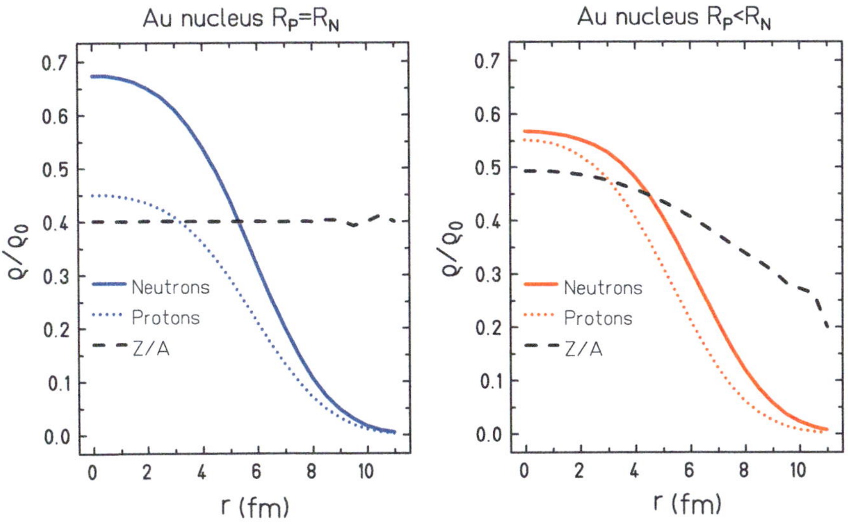

Fig. 15 Density distribution of protons and neutrons and Z/A. Left for $R_P = R_N$, right for $R_P < R_N$

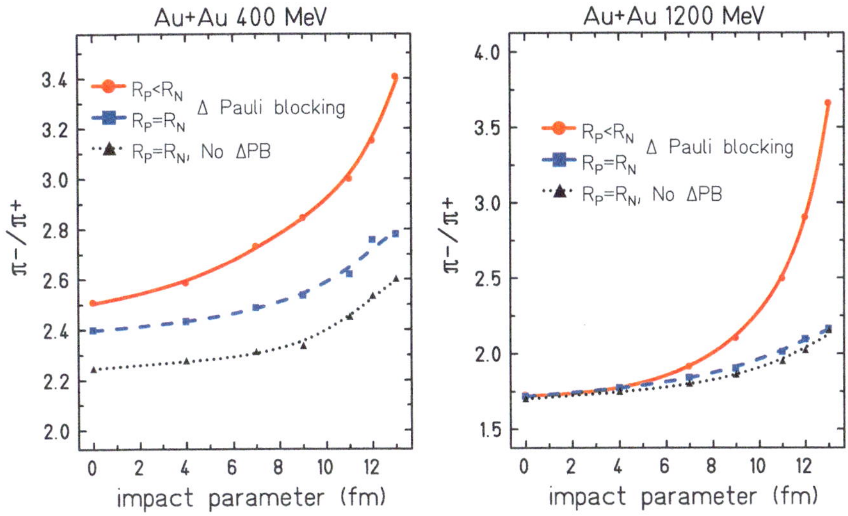

Fig. 16 Impact parameter dependence of the ratio π^-/π^+, left for 400 AMev incident energy, right for 1.2 AGeV

shown on the right hand side. Here the different options yield similar values at central collisions but the neutron skin yields very different values at peripheral collisions, while the Pauli blocking of the delta decay does not play a role any more.

A systematic study of the centrality dependence of the π^-/π^+ ratio at different energies may thus reveal important information about the neutron skin but also on other features, including the equation of state of asymmetric matter [22].

4 Conclusion

As this article is preceded by an unusual introduction, the reader may excuse the author when splitting up the conclusion in an unusual way. It will contain in a first part the (usual) synthesis of the contents of this article devoted to the physics of pions. Next there will be some more general part (but also personal) part on important results from the physics at this energy range and finally some pledge for changing a paradigm.

4.1 *What We Discussed in that Paper*

In this article the properties of pions in heavy ion collisions in the energy range up to about 2 AGeV have been discussed. It is found that the pions, even if initiated at high densities, mostly show signatures of quite low freeze out densities. The pion sideward flow may reflect information on the delta-nucleon interaction. The recombinated mass of the last delta does not show necessarily medium effects of the pole mass but reflects moreover the convolution of different reactions with different available energy. The pions seem to be linked to other particles, especially the K^-, therefore the use of thermal models seems to be working, even if thermalisation cannot be observed. The ratio K^-/K^+ is directly related to the ratio of π/A.

A perspective for future studies might be the analysis of the impact parameter dependence of the π^-/π^+ ratio at different energies. This may help to disentangle information on the nuclear equation of state of asymmetric matter as well as on the neutron skin of nuclei. This may open further insight into the properties of matter by comparing high precision data with models.

4.2 *What Has Marked the Author in that Physics Domain at that Energy*

The author is conscious of the subjectivity of this part and restricts himself to the discussed energy domain of several hundred AMeV to several AGeV, not daring to judge on other fields. But when raising the question "what of all that will remain in the textbooks of physics?" he is remembered of a question Walter Greiner liked to ask experimentalists at a periode when he himself was a young student: "What will you do, if we can pin down the equation of state?" Since that question raised about 3 decades ago, a lot of things have changed but all conclusions converge more and more to a soft equation of state when regarding experiments at GSI. Scaling laws in kaon production yielded model independent predictions which by comparison to KaoS data strongly advocate for a soft eos [23]. Detailed investigations of the squeeze-out of light particles (protons, deuterons, tritons etc separately) performed by the FOPI collaboration confirmed this finding even more precisely [24] and seem to be compatible with astrophysical findings. There are now some interesting ideas to attack the nuclear equation of state of asymmetric matter (asy-eos) in a similar way.

The study of fragments performed by ALaDiN, INDRA and ALaDIN-INDRA indicated a caloric curve of nuclear matter. The experiments HADES allow to get insight into medium effects in matter.

What is very common in a lot of these success stories is, that people dared to do systematics—even if that might be misunderstood by some founding representatives as lack of ideas. Frequently the "devil inside the detail" could only be banned when studying systematically different system sizes, incident energies and different centralities. This leads to the final pledge of this article.

4.3 A Pledge for Changing a Paradigm

This domain of physics—and the author puts himself into the dock as well—has frequently followed one paradigm for doing exciting physics:

- Go to the highest available energy
- Use the largest available system
- Focus on the most central events

Clear insight may frequently only come from a variation of these parameters. A nice example for this has already been shown in Sect. 3.4, where only the combination of different centralities and different energies allow to deconvolute the effect. Nevertheless concerning the system size the systematics have sometimes been reduced to only compare a very large system with a very small one.

Figure 17 shows on the left hand side the time evolution of the central density in a central A+A collision of Au+Au (red line), Ca+Ca (black line) and C+C (blue line). It can be seen that the systems of Ca+Ca and Au+Au obtain nearly the same maximum density. However, the Au+Au system remains longer in a high density phase. When we look on the right hand side, we find that the maximum density (red line) reached a plateau at about $A = 50$ and stays roughly constant afterwards. However, the rescattering of particles—here the pions (blue line)—changes significantly from intermediate to larger systems. This means that when comparing a light system

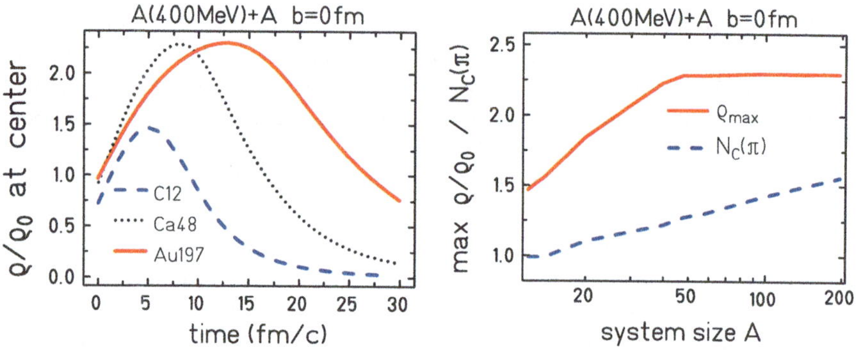

Fig. 17 Left: Time evolution of the density in central Au+Au, Ca+Ca and C+C collisions. Right: Dependence of the maximum density and the number of pion rescattering on the system size A

with an intermediate system we may focus stronger on the direct compression effects with less rescattering. Comparing that to a heavy system will give us more insight on the rescattering in the system. Systematical studies of that kind may help to better disentangle these effects.

Walter Greiner taught us to keep our conviction even if other people disagree strongly with us. Following his advise I call for changing our paradigm: we need more experimental detailed insight and less hunting for spectacular results.

In this context I would like to close with a sentence, that Walter Greiner liked to tell to his students:

It is the privilege of a theoretician to read in the bible of nature
But only an experimentalist is able to turn the pages.

References

1. H. Stöcker, W. Greiner, Phys. Rep. **137**, 277 (1986)
2. H. Kruse, B.V. Jacak, H. Stöcker, Phys. Rev. Lett. **54**, 289 (1985)
3. J.J. Molitoris, H. Stöcker, Phys. Rev. C **32**, R346 (1985)
4. J. Aichelin, H. Stöcker, Phys. Lett. B **176**, 14 (1986)
5. C. Hartnack, L. Zhuxia, L. Neise, G. Peilert, A. Rosenhauer, H. Sorge, J. Aichelin, H. Stöcker, W. Greiner, Nucl. Phys. A **495**, 303 (1989)
6. J. Gosset et al., Phys. Rev. Lett. **62** (1989)
7. Ch. Hartnack, H. Stöcker, W. Greiner, in *Proceedings of the International Workshop on Gross Properties of Nuclei and Nuclear Excitation, XVI*, Hirschegg, Kleinwalsertal, Austria, ed. by H. Feldmeier (1988)
8. J. Cugnon, Phys. Rev. C **22**, 1885 (1980)
9. Ch. Hartnack, S.A. Bass, J. Aichelin, H. Stöcker, W. Greiner, in *International Workshop on Multiparticle Correlations and Nuclear Reactions (Corinne II)*, Nantes, ed. by J. Aichelin, D. Ardouin (World Scientific, Singapore, 1994)
10. C. Hartnack, R.K. Puri, J. Aichelin, J. Konopka, S.A. Bass, H. Stoecker, W. Greiner, Eur. Phys. J. A **1**, 151 (1998), arXiv:nucl-th/9811015

11. S.A. Bass, M. Belkacem, M. Bleicher, M. Brandstetter, L. Bravina, C. Ernst, L. Gerland, M. Hofmann, S. Hofmann, J. Konopka, G. Mao, L. Neise, S. Soff, C. Spieles, H. Weber, L.A. Winckelmann, H. Stoecker, W. Greiner, C. Hartnack, J. Aichelin, N. Amelin, Prog. Part. Nucl. Phys. **41** 225 (1998)
12. W. Cassing, V. Metag, U. Mosel, K. Niita, Phys. Rep. **188**, 361 (1990)
13. J. Aichelin, Phys. Rep. **202**, 233 (1991)
14. S.A. Bass, C. Hartnack, H. Stöcker, W. Greiner, Phys. Rev. C **51**, 3343 (1994)
15. W. Reisdorf et al. [FOPI Collaboration], Nucl. Phys. A **781**, 459 (2007). https://doi.org/10.1016/j.nuclphysa.2006.10.085, arXiv:nucl-ex/0610025
16. S.A. Bass, C. Hartnack, H. Stöcker, W. Greiner, Phys. Rev. C **51**, R12 (1995)
17. S.A. Bass, C. Hartnack, H. Stöcker, W. Greiner, Phys. Rev. Lett. **71**, 1144 (1993)
18. C. Hartnack, H. Oeschler, J. Aichelin, Phys. Part. Nucl. Lett. **8**, 845 (2011)
19. C. Hartnack, H. Oeschler, Y. Leifels, E.L. Bratkovskaya, J. Aichelin Phys. Rep. **510**, 119 (2012)
20. C. Hartnack, H. Oeschler, J. Aichelin Phys. Rev. Lett. **90**, 102302 (2003)
21. Z. Xiao, B.A. Li, L.W. Chen, G.C. Yong, M. Zhang, Phys. Rev. Lett. **102**, 062502 (2009)
22. C. Hartnack, A. Le Fèvre, Y. Leifels, J. Aichelin, arXiv:1808.09868
23. C. Hartnack, H. Oeschler, J. Aichelin Phys. Rev. Lett. **96**, 012302 (2006)
24. A. Le Fèvre, Y. Leifels, W. Reisdorf, J. Aichelin, C. Hartnack, Nucl. Phys. A **945**, 112 (2016)

Die Erste Stunde (The First Hour)

Johann Rafelski

Abstract I recall my "first hour" events following on my meeting in Fall 1968 in the classroom with my academic teacher and thesis mentor Prof. Dr. Dr. h.c. multiple Walter Greiner. My comments focus on the creation of the new "strong fields" domain of physics in Frankfurt. I argue that this was the research field closest to Walter's heart during his lustrous academic career. I will describe the events that led on to Greiner's course books, Walter's actions leading to the rise of Frankfurt School of Theoretical Physics, and show how a stability principle defined his science.

1 Introduction

Walter Greiner arrived in Frankfurt in the mid 1960s. He came in as a reformer, pushing through many changes at the Physics Faculty (Fachbereich Physik) of the Johann Wolfgang von Goethe Universität in Frankfurt. In Fall 1968, Walter's newly approved Theoretical Physics course was offered to physics students in their first semester and attracted many students, including the author. For me and many others in this class the meeting with Walter was a random chance. However, we stayed on because of Walter.

Teaching freshman Theoretical Physics Course was an educational revolution. It was accompanied by another revolution; West Germany was in the midst of a large scale student revolt. I recall that the J.W. Goethe University could not set exams; some courses were even canceled as the zealots focused on particular 'reactionary' lecturers. Other courses were disrupted temporarily by sit-ins organized by idealistic students responding to the crooked but active Soviet propaganda machine operating from East Germany.

Presented at: FIAS International Symposium on *Discoveries at the Frontiers of (Walter Greiner) Science*, Frankfurt, June 26–30, 2017. to appear in proceedings.

J. Rafelski (✉)
Department of Physics, The University of Arizona, Tucson, AZ 85721, USA
e-mail: Rafelski@Physics.Arizona.EDU

J. Kirsch et al. (eds.), *Discoveries at the Frontiers of Science*,
FIAS Interdisciplinary Science Series, https://doi.org/10.1007/978-3-030-34234-0_22

A reader interested in a more general characterization of life and work of Walter Greiner should consult another recent commentary [1]. This article is very different as it describes in form of personal reminiscences an important series of events that occurred mainly between 1968 and 1983, during which 15 year period I interacted strongly with Walter.

1.1 Frankfurt 1968–1971

In Greiner's classroom nobody from within disrupted the lectures. Moreover, when external non-physics revolutionaries tried to stop Greiner's Theoretical Physics freshman class, his students defended the classroom, expelling the non-course students, throwing back the stink bombs and barricading the entrance doors from the inside with chairs and desks. Despite numerous distractions, and the absence of formal examination (for fear of external disruption), there was lots of learning going on. We had classes, regular tutorial study groups, and the teaching program proceeded well. Among the 70 or more freshman students, many made great scientific careers.

I think that this shows that teaching in the challenging way Walter pioneered in Frankfurt leads to success irrespective of situation. And for those going on to academic careers the near complete lack of formal examination may have been an asset: learning was not distracted by examination. Of the 1968 crop of students, quite a few later joined Greiner's research group, and several became tenured theoretical physics full professors.

This concentration of talent around Walter was due to his proactive approach: Walter cared for and developed young men and women working with him. Walter was bent on keeping his talented students. When someone made a move that displeased Walter's vision, he would straighten out the situation, typically offering his assessment on who was who in theoretical physics. His decisive and convincing arguments were without doubt an important reason for the successful birth of the Frankfurt School of Theoretical Physics in the eventful months of 1968–70.

I still remember how Walter reacted when I told him that I won a very coveted and competitive Studienstiftung fellowship to Oxford. Walter was upfront and direct: "... only a!#s are working there." Also, within days of this conversation, shortly after I completed my Diploma, and despite being only 21 years old, I was appointed his Scientific Assistant, see Fig. 1.

A few 'older' (as compared to me) Assistants supported Walter's teaching and research efforts; Burkhard Fricke and Ulrich Mosel come to my mind. Burkhard led my tutorial group, while Ulrich was the primary pillar of classical Nuclear Science and did both his Diploma and Ph.D. with Walter. However, I was Walter's first hire from among the people he taught from the first semester on, beginning in Fall 1968. I in turn introduced Walter to others who became important in the strong field physics formative years, including Berndt Müller, Gerhard Soff, and a year later, Joachim Reinhardt, the future soul of Walter's rapidly expanding strong field research group.

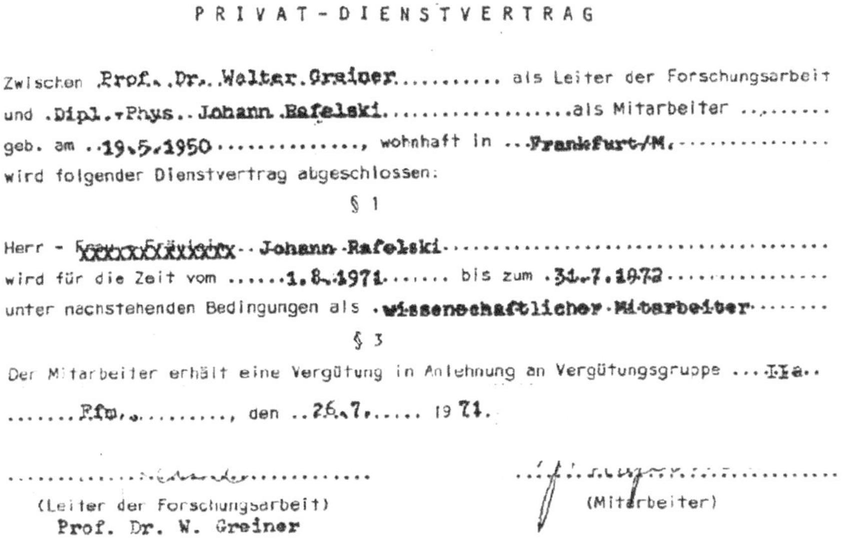

P R I V A T - D I E N S T V E R T R A G

Zwischen .Prof. .Dr. .Walter.Greiner.......... als Leiter der Forschungsarbeit
und .Dipl.-Phys.. Johann .Rafelski.................als Mitarbeiter ....
geb. am ..19.5.1950.............., wohnhaft in ...Frankfurt/M..............
wird folgender Dienstvertrag abgeschlossen:

§ 1

Herr - XXXXXXXXXXXXX.. Johann Rafelski....................................
wird für die Zeit vom ......1.8..1971....... bis zum .31.7.1972...............
unter nachstehenden Bedingungen als .wissenschaftlicher Mitarbeiter........

§ 3

Der Mitarbeiter erhält eine Vergütung in Anlehnung an Vergütungsgruppe ....IIa..

.......Ffm.,........., den ..26..7...... 19 71.

(Leiter der Forschungsarbeit)
Prof. Dr. W. Greiner

(Mitarbeiter)

Fig. 1 Walter Greiner appoints the author as his "Scientific Assistant" in July 1971. *Source* Johann Rafelski archives

1.2 Theoretical Physics Course Books

It is of interest to many to understand how and why Walter Greiner's red-book series, "Theoretische Physik," was created. Before Greiner, nobody dared to teach a Theoretical Physics course in the 1st semester; thus no appropriate textbooks were available for freshman student use in the study of theoretical physics. In order to teach material which previously would appear typically only two years later in the curriculum, Walter clearly recognized the need to simplify, to explain by example, and to offer full solutions of exercises.

To create such a new series for students, Walter realized that much of it had to be co-written by students. His "Assistant," Burkhard Fricke, who set the tutorial exercises (Übungen) for each week, would also collect from student volunteers their class protocols of class lectures, exercises, and solved problems. These were reviewed and edited by Burkhard. Walter would make another set of edits before these notes were typed, figures drafted and all put together into a "Script" by Walter's secretarial staff, supervised by the "Assistant".

The first class course "Scripts" of Classical Mechanics I+II and Electrodynamics were prepared by Burkhard Fricke, as can be read in Fig. 2. Scripts were printed just like a Ph.D. thesis by an in-house printing press maintained by the Theoretical Physics Institute. These scripts were available to any interested party and were popular among students, selling for 2 DM a copy, the cost of a lunch in a student Mensa restaurant.

Walter would use the first edition of his script, repeating three years later the same series of classes. Doing this he caught inevitable errors, and added and expanded the

II

Das ursprüngliche Manuskript der Vorlesungen wurde 1969/70 in Zusammenarbeit mit den damaligen Assistenten Dr. Burkhard Fricke* und Dr. Ulrich Mosel**, von den Studenten P. Bergmann, H. Betz, J. Briechle***, M. Bundschuh, C.v.Charzewski, S. Erlenkämper, J. Hahn, H.J. Lustig, B. Müller, H. Müller, H. Peitz, J. Rafelski, H.J. Scheefer, H. Severin und P. Struckmeier erstellt. Die Neufassung erfolgte dann 3 Jahre später durch W. Betz, D. Blöcker, J. Hofmann, B. Jacob, P. Kurowski, J. Kirsch, H. Leber, A. Mahn, P. Mohring, J. Reinhardt, D. Schebesta, X. Stiebing und J. Wagner.

Über die rege Beteiligung dieser Studenten an der Gestaltung der Vorlesungen und ihrer Ausarbeitung waren wir besonders froh, weil dadurch die speziellen Verständnisschwierigkeiten der Erstsemester besondere Berücksichtigung fanden.

Die Damen M. Knolle, R. Lasarzig, B. Utschig und Herr G. Terlecki haben uns bei der Anfertigung des Manuskripts viel geholfen. Bei ihnen allen bedanken wir uns sehr.

Frankfurt/Main, im August 1974 Walter Greiner

Herbert Diehl

* Jetzt Professor an der Universität Kassel. ** Jetzt Professor an der Justus Liebig-Universität Gießen.

*** Wir gedenken besonders Herrn Jörg Briechle, der am 24.3.1970 bei einem Lawinenunglück gestorben ist.

III

INHALTSVERZEICHNIS

Fig. 2 View on introductory remarks in the W. Greiner and H. Diehl *Electrodynamik* (Verlag Harri Deutsch, Frankfurt 1975, ISBN 3871441856). See text for further comments. *Source* Johann Rafelski archives

material. A second edition of the script was then created, following the path of the first. In the case of *Electrodynamik* shown in Fig. 2, the first script edition was in late 1970; the second in late 1973, and the book version was ready to go to press a year later.

As the fame spread, a publisher, Verlag Harri Deutsch, located just across the street from the institute, became interested. So the 2nd edition of the lecture script was published reprinted directly from the script document. In this way Greiner's script became available beyond Greiner's institute. This in turn opened the path for other German Universities to embark on Greiner's teaching method. The newly developed Theoretical Physics teaching model began to influence all German-speaking Universities.

I selected the pages of the *Electrodynamik* 1st book edition in Fig. 2 to show that Walter credited everyone deserving for the help in creating the main course books: we can read the names of all involved students and Assistants, and Herbert Diehl of 2nd Script edition is coauthor. After 40 years, one would expect that most, if not all, students listed in this page are long forgotten. This is not the case: many embarked on great careers; some returned (Johannes Kirsch for example) after their retirement, contributing to the academic life in the newly formed Institute for Advanced Studies.

Aside of the core stream of books, Walter saw the need and also an opportunity for additional texts offering specialized topics. While I was teaching Special Relativity in 1981, Walter asked me if I wanted to contribute as an author to the new topical series he would publish as "Herausgeber". I agreed and Walter was keen to see my text as soon as possible; it was to be the first in the topical series. Being a young professor I did not at first have an "Assistant" to help me and I was teaching without sources as secondary literature was often inaccurate and misleading. Moreover, I had research priorities. As a result three years passed before the *Spezielle Relativitätstheorie* book was ready [2].

By that time the original publication model for the special topics series was colored by other projects and the question of economic success: Walter Greiner was a successful author, Johann Rafelski (and the authors of other topical projects underway) were not. Verlag Harri Deutsch wanted to see the special series books as Greiner's books. To account for the fact that I did all of the author's work, Walter proposed to give me 3/4 of author compensation, considering that editors of series in general get some fraction of the revenue. Indeed some 15 years later when publishing a book on Relativistic Heavy Ion physics, I was offered the same contract with Cambridge University Press; the difference was that when the CUP book appeared, the authors, and series editors, were clearly set apart. This was not the case with Verlag Harri Deutsch.

My German relativity text was very successful: in 10 years and three editions it sold 6,300 copies. The 2nd and 3rd editions where hard cover bound. However, the used physics book market trades today soft cover reprints of the 3rd edition. Walter used to say that one way to test popularity is to discover that one's books are stolen in libraries. Together we made it to a higher dimension of popularity: the German Special Relativity book is so popular that someone prints softcover wild copies.

Several of Walter's books have been published in English by Springer Verlag (beginning in 1989). Seeing the success of "Relativity" and the fact that after 1983 I was teaching in English, we discussed a possible English edition. However, Walter, I, Verlag Harri Deutsch, not to mention Springer Verlag, never found a contractual solution. Our book went out of press by 1995/6. Finally, after 20 more years, in 2017, I published a new English book on special relativity [3], dedicating this work to Walter who had passed away a few months earlier.

1.3 The Beginning of Strong Field Physics in Frankfurt

Following on my "Vordiplom" in early 1970 I was drawn into a new topic of "strong fields". Walter correctly saw that I would best fit this intellectual adventure that even today still seeks an intellectual home, falling often through the cracks between particle, nuclear, atomic foundational physics domains.

Walter needed someone ready to jump in since by mid-1969 he had convinced himself that we had not understood the physics of atomic electrons bound to superheavy large atomic number Z nuclei. It is best here to let Walter speak, citing from the conference panel discussion printed in the proceedings of the International Conference on Properties of Nuclear States held in Montreal, Canada August 25–30, 1969. This was the premier meeting event where who-was-who in the international nuclear science appeared.

We read (p. 611 in [4]): "*Greiner*: The important thing is that for $Z = 80$ you have $Z\alpha$ ($\alpha \simeq 1/137$ (JR)) less than unity, but for super-heavy nuclei around $Z = 164$ it is suddenly larger than unity and you do not know whether the expansion in $Z\alpha$

converges anymore. You really have to start from a completely different point of view and develop new methods." Responding and addressing challenges by G. E. Brown and D. H. Wilkinson, Walter continues (p. 612): "...the 1s-electron levels are very quickly very strongly bound and dive into the lower continuum. Their binding energy very quickly increases up to 1 MeV. This is the point where the difficulties arise. ...I would like to stress that this quantum electrodynamics problem is very interesting from a purely theoretical point of view."

The last remark shows what attracted Walter to the topic, while the first part merely explains how he came to consider the research program in the first place. Walter knew that the island of nuclear stability at $Z = 164$ would not lead to what we call today supercritical fields, but of course this did not matter to him and as the following developments showed, we had an alternate path, the quasi molecules. More on this later; see Sect. 3.2. Continuing the timeline: already in August 1968 Walter and W. Pieper submitted to Z. Physik results showing the need for $Z > 172$ [5]. This publication was delayed while a partial redo of the computation was carried out [6], in which an editor of the journal was thanked for suggesting the research topic.

The results by Pieper and Greiner confirmed and quantified using realistic nuclear charge distributions the earlier results obtained by Werner and Wheeler [7], who in their 1958 publication abstract say: "Despite Z values substantially higher than 137, the K electrons behave perfectly normally because of the finite extension of the nucleus. Vacuum polarization and vacuum fluctuations are roughly estimated to make relatively minor alterations in the K electron binding—which exceeds mc^2."

The above describes succinctly the state of knowledge before Pieper and Greiner: what Werner and Wheeler *overlooked* is that at sufficiently large $Z > 172$, the problem that earlier was seen for $Z > 137$ reappears and does so in a way that is even less comprehensible. While for the point source of the electromagnetic field the Dirac Hamiltonian becomes non-selfadjoint, meaning that the spectrum of bound states is not complete, for a finite nuclear size when the electron biding exceeds $E > 2mc^2$, a bound state "dives" into the antiparticle solutions of the Dirac equation. While some work (incorrectly) claimed that the old self-adjointness problem related to the $1/r$ singular potential returns, Walter never made this mistake.

I have little doubt that in order to make progress someone as unencumbered by prior thinking as Walter had to become interested in strong binding, who also had to be a person with a wide knowledge of diverse theoretical tools and methods. In particular Walter was aware of the work by U. Fano on embedding of bound states in a continuum [8]. This created the basis for the understanding of the positron autoionization phenomenon, see Sect. 3.1.

As we will describe in Sect. 2, the study of nonlinear limiting field electromagnetism paved the way to the recognition that even in a limiting force theory, there is no way to avoid the phenomenon of electron binding in excess of $2mc^2 = 1.022$ MeV. In colloquial language we call this level crossing into negative energy continuum "diving". This behavior had been described and left unresolved in earlier studies.

Looking back I remember how in the Winter 1971/72, the strong field group met regularly Saturday mornings in Walter's office suite located in the SW corner

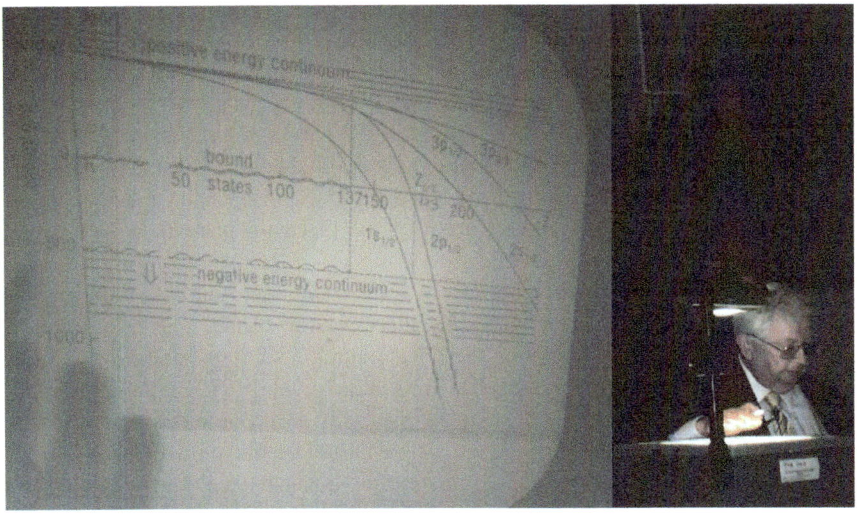

Fig. 3 Walter Greiner showing in March 2006 a slide created with contents created for publication in Spring 1972 [9]. *Photo* Johann Rafelski

of the 5th floor of the Physics building at Robert-Mayer Strasse 10. One Saturday morning, in an spontaneous burst of creativity, Walter adapted Fano's renown work to the case of "diving", calling this process autoionization of positrons, implying that one should interpret "diving" of particle states into antiparticle continuum as a phenomenon where a hole, a vacancy in the bound state turned in the "diving" process into a spontaneously emitted positron; we return to this topic in Sect. 3.1.

I believe that to the end of his life Walter considered this insight his greatest. He never hesitated to tell about this process. For example, at the "International Conference on Strangeness in Quark Matter" (SQM2006), held at the University of California at Los Angeles (UCLA), Walter Greiner gave the UCLA Departmental colloquium, "On the Extension of the Periodic System into the Sector of Strange- and Antimatter."

In this UCLA lecture Walter described to a large and multidisciplinary audience the main domains of research he developed in Frankfurt, beginning in the late 1960s. As the lecture unfolded we sensed Walter's heart beating loudest when he recounted the strong field QED. Water described in detail the physics of "diving" seen in Fig. 3, as if this was still the Spring 1972 when we published this result together [9].

1.4 The Stability Principle

Even though in science dogmatic principles are rarely of value, as the context and ultimate fate of the false Aristotelian Physics proves by example, Walter's adherence to a "stability principle" guided, and may have misdirected, some of his effort. Let me show in a few examples how this worked.

Strong fields

Walter was not willing to embrace his own Pieper-Greiner result unconditionally. In 1970, when he offered me a Diploma Thesis topic, he was searching for a new mechanism to make the critical binding go away. He was searching for improvements such that in the presence of ever larger externally applied fields, the binding energy of an electron should never reach the limit $E \to 2mc^2 \simeq 1\,\text{MeV}$. He followed several paths of which one was the modification of the electromagnetism, which I will describe in technical detail in Sect. 2.

We found that other experimental results limited the opportunity to modify electromagnetism and thus we concluded that there was no chance that this approach could be of relevance for stabilizing the Dirac equation solutions and resolving the electron "diving" problem, see Sect. 3.1. Thus in the pivotal 1972 publication where we quantitatively demonstrate the positron instability, Walter refers to this situation as follows [9]: "At this point the single-particle theory seems to break down. Pieper and Greiner and later Popov have interpreted this to mean that electron-positron pairs are created spontaneously." The word "seems" reflects on his continued hope that that non-perturbative QED many body theory effect could be significant.

I think Walter's change of heart in regard to "diving" behavior occurred with our 1973 work on "Charged Vacuum" [10]. The reinterpretation of the phenomena in terms of vacuum structure and the insight that the structured quantum vacuum acquires localized charge density overwhelmed Walter's adherence to stability.

Into this new context arrived Miklos Gyulassy, freshly minted at Berkeley. Miklos told me, and I agree, that his Frankfurt job and excellent relationship with Walter was a direct outcome of him showing independently of our effort (in a very elaborate numerical work [11]) that the Frankfurt Charged Vacuum theory was right.

Event horizon

Walter posited that gravity should not prevent light from traveling; in other words, an event horizon associated with a black hole solution should not exist. As strong field physics advanced, Walter saw the connection to gravity, and he hoped that there would be a way, using the ideas we developed in relativistic quantum theory of strong fields, to modify Einstein's gravity. He gained this insight in his very first General Relativity (GR) class, which I recall he held in Winter 1971/2. Significant effort went into the understanding of Dirac equation solutions in a strong gravity field [12]. Following in the footsteps of related work on nonlinear EM theory, see Sect. 2, students in Frankfurt worked on what we call today $f(R)$ gravity [13], where action is a nonlinear function of the Ricci tensor R.

When I visited the GSI laboratory in Summer 1977 for three months on the way from US to CERN, Walter wanted me to create a no-event-horizon gravity, asking for my full commitment. In these three months I learned more about gravity than I did in the rest of my life. Aside of me, Berndt Müller became also part of effort. We did not solve the problem, and I must add, Berndt and I challenged Walter, if the existence of an event horizon was really a problem? I believe it was the first time that a clear divisive line opened between Walter and his first hour students. It is worth noting that Walter never relented about black holes, and after some 30 more years he published a no-event-horizon gravity theory [14, 15].

Even if our preoccupation with event horizon did not lead to a good outcome, our effort paid off in a different way. Berndt Müller, Walter and I published on the interpretation of strong external EM field—thus acceleration—in terms of effective temperature [16]. This work implies that some deep connection exists between EM and GR and has influenced my work from the past decade [17].

Quark-gluon plasma

The development of the GSI laboratory near Frankfurt in Wixhausen, now part of the City of Darmstadt, was driven by the hard work and political skill Walter so often displayed. This laboratory today is a renowned center of relativistic heavy ion research. Among the most important physics developments that occurred in late 1970s and early 1980s was the exploration of nuclear matter using beam of relativistic heavy ions. Walter was the pioneer in this field, working with Horst Stöcker, another lustrous student and recent GSI director, on shock waves that nuclear matter should support, see for example [18].

This work began in close cooperation with Erwin Schopper, the founder of (experimental) "Institut für Kernphysik" (nuclear physics). For this work The European Physical Society in 2008 awarded to Walter Greiner and to Schopper's successor, Reinhard Stock, the **Lise Meitner Prize** for nuclear science. Walter's citation reads: "...for his outstanding contributions to the development of the field of relativistic nucleus–nucleus collisions by pioneering the ideas of shock waves and collective flow in nuclear matter, thus inspiring experimental studies of nuclear matter at extreme conditions of density and temperature."

There is no mention in this citation of the new state of matter, the deconfined quark-gluon plasma (QGP) phase of matter, which had perhaps even more consequential impact on nuclear physics. By the late 1970s the recognition grew that this deconfined form of matter could be created in ultra relativistic heavy ion collisions. I reported on the early work on deconfinement, and QGP formation, in the "Hagedorn" volume [19].

The QGP research direction fit both GSI and Greiner's traditional nuclear research program perfectly. He and his group should have been among the pioneers in this new research field. However, in Frankfurt the stability concerns affected the early development of the QGP physics: the idea that nucleons at high temperature could melt and dissolve into the more fundamental constituents of matter, quarks and gluons, did not sit easily in Walter's mind.

Walter attended my inaugural lecture event on June 18, 1980, see Fig. 4, where a contemporary view of these QGP developments was offered based on work I had done at CERN in the prior 2.5 years. I hoped and expected my friend and teacher to sit in the front row nodding approvingly. Instead, he was in the very back of the filled room, not listening, as I observed with some trepidation.

In following years Walter continued in clear and outspoken opposition; several Ph.D. students in his group were working to prove that QGP could not be observable. I departed Frankfurt in 1982, heading back to CERN and later on to Cape Town.

Fachbereich Physik
der Johann Wolfgang Goethe-Universität Frankfurt am Main

Einladung

zu der öffentlichen Antrittsvorlesung des Herrn
Prof. Dr. Johann Rafelski
über das Thema

„Quarkmaterie – Kernmaterie"

Das Verschmelzen der Bestandteile der Protonen und Neutronen – der Quarks – zur Quarkmaterie, einer neuen Phase der Kernmaterie wird aufgrund von neuesten theoretischen Arbeiten in einem experimentell zugänglichen Druck und Temperaturbereich erwartet.

am Mittwoch, dem 18. Juni 1980, 17 Uhr c. t.
im Hörsaal des Instituts für Angewandte Physik
Frankfurt am Main, Robert-Mayer-Straße 2–4

Die Vorlesung findet im Rahmen des Physikalischen Kolloquiums statt.

Der Dekan: Prof. Dr. Werner Martienssen

Fig. 4 The translation of the abstract of author's public Inaugural lecture held in Frankfurt on June 18, 1980 reads: *The fusion of constituents of protons and neutrons—quarks—into quark matter, a new phase of nuclear matter has been studied in recent theoretical work. It is expected to occur in an experimentally accessible domain of temperature and pressure.*" *Source* Johann Rafelski archives

Despite these setbacks, in the ensuing years the QGP effort in Frankfurt grew stronger around some of my students who persevered—Peter Koch deserves to be mentioned and praised for this effort; see for example Ref. [20].

2 Born-Infeld Nonlinear Electromagnetism

Following Max Born's passing in January 1970, his work on nonlinear Born-Infeld (BI) electromagnetism was widely discussed. Walter Greiner, given his adherence to stability, was fascinated by the BI effort to stabilize electromagnetism. In the BI-EM theory, the self-energy of a charged point particle was finite and the acceleration, just like particle speed, had a maximum value.

Walter believed that if the electric field of an atomic nucleus had a limit, there could be a major change in the solutions of the Dirac equation and the critical binding behavior might disappear. This was the project he signed me on for my Diploma, and arranged for assistance from Lewis P. Fulcher. Lewis had just arrived in Frankfurt as a postdoctoral fellow, having graduated with Judah M. Eisenberg from the University of Virginia.

To understand the working of the BI-EM, we need to truly distinguish between EM field and the displacement fields. We note the inhomogeneous Maxwell's equations

$$\nabla \cdot \mathbf{D} = e\rho \ , \quad \nabla \times \mathbf{H} = \frac{\partial D}{\partial t} + \mathbf{e}j \ , \tag{1}$$

connecting with **E** and **B** by means of the first derivative of the Lagrangian density

$$\mathbf{D} = \frac{\partial \mathscr{L}}{\partial \mathbf{E}} \ , \qquad \mathbf{H} = \frac{\partial \mathscr{L}}{\partial \mathbf{B}} \ . \tag{2}$$

This is well known to those who practice electromagnetism in media. Max Born worked with a covariant medium by choosing a nonlinear covariant format of the action. Restricting our study for illustration to electrostatics, where only **E** and **D** do not vanish, we can easily understand Born's idea. Consider

$$\mathscr{L} \equiv -\varepsilon_0 E_{\mathrm{BI}}^2 \left(\sqrt{1 - \mathbf{E}^2/E_{\mathrm{BI}}^2} - 1 \right) \to \frac{\varepsilon_0}{2}\mathbf{E}^2 \ , \tag{3}$$

where the weak field limit is indicated. For the **D** field we obtain

$$\mathbf{D} = \frac{\partial \mathscr{L}}{\partial \mathbf{E}} = \frac{\mathbf{E}}{\sqrt{1 - \mathbf{E}^2/E_{\mathrm{BI}}^2}} \ , \qquad \mathbf{E} = \frac{\mathbf{D}}{\sqrt{1 + \mathbf{D}^2/E_{\mathrm{BI}}^2}} \ . \tag{4}$$

We see that when $|\mathbf{E}| \to E_{\mathrm{BI}}$ the displacement field **D** diverges. Thus for a solution of the Coulomb problem inherent in Eq. (1), the field $\mathbf{D} \propto \hat{r}/r^2$ diverges at the origin for a point source, while at the origin the electrical field **E** reaches its maximum BI value. More generally and independent of the form of the displacement field **D**, we find $\mathbf{E}^2 \leq E_{\mathrm{BI}}^2$.

Considering the Lorentz-force on a particle due to a point (nuclear) source

$$\frac{d\mathbf{p}}{dt} = e\mathbf{E} = -\frac{\hat{r}\, eE_{\mathrm{BI}}}{\sqrt{r^4/(Zr_{\mathrm{BI}}^2)^2 + 1}} \ , \qquad eE_{\mathrm{BI}} \equiv \frac{e^2}{4\pi\varepsilon_0 r_{\mathrm{BI}}^2} \ , \tag{5}$$

we see that there is a limited strength force. The negative sign appears since the electron and the nucleus carry opposite charges.

Born invented (and Born and Infeld improved) a limit to force and this is what attracted the attention of Walter Greiner. Seeing a limit to force one may justly ask if there is a limit on electron binding in the field of a heavy nucleus. However, a classical limit to electric force does not mean that the potential governing solutions of the Dirac equation is is also bounded. The electron atom is characterized by a weak electric field stretching far when compared to the radius of the K-shell electron inside a heavy atom. Therefore to limit the electron binding one must limit the potential depth. $|eV| < 2mc^2$ is required to prevent the electron from "diving" into the negative continuum.

To quantify this we evaluate the radial integral of Eq. (5)

$$eV(r) = \int_r^\infty eE_r dr\ , \qquad eV_{\rm BI}(0) = -\frac{1.8541\alpha Z^{1/2}\hbar c}{r_{\rm BI}}\ . \tag{6}$$

We show the value at origin as $V(r)$ is a complicated hypergeometric function. The BI choice of $r_{\rm BI}$ was made such that the electron mass is accounted for as being the energy content of the electromagnetic field: $r_{\rm BI} = 1.236 r_e = 3.483\,\text{fm}$, where $r_e = e^2/(4\pi\varepsilon_0 mc^2)$ is the classical electron radius and the numerical factor follows from some technicalities, see for example Chap. 28 in Ref. [3].

We thus discover that

$$eV_{\rm BI}(0) = -\frac{Z^{1/2}\,2.666\,\text{MeV fm}}{r_{\rm BI}} = -Z^{1/2} 0.765\,\text{MeV}\ . \tag{7}$$

We see that the depth of the potential is finite but unbounded. $eV_{\rm BI}(0)$ scales with $Z^{1/2}$ instead of the Z we are familiar with in the linear Maxwell theory. One can interpret this result in the context of Maxwell theory, introducing an atomic nucleus effective size $R = Z^{1/2} 2.75\,\text{fm}$. This size is, however, much too large. In order to be compatible with atomic physics data, a much larger BI limiting field is required [21–24].

With a larger $E_{\rm BO}$, according to Eq. (5), we would need a smaller $r_{\rm BI}$, perhaps $r_{\rm BI} \to r_{\rm BI}/5$. Since the EM mass of the electron scales with $1/r_{\rm BI}$, such a field implies an electron EM mass well in excess of experiment. To summarize, the conclusion is that there is on one hand no stability of atomic orbitals, and on the other, the key attractions of BI theory is invalidated: the EM field energy for an electron is clearly too large after the BI limiting field parameter is adjusted to agree with atomic experimental constraints. This was the result that made Walter search for the understanding of what happens when a Dirac state as a function of parameter such as nuclear charge Z mutates from being an electron into new existence of a positron. In colloquial language we call this "diving".

3 Positron Production and Charged Vacuum

For an uninitiated reader the first necessary insight is an explanation of why we call the Coulomb potential that is capable of binding an electron by more than $2m_e c^2$ "supercritical". To answer this question, let us consider the electron-positron e^-e^+-pair production process. The minimum energy required is $2m_e c^2$. However, in the presence of a nucleus of charge Ze, it is possible that we do not require this vacuum energy, since there is an electronic bound to the nucleus, and the binding reduces the pair energy threshold.

3.1 Dirac Equation and "Diving"

The threshold for pair conversion of a γ-ray to an e^-e^+-pair in the presence of a nucleus is

$$E_T^\gamma = m_e c^2 + \varepsilon_n \ , \quad (8)$$

where ε_n is the energy of the bound electron (always including its rest mass) in the eigenstate n. Considering the Pauli principle we recognize that this is only possible if such a state has not been occupied by another electron. The above energy balance for the γ-conversion to e^-e^+ pair implies the following statement:

> When $\varepsilon_n \to -m_e c^2$, the minimum energy required to create an e^-e^+-pair approaches zero: $E_T^\gamma \to 0$. At the critical point $\varepsilon_n = -m_e c^2$, the energy of the ionized atom is equal to the energy of the atom with a filled 1s-electron state and a free positron of nearly zero kinetic energy.

It is important to consider carefully what happens if and when a metastable bound state $\varepsilon_n \to \varepsilon_R < -m_e c^2$ could exist. In such a situation the energy of a fully ionized atom without the $1s$-electron(s) is higher than the energy of an atom with a "filled" K-shell and free positron(s). Thus a bare supercritical atomic nucleus cannot be a stable ground state and therefore the neutral (speaking of electro-positron) vacuum cannot be a stable ground state either.

We conclude that for super-critical binding where a quasi-state dives into the negative energy sea as we see in Fig. 3, the supercritical bare atomic nucleus will spontaneously emit a positron e^+ (or two e^+, allowing for spin), keeping in its vicinity the accompanying negative charge which thus can be called the real vacuum polarization charge. The state that has an undressed atomic nucleus is the "neutral vacuum" (vacuum for electrons, positrons), and beyond the critical point is not the state of lowest energy. The new state of lower energy, called the charged vacuum [10], is the dressed atomic nucleus; that is a nucleus surrounded by the real vacuum polarization charge.

The physics understanding here described was created in early 1972 [9]; this was the great insight Walter gained one Saturday morning, based on the Fano resonance embedment method. However, when viewed in hindsight, we could arrive at this result solving the Dirac equation to determine phase shifts of positron scattering states, which we did a year after [25]. The phase shift analysis shows resonant scattering and allows the determination of the width of the resonance reliably.

3.2 Quasi Molecules

Critical binding requires a super-superheavy nucleus containing $Z \geq 173$ protons within a realistic nuclear volume. On the other hand there is a fundamental interest in seeing the vacuum decay predicted. In 1971 we recognized that in heavy-ion collisions the relativistic deeply bound electrons were moving fast enough to form quasi-molecular states around the two slowly moving nuclear Coulomb potential centers.

This means that the collision of two extremely heavy nuclei could be used to probe the supercritical fields; in the second paragraph in Ref. [21] we state: "Even if superheavy elements cannot be readily produced, enough information could possibly be gathered in the collisions of heavy ions, such as Pb on Pb or Cf on Cf, to decide if this limit exists. In these collisions the adiabatic approximation should have some validity since the velocity of the electrons in the $1s$ and $2p$ atomic orbitals is much faster than the relative nuclear velocity. Hence, as far as the electrons in the lower atomic orbitals are concerned, the collisions of Pb on Pb and of Cf on Cf may simulate superheavy electronic molecules with $Z = 164$ and $Z = 196$, respectively."

The relatively slowly moving heavy-ions with energies chosen to stop the collision at the Coulomb barrier provide a common field for a shared quasi-molecular electron cloud. These electron eigenstates could be computed in a good approximation using the combined Coulomb field corresponding to a super-heavy nucleus of charge $2Z$, with a quasi potential formed by an effective nuclear charge distribution with a diameter $2R_N = R_{12}$, corresponding to the distance R_{12} between the two nuclei [26].

This "monopole" approximation can be justified by averaging the two lowest terms in the multipole expansion. Adopting such an effective radial form of the potential to simulate the effect of axially symmetric potential implements the idea of quasi-molecular states where the electrons circle around the two centers, or seen in reverse, the two nuclear charges circle around each other, and the electron is observing the so obtained averaged potential. The shape of the adopted effective monopole radial potential is

$$V_0(r) = \begin{cases} -\frac{3}{2}\frac{Z\alpha}{(R_{12}/2)}\left(1 - \frac{r^2}{3(R_{12}/2)^2}\right) & \text{for } 0 \le r \le R_{12}/2 \\ -\frac{Z\alpha}{r} & \text{for } R_{12}/2 < r < \infty\,. \end{cases} \tag{9}$$

In Fig. 5 the exact two center potential following the axis connecting the two nuclei (dashed line) is compared to the monopole approximate potential (solid line) for the case of a Uranium–Uranium collision. We show both potentials for $R_{12} \equiv R = 38.6\,\text{fm}$, the critical separation between the two Uranium nuclei. This shows that the electrons experience attractive forces similar to those of a super-heavy nucleus with $Z_{\text{eff}} = 184$, protons.

This simple approximation was tested extensively later, using the numerical methods that were developed in Ref. [28], and found to be a very accurate and useful tool in understanding the physics of strong fields in heavy-ion collisions at sub- and near-Coulomb barrier collisions.

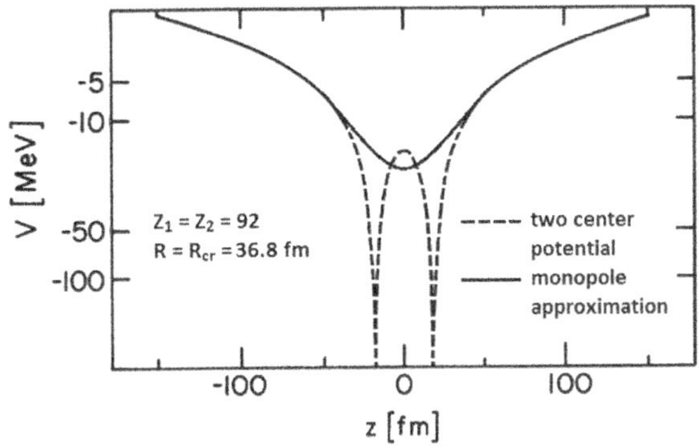

Fig. 5 Solid line: the (averaged) monopole potential that can be used to compute the electron binding in presence of colliding heavy-ions, see text; dashed line: the two center potential cut along the axis connecting the two nuclei. *Source* Ref. [27]

3.3 Experiments on Positron Production

The following experimentally observable effect emerges as a consequence of the supercritical binding: in collisions of high Z heavy ions an empty $1s$-state can be bound by more than $2m_ec^2$. Subsequently, a positron is emitted spontaneously. When the heavy ions separate again, the previously empty $1s$-state is now occupied by an electron; thus we effectively produced a pair by spontaneous vacuum decay. The theoretical treatment of the process is greatly facilitated by the large mass of the two nuclei: the Sommerfeld parameter $\eta = Z_1Z_2\alpha/v > 500$. Hence the classical approximation to the nuclear motion is adequate, and only the electrons have to be treated quantum mechanically.

The actual physical situation is not that simple: the heavy-ion collision is a time-dependent process; thus there may not always be enough time to emit a positron. Moreover there are several processes driven by time dependence of the collision, see Fig. 6. For the positron production to involve the tightly bound eigenstate we need to remove electrons still present in the K-shell quasi-molecular states, see processes a, b. The motion of the ions can induce positron production in the processes d, e, there can be furthermore direct free pair production process f. Coherently superposed to processes d, e, f is the spontaneous positron emission process c. Detailed discussion of the extensive 1970–1981 study of the theoretically anticipated effects can be found in [29].

The rather short lifetime of a supercritical K-shell vacancy against positron emission, $\tau_{e^+} \simeq 10^{-18} - 10^{-19}$ s implies that the supercritical system needs to live only for such a short period of time. An estimate of the order of magnitude of collision time shows that this is indeed feasible: the typical collision time of two nuclei at energies just below the Coulomb barrier is

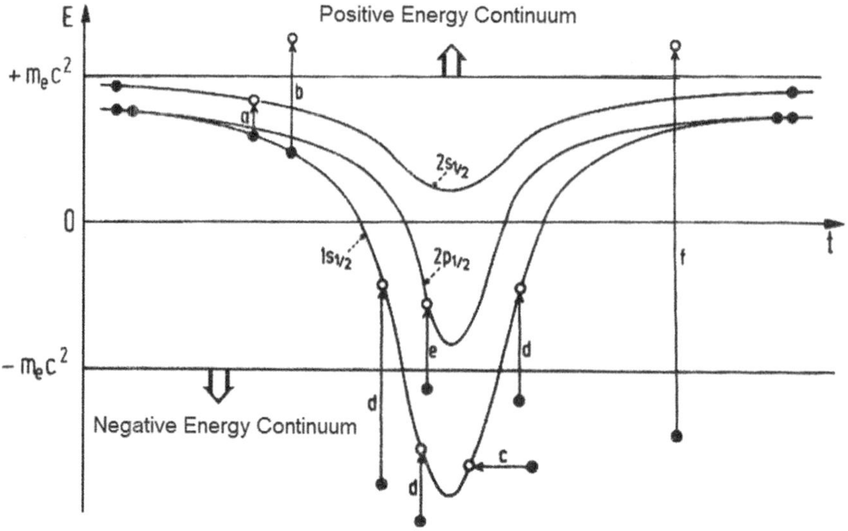

Fig. 6 Schematic representation of pair-production processes in heavy-ion collision as a function of time. We see most tightly bound eigenstates and relevant processes: a, b-ionization; c-spontaneous and d, e-induced vacuum decay, f-continuum pair production. *Source* Ref. [27]

$$\tau_{\rm coll} \simeq \frac{2R_{\rm cr}}{v} \simeq 0.25 \times 10^{-20}\,{\rm s}\,, \tag{10}$$

with $R_{\rm cr} \simeq 35$ fm (see below). The emission time for positrons is typically 100 times longer such that one expects a yield of roughly 1% in this reaction.

The eigenstate energy of of most tightly bound electrons increases as ions approach and at $R_{\rm cr} \simeq 35$ fm, it equals $-2m_e$ a for the $1s_{1/2}\sigma$ electron state in U+U collisions [30]. The quasi-molecule is rendered supercritical in just the same way as the super-heavy atom was at $Z > Z_{\rm cr}$.

A lot of effort went into the experimental search for spontaneously emitted positrons. A contemporary discussion of the experimental results has been recently presented [27]. In a nutshell, the consensus view today is that positrons observed were due to system dependent nuclear excitations converting into pairs. Walter Greiner was deeply marked by these disappointing experimental developments. In fact the word “disappointment” does not even come close to describing his feelings. While today there is no ongoing heavy ion positron production experiment, many regret that the experimental effort ended without an experimental result addressing strong field physics.

Fig. 7 Walter Greiner prepares to become witness at the marriage of Johann Rafelski (in picture) and Helga E. Betz, August 1973 at the Frankfurt Römer. *Photo* Johann Rafelski

Fig. 8 Walter Greiner with the Rafelski Family in Tucson, 1988. *Photo* Johann Rafelski

4 Our Life

These comments about the life and work of my teacher are best concluded with a few pictures that tell more about the lasting relationship Walter Greiner enjoyed with the author and his family (Figs. 7, 8 and 9).

Fig. 9 Walter Greiner and the author in March 2006 Dinner at SQM2006, UCLA. *Photo* Johann Rafelski

Acknowledgements The author thanks FIAS and Horst Stöcker for support making this presentation possible.

References

1. T. Biró, C. Greiner, B. Müller, J. Rafelski, H. Stöcker, Topical issue on frontiers in nuclear, heavy ion and strong field physics. Eur. Phys. J. A **54**, 31 (2018). https://doi.org/10.1140/epja/i2018-12477-6
2. W. Greiner, J. Rafelski, Spezielle Relativitätstheorie (Verlag Harri Deutsch, Thun and Frankfurt am Main, 1984), 322 pp. ISBN 3-87144-711-0
3. J. Rafelski, *Relativity Matters : From Einstein's EMC2 to Laser Particle Acceleration and Quark-Gluon Plasma* (Springer, Berlin, 2017), 468 pp. https://doi.org/10.1007/978-3-319-51231-0
4. G.E. Brown, W. Greiner, D.H. Wilkinson (those only contributing to the pertinent topic of panel discussion), Panel discussion: new directions in nuclear spectroscopy, in *Proceedings of International Conference on Properties of Nuclear States*, 25–30 August 1969 (University of Montréal Press, Montréal, 1969)
5. W. Pieper, W. Greiner, Interior electron shells in superheavy nuclei. Z. f. Phys. **218**, 327 (1969). https://doi.org/10.1007/BF01670014
6. D. Rein, Über den Grundzustand überschwerer Atome (translated: On the groundstate of superheavy atoms). Z. f. Phys. **221**, 423 (1969). https://doi.org/10.1007/BF01393233
7. F.G. Werner, J.A. Wheeler, Superheavy nuclei. Phys. Rev. **109**, 126 (1958). https://doi.org/10.1103/PhysRev.109.126
8. U. Fano, Effects of configuration interaction on intensities and phase shifts. Phys. Rev. **124**, 1866 (1961). https://doi.org/10.1103/PhysRev.124.1866
9. B. Müller, H. Peitz, J. Rafelski, W. Greiner, Solution of the Dirac equation for strong external fields. Phys. Rev. Lett. **28**, 1235 (1972). https://doi.org/10.1103/PhysRevLett.28.1235
10. J. Rafelski, B. Müller, W. Greiner, The charged vacuum in over-critical fields. Nucl. Phys. B **68**, 585 (1974). https://doi.org/10.1016/0550-3213(74)90333-2
11. M. Gyulassy, Higher order vacuum polarization for finite radius nuclei. Nucl. Phys. A **244**, 497 (1975). https://doi.org/10.1016/0375-9474(75)90554-0
12. M. Soffel, B. Müller, W. Greiner, Dirac particles in rindler space. Phys. Rev. D **22**, 1935 (1980). https://doi.org/10.1103/PhysRevD.22.1935

13. A. Müller, U. Heinz, B. Müller, W. Greiner, On gravitation theories with limiting curvature. J. Phys. A **11**, 1781 (1978). https://doi.org/10.1088/0305-4470/11/9/012
14. P.O. Hess, W. Greiner, Pseudo-complex general relativity. Int. J. Mod. Phys. E **18**, 51 (2009). https://doi.org/10.1142/S0218301309012045
15. P.O. Hess, M. Schäfer, W. Greiner, *Pseudo-Complex General Relativity*. FIAS Interdisciplinary Science Series (Springer, Switzerland, 2016), p. 235
16. B. Müller, W. Greiner, J. Rafelski, Interpretation of external fields as temperature. Phys. Lett. A **63**, 181 (1977). https://doi.org/10.1016/0375-9601(77)90866-0
17. L. Labun, J. Rafelski, Acceleration and vacuum temperature. Phys. Rev. D **86**, 041701 (2012). https://doi.org/10.1103/PhysRevD.86.041701
18. H. Stöcker et al., Nuclear fluid dynamics versus intranuclear cascade-possible evidence for collective flow in central high-energy nuclear collisions. Phys. Rev. Lett. **47**, 1807 (1981). https://doi.org/10.1103/PhysRevLett.47.1807
19. J. Rafelski, *Melting Hadrons, Boiling Quarks: From Hagedorn Temperature to Ultra-Relativistic Heavy-Ion Collisions at CERN – With a Tribute to Rolf Hagedorn* (Springer, Switzerland, 2016). https://doi.org/10.1007/978-3-319-17545-4, ISBN 9783319175454 (online), 9783319175447 (print)
20. C. Greiner, P. Koch, H. Stöcker, Separation of strangeness from antistrangeness in the phase transition from quark to hadron matter: possible formation of strange quark matter in heavy ion collisions. Phys. Rev. Lett. **58**, 1825 (1987). https://doi.org/10.1103/PhysRevLett.58.1825
21. J. Rafelski, L.P. Fulcher, W. Greiner, Superheavy elements and an upper limit to the electric field strength. Phys. Rev. Lett. **27**, 958 (1971). https://doi.org/10.1103/PhysRevLett.27.958
22. J. Rafelski, Die Konsequenzen nichtlinearer elektromagnetischer Feldtheorie in überschweren Elementen. Diploma thesis, Institut für Theoretische Physik der Universität Frankfurt am Main, June 1971, http://inspirehep.net/record/1468114/files/Thesis-1971-Rafelski.pdf
23. J. Rafelski, W. Greiner, L.P. Fulcher, Superheavy elements and nonlinear electrodynamics. Nuovo Cim. B **13**, 135 (1973). https://doi.org/10.1007/BF02726701
24. J. Rafelski, G. Soff, W. Greiner, Lower bound to limiting fields in nonlinear electrodynamics. Phys. Rev. A **7**, 903 (1973). https://doi.org/10.1103/PhysRevA.7.903
25. B. Müller. J. Rafelski, W. Greiner, Electron wave functions in overcritical electrostatic potentials. Nuovo Cim. A **18**, 551 (1973)
26. J. Rafelski, B. Müller, W. Greiner, Superheavy electronic molecules. Lett. Nuovo Cim. **4**, 469 (1972). https://doi.org/10.1007/BF02757147
27. J. Rafelski, J. Kirsch, B. Müller, J. Reinhardt, W. Greiner, *Probing QED Vacuum with Heavy Ions*, Springer FIAS Interdisciplinary Science Series (Springer, Berlin, 2017), p. 211. https://doi.org/10.1007/978-3-319-44165-8
28. J. Rafelski, B. Müller, Magnetic splitting of quasimolecular electronic states in strong fields. Phys. Rev. Lett. **36**, 517 (1976). https://doi.org/10.1103/PhysRevLett.36.517
29. J. Reinhardt, U. Müller, B. Müller, W. Greiner, The decay of the vacuum in the field of superheavy nuclear systems. Z. Phys. A **303**, 173 (1981)
30. J. Rafelski B. Müller, The critical distance in collisions of heavy ions. Phys. Lett. **65B**, 205 (1976). https://doi.org/10.1016/0370-2693(76)90163-5

In Memoriam Walter Greiner 1935–2016. Was Bleibt?

FIAS International Symposion on Discoveries of Science. Summary Talk

Amand Faessler

1 From Neuenbau to Frankfurt

Walter Greiner is born in Neuenbau near Sonnenberg in the Thueringer Wald on October 29th. 1935. Neuenbau is a small village of about 200 inhabitants. When Walter was born, it was almost in the center of Germany. After the war in 1945, Neuenbau was situated in Communist East Germany, the DDR, only about 2 km to the border to West Germany in the restricted area, which one could only visit with special permission. Walter's parents, his father was a shoemaker, realized that Neuenbau could not provide an adequate education to the gifted young boy. In 1947, they did send Walter to his grandparents near Frankfurt/Main. Finally the total family moved to the area of Frankfurt. In the chemical company, Farbwerke Hoechst, Walter made an apprenticeship as metalworker. He obtained the Abitur, which in Germany is needed for the admission to the University, in the Abend-Gymnasium. There he also met his later wife Baerbel Chun. After the Abitur, Walter studied Physics at the Universities of Frankfurt and the Technical University of Darmstadt. In Darmstadt, he obtained in 1960 with Prof. Scherzer the diploma degree with work on plasma physics for fusion reactors. In Darmstadt, he also met Prof. Hans Marschall from the University of Freiburg, who had an offer to move to Darmstadt. He asked him that if he could do his Ph.D. with him and get an "Assistant" position. When Hans Marschall decided to stay in Freiburg, he offered him to join him in Freiburg for his doctoral degree.

A. Faessler (✉)
Institute of Theoretical Physics, University of Tuebingen, Tübingen, Germany
e-mail: faessler@uni-tuebingen.de

J. Kirsch et al. (eds.), *Discoveries at the Frontiers of Science*,
FIAS Interdisciplinary Science Series, https://doi.org/10.1007/978-3-030-34234-0_23

2 Ph.D. in Freiburg and the Rotational-Vibrational Model for Nuclei

I met Walter when he came to Freiburg to Hans Marschall (1913–1986) in May 1960. In summer after he obtained the position as “Assistant” and had the feeling that he could support a family, he married Baerbel Chun from Frankfurt-Hoechst. Baerbel and Walter (Fig. 1) have two sons: Martin Greiner, born 1963, is Professor at the University of Aarhus in the Engineering Department. He has three daughters. Carsten Greiner, born 1964, is Professor of Theoretical Physics at the University of Frankfurt. He has two sons.

Walter started his doctoral thesis on polarization of Lead 208 by a bound muon. In lead, the nuclear radius is about 7 fm, and the Bohr radius of the μ^- is 3 fm. The muon moves inside the nucleus and tries to suck protons within its Bohr radius. For the nucleus, he used the Steinwedel–Jensen model developed for the giant dipole resonances with a proton and neutron liquid. The relativistic 1s muonic wave function in Lead 208 he could take from the literature (Fig. 2).

The shift of the 1s muonic level in Lead 208 turned out to be less than 1 keV, which was at that time below the resolution for X-rays detection at the relevant energies. After Walter had finished the analytic formulation of the problem, he asked me to program the numerical calculations on the “ZUSE” with only 16 electronic storages

Fig. 1 Photo Walter Greiner

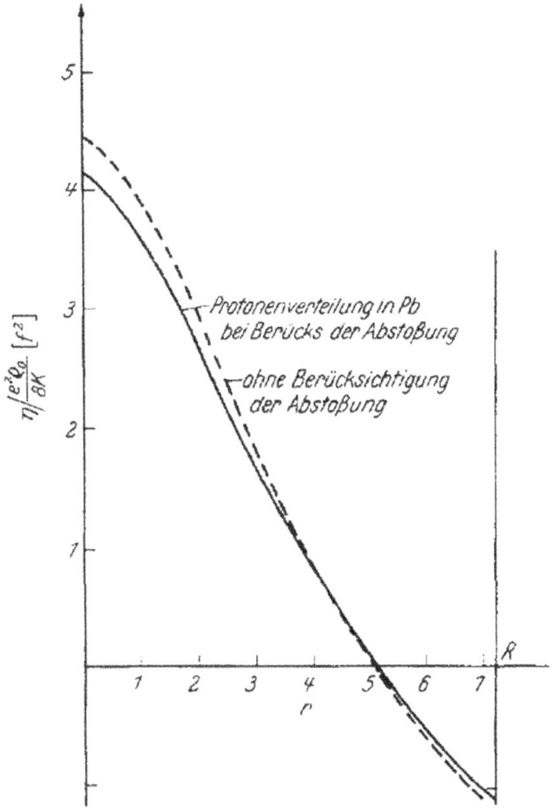

Fig. 2 Change of the proton density in Lead 208 due to polarization by an 1s muon. Proton density with the polarization minus the proton density without it. The solid line includes while the dashed lines neglects the increased repulsion between the protons due to the increased density induced by the polarization inside the 1s muonic Bohr radius

and 8192 storage places on a drum. After Walter finished his Ph.D. [1], he gave me for the numerical calculations a copy of the book "Introduction into Quantum Field Theory" by F. Mandl. Later on, Walter himself wrote several popular books for quantum field theory. But I know he learned this field from "Mandl", as I also did. I still can recommend this book for beginners in the field. It is with around 180 pages, which is quite short, and explains everything easy to understand. After Walter finished his Ph.D. and I my diploma degree with work on the Sternheimer effect, we started to work together on improvements of the Bohr–Mottelson model and moved from the old quite narrow physics building into the "Neubau" only about 100 m away in the Hermann-Herder-Strasse 3. Walter and I moved in the seventh floor into the same room with a view directly to the gate of the largest prison of the State of Baden-Wuerttemberg. From the seventh floor, we had a nice view into the prison yard, where the prisoners played soccer every day, in German Fussball. When Walter wanted a rest, he watched often the football game.

Around 1960 was a very important time for Nuclear Physics: In 1959, van der Graff had published the paper on the Tandem accelerator principle. He had not only the idea for the Tandems but also the company "National Electrostatic Corporation" not far from MIT. The big advantage of the Tandems over the cyclotrons was the much better energy resolution. The cyclotron could for higher energy excitation not resolve single states. With the Tandems, one could now study new physics at higher excitation energies. One class of such states was higher spin rotational states for deformed nuclei described by the Bohr–Mottelson model. The approximate solution of the Bohr–Mottelson model used at that time to describe the high spin states in deformed nuclei mainly in the rare-earth region showed several deficits.

The E2 branching ratio $B(E2, 2^{+}\gamma \rightarrow 2^{+}ground)/B(E2, 2^{+}\gamma \rightarrow 0^{+}ground)$ in the simple version of the Bohr–Mottelson model is a constant and given by the Alaga rule, the square of the ratio of two Clebsch–Gordan coefficients shown in Fig. 3. Walter Greiner and I took into account the changes of the moment of inertia of the deformed rotational nuclei due to the β and γ vibrations [2]. The improvement relative to the experiment was tremendous (see Figs. 3 and 4 and Ref. [2]).

In spring 1962, Walter and I went to our first conference to the meeting of the German Physical Society in Bad Pyrmont about 50 km north of Goettingen. Walter spoke about our results of the rotational-vibrational (RotVib) model and I on the results of my diploma thesis on the Sternheimer effect (screening of the nuclear quadrupole moment by the inner electrons for the valence electron, a correction for the determination of the nuclear quadrupole moment by electron spectra). There we also met Max Born (1882–1970), who returned after the war from Edinburgh

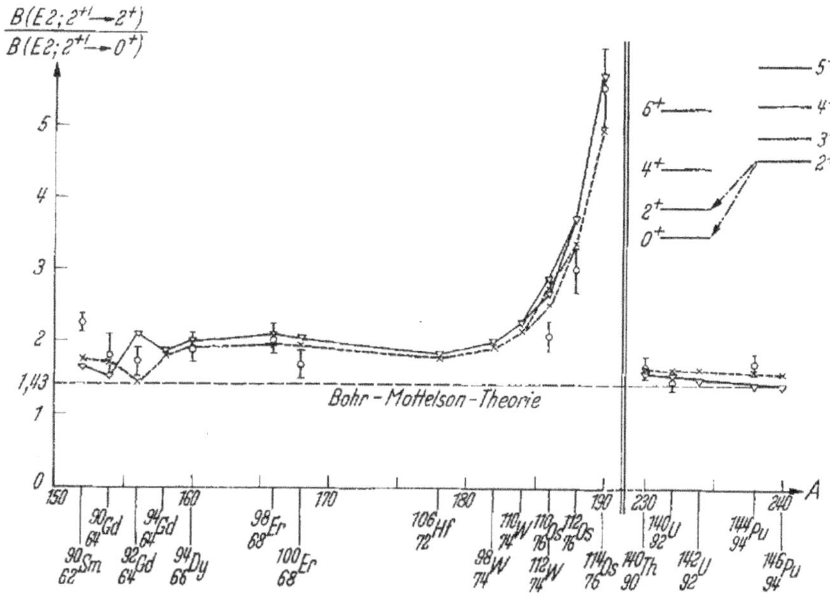

Fig. 3 Branching ratio of the reduced electric quadrupole transitions of rare-earth and transuranic nuclei: $B(E2, 2^{+}\gamma \rightarrow 2^{+}ground)/B(E2, 2^{+}\gamma \rightarrow 0^{+}ground)$. The solid and short dashed lines are different approaches to fit the three parameters: (1) moment of inertia, (2) $\beta-$, and (3) $\gamma-$ vibrational energies of the RotVib model for each nucleus

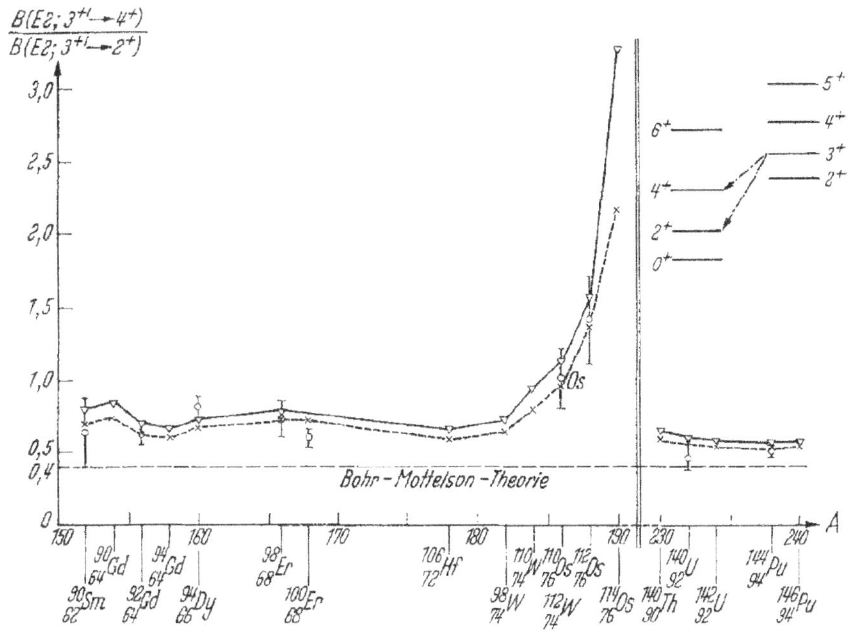

Fig. 4 Branching ratio of the reduced electric quadrupole transitions of rare-earth and transuranic nuclei: $B(E2, 3^+\gamma \rightarrow 4^+ground)/B(E2, 3^+\gamma \rightarrow 2^+ground)$. For more explanations see Fig. 3

to Germany for spending his retirement in Bad Pyrmont. At that time, we did not know that Walter will receive the Max Born price in 1974 and I in 1984 from the Institute of Physics, the Physical Society of the United Kingdom. We also did not know that we are in the scientific genealogy the grand-grand sons of Max Born. Siegfried Fluegge had done his Ph.D. with Max Born and Hans Marschall with Siegfried Fluegge, and Walter and I both with Hans Marschall. We also did not know that our scientific genealogy Ph.D. by Ph.D. goes back to Leibniz (1646–1716), the inventor of differential and integral calculus with the nomenclature, which we still use today.

The same improvement as in Fig. 3 is found by the inclusion of the rotational-vibrational interaction for the branching ratio of the reduced electric quadrupole transitions of rare-earth and transuranic nuclei: $B(E2, 3^+\gamma \rightarrow 4^+ground)/B(E2, 3^+\gamma \rightarrow 2^+ground)$ (see Fig. 4 and Ref. [2]).

3 Walter in Maryland and in Tallahassee/Florida

In fall 1962, Walter went till 1964 to the States as Assistant Professor at the University of Maryland near Washington to work with Mike Danos from the National Bureau of Standards. Mike Danos a former student of Hans Jensen, still in Hannover before

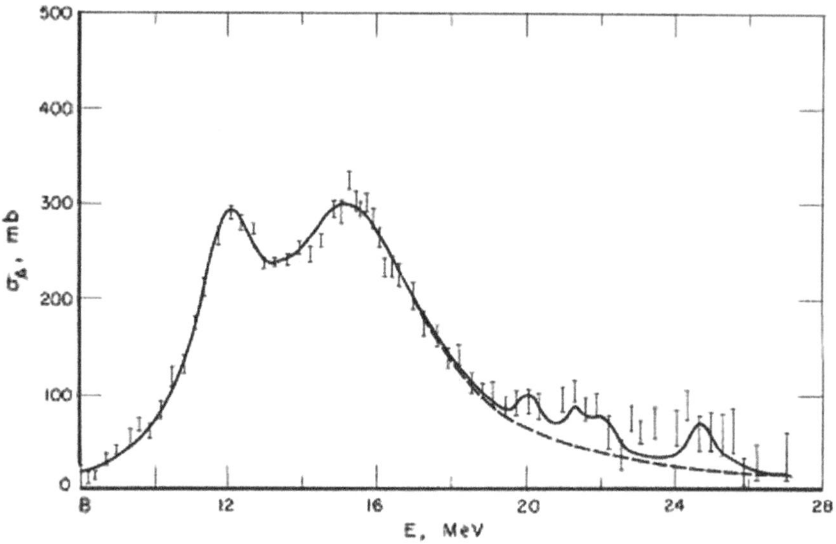

Fig. 5 Giant dipole and also possibly giant quadrupole resonance in ^{165}Ho [3]. For a detailed discussion see text

Jensen moved to Heidelberg, was for about half a year visiting Hans Marschall in Freiburg, where Walter got to know him. In Maryland, Walter worked with Mike Danos on giant resonances. Specifically, they included the coupling of the giant dipole resonance to the quadrupole vibrations of the RotVib model. Figure 5 shows the experimental giant resonance region of ^{165}Ho and the calculation of Ligenza, Danos, and Greiner [3]. For the giant dipole resonance, the lowest excitation is due to a_0 (isovector vibrations along the symmetry axis), and the a_2 (asymmetric isovector γ-vibrations) is higher in energy and twice as strong. The excitation around 25 MeV is the giant quadrupole resonance, the lowest maximum corresponds to the a_0 degree, the second lowest to the a_1 and a_{-1} excitations. It splits due to the coupling to the low lying RotVib excitations. The highest energy is due to the quadrupole excitations a_2 and a_{-2}. The detection of the giant quadrupole resonance is today attributed to the work of Thomas Walcher and coworkers in 1974 by electron scattering in Darmstadt. But looking back the investigation of Ligenza, Danos, and Greiner has probably seen this resonance already 8 years earlier in 1966, although at that time they could not convince the community that they have seen the giant quadrupole resonance.

As mentioned already above, the better energy resolution of the Tandems allowed to study high spin states in deformed nuclei. The first Tandem was delivered to Chalk River in Canada. At that time, it is one of the hot spots for Nuclear Physics. The second Tandem went to Yale University in New Haven with Alan Bromley, original from Canada, as the leading figure. The third Tandem was located in Tallahassee/Florida. The Heidelberg Tandem was number seven. In Tallahassee, one studied with the Tandem the newly detected isobaric analogue states (experiments: J. D. Fox, Fred

Moore; theory: Donald Robson) and the high spin rotational states (Ray Sheline). Ray Sheline was a nuclear chemist who was already active in the Manhattan Project. He came back to Tallahassee from a longer stay in the Niels Bohr Institute in Copenhagen and was familiar with the Bohr–Mottelson model. He had learned about our work on the RotVib model [2], although till then we had only published in German language in Zeitschrift fuer Physik. With the Tandem in Tallahassee and a magnet spectrometer and a "detector" based on a photo film in the focus of the spectrometer and girls scrutinizing these films, he measured with reactions like (p, d) or (d, p) high spin states mainly in the rare-earth region. He also had the gift to copy from slides, which were shown at Gordon Research Conferences, in the dark very fast the newest data on high spin states. So Ray had at that time the most complete information on high spin data, partially his own and partially from other groups.

For the summer 1963, Ray invited Walter from Maryland down to Tallahassee/Florida. Since both Walter and Ray were not familiar with numerical calculations on computers, Ray invited me also to Tallahassee. Walter and Ray had no idea about the computer available at Tallahassee, and thus, they did also not bother to inform me in advance about the computer facility at Florida State University. When I arrived on a Friday evening, I realized that my internal machine code for the "Siemens 2002" running in Freiburg was useless for Tallahassee. They had the fastest and best computer IBM-709 available at that time, still running with tubes. The code was FORTRAN, totally unknown to me at that time. So I picked up just after arrival on Friday evening from the local library a book to learn FORTRAN, and I translated statement by statement my code into FORTRAN. I worked through till Monday morning. When Walter and Ray came into the Institute at the beginning of the week, the program was running with FORTRAN on the IBM-709. We could start explaining the data of Ray Sheline. The model had for each nucleus three parameters: (1) the moment of inertia, (2) the beta, and (3) the gamma vibrational energies, which had to be fitted in each nucleus. We concentrated first on the transitional region Osmium, where the usual approximate solution of the Bohr–Mottelson model showed the largest discrepancy with the data [4] (see Fig. 6). The publication [4] was the first for the RotVib model in English language and in the most read and popular journal, Physical Review. Walter and I received after that many offers for positions. I do not know how many Walter received. One offer to him was the Full Professorship from Frankfurt, which he accepted in 1965. I received six offers: I made the error to accept the first offer, which was only an Assistant Professorship at UCLA. The others were Associate and Full Professorships. After 1 year at UCLA, I accepted the offer for a Full Professorship at the University of Muenster in Germany in 1967. At that time, one was still appointed in Germany as Director of the Institute of Theoretical Physics. In 1971, I was appointed as Director at the "Institut fuer Kernphysik" in the KFA Juelich and Professor at the University of Bonn. In 1980, when I moved to the University of Tuebingen, I was appointed only as Professor. So according to the titles, I moved always downwards.

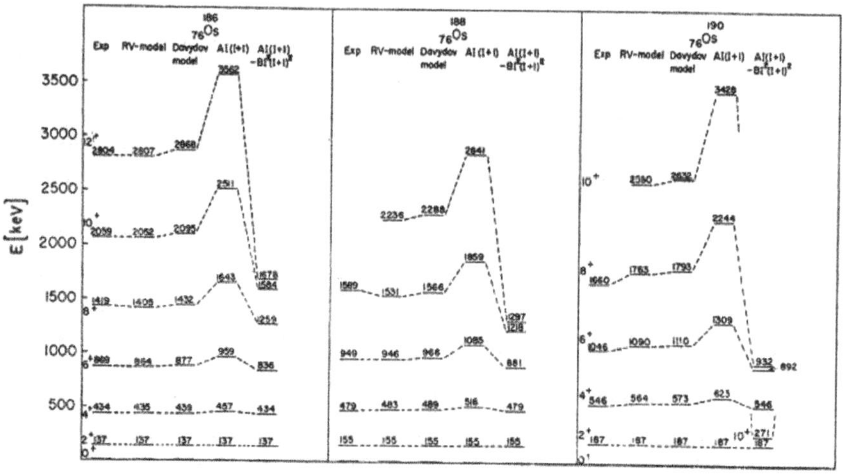

Fig. 6 Experimental energies, RotVib model, Davydov model, $AI(I+1)$ and $AI(I+1) - BI^2(I+1)^2$ for ^{186}Os, ^{188}Os, and ^{190}Os transitional nuclei at the end of the rare-earth region [4]

4 Walter Accepts the Offer for a Full Professorship in Frankfurt

As soon as Walter was in Frankfurt in summer 1965, he started to push for a center for heavy ion physics with the corresponding accelerator and access of neighboring universities to this lab, called later on "Gesellschaft fuer Schwerionenforschung" (GSI). Rudolf Bock, one of the founding fathers of the GSI, showed in his talk a note of the secretary of Prof. Peter Brix (1918–2007), Darmstadt, from February 22, 1966 at 17:44 h about a phone call of Walter Greiner, in which Walter asked that Peter Brix should come the next day at 18:30 h into his office in Frankfurt to speak about the plans for the GSI. Walter was at that time 30 years old and only since half a year Professor in Frankfurt, and Peter Brix was one of the leading nuclear physicists in Germany (See Fig. 7).

5 The Positron Lines

A very exciting time in Frankfurt and in the whole nuclear physics community at the beginning of the seventies was the prediction of the positron lines in supercritical heavy ion collisions by Walter and coworkers [5]. The relativistic binding energy of an electron in a pointlike Coulomb field with the charge Z is

$$E_{1s} = m_e c^2 [1 - (Z/137)^2]^{1/2} \tag{1}$$

Fig. 7 Note of the secretary of Peter Brix, at that time in Darmstadt, about a phone call of Walter Greiner on February 22, 1966 inviting (ordering) Brix the next day at 18:30 h into his office in Frankfurt for a discussion of plans for a lab for heavy ion physics, the future GSI (this note was shown in the talk of Rudolf Bock about the foundation of the GSI)

If the charge gets larger than Z = 137, this binding energy is imaginary. What does this mean: The electron orbit dives down into the lower Dirac sea and, if ionized, this corresponds to a positron, which can escape. Since one knows no superheavy nuclei with a charge larger than Z = 137, Greiner, Mueller and Rafelski [5] proposed to produce such a large charge in heavy ion collisions, e.g., in lead-on-lead $Z_{united} = 82 + 82 = 164$. During the touching configuration, the energy of the positron is fixed and it can escape. This is the heavily discussed and searched positron line (Figs. 8 and 9).

The main experimental groups searching for these lines were the ones of Paul Kienle (TU Muenchen) [6], of Dirk Schwalm (MPI Heidelberg) and Jack Greenberg (Yale University) [7], and of E. Kankeleit (TU Darmstadt) [8]. Since the electron orbit forming the positron had to be ionized, all groups finally tried to enlarge the sensitivity by requesting an electron–positron coincidence. By different cuts through the electron-positron spectrum, one could produce positron lines. But the energies varied from experiment to experiment and even by varying the cuts. This gave finally the “coup de grace” to the positron lines in supercritical heavy ion collisions. Paul Kienle insisted up to his death, that his data are correct and he did no massaging of the spectra.

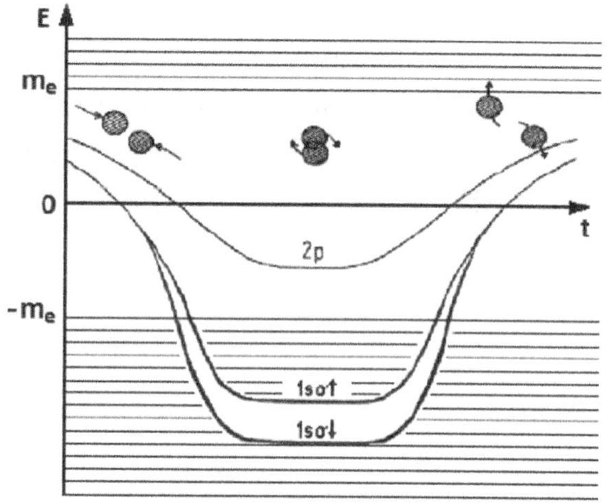

Fig. 8 Positron lines in supercritical heavy ion collisions. The electron levels, e.g., in a lead-on-lead collision $Z_{united} = 82 + 82 = 164$ dive down into the lower Dirac sea. If the level gets ionized during the collision, this corresponds to a positron which can escape with definite energy determined by the touching configuration. The united system rotates and produces a strong magnetic field, which splits the positron line into positron-spin parallel and antiparallel to the magnetic field

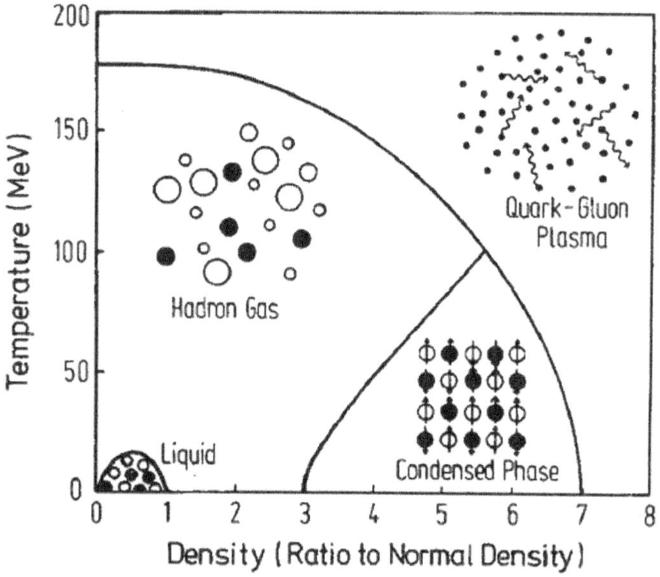

Fig. 9 The different phases of nuclear matter and the equation of state (EoS)

6 The Different Phases of Nuclear Matter and the Description of Heavy Ion Reactions

Walter was already in the 60s interested in the different phases of nuclear matter and its equation of state (EoS). He realized that one needs for the extraction of the different phases of nuclear matter and its EoS a reliable description of heavy ion collisions. Three different descriptions of heavy ion collisions were pushed forward in Frankfurt and related groups:

- Hydrodynamics (Hydro; Talks of M. Gyulassy, Barbara Jacak, and Uli Heinz).
- Quantum Molecular Dynamics (QMD; Talks of Hatnack and Aichelin): The relativistic version was pioneered by Sorge. This approach was used by Aichelin and coworkers, by the Tuebingen group, and by Herman Wolter and Fuchs in Muenchen.
- Boltzmann-Uehling-Uhlenbeck (BUU): Used in Frankfurt and in Giessen by the Mosel group.

At the beginning, Boltzmann-Uehling-Uhlenbeck (BUU) had a head start due to his long use and justification in solid-state physics and in gas dynamics. QMD was second and Hydro had first a bad reputation, since it required local equilibrium and one doubted, that this could be achieved in a few collisions. But soon it turned out that at lower heavy ion (HI) collision energies, all three approaches yielded the same or at least very similar results. Even time-dependent Hartree–Fock (TDHF) was very similar to the Vlasov approach (BUU without the collision term) and QMD omitting collisions. In general, BUU and QMD yield very similar results. But as a surprise, the data from RHIC (Relativistic Heavy Ion Collider at Brookhaven) could be best described by Hydro. Why is the hydrodynamic so good at high energies? I hoped to learn this at this symposion from the talk of Barbara Jacak and Uli Heinz. But both showed only that Hydro describes the high energy data very well, but not why. Why is hydrodynamics working so well, perhaps even without a local equilibrium? At this point, most practitioners adopt the "medical principle": What helps is correct, even if we do not understand why.

Uli Heinz showed in his talk hydrodynamic results, which agree with the data on high energy central collisions for proton–proton at 13 TeV, for proton–lead at 5.02 TeV, and for lead–lead at 5.02 TeV (see Fig. 10). Zabrodin gave in his talk a hint, why Hydro works so well: He investigated as a function of time in UrQMD (ultrarelativistic QMD) and in QGSM (quark-gluon string model) for different energies at RHIC the three pressures: total pressure, the corresponding pressure in beam direction $3 * P_z$, and perpendicular to beam direction $3 * P_x$. An equilibrium $P = 3 * P_z = 3 * P_x$ is already reached after $7\, fm/c = 2 * 10^{-23}$ s. The total reaction time is about $30\, fm/c$. Steffen Bass presented in his talk a very interesting approach: He fitted with his collaboration with Hydro and for fragmentation with ultrarelativistic QMD the hydro parameters to a set of sample data and then used these parameters to describe other HI reactions for lead-on-lead at 2.76 TeV and at 5.02 TeV. Cernay from Bergen presented a nice variant of Hydro: Particles in Cell Relativistic = PICR.

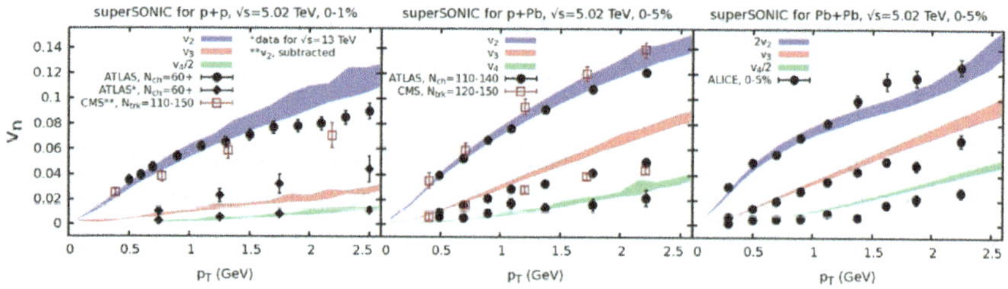

Fig. 10 v_n from the angular distribution $dN/d\Phi \propto v_n cos(n\phi)$ for central collisions p+p, for p+Pb, and for Pb+Pb at 5.02 TeV from ATLAS, CMS, and ALICE at LHC

Hydro describes the motion and distortion of the cell and the particles in the cell are dragged along. A. Murungo from Cape Town showed that Hydro can not only be used for HI collisions, but that it can also describe astrophysical processes.

Gorenstein studied in his talk the Hagedorn limiting temperature with and without strings. The Hagedorn limiting temperature arises, because it is cheaper to produce more and more particles due to the exponentially increasing spectral density, then to increase the temperature. With strings, he finds a limiting temperature of about 120 MeV, and without strings, the limiting temperature is increased to about 180 MeV without showing a saturation. He extracts an experimental limiting temperature from different reactions $p + p,\ p + \bar{p},\ K^+ + p,\ \pi^+ + p,\ e^+ + e^-$ around $170 + 40 - 20$ MeV.

Marc Strikman, Penn State University, spoke about the difference of color fluctuations in vacuum and in nuclear Matter. Marc got famous by the theoretical prediction of color transparency. At higher energies, the wavelength of the projectile is getting shorter and shorter. For larger objects, the positive and negative contributions interfere to zero. Only for small objects, the interaction is still strong. Since all the objects have a minimal size interaction interfers to zero for high energy projectiles with smaller and smaller wavelength. This is the color transparency. Color fluctuations single out different size objects of a wave function.

$$|Nucleon\ in\ vacuum >= |r, b, y >_{R,3q} + |r, b, y, gluon >_{R,3q+gluon} + \\ |r, b, y, meson >_{R,3q+meson} + \cdots \tag{2}$$

$$|Nucleon\ in\ Nuclear\ Matter = |NM >= |r, b, y >_{R1,NM} + \\ |r, b, y, gluon >_{R2,NM} + |r, b, y, meson >_{R3,NM} + \cdots \tag{3}$$

The different components of the nucleon wave function are admixtures with three colored quarks red, blue, and yellow without an additional admixture; a part with admixtures of gluons; and one with admixtures of mesons (see Eq. 2). The different components of a nucleon wave functions represent probabilities and the changing results due to these probabilities, if measured, constitute the color fluctuations. These

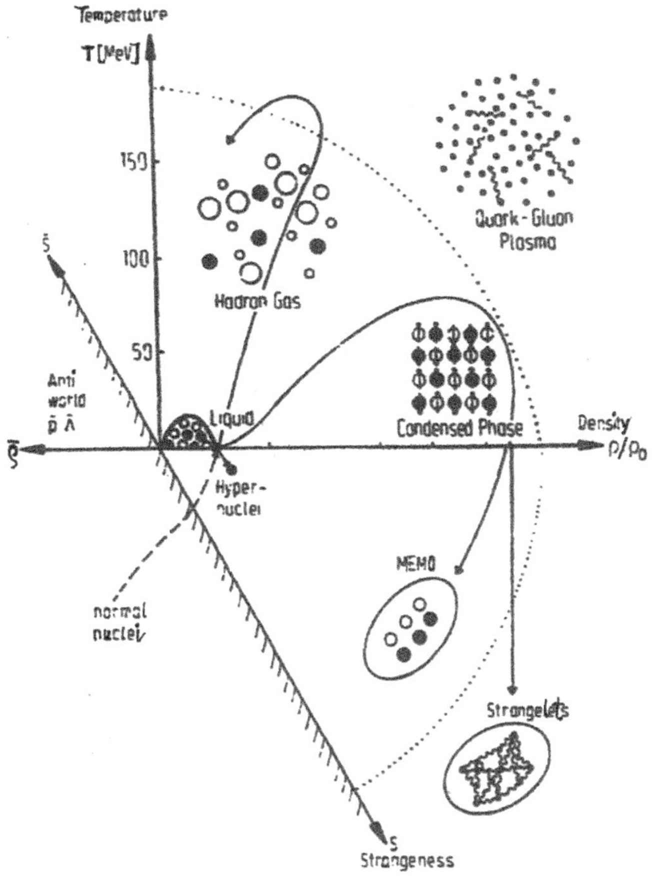

Fig. 11 The different phases of nuclear matter including density, temperature, and as additional degree also strangeness [9]

fluctuations change if the nucleon is in nuclear matter (inside a nucleus). The different objects are larger in nuclear matter, and thus, the color fluctuation is changing relative to a single nucleon (see Eq. 3). This was the topic of the talk of Marc Strikman.

Walter was also always interested to extend the nuclear EoS beyond density and temperature; he wanted to include also as an additional degree of freedom Strangeness. (See Figs. 11 and 12 and Ref. [9].)

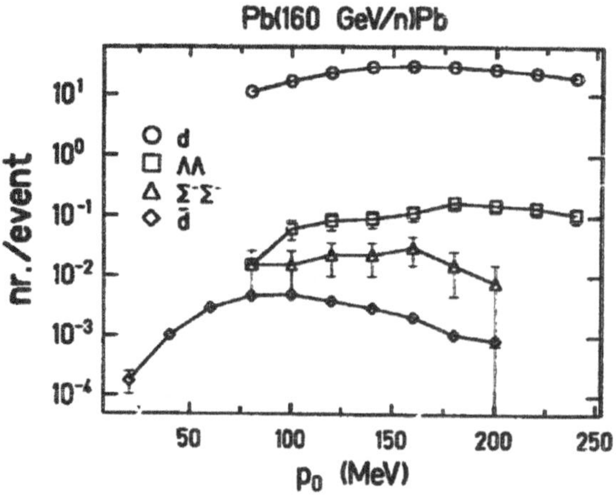

Fig. 12 Number of events with deuteron d, $\Lambda - \Lambda$, $\Sigma^- - \Sigma^-$ and the antideuteron $\bar{d}$ per Pb on Pb collision at 160 GeV/n [9]

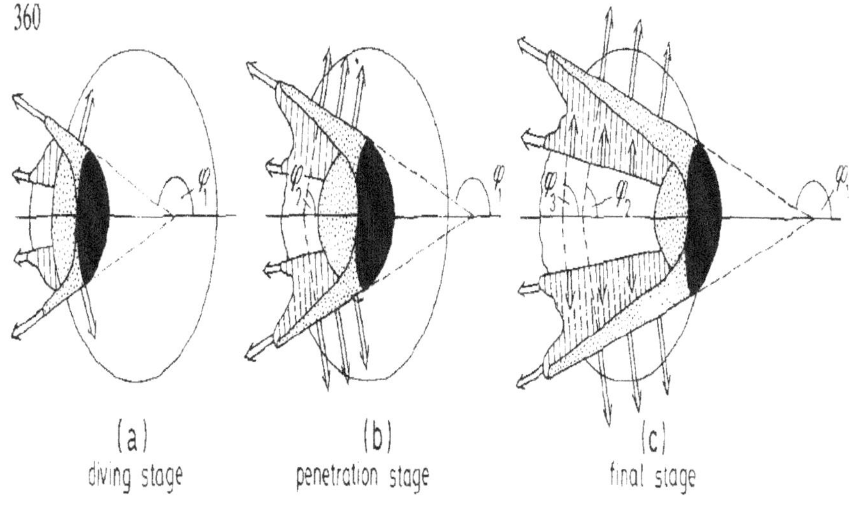

Fig. 13 Principle of a Mach shock in a collision of a nucleus with nuclear matter. The projectile comes from the left and hits in the experiment photo emulsion with silver and chloride nuclei

7 Shock Waves and Density Isomers

In the second half of the 70s, Walter Greiner and his collaborators [10] were interested in density isomers produced, for example, by pion condensation, Kaon condensation or Delta Matter. The idea was to compress nuclei by shock waves. Figure 13 shows the principle of a Mach shock in the collision of two nuclei. Figure 14 presents the data of the search the Mach shock by the experimental group of Schopper and Stock. With

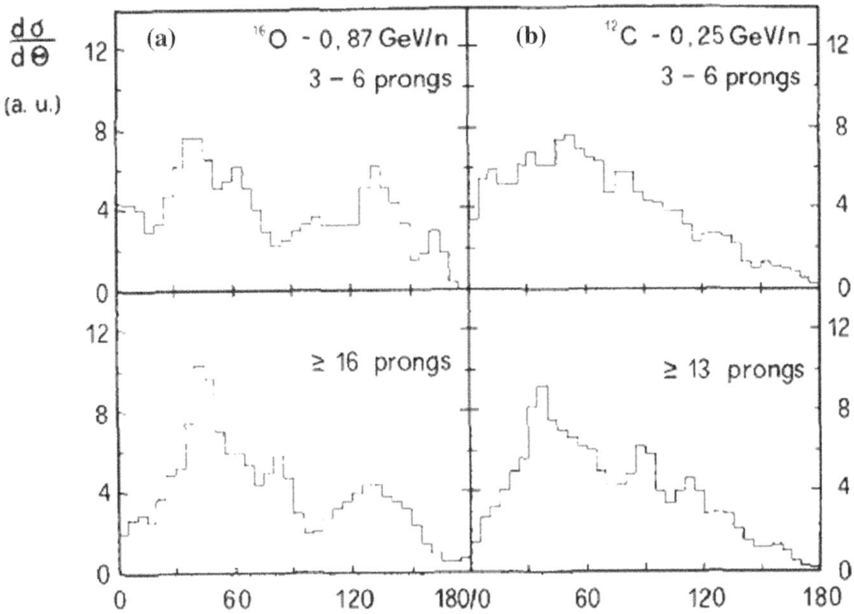

Fig. 14 The left-hand side of the image shows angular distribution in arbitrary units $d\sigma/d\theta$ of particles from collision of different projectiles, ^{16}O 2.1 GeV/n, ^{16}O 0.87 GeV/n, and ^{12}C 0.25 GeV/n [10] with photo emulsion of silver and chloride. On the right-hand side, the evaporation component is subtracted. The shock maximum moves to smaller angles at higher beam energies as expected from a Mach cone

increasing energy, the shock maximum moves to smaller angles defined in Figs. 13 and 14. A density isomer shows up as a pocket in the binding energy per particle in nuclear matter as a function of increasing density. This produces an anomaly for the shock angle: It shows an increase with higher energy. No such anomaly and thus no density isomer were found.

8 New Results from ALICE at LHC

Peter Braun-Munzinger Fig. 15 and Johanna Stachel Fig. 16 showed new results from the ALICE detector at LHC/CERN. It is surprising how well the relative abundance of the different particles can be described by the simple grand canonical statistical model $exp[-M/T]$ with the temperature 156 MeV (see Fig. 15). Johanna Stachel showed the ALICE result for the D-meson $D(c, \bar{q})$, $m_D = 1.869$ GeV production in Pb + Pb at 2.76 TeV.

$$\sigma(AA,\ D,\ m_D\ =\ 1.869\ \text{GeV})/[\sigma(pp,\ D) * N_{Collisions}] \tag{4}$$

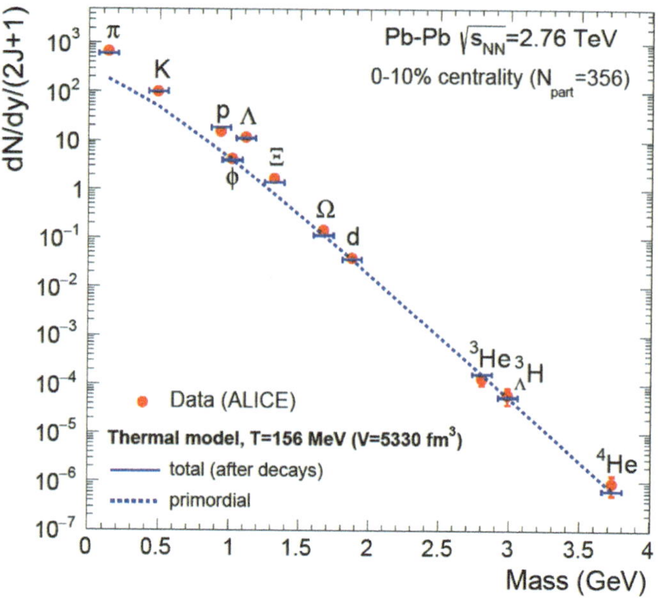

Fig. 15 Relative abundance of elementary particles $dN/dy * (2J + 1)$ for Pb+Pb at 2.76 TeV measured by the ALICE detector at the LHC. The theoretical results are statistical probabilities $exp[-M/T]$ for the temperature 156 MeV

Fig. 16 D-meson $D(c, \bar{q})$, $m_D = 1.869$ GeV production in Pb+Pb at 2.76 TeV presented by Johanna Stachel as results from ALICE

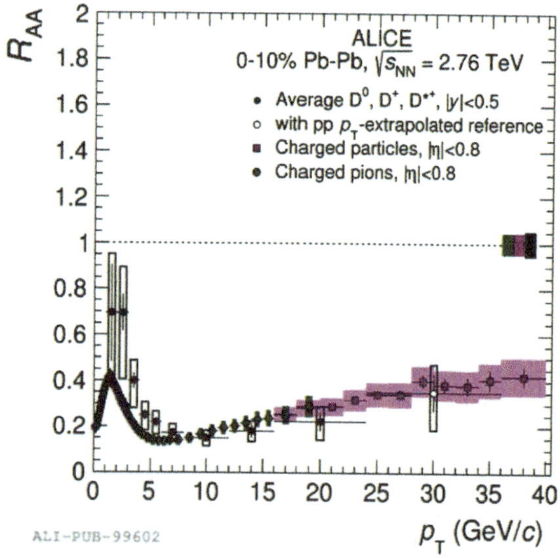

Joerg Aichelin explained in his talk the probability for the D-meson production as a function of its transversal momentum with perturbative QCD with known coupling constants and infrared regulators. The reduction at low p_T is due to shadowing in central collisions, while it is obvious that the probability for D-meson production at high p_T must decrease (Fig. 17).

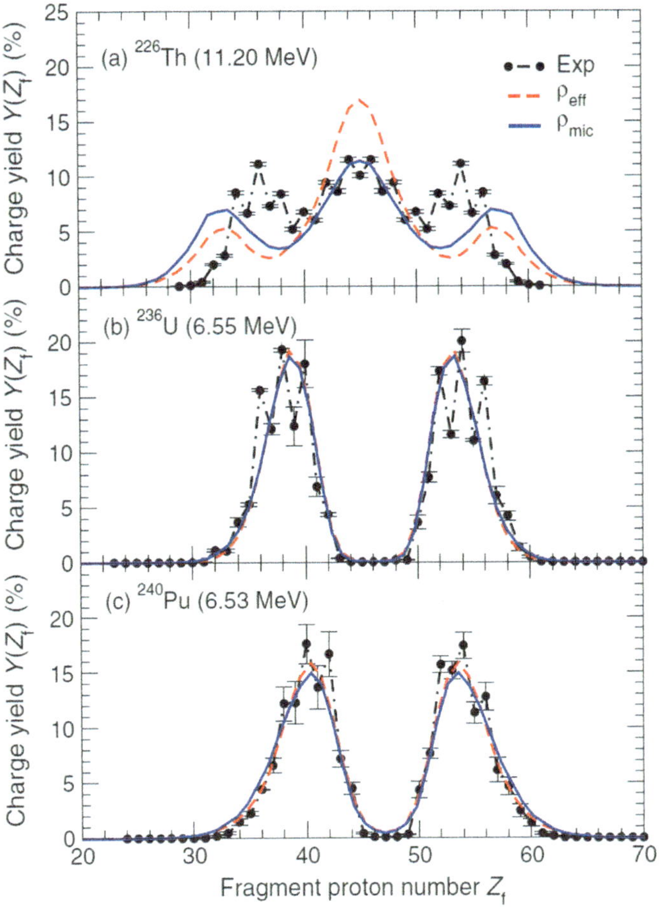

Fig. 17 Charge fragment distributions for the fission fragments of ^{226}Th, ^{236}U, and ^{240}Pu calculated with the method of Randrup and compared to experiment. The odd-even effects of the level density is here not included in this figure, but Randrup showed also results including this effect, which reproduce the fine structure of the data

9 Superheavy Nuclei

The production and investigation of superheavy nuclei was one of Walter's favorite topics since the 60s. He and his coworkers collaborated closely with the experimental superheavy group around Sigurd Hofmann predicting cross sections and optimal projectile target combinations. Recently, he was also closely connected with the Dubna group for the production of superheavy nuclei [11].

At this symposion, we had talks by Sigurd Hofmann/GSI and by M. Itkis, the Head of the Heavy Ion Physics Laboratory in Dubna, on superheavies.

The present status of the superheavy field is listed below:

- 104 Rutherfordium, Rf (Berkeley).
- 105 Dubnium, Db (Dubna).
- 106 Seaborgium, Sg (Berkeley).
- 107 Bohrium, Bh (Darmstadt).
- 108 Hassium, Hs (Darmstadt).
- 109 Meitnerium, Mt (Darmstadt).
- 110 Darmstadtium, Ds (Darmstadt).
- 111 Roentgenium, Rg (Darmstadt).
- 112 Copernicium, Cn (Darmstadt).
- 113 Nihonium, Nh (RIKEN, Japan)
- 114 Flerovium, Fl (Dubna).
- 115 Moscovium, Mc (Dubna).
- 116 Livermorium, Lv (Dubna).
- 117 Tennessine, Ts (Dubna).
- 118 Oganesson, Og (Dubna).

Dubna has good chances to establish element 119:

$$^{48}Ca + {}^{254}Es \rightarrow {}^{299}119 + 3n \ (\sigma = 0.3\, pb) \tag{5}$$

They identified already a decay chain of five alpha particles.

Itkis told us that Dubna, under the leadership of the Flerov Laboratory, has erected a special building with a DC-280 cyclotron and a new gas-filled recoil separator for superheavy ion research. The first experiments are planned for 2018. Dubna welcomes collaborations.

Fission is the biggest enemy for the detection of superheavy elements. Randrup presented in his talk a novel approach to calculate the mass and the charge distribution of the fission fragments. He parameterizes the shape of the fissioning nucleus in the way of Ray Nix and calculates for each set of shape parameters the total energy by the liquid drop model and the Strutinsky approach for the shell corrections. Then he chooses a Metropolis statistical walk through the different shapes of the nucleus to fission. The steps of the random walk are selected statistically with the weight:

$$P_{i \rightarrow f} \propto exp[-(E_f - E_f)/T] \tag{6}$$

After a larger number of random walks to fission, he obtains an excellent agreement with the experimental charge and mass distribution of the fission fragments. P. G. Reinhard presented a very interesting method to determine the fission barrier even with an error bar. The first step is for a known barrier height of 5.5 MeV in ${}^{266}_{108}Hs$ to calculate the barrier in parameter space of effective mass m^*, of symmetry energy, and of pairing strength for different nucleon–nucleon forces: Gogny, Skl3, Sly6, SkT6, SkM, SkP, BSK1, SV-bas, SV-min, UNDF2, NL-Z2, DD-PG, and DDME.

$$Barrier_{th} = Barrier(m^*/m,\ pairing\ gap,\ symmetry\ energy) \quad (7)$$

Then on determines for each force and set of parameters χ^2 (see Fig. 18).

$$\chi^2 = [Barrier_{th} - Barrier_{exp}]^2 \quad (8)$$

Figure 19 shows how the pairing gap (in percent of the experimental value) influences the barrier heights if the other parameters are optimized. The same is shown for the effective mass m^*/m $(in\ units\ of\ m = 1)$ and the symmetry energy J in [MeV] (Fig. 20).

Walter was not only interested in the nuclear physics of superheavy nuclei but also in atomic physics and in the chemistry of these elements [12]. Burkhardt Fricke told me that the paper [12] was already in 2007 referenced 4961 times, probably because it appeals also to chemists. It is the most referenced publication, which I know.

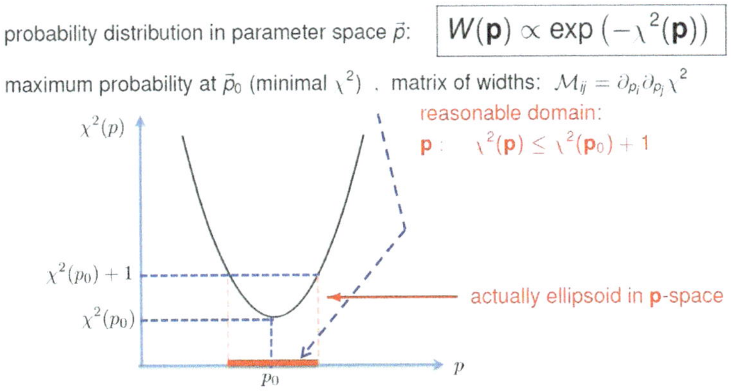

Fig. 18 χ^2(parameter space) is approximated for each parameter m^*/m, symmetry energy, pairing strength by a parabola. The width at $\chi^2 + 1$ gives the uncertainty of this parameter

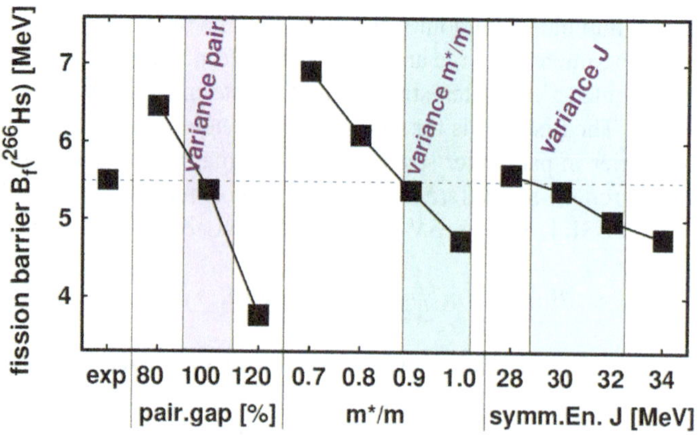

Fig. 19 Variation of the fission barrier ($exp = 5.5$ MeV) with variation of the parameters pairing gap, in percent of the experimental value, the effective mass m^*/m, and the symmetry energy J [MeV]

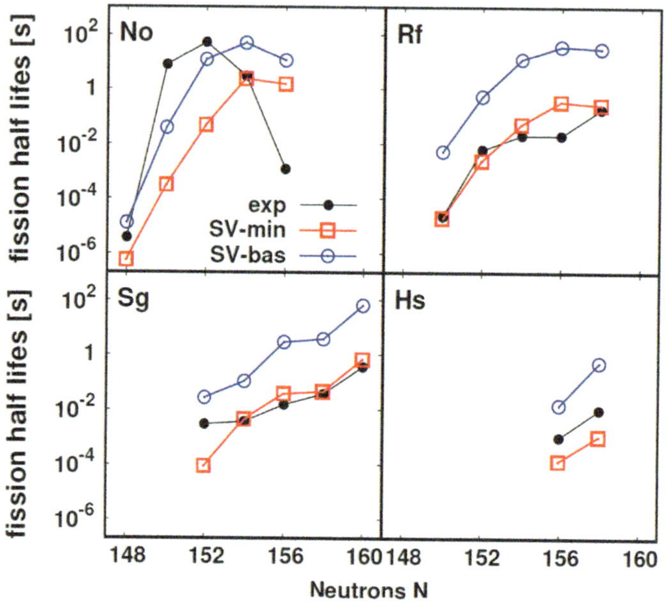

Fig. 20 Results for fission half-lives with Reinhard's method for different isotopes of nobelium Z = 102, rutherfordium Z = 104, seaborgium Z = 106 and hassium Z 108 calculated with the parameters of Fig. 19 for the NN forces SV-min, SV-bas compared with experiment

Fig. 21 Black hole with surrounding. In pc-GR (pseudo-complex-general relativity), the yellow rim is 100 times more intense than in GR (general relativity) of Einstein. This test will be measured soon as reported by Thomas Boller

10 Pseudo-complex General Relativity and Black Holes

Walter recent favorites were black holes and the "pseudo-complex general relativity (pc-GR)", a generalization of GR of Einstein [13, 14]. Among others, pc-GR predicts for the yellow rim of black holes a 100 times higher intensity (see Fig. 21); then GR. K. H. Kampert mentioned in his talk that the missing suppression of the GRZ = Greisen–Zatsepin–Kuzmin effect for Cosmic Rays at about 10^{19} eV normally explained by the violation of Lorentz-invariance can be described in pc-GR without such a Lorentz-invariance violation.

11 Frankfurt School

The strong Nuclear Physics activities of the group of Walter Greiner established the expression "Frankfurt School". I am listing here some of the earlier Ph.D. students of Walter Greiner, who hold Professorships at different places:

- 1967 Christian Toepffer (Erlangen), Werner Scheid (Giessen).

- 1968 Ulrich Mosel (Giessen).
- 1970 Burkhardt Fricke (Kassel), Paul-Gerhard Reinhard (Erlangen).
- 1973 Joachim Maruhn (Frankfurt), Berndt Mueller (Duke University and Associate Laboratory Director at Brookhaven National Lab.), Johann Rafelski (Tucson).
- 1977 Volker Oberacker (Vanderbilt Univ.), Gerhard Soff (Dresden).
- 1979 Horst Stoecker (Frankfurt and GSI Darmstadt).
- 1980 Ulrich Heinz (Ohio State Univ.), Peter Hess (University of Mexico, Mexico), Joachim Reinhardt (Frankfurt).
- 1985 Andreas Schaefer (Regensburg).
- 1993 Dirk Rischke (Frankfurt).
- 1994 Juergen Schaffner-Bielich (Heidelberg).
- 1999 Markus Bleicher (Frankfurt).

12 Walters Textbooks and Awards

Walter Greiner in addition to his more than thousand journal articles published many popular textbooks (see Fig. 22). Amazon offers 16 different textbooks from Walter. Some of them had eight editions.

Walter Greiner is also one of two founders of the "Frankfurt Institute of Advanced Studies (FIAS)". The five-story building for physics was built mainly from private donations, which Walter collected from different sources. More than 100 collaborators in FIAS for physics are supported by grants from the State of Hessen, the

Fig. 22 Walter in front of his bookshelf in the office. Practically, all books in the middle shelf are textbooks written by Walter with some coauthors

Deutsche Forschungsgemeinschaft, the Bundes-Ministerium for Research and Technology (BMFT), and other sources.

Walter has also an almost uncountable number of awards, among them 11 Honorary Doctor Degrees. The following is a list of the main awards:

- 1974 Max Born Prize, Institute of Physics, London, and the German Physical Society.
- 1982 Otto Hahn Prize (Frankfurt/Main, Germany).
- 1986 First Professor holding the “Frankfurt chair” at the University of Tel Aviv/Israel.
- 1987 Fellow of the Royal Society of Arts (FRSA), London.
- 1989 Honorary Member of the Roland Eötvös Society of Hungary (Budapest).
- 1993 Honorary Member of the Rumanian Academy, Bucharest.
- 1998 Alexander von Humboldt Medal.
- 1999 Officier dans l’Ordre Palmes Académiques.
- 2001 Member of Academia Gioenia di Catania.
- 2001 Honorary Professor, Jilin University, China.
- 2004 Fellow of Institute of Physics (FINSP), London.
- 2006 1st Degree Physical Medal, Czech Physical Society.
- 2006 Chairman of the Program Advisory Board of the Flerov Laboratory at Dubna.

13 Radiation Therapy with Heavy Ions

Walter always supported in committees and with his authority the development of heavy ion radiation therapy with ^{12}C at GSI for tumors in the brain (see Fig. 23). This therapy was later realized at the “Deutsches Krebsforschungszentrum” in Heidelberg. When Walter himself had a tumor in the brain, he was sure that ^{12}C irradiation at Heidelberg could help him. But he lost his final battle.

14 Was Bleibt? What Remains?

Everyone asks himself, what remains when I die. We ask this here specifically for Walter Greiner.

- First: Baerbel his wife and the two sons Martin and Carsten with the five grandchildren: three girls and two boys,
- his former students, his successful “Schueler”,
- his popular textbooks,
- his publications, and
- the momentum he gave to Nuclear Physics.

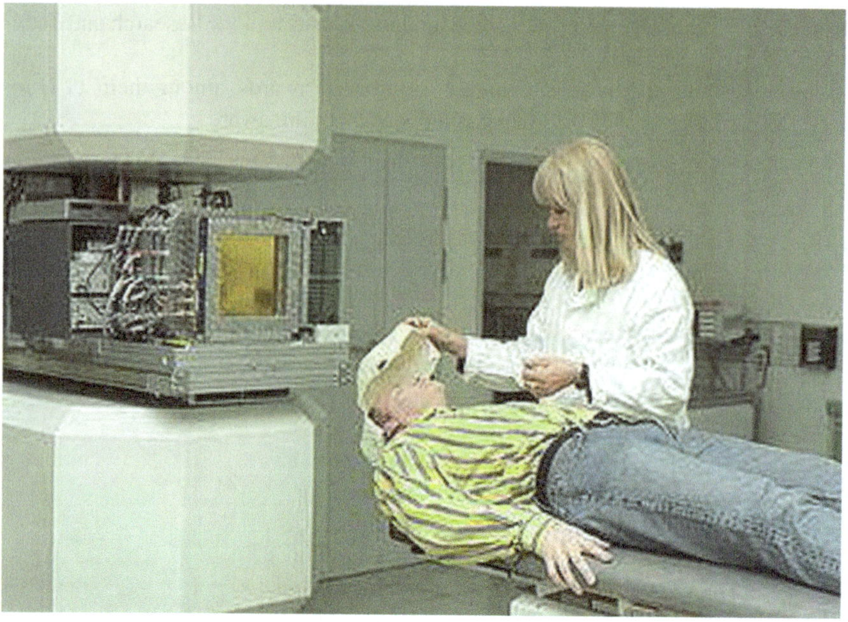

Fig. 23 Heavy ion (^{12}C) radiation therapy for tumors in the brain as tested at the GSI

Acknowledgements We colleagues will be always thankful to Walter for his inspiring personality and his scientific work. Thank you, Walter.

References

1. W. Greiner, Z. fuer Phys. **164**, 374–388 (19nd. 1966 at 17:44 h. 61)
2. A. Faessler, W. Greiner, Z. fuer Phys. **168**, 425–464 (1962)
3. R. Ligenza, M. Danos, W. Greiner, Phys. Rev. Lett. **16**, 364 (1966)
4. A. Faessler, W. Greiner, R.K. Sheline, Phys. Rev. **135**, B591 (1964)
5. W. Greiner, B. Mueller, J. Rafelski, Z. fuer Phys. **257**, 62 (1972)
6. P. Kienle et al., Phys. Lett. **137B**, 41 (1984) (1986)
7. J.S. Greenberg, D. Schwalm, Phys. Rev. Lett. **56**, 445 (1986)
8. E. Kankeleit et al., Phys. Rev. Lett. **40**, 1443 (1978) and Phys. Lett. B **245**, 17–20 (1990)
9. Ch. Hartnack, M. Berenguer, A. Jahns, A. V. Keitz, R. Matiello, A. Rosenhauer, J. Schaffner, Th. Schoenfeld, H. Sorge, L. Winckelmann, H. Stoecker, W. Greiner, Nucl. Phys. A **538**, 53c (1992)
10. H.G. Baumgardt et al., Z. fuer Phys. A **273**, 359 (1975)
11. V.I. Zagrebaev, A.V. Karpov, W. Greiner, Phys. Rev. C **85**, 014608 (2012)
12. B. Fricke, W. Greiner, J.T. Waber, Theor. Chim. Acta **21**, 235–260 (1971)
13. P.O. Hess, M. Schaefer, W. Greiner, *Pseudo-Complex General Relativity*. FIAS Interdisciplinary Science Series (Springer, 2015)
14. G. Caspar, T. Schoenenbach, P.O. Hess, M. Schaefer, W. Greiner, Int. J. Mod. Phys. E **21**, 1250015 (2012)

CPI Antony Rowe
Eastbourne, UK
March 18, 2020